THE ILLUSTRATED ALMANAC OF

SCIENCE, TECHNOLOGY, AND INVENTION

DAY BY DAY
FACTS, FIGURES, AND THE FANCIFUL

The Illustrated Almanac of Science, Technology, and Invention

Day by Day Facts, Figures, and the Fanciful

RAYMOND L. FRANCIS

PLENUM TRADE • NEW YORK AND LONDON

ISBN 0-306-45633-8

Plenum Press is a Division of Plenum Publishing Corporation
233 Spring Street, New York, N.Y. 10013-1578
http://www.plenum.com

10 9 8 7 6 5 4 3 2 1

Printed in the United States of America

PREFACE

How this book began, how it exploded, how to use it, and who made it possible.

One sunny day in the summer of 1990, sitting on the banks of St. Andrew Bay in northern Florida, I happened to read that Marie Curie's nomination to the French Academy of Sciences was voted down by its all-male membership on January 23, 1911. On that date Marie Curie already had one Nobel Prize, and she would soon become the first person to win a second Nobel Prize. She was one of history's greatest scientists, eventually to die as a direct result of years of tireless research on radiation. Her snub by the French Academy was an incident filled with irony, injustice, prejudice, and tension, and it quickly sparked the thought that such lesser-known tidbits, as well as well-known historic incidents, might make a fun calendar that would be valuable to students, teachers, and interested laymen.

The calendar was originally limited to the history of biology. Soon there was an explosion in its scope, when I realized that events in the history of ecology (including great environmental tragedies) belonged in the calendar, in hopes that someone might read them and be inspired to action to improve Earth's health.

Another explosion in the material occurred with the inclusion of incidents that were downright humorous or attractive because of their entertainment value, like the world's longest sneezing fit, the invention of a food slicer that doubled as a mouse trap, or the 1985 announcement that Coca-Cola was changing the formula of history's best-selling soft drink.

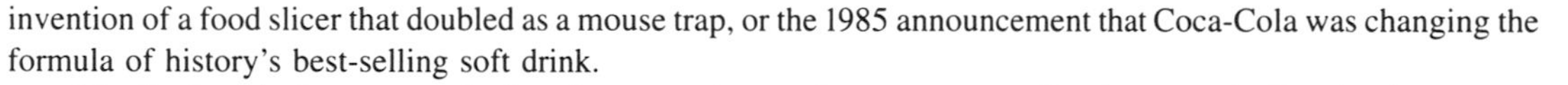

The final explosion occurred when the research was thrown open to all sciences and all technology.

The result seems to be the first historic almanac devoted to science; in fact, the first devoted to any specialized area of human endeavor. It was prepared for those who appreciate history, science, or trivia, for those who want to know what happened on any particular day, and for teachers who want to spice up a day's lesson by discussing with students what happened on that day.

The sources for the entries in this book are easily obtainable—encyclopedias, biographies, newspapers, science histories, etc. The reader who wants to know a specific reference or source can contact me directly through the publisher, and I will get the information to you.

Similarly, I would be grateful to hear from readers who discover errors in the text, who have suggestions, or who know of especially important or inspirational events that did not make it into this edition of this book.

Acknowledgments. A number of people made important contributions to this work. Any mistakes in the text are their fault.

That, of course, is a joke; my gratitude to them is not. I want to express my heartfelt thanks to Michael Snell, the agent who sold the book to Plenum; to Vanessa Tibbits and Linda Regan, editors at Plenum who worked so hard to prepare the text; to the members of the business management team at Plenum, for making the company and this book possible; and to Mary Curioli and the rest of the production team at Plenum who did such a fine job making the text into a book. Mary Curioli devoted many hours to selecting the hundreds of line art figures, and Susanne Van Duyne did an excellent job of preparing these figures for inclusion in the text.

The spectacular oil paintings of historic scenes appear with the generous permission of the Parke-Davis Division of Warner-Lambert. They were painted by Robert A. Thom, who collaborated with George A. Bender on the years of research. The lineart appears courtesy of Zedcor, Inc.

I wish to thank the trustees, the administrators, and my colleagues at Gulf Coast Community College in Panama City, Florida, for personal and professional support of this project. Much of the research and writing was done at the school's fine library. I have also received valuable and courteous assistance from the staff at the Bay County Public Library in Panama City.

I want to express my gratitude to Dr. Jane Goodall, the historic primatologist who now devotes herself to the welfare of captive and threatened chimpanzees; her research into chimpanzee behavior has inspired many. Thanks also to my family, Kathy and Pete Belovary, Charlie and Carolyn Francis, and Velma Francis, a word artist and pioneer in her own right.

Raymond Francis
Bayou George, Florida

The Illustrated Almanac of Science, Technology, and Invention

Day by Day
Facts, Figures, and the Fanciful

1600 Two legendary astronomers begin a short, but monumental collaboration, when Johannes Kepler (a 28-year-old German, still reeling from a financial crisis in Austria) arrives in Prague at the observatory of Tycho Brahe (a 53-year-old Dane, in poor health and about to die). Kepler was a supreme mathematician, Brahe a supreme observer who had compiled years of very accurate data on the movements of planets, comets, the sun, stars, and other heavenly bodies. But massive as the data were, they could not be explained or connected theoretically, until Kepler, well after Brahe's death, derived a series of mathematical laws to explain Brahe's figures. Foremost of Kepler's insights were that planets' orbits are ovals, not circles, and that the theory that planets orbit the sun was correct.

1739 The tiny, frostbitten island of Bouvetøya is discovered in the south Atlantic by J. B. C. Bouvet de Lozier, a young French explorer who was actually searching for the tropical paradise of Madagascar at the time. Bouvet did not realize that he had stumbled on the most remote island on Earth. No island is farther from land than Bouvetøya, now a Norwegian dependency. The nearest land is the uninhabited Queen Maud Land in the Antarctic, 1050 miles away.

JANUARY 1

1790 We now understand heat to be a form of motion, largely because of the theories and researches of the American-British physicist Benjamin Thompson (Count Rumford). On this date, the 36-year-old Thompson (under commission of Elector Karl Theodor of Bavaria) leads a troop of soldiers that clears the Munich streets of beggars and homeless, who are then forced to work in Thompson's uniform factory. The factory prospers, affording Thompson the funds and time to conduct historic experiments on a variety of topics, including food, light, and the nature of heat. Among Thompson's inventions were a drip coffee pot, a kitchen range, a double boiler, and improved fireplaces and chimneys.

1800 Louis-Jean-Marie Daubenton, a pioneer in both anatomy and paleontology, dies at 83 in Paris. Daubenton was a 26-year-old medical student when naturalist Georges Buffon invited him to prepare anatomical descriptions for *Histoire naturelle*, Buffon's multivolume comprehensive survey of Earth's natural history. The two had a stormy relationship, and parted company on several occasions, but were nonetheless able to collaborate on describing and cataloguing many forms of existing and extinct animals, plants, and minerals. When France's national Museum of Natural History was established in 1793, Daubenton was its first director.

1801 In his observatory in Palermo, Sicily, Giuseppe Piazzi discovers the asteroid Ceres, named after the Roman goddess associated with Sicily. Ceres was the first and largest asteroid ever discovered.

1800 Members of the du Pont family arrive from France in the United States, where they establish an empire in the chemical industry that has made them the wealthiest family in history.

1905 Clifford Beers begins writing *A Mind That Found Itself*, an autobiographical exposé of the brutal treatment of mental patients in asylums. The book started a campaign that changed forever the treatment of the insane. After two days Beers had written 15,000 words; his enthusiasm so upset his family that they had him committed temporarily to an asylum.

1939 The nation's first flea laboratory opens (in San Francisco, as part of the University of California's Hooper Foundation for Medical Research).

1964 Open burning of all materials is banned in New York City, as an attempt to control air pollution.

1984 AT&T is broken up. The communications giant is divested of its 22 Bell Systems companies according to an antitrust agreement.

1992 Five thousand U.S. hospitals begin to ban smoking on their premises. Smoking is now permitted only in restricted areas; it will be totally outlawed in two years.

1993 At the stroke of midnight thousands more Americans officially have AIDS than just seconds before. The increase in numbers is triggered by a new definition of the disease, which now includes pulmonary tuberculosis, recurrent pneumonia, invasive cervical cancer, and a severe drop in the level of the body's master immune cells, called CD4s. The new definition involves diseases that are especially common to women, whereas the old definition involved diseases most common to men; the shift represents an expanded knowledge of AIDS.

1403 Scholar John Bessarion is born in Trabzon, Turkey. He accumulated a large library of manuscripts and books of the ancient Greeks, and made these accessible to other scholars. He was the first non-Arab to translate the works of Aristotle.

1729 German astronomer Johann Daniel Titius is born the son of a draper in Konitz, Prussia. He was the first to state Bode's law (in 1766), a single, simple equation describing the distances of our planets from the sun. The formula fit the data for the known six planets at the time, and when Uranus was discovered in 1781, it also fit the magical formula. But ironically, the greatest success of the "law" came from its failure. The formula predicted the presence of a yet-undiscovered planet between Mars and Jupiter, and a group of German scientists (unofficially known as the "celestial police") spent years in the early 1800s systematically searching for the planet. It was never discovered, and apparently does not exist, but the search did lead to the discovery of Pallas, Juno, and Vesta. These are three of the four so-called minor planets, or asteroids, which follow regular orbits around the sun, but which are too small to be called true planets. Bode's law was buried forever in 1846 when Neptune was found, and shown not to follow the law, but by then it had served science well.

1822 Physicist Rudolf Clausius is born the son of a schoolmaster in Köslin, Germany (now in Poland). He is best known for coining the term "entropy" to describe the fraction of energy that is absolutely wasted when one form of energy is converted into another (such as the conversion of heat energy into mechanical energy by an engine), and for theorizing that entropy is constantly increasing in the universe because there is always some wastage during energy conversions. This process of increasing entropy is now called "the second law of thermodynamics."

1897 The Swedish newspaper *Nya Dagligt Allehanda* publishes the will of Alfred Nobel, the document that created the Nobel Prizes. One commentator at the time headed his column, "Nobel's will—Magnificent Intentions–Magnificent Blunder."

1920 Isaac Asimov, one of history's great science writers, is born in Petrovichi, Russia, the son of a Jewish candy-store keeper. At age 3 his family emigrated to the United States. Asimov was awarded his Ph.D. in chemistry at Columbia University, then went on to teach biochemistry at Boston University. He soon began supporting himself by a prolific output of books and articles. At his death in 1992, he had written nearly 500 books. His range was staggering: biology, astronomy, science fiction, mathematics, the Bible, Shakespeare, physics, and humor. Carl Sagan called him "one of the master explainers of the age ... I think millions of people owe their knowledge of science, their familiarity with some scientific fact, to reading either the fact or fiction of Isaac Asimov."

1925 The APA (American Psychological Association) incorporates in Washington, D.C. "The object of this society shall be to advance psychology as a science." The APA was the country's first national psychology society, and remains its largest and most influential. Organized at Clark University in Worcester, Massachusetts, the APA's first president was Granville Stanley Hall, a pioneer in educational and child psychology.

1929 Canada and the United States reach agreement on joint protection of Niagara Falls, the honeymooners' paradise and now a major source of hydroelectric power.

1953 Double Nobel laureates Linus Pauling and Robert Corey send an article to *Nature* suggesting a structure for DNA. Less than four months later, the race to describe this molecule ends, when Watson and Crick publish the now-accepted version in the same journal.

1959 The first Russian space probe, *Daydream*, is launched toward the moon. Although it missed its target by 4600 miles and went into orbit around the sun, it was the first probe with enough power to escape Earth's gravitational pull.

1988 In one of history's largest inland oil spills, a tank rupture sends 3,800,000 gallons of oil into Pennsylvania's Monongahela River.

1871 The first patent for margarine is awarded to Henry W. Bradley of Binghamton, New York, on a "compound for culinary use" containing lard, vegetable butter, or shortening.

1872 The *Official Gazette of the U.S. Patent Office* is issued for the first time. The publication provides details of each week's patents (patent numbers, titles, claims, and the names and addresses of the inventors), and is still published today.

1873 In the Atlantic Ocean off the coast of Lisbon, scientists on the HMS *Challenger* measure the ocean depth in five places, and they haul up three samples of ocean floor. It is the first full day of data gathering on the historic 3½-year voyage of the *Challenger*, acknowledged as the birth of scientific oceanography. Thousands of new species are discovered, and the sea bottom is found to contain enormous mountains, plains, and caverns.

JANUARY 3

1888 The drinking straw is patented. Marvin C. Stone of Washington, D.C., invented the paraffined manila paper object to replace natural rye straws, which were unsanitary and cracked. The artificial straws were hand-rolled until 1905, when a machine was created by the Marvin C. Stone Estate. Stone's patent is No. 375,962.

1906 Astronomer William Wilson Morgan is born in Bethesda, Tennessee. His first major contribution, published in 1943, was discovering a relationship betweeen the pattern of light that a star emits and its distance from Earth. In the late 1940s he studied the large blue-white stars of the Milky Way, and from the characteristics of light they emitted (i.e., their "spectral emissions") he discovered that our galaxy has a spiral shape. In 1956 he created a system of classifying galaxies according to the characteristics of the stars in them.

1938 The March of Dimes, dedicated to fighting polio, is organized.

1957 The electric watch is introduced to the U.S. public, by the Hamilton Watch Company of Lancaster, Pennsylvania. Powered by a small battery guaranteed to last a year, the watch eliminated 35% of the parts of a conventional watch, including the mainspring.

1961 An accident at an experimental nuclear reactor at a federal installation near Idaho Falls, Idaho, releases radiation throughout the plant, and three workers die. These are the first three deaths at a U.S. reactor.

1961 Los Angeles sets a record for air pollution, when the concentration of nitrogen oxide molecules is measured at 3.93 parts per million (ppm). Current EPA standards specify that a "public alert" be issued when nitrogen dioxide concentration reaches 0.6 ppm, a "public warning" be issued at 1.2 ppm, and a "declaration of public emergency" be issued at 1.6 ppm.

1970 A record in heart-transplant longevity begins at the Stanford Medical Center in Palo Alto, California, when Dr. Edward Stinson leads a team that transplants an anonymous donor's heart into William George van Buuren, who survives for 21 years 10 months 24 days.

1980 Joy Adamson, conservationist and *Born Free* author, is killed by an employee in a wage dispute in northern Kenya.

1987 The world's most powerful atom smasher is the *Tevatron*, a 1.25-mile-diameter proton synchrotron at the Fermi National Accelerator Laboratory (the "Fermilab") near Batavia, Illinois. On this date it achieves a center-of-mass gravity of 1.8 TeV (1.8×10^{12} eV) by colliding protons with antiprotons. It is the greatest quantity of energy generated, measured, and reported in such a device.

1993 A bald eaglet hatches at the Audubon Society's Birds of Prey Center in Maitland, Florida. A week before, the egg survived a 40-foot fall, when the pine tree it was in was felled with a chain saw. Today is the first time in 50 years that an eagle egg successfully hatched in captivity after being laid in the wild. Who committed the crime of endangering the bird (a protected species) is not known, but authorities believe the tree was felled to remove bald eagles from the area. In the presence of a protected species, housing construction and other forms of "development" of the land would be hindered by laws protecting wildlife.

JANUARY 4

1643 Isaac Newton is born premature and sickly in Woolsthorpe, England. At age three he was given to his grandparents to be raised. This uprooting, and the hate he felt for his stepfather, are often cited as reasons for his lifelong emotional disorders. Newton was a slow student, and did not begin to excel in his studies until his teens.

1737 Baron Louis Bernard Guyton de Morveau is born in Dijon, France. As well as helping to modernize chemical nomenclature (so that the name of each chemical compound reflects its composition), he changed the face of aviation. A balloonist himself, Guyton de Morveau convinced the French revolutionary government to establish a fleet of balloons that facilitated several victories in the 1790s. It was history's first military air force.

1746 Benjamin Rush, a founder of psychiatry, is born in Byberry, Pennsylvania.

1761 Stephen Hales, a lifelong clergyman and the founder of plant physiology, dies at 83 in Teddington, England. Hales was the first to accurately measure transpiration in plants and the pressure on sap as it moves within plants. He was also the first to measure human blood pressure.

1797 Wilhelm Beer, the first to map both the moon and Mars in detail, is born in Berlin, Germany.

1809 Louis Braille, inventor of the universal reading system for the blind, is born in Coupvray, France. At age 3 he blinded himself accidentally with an awl in his father's harness shop. He adapted to the tragedy, however, and at age 15 created the alphabet that bears his name.

1885 Dr. William W. Grant performs the first successful appendectomy in medical history, on 22-year-old Mary Gartside in Davenport, Iowa. The operation probably saved her life, and she died 34 years later of an unrelated illness.

1887 History's first bicycle trip around the world ends in San Francisco, where it began on April 22, 1884. Thomas Stevens first rode his 50-inch cycle (the diameter of the front wheel) to Boston, then across Europe and Asia.

1896 German physicist Wilhelm Conrad Roentgen reports his discovery of X rays to the Berlin Physical Society. The next day a newspaper reports the discovery.

1912 This century's closest pass between Earth and the moon occurs; the distance is 221,441 miles.

1940 Physicist Brian David Josephson is born in Cardiff, Wales. He won the 1973 Nobel Prize at just age 33 for his theories on superconductivity (i.e., the ability of some materials to pass an electric current with no resistance when they are supercooled to temperatures near absolute zero). While a 22-year-old graduate student at Cambridge, Josephson theorized/predicted that an electric current would spontaneously flow between two pieces of superconducting material, and that if a power source was applied to this system, the current would stop flowing. This is now called the "Josephson effect," and subsequent experiments by others proved Josephson correct.

1958 The first man-made object in space, *Sputnik I* (meaning "fellow traveler" in Russian), falls out of orbit and disintegrates on its way back to Earth. It was a 184-pound satellite containing a radio beacon and a thermometer, and it had been orbiting for 92 days.

1966 A record-setting sneezing fit begins in Miami. Seventeen-year-old June Clark starts sneezing in Jackson Memorial Hospital while recovering from kidney trouble; she stops 155 days later through the use of electric shocks. Flying particles were clocked at 103.6 mph.

1794 Dr. Benjamin Rush fights the yellow fever epidemic in Philadelphia. Rush heroically remained in the city while many physicians fled and hundreds of citizens died daily. Born on January 4, 1746, Rush was his nation's first great physician and a founder of psychiatry.

1786 Naturalist Thomas Nuttall is born in Long Preston, England. After spending years as a journeyman printer for his uncle, he moved to Philadelphia at age 22 and studied botany. He made numerous field trips throughout North America, collecting and identifying a variety of plant species. He then turned to the study of birds after joining the faculty of Harvard in 1822.

1794 Edmund Ruffin, a founder of soil science, is born in Prince George County, Virginia. A leading secessionist, Ruffin fired one of the first shots of the Civil War at Fort Sumter. At age 19 he took over his deceased father's unproductive tobacco plantation, and, with no scientific background, Ruffin discovered that crop rotation and fertilization made the soil fertile again.

1838 French mathematician Camille Jordan is born in Lyon. He made important contributions to theories in geometry and algebra, but he is best known to the layman for the Jordan's curve, which is a circle that is twisted out of shape to form a maze.

1874 Physiologist Joseph Erlanger is born the son of a German immigrant in San Francisco. He applied the newly invented oscilloscope to the study of nerves. Erlanger made a host of discoveries, and shared the 1944 Nobel Prize for Physiology or Medicine with collaborator and ex-student Herbert Gasser.

1885 The first U.S. "piggyback" railroad operation, in which railroad cars carried other forms of transportation, begins at Albertson's Station, Long Island, New York, with the departure of a produce train, consisting of eight flatcars carrying farmers' wagons, eight cars for horses, and one coach for teamsters.

1896 *Wiener Presse* in Vienna carries the first newspaper report of the discovery of X rays by German physicist Wilhelm Roentgen.

1903 The first cable across the Pacific Ocean, from San Francisco to Honolulu, opens to the public for the transmission of telegrams.

1933 Construction officially begins on the Golden Gate Bridge in San Francisco. It is the nation's first bridge with piers sunk in the open ocean and the first bridge built across the outer mouth of a major ocean harbor.

1943 Educator-agricultural scientist George Washington Carver dies at 81 in Tuskegee, Alabama. In his early 30s, Carver was hired by the Tuskegee Institute to head the school's newly formed department of agriculture. He created a research program that ultimately developed 300 derivative products from peanuts, including cheese, inks, and plastics, and another 180 products from sweet potatoes. The peanut was not even recognized as a crop when Carver began at Tuskegee, but by his death peanuts were one of the six leading crops in the United States and the second cash crop (following cotton) in the South. Carver's birthdate is unknown. He was born a black slave in Missouri. Before his first birthday he and his mother were stolen by slave rustlers and taken to Arkansas. The fate of his mother is unknown, but Carver was traded back to his original owner, Moses Carver, for a $300 racehorse.

1955 Hurricane Alice finally weakens, and is downgraded to a tropical storm. She is no longer a hurricane, but during her brief life she set two, apparently contradictory records. She became a hurricane very late in one year (December 30, 1954), and remained a hurricane very early in the next; thus, Alice holds the record as both the earliest and latest Atlantic hurricane.

1968 Navy pilot Ralph E. Foulks, Jr., is shot down over North Vietnam. Twenty-five years later he becomes the first Vietnam soldier whose remains were identified solely by DNA analysis.

1991 A speed record for a solar battery-powered car is set by Manfred Hermann in Star Micronics's *Solar Star* at an air force base in Richmond, New South Wales, Australia, at 83.88 mph.

1993 The Food and Drug Administration publishes its first regulations for bottled water.

1993 "One of Britain's most vulnerable sites for marine wildlife" is devastated when the single-hull tanker *Braer* runs aground in the Bay of Quendale, Shetland Islands, and spills nearly 25 million gallons of oil. This is more than twice the amount spilled in the *Exxon Valdez* holocaust.

JANUARY 6

1639 Surplus crops are ordered to be destroyed for the first time in U.S. history. The Virginia General Assembly orders that "the tobacco of that year be viewed by sworn viewers and the rotten and unmerchantable, and half the good to be burned."

1655 Jakob Bernoulli is born in Basel, Switzerland. He coined the term "integral" to analyze a curve in calculus, and was one of the first to apply the newly invented calculus to real-world problems—design of bridges.

1745 Jacques Montgolfier is born in Vidalon-les-Annonay, France, one of 16 children of a prosperous paper manufacturer. He and one of his brothers invented the hot air balloon.

1795 Chemist Anselme Payen is born in Paris. He discovered and named "diastase"; this was the first enzyme prepared in concentrated form, and in its naming Payen established the tradition of using the suffix "ase" to name enzymes. Payen also discovered and named "cellulose"; this name established the tradition of using the suffix "ose" to indicate a carbohydrate.

1838 Samuel Morse first publicly demonstrates his telegraph, in Morristown, New Jersey.

1844 Patent No. 3,605 is issued to S. Broadmeadow of Woodbridge, New Jersey, for a method "to obtain malleable iron direct from iron ore." It is the first iron-related patent in U.S.history. The first zinc patent in U.S. history was issued on this date in 1857 to Samuel Wetherill of Bethlehem, Pennsylvania; it was patent No. 16,362, describing a method to obtain usable zinc from zinc ore.

1852 Louis Braille, creator of the world-famous Braille alphabet for the blind, dies of tuberculosis in Paris two days after his 41st birthday. His invention of the Braille alphabet stemmed from an accident in childhood that took his eyesight.

1858 The first-ever telephone call from a submerged submarine is placed by inventor Simon Lake (who is sitting at the bottom of the Patapsco River) to the mayor of Baltimore, William Talbot Talster. The submerged Lake then places calls to New York and Washington.

1884 Suffering from dropsy, obesity, and a long political fight against taxation of churches, geneticist Gregor Mendel dies at 61 in the monastery at Brünn, Bohemia. The entire town attended the funeral of the beloved monk and teacher. None of the grievers realized they were mourning the loss of the discoverer of the basic laws of genetics.

1939 Otto Hahn and Fritz Strassmann publish experimental results from their Berlin lab in which they bombarded uranium with neutrons. The element barium was produced in the reaction, a result that was neither anticipated nor fully understood by Hahn and Strassmann. It took physicist Lise Meitner (a longtime collaborator of Hahn's who had recently fled Germany and the Nazis) to realize and announce that nuclear fission had been achieved. The atom had been split. It was the critical step in later producing an atomic bomb, and it took place only a few blocks from Hitler's headquarters.

1942 Pan American Airways's *Pacific Clipper* lands in New York at the end of the first round-the-world flight by a commercial airplane.

1986 A tank of nuclear material explodes after being improperly heated at the Kerr-McGee plant in Gore, Oklahoma. Radioactive molecules are scattered throughout the atmosphere. One worker dies and one hundred are hospitalized. The incident is one of the major mishaps in U.S. nuclear history.

1995 A coalition of government and private organizations in Tallahassee, Florida, calls for tougher enforcement of speed limits for motorboats, shortly after the State announced that 1994 was the second-worst year ever for the manatee, a gentle, aquatic mammal, nicknamed the "sea cow." Each year brings the species closer to extinction in the wild. In 1994, 192 manatees perished, which is about 10% of the total remaining on Earth. Of the 192, 70 were killed by pleasure boats and other man-made causes.

1610 The great Italian scientist Galileo discovers three moons orbiting Jupiter; within a few weeks he discovered a fourth. Galileo was using one of history's first telescopes, which he designed and constructed in 1609. The four moons were the first bodies discovered to regularly orbit some object in space other than Earth, and were important evidence that Earth is not the center of the universe.

1745 Danish entomologist-naturalist Johann Christian Fabricius is born in Tøndern.

1784 David Landreth of High Street, Philadelphia, organizes the country's first significant seed business whose seeds were not imported from Europe. The location of the business is now covered by buildings at 1210 and 1212 Market Street.

1794 Eilhardt Mitscherlich is born the son of a minister in Neuende, Germany. He discovered chemical isomorphism, a condition in which compounds of similar composition crystallize together.

JANUARY 7

1834 German physicist Johann Philipp Reis, the first to build an electric telephone, is born in Gelnhausen. Like Alexander Graham Bell, Reis became interested in the electrical transmission of speech as a way to help the deaf. Bell modified Reis's devices to invent the first practical and successful telephone.

1853 Just after 9 AM a gunpowder factory explodes in Concord, Massachusetts, rocking the countryside for miles around. Author Henry David Thoreau hears the blast at his tranquil homesite on nearby Walden Pond. He visits the accident scene that day, and is filled with "anger and heartsick" at the loss of life, the destruction of the rustic beauty, the conflict between nature and factories. Several days later Thoreau begins writing the final draft of his famous book *Walden*, which he called a "cockcrow" to awaken mankind to the realities, drawbacks, and ugliness of industrialization.

1860 The second edition of Charles Darwin's *Origin of Species* is published.

1871 Russian chemist Dmitry Mendeleyev publishes an updated version of his periodic table of the elements. This time he takes the bold step of predicting that yet-unknown elements will be discovered.

1892 Physiologist Ernst von Brücke dies at 94 in Vienna. Best known as an early influence on Freud, Brücke was an eminent researcher into vision, speech, and muscle. He was an early advocate of animal experimentation in medical research, and he urged that biology be rooted in physics and chemistry.

1894 The copyright application of the film, "The Sneeze," is dated. The movie was shot at the studio of Thomas Edison (see February 2, 1893) and was the first American movie copyrighted.

1927 Commercial transatlantic telephone service is inaugurated between London and New York.

1949 The first-ever photograph of genes is announced by Drs. Daniel Chapin Pease and Richard Freligh Baker of the University of Southern California.

1953 President Truman announces the United States has developed the hydrogen bomb.

1963 Rachel Carson receives the Schweitzer Medal of the Animal Welfare Institute. Carson was inspired by Albert Schweitzer, and dedicated *Silent Spring* to him.

1992 Spain and Australia become the first countries to ban silicone breast implants.

1993 The Environmental Protection Agency releases a report that identifies secondhand smoke as a proven carcinogen.

1785 French balloonist (and inventor of the parachute) Jean-Pierre-François Blanchard makes the first aerial crossing of the English Channel. He is accompanied by U.S. physician John Jeffries. The pair were forced to throw all equipment overboard except for a package of mail; this package was the first international airmail delivery.

JANUARY 8

1587 Astronomer Johannes Fabricius is born in Resterhafe, the Netherlands. His father was the famous astronomer David Fabricius, who joined Galileo as one of the first to use the telescope. In 1611 Johannes was the first to report sunspots, mysterious great blotches on the face of the sun that move around erratically and are seen in great numbers only once every 11 years.

1642 Having spent the last eight years under house arrest for maintaining that Earth orbits the sun, the great astronomer Galileo dies at 77 in Arcetri, Italy.

1823 Alfred Russel Wallace (who devised the theory of evolution independently of Charles Darwin) is born in Usk, Wales, the eighth of nine children. His childhood was spent in poverty, with his family moving to a series of increasingly rural surroundings; this exposure to the country may have been critical in turning Wallace to a life as a naturalist. This came after failures at several other careers, just as Darwin had fumbled through several professions as a young adult. Again like Darwin, Wallace hit on the theory of evolution after reading *Essay on Population* by Thomas Malthus. Darwin struggled for 20 years to conceive and refine the theory, whereas Wallace wrote it out in two days during a bout of malaria in the jungles of Borneo in Southeast Asia.

1851 At 2 AM in the basement of his Paris home, physicist Jean Foucault, 31, becomes the first to actually demonstrate that Earth rotates. His device, now called the Foucault pendulum, is a heavy weight suspended by a long wire; as the weight swings back and forth through a fixed plane of motion, the Earth's spin below is revealed by marks of the weight in the sand. Foucault later erected an enormous pendulum in the Pantheon in Paris, and became famous for his demonstration.

1891 Physicist Walther Bothe is born the son of a merchant in Oranienburg, Germany. He won the 1954 Nobel Prize for Physics for inventing a "coincidence counter," which detects subatomic particles (i.e., electrons, protons, and other tiny pieces that make up the atom) and measures atomic events that occur at the incredibly fast speed of a billionth of a second or less.

1894 Parasitologist Pierre-Joseph van Beneden dies at 84 in Lowain, Belgium. He is most famous for a 15-year study of tapeworms, which resulted in the first complete description of their complicated life cycle. (The pork tapeworm, for example, typically spends part of its life in a human and part of it in a pig.) His son Edouard also studied worms, but became famous for discoveries in the field of genetics.

1935 The spectrophotometer is patented. This device, very useful to the study of light, can detect 2 million shades of color. Its inventor, Professor Arthur Cobb Hardy of Wellesley, Massachusetts, receives patent No. 1,987,441 for his "photometric device."

1942 Cosmologist-physicist Stephen William Hawking is born in Oxford. Within science circles he is best known for theories relating to the birth and death of the universe, and in the 1980s became famous worldwide for his best-selling *A Brief History of Time*, a popular treatment of these topics.

1995 Wire services report findings by the U.S. Fish and Wildlife Service on the endangered status of the Key deer, a small animal of maximum height 28 inches, found only in lower Florida. Of its estimated population of just 250–300, there were 67 deaths in 1994, 47 caused by vehicles. U.S. Route 1, the main highway through the Florida Keys, is the main site of fatal collisions. Dogs, fences, disease, and stomach obstructions account for the rest of the species's mortality in 1994.

1656 Issue No. 1 of *Weeckelyche Courante van Europa* is published in Haarlem, the Netherlands. A copy still remains, making it the oldest still-existing commercial newspaper.

1324 Italian explorer Marco Polo, known for his travels across central Asia to China, dies at about 70 in his birthplace, Venice.

1778 Philosopher-metaphysician Thomas Brown is born in Kilmabreck, Scotland. He is remembered for founding a branch of the "common sense" school of philosophy that stressed the importance of sensory perception in understanding human nature.

1793 French balloonist Jean-Pierre Blanchard becomes the first man to fly over U.S. soil, in a hydrogen balloon trip between Philadelphia and Woodbury, New Jersey.

1868 Danish chemist Søren Sørensen, creator of the term "pH" and of the pH scale, which is now the standard measure of acidity in chemical solutions, is born the son of a farmer in Havrebjerg.

JANUARY 9

1869 Physical chemist Richard Wilhelm Heinrich Abegg is born in Danzig, Prussia. His major contribution was understanding the capacity of atoms to bond with each other, known as "valence."

1878 Psychologist John B(roadus) Watson is born in Greenville, South Carolina. His 1913 article "Psychology as a Behaviorist Views It" is regarded as the founding of behaviorism, a school of social science that considers observable behavior to be the only proper object of study; emotions, thoughts, and other mental processes are ignored by strict behaviorists.

1903 Wind Cave National Park (South Dakota) is authorized.

1922 Indian-U.S. biochemist Har Gobind Khorana, winner of the 1968 Nobel Prize for helping to explain how the information in DNA is translated into protein manufacture in cells, is born in Raipur, India.

1929 The Seeing Eye, the nation's first institute to train guide dogs for the blind, is incorporated in Nashville, Tennessee, as a nonprofit organization. Mrs. Harrison Eustis is the first president. "Buddy," the first Seeing Eye dog in the United States, was imported from Switzerland in April 1928.

1938 The first patent reissue in U.S. history is granted. A reissue involves changing the wording in a patent so that the inventor has more legal protection for his or her invention. Julius Hatch of Great Bend, Pennsylvania, receives Reissue No. 1 on his "machine for sowing plaster, ashes, seed and other separable substances."

1940 In London, legendary biologist J.B.S. Haldane is asked by his government to study means of escape from sunken submarines. The study soon becomes research into breathing and other bodily reactions under extreme conditions: extreme cold, enormous pressure, the relative absence of oxygen, and all of these conditions combined. The lives of the small group of volunteer subjects were regularly endangered by the experiments; Haldane used himself most often as a subject. It was the first time science had looked carefully at functioning under such rigorous conditions.

1950 "The grand aim of all science is to cover the greatest number of empirical facts by logical deduction from the smallest number of hypotheses or axioms."—Albert Einstein, quoted in *Life* magazine.

1992 The federal government restricts logging on 6.9 million acres of ancient northwestern forest considered critical to the survival of the northern spotted owl.

1995 The Associated Press reports that a new museum of prehistoric art has just opened on the banks of Skunk Creek in the Deer Valley outside Phoenix, Arizona. "The Deer Valley Rock Art Center" consists of 500 boulders on which prehistoric peoples carved some 1500 patterns and figures called "petroglyphs." The site is under the protection of the Army Corps of Engineers and Arizona State University. The artwork dates as far back as 5000 BC.

1968 *Surveyor 7* lands on the moon, having been launched from the United States on January 7. This is the last "soft landing" (i.e., the ship did not crash, but landed carefully and remained functional once on the moon) of an unmanned vehicle prior to the manned *Apollo* flights that put men on the moon. *Surveyor 7* is the first ship to land in the lunar highlands, near the crater Tycho, and the first spacecraft to observe artificial light from Earth.

1778 Famed botanist-explorer Carolus Linnaeus dies at 70 in Uppsala, Sweden. His fascination with flowers blossomed early in life; his childhood nickname was "the little botanist." His *Systema Naturae*, published in 1735, presented the modern system of classifying and naming plants and animals.

1833 New rules for the Massachusetts Lunatic Hospital take effect. Thought to be written by educational reformer Horace Mann, the regulations were the most progressive in their day.

1853 Astronomer John Martin Schaeberle is born in Württemberg, Germany. He was the second to discover a "dim companion" of a known star. He discovered the body in the vicinity of the star Procyon in 1896. Such companions are now known to be common. They lie near huge stars and emit light themselves, but their small size makes them seem dim in comparison.

1864 Alfred Nobel dates his patent application for dynamite.

1911 A photograph is taken from an airplane for the first time in U.S. history. Major H.A. "Jimmie" Erickson photographs San Diego from a Curtiss biplane piloted by Charles Hamilton.

1916 The Supreme Court upholds the Sherley Amendement of 1912, which stated that the labels on foods and drugs could not make false claims about the product's powers, effectiveness, or properties.

1936 Radio astronomer Robert Woodrow Wilson, discoverer of evidence for the big bang theory of the universe's creation from an enormous explosion, is born in Houston, Texas. In collaboration with Robert Penzias at the Bell Telephone Laboratories, Wilson discovered a background radiation that permeates the cosmos uniformly, and is now thought to be a remnant of the big bang. The pair shared the 1978 Nobel Prize for Physics.

1946 A radar signal is bounced off the moon for the first time. Supervised by Lt. Col. John H. DeWitt, members of the U.S. Army Signal Corps at the Evans Signal Laboratories in Belmar, New Jersey, shoot a brief signal at the moon, and receive an echo 2.4 seconds later.

1950 Camilo and Catherine Lyra are found bludgeoned to death in their apartment in Brooklyn, New York. Thus begins a historic case in the use of hypnosis in the court system. The Lyras' son confesses to the crime under hypnosis, and is sentenced death—three times. His appeal is upheld after each trial and conviction, and he is eventually released because he was tricked into being hypnotized.

1978 Two Russian cosmonauts are launched in a *Soyuz* capsule for a rendezvous with the orbiting *Salyut VI* space station, on which other cosmonauts have been living for a month. *Salyut VI* was the first in a new generation of space stations, featuring an advanced refueling system, docking ports in both the front and the back, and more comfortable living quarters.

1994 Called "a hopeful and historic breakthrough" by President Bill Clinton, Ukraine agrees in Brussels to eliminate all of its nuclear weapons in about seven years. Among its arsenal are some 1500 warheads aimed at the United States. In return, we have pledged cash and aid in refurbishing Ukraine's ailing nuclear power program, including the infamous Chernobyl reactor.

1995 A team of Japanese and U.S. astronomers report discovering evidence of the largest known black hole in the universe. Using the Very Long Base Array radio telescope, which is actually a coordinated set of ten telescopes spread one-third of the way around the world, the scientists located an enormous disk, rotating at 650 miles per second, in a galaxy 21 million light-years from Earth. The black hole sits in the middle of this disk.

1863 The world's first subway, the "Metro" in London, opens to the public.

1638 Nicolaus Steno is born the son of a prosperous goldsmith in Copenhagen. As a geologist he devised the first law of crystallography (also called Steno's law, it states that the crystals of specific substances have fixed, characteristic angles at which the crystal faces meet), proposed that layers of earth that are now tilted were once horizontal, and was the first to realize that Earth's crust reveals geological history, which led him to conduct the first detailed studies of rock layers. He made the radical suggestion that fossils were once living creatures (revising a notion of a few ancient philosophers), and proposed that many rocks resulted from sedimentation. As an anatomist, he discovered that muscles are made of fibers, identified the pineal gland in animals other than man (which shot down Descartes's widely held notion that the pineal gland was the seat of human thought), and discovered the duct of Steno, between the mouth and the parotid salivary gland. Much of his scientific findings clashed directly with church dogma. At age 29 he abandoned science completely, became a Catholic, and was in time made a bishop.

1770 Rhubarb is first shipped to the United States, from London. Benjamin Franklin dispatches a quantity to John Bartram in Philadelphia. Through the ages, rhubarb has been used in landscaping, as a food, and as a medicine, especially as a laxative, despite the fact that the roots and leaves are poisonous.

JANUARY 11

1786 J.J. Lister, inventor of the achromatic, or blur-free, lens for microscopes, is born the son of a Quaker wine merchant in London. His innovation came in 1830, and is considered a milestone in modern microscopy. With it, Lister was the first to see the true shape of red blood cells. Despite his accomplishments (which won Lister membership in the Royal Society), his place in history is overshadowed by his son, surgeon Joseph Lister, who founded antiseptic medicine.

1840 The Antarctic ice cap is reached by the naval expedition of Charles Wilkes (the first scientific expedition sponsored by the U.S. government). The voyage later determined that Antarctica is a continent, which Wilkes named.

1842 William James, a groundbreaker in both psychology and philosophy, is born the son of a philosopher in New York City. His brother Henry was the distinguished novelist. James was plagued by poor health throughout life, and he underwent a suicidal breakdown in his mid-20s. He had trouble deciding on a career, studying art, then medicine, then natural history. In 1872 he taught physiology at Harvard, but in 1876 he took the radical step of switching fields to psychology, then in its inception and firmly tied to philosophy and theology. James's monumental *The Principles of Psychology* (1890) was revolutionary because it treated mental activity as a biological event, and stressed scientific experimentation in psychology.

1882 Theodor Schwann, the founder of modern biology for his identifying cells as the basic unit of life, dies at 71 in Cologne, Prussia. Schwann was just 24 when he made the major discovery that digestion does not occur only by acid, but through acid plus another substance, which he named "pepsin." It was the first time that an enzyme was prepared from animal tissue, and was a milestone in biochemistry. But Schwann's greatest achievement came in 1839 when he synthesized his own and others' work into one, grand doctrine, which he called "the cell theory," which states that all plants and animals are composed of cells. It is now a universally accepted law. Schwann was also the first to identify eggs as single cells, coined the term "metabolism" to describe chemical processes in living things, and discovered "Schwann cells," which form the sheath around nerve processes that enables nerve impulses to move very rapidly through the body.

1889 Geneticist Calvin Bridges is born in Schuyler Falls, New York. His research helped prove the fledgling "chromosomal theory of heredity," which states that hereditary information is carried in chromosomes. Working with Thomas Hunt Morgan at Columbia University in the early 1900s, Bridges showed that inherited variations in an animal are related to chromosome structure. This led to the construction of "gene maps," in which specific traits are identified with specific chromosome segments. Morgan and Bridges did their work on the fruit fly, making this animal one of the favorite subjects of genetics research and teaching ever since.

1993 When a young child eats a hamburger at a Jack in the Box restaurant in Seattle and eventually dies (on this date), it is found that improper cooking failed to kill one strain of the dangerous bacterium *E. coli*. Eventually hundreds of people fell ill, leading to new cooking standards in the fast-food industry.

1580 Jan Baptist van Helmont, the father of biochemistry and originator of the term "gas," is born into a noble family in Brussels. Although he was an alchemist and his explanations often involved mysticism, he made an everlasting contribution to science by being the first to conduct biological experiments in which the conditions and results were numerically measured. He performed one of the most famous experiments in biology by growing a willow tree in a precise quantity of soil, to which only water was added; after five years the tree had gained 164 pounds but the soil had lost only 2 ounces. He erroneously concluded that the water was the major source of the tree's nourishment. Ironically, air's carbon dioxide (which Helmont also discovered) contributes most to the tree's growth.

1729 Italian physiologist Lazzaro Spallanzani is born the son of a distinguished lawyer in Modena. He performed groundbreaking experiments on many phenomena, including the regeneration of organs, the spontaneous generation of organisms (which paved the way for Pasteur's work), the sonar of bats, the circulation of blood, the components of digestion, and the role of sperm in artificial insemination.

1773 America's first public museum is organized, at the annual anniversary meeting of the Charleston [South Carolina] Library Society. The first curators of the Charleston Museum are Charles Cotesworth Pinkney, Thomas Heyward, Dr. Alexander Baron, and Dr. Peter Fassoux.

1861 James Mark Baldwin, heavily influential in moving psychology from the realm of philosophy to the realm of science, is born in Columbia, South Carolina. He studied in Leipzig, Germany, where he became acquainted with the founder of experimental psychology, Wilhelm Wundt. Baldwin's *Handbook of Psychology*, published in two volumes in 1889–1891, was the first treatment in English of the emerging science of psychology. He then held positions as a professor of philosophy in Toronto and Princeton, where he established psychology laboratories. His *Mental Development in the Child and the Race* and *Social and Ethical Interpretations in Mental Development* treated mental phenomena from a new perspective: as biological events that are influenced by the processes of evolution. In 1894 he collaborated with James McKeen Cattell to found *Psychological Review*, which spawned *Psychological Index* and *Psychological Bulletin*, which is still going strong today.

1896 The first X-ray photograph in U.S. history is taken by Dr. Henry Louis Smith, professor of physics and astronomy at Davidson College, Davidson, North Carolina, who fires a bullet into the hand of a corpse, then takes a 15-minute X-ray exposure of it. The bullet's position is revealed exactly when the photographic plate is developed.

1899 Paul Hermann Müller, who discovered that DDT could be an exceptional insecticide, is born in Olten, Switzerland. He won the 1948 Nobel Prize for Physiology or Medicine for this discovery. The son of a civil servant, Müller originally worked as a dye chemist until he began searching for the "ideal" insect-killer—cheap, long-lasting, not harmful to mammals or plants, fast-acting, and toxic to many insect species. After four years of intense but fruitless effort, Müller found it: DDT. He first tried this substance (which was actually created by Othmar Zeidler in 1874) in September 1939, coincident with the start of World War II. Use of the chemical spread quickly throughout the world, but after 25 years insidious environmental effects were discovered, eventually leading to the banning of DDT in many places.

1986 The shuttle *Columbia* blasts off with the first Hispanic American in space, Dr. Franklin R. Chang-Diaz.

1995 The wolf is reintroduced to Yellowstone National Park. By 10:45 PM the cages of eight gray wolves (*Canis lupus*) are opened in remote spots in the Park. The lupine pioneers are to roam one-acre pens for about a month, after which they will be set free. Once common, wolves were hunted to extinction in the area by settlers 60 years ago.

1944 "N'gagi," the heaviest gorilla ever kept in captivity, dies at age 18 in the San Diego Zoo. In 1943, N'gagi set the record at 683 pounds. Another record animal, the oldest age-verified rodent, dies on this day in 1965 in the National Zoological Park in Washington, D.C.; the animal was a 27-year-old Sumatran crested porcupine.

1843 Brain scientist Sir David Ferrier is born in Aberdeen, Scotland. He obtained a medical degree, but did not like general practice, and devoted himself to neurological research (the study of the nervous system). Using living animals (mainly primates, which resulted in a cruelty-to-animals lawsuit against him in 1882), he showed that the brain's cortex contains sensory regions, which receive information from the senses, as well as motor regions, which control the actions of muscles and other organs.

1854 The nation's first accordion patent is No. 11,062, issued on this date to Anthony Faas of Philadelphia.

1858 Physiologist Oskar Minkowski is born in Aleksotas, Russia (now Kuanas, Lithuania). In 1889 he and collaborator Joseph Mering were able to produce a diabetic condition in dogs by removing the pancreas. Minkowski guessed that the pancreas produces some "antidiabetic" substance, a theory that later led the team of Sir Frederick Banting and Charles Best to discover insulin, a hormone that successfully treats the disease when injected into the body. Minkowski's younger brother Hermann was a world-famous mathematician-physicist whose theories laid the groundwork for Einstein's theory of relativity.

JANUARY 13

1864 Physicist Wilhelm Wien, who explained why objects glow red when they are heated, is born the son of a landowner in Gaffken, Prussia. Using a special chamber with one tiny hole in it that allowed Wien to analyze the light that was given off as the inside of the chamber was heated to different temperatures, he discovered that many wavelengths are emitted at elevated temperatures, but each temperature produced a characteristic emission peak. As the temperature increased, the peak shifted from the dull red to the bright red to the yellow-white and finally to the blue-white wavelengths. Wien received the 1911 Nobel Prize for Physics for this work.

1870 Ross Granville Harrison, the first to successfully grow isolated animal tissue in the lab, is born in Germantown, Pennsylvania. During his first year as a Yale professor Harrison was able to cultivate sections of tadpole; this and subsequent innovations have been widely applied to disease-fighting and to organ transplantation and grafting.

1906 A radio is first advertised for sale, in the United States in the current issue of *Scientific American*. Its price is $7.50 and it "will work up to one mile."

1942 Henry Ford of Dearborn, Michigan, patents the construction of plastic automobiles.

1976 A machine that reads printed material aloud is first demonstrated publicly in the United States. Raymond Kurzweil of Cambridge, Massachusetts, invented the device to assist the blind; pages are scanned by a camera connected to a computer, which analyzes the letters and provides an audio output. It operates at 150 words per minute.

1910 The latest radio innovations of Dr. Lee De Forest are demonstrated by the great opera star Enrico Caruso, who is performing *I Pagliacci* at the Metropolitan Opera House in New York City. It is the first opera broadcast in the United States.

1981 Twelve-year-old Donna Griffiths of Pershore, England, begins sneezing. She does not stop sneezing for 978 days (in September 1983), which is a record. The fit contained an estimated one million sneezes in its first year.

1994 The Hubble Space Telescope is finally functional after a series of in-space repairs. "It's fixed beyond our wildest expectations," says Ed Weiler, NASA's Hubble program scientist. The bus-sized, orbiting Hubble was launched in 1990 with a number of problems, not the least of which was a misshapen mirror that blurred all of its images. During the latest 11-day shuttle mission to the Hubble, astronauts fixed all of the problems, and now the telescope is poised to send back the clearest-yet views of very old and very distant objects throughout the universe.

1994 The woman who founded MADD (Mothers Against Drunk Driving) announces in Sacramento that she is now a paid lobbyist for the liquor industry. Candy Lightner began the world-famous MADD in 1980 after her 13-year-old daughter was struck by a hit-and-run drunk driver, but in 1985 she was fired by the group's board over money disagreements. Now she is working for the American Beverage Institute in their fight against laws to lower the legal standard for drunken driving (from 0.10% to 0.08% blood alcohol content).

1791 Midget Calvin Phillips is born in Bridgewater, Massachusetts. His maximum height was 26½ inches, making him history's smallest adult male, a title he held for nearly two centuries, when Gul Mohammed was born in New Delhi; Mohammed's maximum adult height was 22½ inches.

1794 Elizabeth Hog Bennett becomes the first U.S. woman to successfully deliver a child by Cesarean section, in Edom, Virginia. The surgeon is her husband, Dr. Jessee Bennett, who had asked Dr. Alexander Humphries of Staunton, Virginia, to assist. Humphries refused, stating that the risk of failure was too great. Bennett got help from two fieldhands, who held the woman on a wooden table.

1801 Adolphe-Théodore Brongniart, the founder of modern paleobotany (the study of extinct plants), is born in Paris.

JANUARY 14

1806 Matthew Fontaine Maury, author of the first textbook on oceanography (*The Physical Geography of The Sea*, published in 1855), is born the son of a small planter near Fredericksburg, Virginia.

1875 "Reverence for life" became the driving ethic of Albert Schweitzer, born this day in Kayersberg, Alsace, the eldest son of a Lutheran pastor.

1890 English geologist Arthur Holmes is born in Hebburn on Tyne. His lifework was dating the age of rocks through analysis of radioactive decay. At the beginning of his career, the predominant estimate of Earth's age was about 100 million years (derived by the eminent Lord Kelvin); Holmes's work reset this to about 4.6 billion years. The exact age of Earth is currently unknown.

1938 The nation's first organization to promote legal euthanasia is formed in New York City by Reverend Charles Francis Potter (president), Dr. Harold Hays (secretary), and Charles Edward Nixdorff (treasurer). Originally called the National Society for the Legalization of Euthanasia, the group incorporated on November 30, 1938, as the Euthanasia Society of America.

1947 The first Pan African Congress on Prehistory convenes in Nairobi, under the leadership of Louis Leakey.

1984 Ray(mond Albert) Kroc, founder of the McDonald's empire, dies at 81 in San Diego. Kroc was an ambulance driver at age 15 in World War I, and then held a number of postwar jobs, including jazz pianist, real estate salesman, and paper cup merchant. While selling blenders in 1954, Kroc learned of a restaurant owned by brothers Maurice and Richard McDonald in San Bernardino, California, that used an assembly line to market burgers, fries, and milk shakes; Kroc was impressed because the brothers needed an exceptional number of his blenders to keep up with the orders. Kroc decided to establish a chain of drive-in restaurants based on this model, and agreed to pay the McDonalds 0.5% of the gross receipts. The first McDonalds as we know them opened on April 15, 1955, in Des Plaines, Illinois. The rest(aurant) is history.

1994 Remains of a whale that once walked on land have been discovered in Pakistan, reports paleobiologist J.G.M. Thewissen in the journal *Science*. "This critter is a missing link between land animals and modern whales," said Thewissen, who named the species *Ambulocetus natans*. It was about the size of a sea lion, weighing 600 to 700 pounds, and moved on land like seals and sea lions, by lurching forward and bumping along on its chest. It also had the teeth of a meat eater, but probably was unable to catch prey on land. It apparently died out 50 million years ago, some 10 million years before seabound whales evolved.

1914 Ford Motor Co. greatly improves efficiency in its car factory by adding a chain to move car bodies along the assembly line. This innovation is now standard in many factories worldwide.

1609 The first weekly newspaper, *Aviso Relation oder Zeitung*, debuts. It is published by Julius Adolph von Sohne in Wolfenuttel, Saxony, Germany.

1759 The great British Museum opens to the public in London.

1767 Swedish chemist Anders Gustav Ekeberg, discoverer of the element tantalum, is born in Stockholm.

1784 English chemist Henry Cavendish ignites hydrogen and produces water. This demonstrates that water is a combination of gases, and it overthrows the ancient Greek belief that the world is composed of four elements (air, water, earth, and fire).

1785 Chemist-physiologist William Prout is born the son of a farmer in Horton, England. In 1824 he identified the acid in stomach secretions as hydrochloric acid, which was already known as a substance that could dissolve metal. He was the first to divide food into the familiar groupings now known as fats, proteins, and carbohydrates (in 1827), and the first to point out that the atomic weight of each element is some multiple of hydrogen's atomic weight (in 1815).

JANUARY 15

1842 Physician Josef Breuer, destined to play a pivotal role in Sigmund Freud's development of psychoanalysis, is born the son of a Jewish religion teacher in Vienna. Breuer became an outstanding physician, and in partnership with Ewald Hering discovered breathing reflexes and the role of semicircular canals in balance. From 1880 Breuer began treating the hysterical symptoms of a woman known to science history as "Anna O." He found that hypnosis and hysteric outbursts were helping to ease the woman's distress. Breuer discussed the case with Freud, then in his mid-20s, who took over care of Anna O. She was the first to receive psychoanalysis, which Freud was inventing as he was treating her.

1877 Lewis M(adison) Terman, developer of the modern intelligence test, is born in Johnson County, Indiana. In 1916 he published *The Measurement of Intelligence*, which presented the Stanford–Binet test. At its core is "IQ," (intelligence quotient), which compares the chronological age and the mental age of an individual. In 1921 Terman launched a study of "gifted" persons; it is expected to end in 2010.

1895 Biochemist Artturi Virtanen is born in Helsinki. He won the 1945 Nobel Prize for discovering that adding acid to cattle fodder prevents rotting without reducing nutritional value.

1907 The three-element radio tube, the device that revolutionized human communication, is patented by Lee De Forest. Patent No. 841,387 is for a "device for amplifying feeble electrical currents." Radio finally allowed complex messages to be sent over great distances without the use of wires.

1908 Physicist Edward Teller, the father of the H bomb, is born in Budapest.

1919 An enormous flood of molasses swamps the waterfront section of Boston. A poorly constructed storage tank bursts in unseasonably warm temperatures; 2.3 million gallons of crude molasses is unleashed in a 30-foot-high wave. A firehouse is tipped over, homes are ripped from foundations, and 21 people die.

1939 Workers in the Copenhagen laboratory of famed physicist Niels Bohr split the atom, confirming the feat first accomplished by Hahn and Strassmann weeks before.

1870 A donkey is used to represent the Democratic political party for the first time, in a Thomas Nast cartoon in *Harper's Weekly*.

1943 The nation's first building to contain 6.5 million square feet of usable space, the Pentagon, is completed.

1952 Boston Red Sox baseball star Jimmy Piersall suffers a complete and violent nervous breakdown in the lobby of the Sarasota Terrace Hotel in Florida. When recovered, Piersall wrote *Fear Strikes Out*, a book that looked into his mind as he went insane and as he recovered. The book is now recognized as a pioneering insight into mental disease. It became a movie starring Anthony Perkins.

1477 Geographer Johannes Schöner is born in Karlstadt, Germany. In 1515 he produced the first globe that included the lands discovered by Columbus, and it was the first globe to use the name "America."

1831 Off Praia, Cape Verde Islands, 21-year-old Charles Darwin first sees tropical terrain on the fateful journey of the *Beagle*, during which the seeds for the theory of evolution were planted in his mind.

1840 The first scientific expedition sponsored by the U.S. government sights the continent of Antarctica. The expedition is the four-ship convoy under the command of Charles Wilkes. The area of land seen is now called Wilkes Land, and Charles Wilkes gave the continent its name.

1874 Chang and Eng, history's original Siamese twins, begin dying in the middle of the night in Mt. Airy, North Carolina, at age 62. The moody and alcoholic twin, Chang, had suffered paralytic stroke several years before, and preceded his brother in death by three hours.

1874 German zoologist-cytologist Max Schultze dies in Bonn, Germany, at 48. He was the first to realize that the watery gel that fills most cells, now called "protoplasm," is a critical and common substances in all plants and animals.

1875 Leonor Michaelis, the biochemist who made the home permanent possible, is born in Berlin. He discovered that keratin (a protein that is the chief constituent of hair) partially dissolves in certain acids. This led others to create a revolution in cosmetics, with products that allowed the hair to be curled and shaped. Of more profound importance to science is the Michaelis–Menten equation, which describes the speeds at which enzymes will combine or split apart chemical compounds.

1881 Radar developer Sir Arthur Percy Morris Fleming is born in Newport, Isle of Wight.

1919 Nebraska becomes the critical 36th state to ratify the 18th Amendment to the Constitution, Prohibition. Wyoming and Missouri also ratify the amendment today. The Secretary of State proclaims that the amendment will become law the become law the following January, which it did on this date in 1920.

1936 The first photo-finish camera begins operation at a U.S. racetrack, the Hialeah Race Course in Hialeah, Florida.

1939 Two historic papers on atom-splitting are completed and dated in Copenhagen, and soon will be sent to *Nature*. The first is by Lise Meitner and Otto Robert Frisch (Meitner's nephew); it coins the term "nuclear fission" and is the first to suggest that Hahn and Strassmann have, indeed, split the atom. The second paper, by Frisch, is a modest effort of about 500 words reporting experiments that confirm this conclusion. These papers mean that an atomic bomb is possible.

1957 The first nonstop flight around Earth begins. Three B-52s depart Castle Air Force Base, California, and stay in the air for 45 hours 10 minutes.

1963 Soviet Premier Nikita Khrushchev announces in Berlin that the first 100-megaton bomb is ready.

1967 On her 35th birthday, Dian Fossey begins her historic field studies of gorillas in Rwanda's Virunga Mountains.

1969 Two Russian *Soyuz* spaceships become the first to dock and exchange personnel in outer space.

1995 Keiko (the whale from the movie *Free Willy*) will be getting a new home, announce wire services throughout the world. Keiko languishes in a pool in Mexico, losing weight and suffering from a skin disease. Some $5 million had been raised, however, allowing construction of a larger enclosure in Oregon.

1932 Primatologist-conservationist Dian Fossey is born in San Francisco.

1501 Leonhard Fuchs, author of the first important botany dictionary, is born in Wemding, Bavaria. His botany writings included detailed descriptions of many plants. A group of shrubs and the color of its flower (now called fuchsia) are named for him.

1834 Biological theorist (forced to theory-making because of poor eyesight) August Weismann is born the son of a classics professor in Frankfurt am Main, Germany. Ahead of his time, he suggested that chromosomes carry hereditary information and he devised the "germ plasm theory," which states that some portion of organisms lives on from generation to generation.

1893 Hawaii's monarchy ends. Queen Liliuokalani is forced to abdicate by a coalition of white businessmen and sugar planters.

1911 Explorer-anthropologist-eugenicist Sir Francis Galton dies at 89 in Haslemere, England (see February 16 for details of his life).

JANUARY 17

1912 Alexis Carrel, in his lab at Rockefeller Institute for Medical Research in New York City, begins a historic experiment in which isolated fragments of a chick embryo heart were kept alive for more than 30 years, much longer than the life span of the chicken itself. Carrel won this year's Nobel Prize for Physiology or Medicine for developing methods of suturing blood vessels and for research into organ transplantation.

1928 The nation's first fully automatic film-developing machine is patented (No. 1,656,522) by Anatol M. Josepho, who built the first model in a loft on 125th Street in New York City. He is said to have received $1 million for his invention. The first "Photomaton" studio opened to the public at 1659 Broadway, New York City, in September 1926 (a year and a half after Josepho applied for a patent).

1955 The submarine USS *Nautilus* is launched on its first nuclear-powered test run from its berth in Groton, Connecticut. In 1957 it became the first submarine to remain submerged for two weeks.

1962 The nation's first indoor ski slope, the Ski-Dek Center, opens in Buffalo, New York. It is built in a former movie theater, and involves endlessly moving tracks that are treated to simulate two inches of powdered snow on a firm base. It has nine separate runs, and can accommodate 144 skiers.

1994 A new strategy in the fight against heart disease is unveiled by Dr. Stephen Epstein (National Institutes of Health) at a meeting of science writers in Clearwater, Florida. Eight years earlier, Epstein had begun experimenting with the growth of new blood vessels to compensate for the clogging of old ones. Now his work has paid off. He has been able to genetically engineer a naturally occurring substance called "fibroblast growth factor." As little as seven days' treatment triggered a 50% increase in the growth of new coronary arteries in dogs. If adaptable to humans, the procedure promises to be a noninvasive alternative to angioplasty and surgery. But it does have a dark side: The substance could spur unwanted growth in various body parts, including growth of dormant tumors.

1994 An earthquake of magnitude 6.6 on the Richter scale strikes Los Angeles. Several bridges collapse, but three special highway ramps near Dodger Stadium remain standing. It is a triumph for a new composite material, partly glass and partly plastic, with which the ramps were reinforced two years ago. It is the same composite material that was once kept secret because of its use in the Stealth Bomber. Even though it is very expensive, engineers see it as a revolution in highway construction. Exactly one year after this test (1995) a huge earthquake roars through Kobe, Japan, shattering that country's confidence in their antiquake building technology.

1706 Benjamin Franklin is born in Boston, the 15th of 17 children of a poor candlemaker. Although he only had two years of formal education, Franklin achieved fame as an inventory, scientist, and philosopher.

1733 A white bear is exhibited for the first time in the United States. "Ursa Major," a 9-month-old cub, is shown in a cage at Clark's Wharf, in Boston's North End. The cub had been captured in Greenland.

1734 German physiologist Caspar Friedrich Wolff is born the son of a tailor in Berlin. Regarded as the founder of modern embryology, his 1759 book presented the notion that the organs of living things take shape gradually from nonspecific tissue (conventional wisdom of the time started that living things came into this world fully formed, but small). The book was largely ignored for a half-century. Wolff's name lives on in the Wolffian body and the Wolffian ducts; both are part of the kidneys of embryos.

1778 James Cook "discovers" the Hawaiian Islands, and exactly 134 years later (1912) Robert Scott reaches the South Pole. Both were great English explorers. Cook perished in the Islands about a year later, and Scott died on the return trip from the Pole.

1788 At Botany Bay, European settlers first reach Australia.

1825 Sir Edward Frankland, a pioneer in both structural chemistry and the study of pollution, is born illegitimately in Churchtown, England. He taught himself chemistry while a druggist's apprentice. Frankland was the first to study the organometals, announced the fundamental theory of a valence in 1852 (i.e., that each atom has a fixed capacity for combination with other atoms), and in 1868 was appointed to a commission to study contamination in English rivers.

1858 Daniel Hale Williams, credited with performing the first successful heart surgery, is born in Hollidaysburg, Pennsylvania. Frustrated at the lack of opportunity for black physicians, Williams founded his own hospital, Provident Hospital in Chicago. The nation's first interracial hospital, it was there on July 10, 1893, that he opened a patient's chest (without the use of modern anesthetics, antibiotics, or blood transfusion) and repaired the sac that surrounds the heart.

1861 Hans Goldschmidt is born in Berlin. Working in his family's metallurgy business, he invented a highly efficient way of extracting pure metals from their oxides. The process is called the aluminothermic process, or the Goldschmidt reduction process. Like Frankland (above), Goldschmidt was a student of the great chemist Robert Wilhelm Bunsen, who is most famous for the Bunsen burner, which he did not invent.

1896 An X-ray machine is first exhibited in the United States, at the Casino Chambers in New York City. Spectators are charged 25 cents to view the device, nicknamed the "Parisian sensation."

1908 Jacob Bronowski, mathematician, humanist, and popularizer of science, is born in Poland. During World War II, with a Ph.D. in math from England's Cambridge University, Bronowski pioneered the field of "operational research" in his efforts to enhance the efficiency of Allied bombing missions. His 1945 visit to the nuclear site of Nagasaki, however, made him abandon military research and devote his life to humanistic applications of science, which he pursued as an author and television host.

1911 An airplane lands on a ship for the first time. Traveling at 35 mph, civilian pilot Eugene B. Ely flies his craft onto a sloped platform on the cruiser *Pennsylvania* in San Francisco Bay; ropes and sandbags are used to stop the plane. Ely was also the first to take off in a plane from a ship, in 1910 off the coast of Virginia.

1994 The U.S. Department of Energy announces a major breakthrough in solar energy. United Solar Systems Corp. of Troy, Michigan, has produced solar panels that are nearly twice as efficient as current panels in converting sunlight into electricity. Current panels are able to harness about 6% of the available solar energy, whereas the new panels have achieved a record 10.2%.

1994 Brutalis, a 2.3-ton white rhinoceros, is tranquilized and taken from Givkud Lion Park in Denmark, after suffering a nervous breakdown in captivity. He went on a rampage and destroyed $22,000 in fencing and cages. Brutalis was finally shipped to a wildlife sanctuary in northern Namibia.

1943 A wartime ban on sliced bread goes into effect in the United States, in an effort to reduce demand for metal parts by bakers.

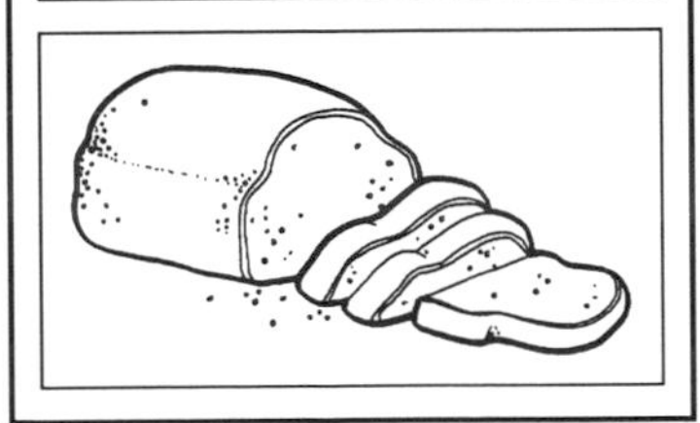

1747 Astronomer Johann Elert Bode, popularizer of Bode's law (also called the Titius–Bode law, it is a simple rule giving the approximate distances of our planets from the sun), is born the son of a schoolteacher in Hamburg, Germany. The law was eventually found to be invalid, but by then it had been a boon to astronomy because it stimulated many astronomers to search the skies for missing planets.

1813 Metallurgist-inventor Sir Henry Bessemer is born the son of an engineer in Charlton, England. In his teens Bessemer invented a new method for stamping deeds; the British government thought so highly of the process that they immediately adopted it—without compensating its inventor. Bessemer had learned his lesson about getting patent protection. His most famous invention was the Bessemer process, the first way discovered for mass-producing steel cheaply.

1825 The first canning patent in the United States is issued to Ezra Daggett and his nephew, Thomas Kennett, for a way to "preserve animal substances in tin." The pair began canning salmon, oyster, and lobsters in New York City in 1819. It was not until after the Civil War that canning became a major industry.

1851 Educator-naturalist-philosopher David Starr Jordan is born in Gainesville, New York. Educated as a doctor, Jordan became a world-class authority on fish; he was the president of Indiana University in the 1880s and the president of Stanford from 1891 to 1913. He devoted the remainder of his life to world peace and progressive legislation.

1851 Dutch astronomer Jacobus Cornelius Kapteyn, known for using photography and statistics to determine the characteristics of our galaxy, is born the son of a schoolmaster in Barneveld. His achievements include a catgalogue of nearly 500,000 southern stars, discovery of star streaming (i.e., that star movement is not random, but follows two cosmic streams), and the first description of the shape of our galaxy (announced the year he died, 1922).

1879 The tallest married couple on record (Anna Hanen Swan of Nova Scotia and Martin van Buren Bates of Whitesburg, Kentucky; 7 feet 5½ and 7 feet 2½ inches, respectively) give birth to one of the biggest babies ever: 23¾ pounds.

1929 The country's first national park east of the Mississippi, and its first national park on an ocean, is given its third (and current) name: Acadia National Park. The 27,871-acre Maine preserve was established in July 1916 by President Wilson as "Sieur de Monts National Monument." In February 1919, it became "Lafayette National Park." The area is located on Mount Desert Island, about a mile south of Bar Harbor.

1963 History's first spleen transplant is performed in Denver, Colorado, by Dr. Thomas Starzl on a hemophiliac child.

1977 The largest-ever gathering of humans occurs when 12,700,000 meet for the Hindu feast of Kumbha Mela in Allahabad, India.

1991 The U.S. Fish and Wildlife Service begins trapping Florida panthers in the wild in a desperate effort to save the species from extinction. Those trapped will be put in four Florida zoos, and their offspring will be released in the wild. Only 30–50 of the species remain, and that number is reduced every year by speeding motorists. Growing mercury contamination in the area, shrinking habitat, and increased inbreeding caused by a drop in population all put further pressure on the panther.

1995 At 80 years young, fitness guru Jack La Lanne is appointed to California's Council on Physical Fitness and Sports. His exploits and demonstrations of physical stamina are legendary, but he led a troubled life until age 14. "I was eating sugar, sugar, sugar. I was a troublemaker in school. I had an uncontrollable temper. But then I became a vegetarian and got involved in physical fitness. It changed my life."

1736 Engineer James Watt, inventor of the steam engine, which drove the Industrial Revolution, is born in Greenock, Scotland. He was so sickly as a child that he never went to formal school. He suffered tremendously from chronic migraine headaches, and was thought mentally retarded as a youth.

1809 The first significant U.S. geology book, *Observations on the Geology of the United States* by William Maclure, is presented to the American Philosophical Society in Philadelphia. It was revised and finally published in 1817. It contained the first geological map of the eastern United States, and the first complete geological map of the entire country.

1815 Horace Wells, the first to use gas as an anesthetic, is born in Hartford, Vermont. He first noticed the pain-killing properties of nitrous oxide (laughing gas) during the visit of a traveling road show to Hartford, Connecticut, where he had a dental practice, and thereafter he used the gas to perform painless operations. In January 1845, he gave a demonstration at the Massachusetts General Hospital, but the patient was unresponsive to the gas, covering Wells in ridicule. In October 1846, William Morton (Wells's former partner) gave a successful public demonstration of the same techniques, and Morton was hailed as the innovative hero. Wells's life turned sour. He began to self-experiment with a number of chemicals suspected of having anesthetic properties, and his personality was wrecked. In 1848 he was arrested in New York City for throwing acid at a passerby. He committed suicide in jail, just as the Paris Medical Society proclaimed him discoverer of anesthetic gases.

1896 Several months after their discovery, X rays are first used in a clinical setting—on two different continents on the same day! In Dartmouth, New Hampshire, they are used during the diagnosis and setting of a broken arm of one Eddie McCarthy. In Berlin, Germany, they are used to locate a glass splinter in a finger.

1929 The movie *In Old Arizona* (actually filmed in Utah and California) is released. It is the first full-length talking picture filmed outdoors.

1930 Edwin E. "Buzz" Aldrin, the second man to walk on the moon, is born in Glen Ridge, New Jersey. In 1963 Aldrin is both selected as an astronaut and takes his doctorate in aeronautics from MIT; his thesis on orbital mechanics and rendezvous establishes techniques that make lunar landings possible.

1982 The Big Five in video equipment (Hitachi, JVC, Matsushita, Sony, and Philips) sign an agreement to cooperate on construction of a camera with a built-in videocassette recorder.

1994 Australian researchers report in *Nature* the discovery of a genetic defect in humans that predicts osteoporosis. This disease is a weakening of bones by excessive loss of tissue, causing an estimated 1.5 million broken bones annually in the United States. The new gene test means that diagnosis could be made in childhood, thereby allowing precautionary measures to be taken long before the disease ever begins.

1995 Officials at the Smithsonian Institution announce the end of public access to its collection of research photographs of college students. Why? Because the photographs are of some of the nation's best-known personalities—naked. George Bush, Diane Sawyer, and Hillary Rodham Clinton are among those photographed when they were students. The pictures were taken to study posture, and to test the theory of W. H. Sheldon that body shape is linked to intelligence and other aspects of personality. Both Sheldon and his theory are now dead.

1573 Astronomer Simon Marius, a rival of the famous Galileo, is born in Gunzenhausen, Germany. Both men claimed priority in discovering the four moons of Jupiter. It is possible that both made the discovery independently around 1610, as both were among the first to use telescopes. Although their discovery is forever linked with Galileo's name, it was Marius who first prepared tables of their motions, and it was Marius who named them Io, Europa, Ganymede, and Callisto, four mythical beings associated with Jupiter. A discovery that is credited solely to Marius is the spiral nebula, a mass of stars and gases in the Andromeda galaxy.

1633 Galileo, 68, leaves his home in Florence to face the Inquisition in Rome. On June 22 of this year, after intense questioning and threats by the Inquisition, Galileo renounced what he knew to be true, that Earth revolved around the sun.

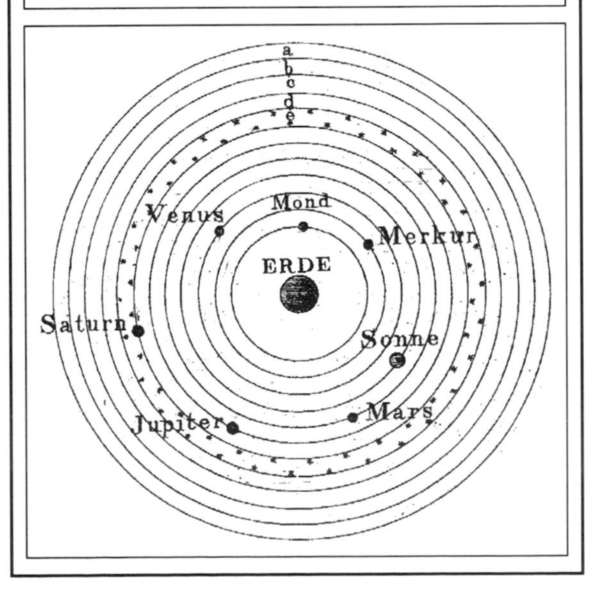

1506 The oldest still-existing army in the world (the 80- to 90-person Pontifical Swiss Guard in Vatican City) is founded.

1743 The unsung and unfortunate inventor John Fitch is born in Windsor, Connecticut. He had little schooling, a tyrannical father, and a nagging wife; he ran a gun factory during the American Revolution, but was paid in worthless currency and taken prisoner; his death is thought a suicide. His claim to fame is inventing a working steamboat before Robert Fulton, but Fulton is famous and Fitch remains unknown.

1853 A day in the history of junk mail—the country's first successful envelope-folding machine is patented. Dr. Russell L. Hawes of Worcester, Massachusetts, receives patent No. 9812 for the device, which was not self-gumming but nevertheless allowed three workers to produce over 2000 envelopes in an hour.

1887 The mental power of insight was discovered in apes by psychologist Wolfgang Köhler, born on this day in Revel (now Tallinn), Estonia. Köhler spent World War I studying chimp behavior in the Canary Islands. Perhaps his most famous experiment involved suspending bananas above a few scattered boxes; through insight and imagination, chimps piled up boxes to form a ladder to the fruit. Köhler was also a founder of the Gestalt school of psychology.

1908 New York City's Board of Aldermen passes a law prohibiting women from smoking in public.

1912 Biochemist Konrad Bloch is born in Neisse, Germany. In 1964 he received the Nobel Prize for Medicine or Physiology for determining how the body makes cholesterol, a process requiring over 30 chemical steps.

1926 Physiologist-cytologist Camillo Golgi dies at 82 in Pavia, Italy. Originally planning a career in psychiatry, Golgi switched to researching the cell after reading the works of pathologist Rudolph Virchow. Golgi's greatest contribution to science came in 1873 when he developed a new way of staining material for microscope viewing. Previous stains used organic dyes, whereas Golgi used silver salts for staining. Never-before-seen details of all kinds of tissue were suddenly visible. Golgi, of course, was the first to benefit from this innovation. He discovered cell parts that still bear his name: the Golgi complex and Golgi bodies (which he first detected in the brains of barn owls in 1898). He further confirmed that nerve cells do not actually touch, but are separated by small spaces (synapses). He also discovered that several different parasite species are responsible for different varieties of malaria. Golgi was awarded the 1906 Nobel Prize for Physiology or Medicine.

1954 The first atomic-powered submarine, the *Nautilus*, is launched in Groton, Connecticut. The occasion is the first time a U.S. submarine is christened by a president's wife, Mamie Eisenhower.

1970 The first jumbo jet, the Boeing 747, is put into service by Pan American Airways.

1994 Although no cause is known for the "Persian Gulf War syndrome," or even if it exists, U.S. government officials announce the creation of the Persian Gulf Veterans Coordinating Board, which will administer compensation to U.S. veterans who fought in the Persian Gulf and who exhibit unexplained symptoms of the mysterious ailment, including headaches, fatigue, and pain. Parasites, petrochemicals, the multitude of oil well fires, microwaves, uranium-rich U.S. tank shells, and chemical and biological weapons have all been listed as possible causes of the "disease." "Certainly, this was a very environmentally dirty war," said Dr. Robert Roswell, an official with the new Board.

1994 The *Miami Herald* reports that multimillionaire Paul Tudor Jones II, currently an active supporter of the move to cleanup the Florida Everglades, was once fined $2 million for illegally filling-in Maryland wetlands (although this was actually part of a plan to create a wildlife refuge). Another twist: Jones made his fortune as a commodities trader, and he is now urging that the environmental clean up be paid for by taxing a commodity, sugar.

1976 The first supersonic passenger plane, the British–French Concorde, is put into regular service simultaneously by Air France and British Airways.

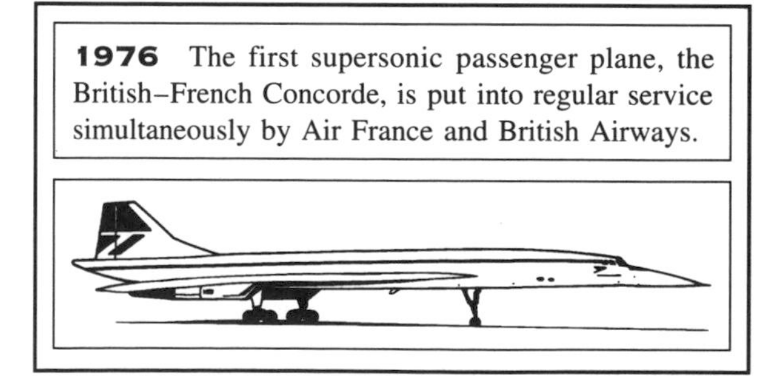

1561 Francis Bacon is born in London into a family prominent in the English court. Although known as a philosopher, Bacon made an important contribution to science with his book, *Advancement of Learning* (1605), which advocated logic and experiments in place of mysticism, and which was partially responsible for the founding of the Royal Society, one of the world's great scientific organizations.

1592 French philosopher Pierre Gassendi is born into poverty in Champtercier. Like Bacon (above) he advocated scientific experimentation. Although an ordained priest, Gassendi supported Galileo's theories; he was the first to observe the transit of a planet across the disk of the sun (Mercury, in 1631), and he described and named the aurora borealis (1621).

1673 Mail is carried over an organized postal route (New York-to-Boston) for the first time in U.S. history.

1775 André-Marie Ampère, the founder of the science of electrodynamics, is born the prodigy son of a prosperous merchant in Lyon, France. In 1820 Ampère jumped on Oersted's discovery that a magnetic needle moves near an electric current, and quickly produced rules and theories to explain the phenomenon. Ampère's law mathematically describes the relationship between a magnetic field and the electric current that produces it. He was the first to build equipment to measure electric current, and the unit of current is now called the ampere or amp.

1840 Johann Friedrich Blumenbach, who coined the terms "Caucasian" and "Mongolian" to describe human races, dies at 87 in Göttingen, Germany. Trained in medicine and comparative anatomy, Blumenbach presented one of the first classification systems of mankind, based on an extensive study of cranial measurements. He has since been regarded as the father of physical anthropology.

1855 German physician Albert Ludwig Neisser is born the son of a physician in Schweidnitz. Neisser discovered the organisms that cause gonorrhea and leprosy. He then tackled syphilis, but his vaccination against the disease may have spread it instead.

1865 Physicist Louis Paschen is born into a military family in Schwerin, Germany. He specialized in the study of light emitted by heated gases. In 1895 he analyzed the spectrum of the newly discovered gas, helium, and found it identical to helium around the sun. The Paschen lines in the hydrogen spectrum are named for him.

1908 Lev Davidovich Landau, a founder of theoretical physics in the Soviet Union, is born in Baku. Landau received the 1962 Nobel Prize for his theories explaining the unique properties of liquid helium, but his interests were many, as are the number of technical terms bearing his name (e.g., Landau diamagnetism, Landau levels in solid-state physics, Landau damping in plasma physics, Landau energy spectra, and Landau cuts in high-energy physics).

1943 The most drastic rise in temperature ever recorded occurs in Spearfish, South Dakota. The temperature jumps from −4°F at 7:30 AM to 45°F at 7:32 AM—a 49° shift in two minutes, which has never been equaled.

1947 The nation's first commercial television station west of the Mississippi, KTLA begins operations in Hollywood, California.

1970 History's first jumbo jet, the Boeing 747, completes its maiden regularly scheduled flight between New York and London.

1973 The Supreme Court announces the *Roe v. Wade* decision.

1617 King James I grants a charter to the "Master, Wardens and Society of the Art and Mystery of the Apothecaries of the City of London." This was the first organization of pharmacists in the English-speaking world. Its formation was bitterly opposed by the Guild of Grocers, which heretofore had controlled all trade and activities relating to spices and drugs. The sponsorship of court insider Francis Bacon (born on January 22, 1561) was critical to James's acceptance of pharmacy as a distinct specialization.

JANUARY 23

1796 Russian chemist Karl Karlovich Klaus, discoverer of the element ruthenium (in 1844), is born in Dorpat (now Tarta, Estonia), Ruthenium was the last dense, inert, platinumlike metal to be found. Klaus took the name from the Latin word for Russia.

1849 Elizabeth Blackwell, the first female physician in modern medicine, receives her M.D. degree from the Geneva (New York) Medical College. The school is now Hobart College.

1857 Yugoslav geologist Andrija Mohorovicic, who discovered the boundary between Earth's outermost crust and its mantle, is born in Volosko, Croatia. The boundary (called the Mohorovicic discontinuity, or simply the "Moho") occurs about 22 miles under continents and just 4.3 miles under oceans. It has never been reached by man, but a penetration was planned in the 1960s, known as the "Mohole."

1862 Mathematician David Hilbert is born the son of a judge in Königsberg, Prussia. Hilbert's *Foundations of Geometry* (1899) presented the first complete set of geometry axioms since Euclid, 21 centuries before. Hilbert's 1909 work on integral equations founded the branch of mathematics called functional analysis (in which functions are studied as groups), and established the basis for consideration of infinite-dimensional space, now known as Hilbert space in quantum physics.

1872 French physicist Paul Langevin is born in Paris, the son of an appraiser and the great-great-grandnephew of the renowned psychiatrist Philippe Pinel. Langevin popularized Einstein's work, and developed the first practical sonar systems.

1875 Charles Kingsley, one of the first churchmen to accept Darwin's theory of evolution, dies at 55 in Eversley, England. His rearing in a country setting encouraged his interest in nature and science. Although frail of constitution, he maintained an independent, crusading mind through most of his life. He wrote a series of novels intended to raise the conscience of England's upper class to health, education, and welfare troubles of the poor. He also wrote historical novels (including *Westword Ho!* in 1855) and the popular *The Water-Babies* (1863), which was based on Darwin's writings. He was made chaplain to Queen Victoria in 1859, professor of history at Cambridge in 1860, and canon of Westminster in 1873.

1911 Marie Curie's nomination to the French Academy of Sciences is voted down by its all-male membership, even though she already had one Nobel Prize and would soon win her second.

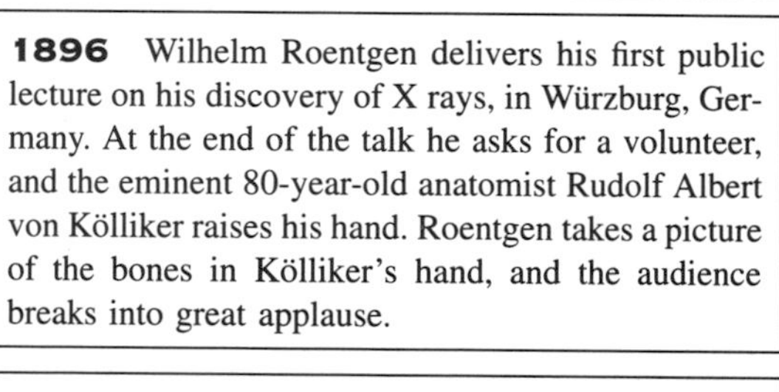

1896 Wilhelm Roentgen delivers his first public lecture on his discovery of X rays, in Würzburg, Germany. At the end of the talk he asks for a volunteer, and the eminent 80-year-old anatomist Rudolf Albert von Kölliker raises his hand. Roentgen takes a picture of the bones in Kölliker's hand, and the audience breaks into great applause.

1960 "Here, in an instant, was the answer that biologists had sought for decades. Could life exist in the greatest depths of the ocean? It could!" This was the summary of Jacques Piccard of his voyage today (with Donald Walsh in the special submarine *Trieste*, which Piccard's father had designed and built) to the floor of the Challenger Deep in the Marianas Trench in the Pacific Ocean. The craft touches bottom at 1:06 PM at a depth of 6.8 miles, a record that still stands. Within minutes of landing, the explorers see a foot-long fish that looks like a flounder with both eyes on the same side of its head.

1985 Debate in England's House of Lords is televised for the first time.

1995 The nation's first all-disability radio show has begun broadcasting from New York City, announces the Knight-Ridder wire service. It is sponsored by the National Multiple Sclerosis Society and National Public Radio, and deals with all disabilities.

1828 Ferdinand Cohn is born in Breslau, Prussia. Raised in a Jewish ghetto in an anti-Semitic society, he was hard of hearing and socially inept, but his merchant father lavished him with education. Cohn obtained his Ph.D. at the young age of 19, and began his career as a botanist. His interest in algae, one-celled plants, led to his demonstration that protoplasm in plant cells was essentially the same as protoplasm in animal cells, therefore creating one of the first major bridges between the two types of life. In the 1860s Cohn moved onto another simple form of life: bacteria; his 1872 three-volume treatise on the subject is recognized as the birth of bacteriology as a separate branch of science. He was the first to separate different bacterial types into species and genera. Cohn is also remembered as the discoverer and mentor of Robert Koch, later a Nobel laureate and founder of medical bacteriology.

1847 French chemist Joseph-Achille Le Bel is born in Merkwiller-Péchelbronn. His family was wealthy in petroleum holdings, which allowed Le Bel to build his own laboratory. He was first to present a theory on the relationship between the structure of molecules and how they reflect or absorb light.

1848 Three days after his 33rd birthday, Horace Wells dies by his own hand in a New York City jail—just as the Paris Medical Society proclaims him the originator of surgical anesthesia.

1848 James W. Marshall discovers a gold nugget at Sutter's Mill in northern California, leading the gold rush of 1849.

1872 English chemist Morris W. Travers, discoverer of krypton (May 1898, while working in the London lab of Sir William Ramsay), is born the son of a physician in London.

1902 Economist-mathematician Oskar Morgenstern, creator of "game theory" (in which many phenomena, including the behavior of men and mice, are analyzed according to win–loss strategies), is born in Görlitz, Germany.

1922 The Eskimo Pie is patented. Christian K. Nelson of Onawa, Iowa, receives patent No. 1,404,539 on his invention of an ice cream center encased in chocolate.

1935 Beer in cans is sold retail for the first time. Packaged by the Krueger Brewing Co. of Newark, New Jersey, the beer is introduced in Richmond, Virginia.

1948 The IBM "SSEC," Selective Sequence Electronic Calculator, is unveiled in New York City. It has been called "the first true computer" because it handled both data and instructions. It contained 13,500 vacuum tubes and 21,000 electronic relays.

1964 The world's first heart transplant takes place, at the Jackson (Mississippi) University Hospital. Dr. James D. Hardy transfers the heart of a chimpanzee into 58-year-old Boyd Rush, who lives for three more hours.

1972 The nation's first comprehensive legislation to control noise pollution is signed by New Jersey Governor William Thomas Cahill. The Noise Control Act of 1971 created a Noise Control Council, allocated $100,000 for the job, and gave the state's Department of Environmental Protection the power to create the necessary codes and regulations.

1978 A nuclear-powered Soviet satellite disintegrates on its return to Earth, scattering radioactivity over northern Canada.

1984 The Apple Macintosh personal computer is unveiled.

1986 Science gets its best look yet at Uranus, as the *Voyager 2* space probe comes within 50,679 miles of the planet and snaps dozens of photographs.

1991 *Nature* reports that scientists in Texas have reproduced a gene that, when missing, causes muscular dystrophy. Discovered several years ago, the gene is responsible for instructing cells to produce the protein dystrophin. It is the largest known human gene, and its size thwarted previous attempts to clone it. Scientists have now done that, using recycled bits of mouse genes. The synthetic gene has worked in isolated cells, causing them to manufacture dystrophin. Getting the gene to function in a living organism is the next hurdle. Nevertheless, today's report is called a "milestone" in the struggle to save "Jerry's kids."

1627 Physicist-chemist Robert Boyle is born a prodigy in Lismore Castle, County Waterford, Ireland. By age eight he spoke Latin and Greek, and soon began reading the works of Galileo, who had just died. Boyle is known mainly for studying gases (he was the first to collect a gas, and he discovered Boyle's law, which describes the relationship between volume and pressure). He also was the first to carefully describe experiments so that others could reproduce them (now standard practice in all sciences), and he helped found the Royal Society.

1736 Joseph-Louis Lagrange, the greatest mathematician-physicist of his day, is born the youngest of 11 children (and the only one to survive to adulthood) in Turin, Italy.

1798 Benjamin Thompson defines the nature of heat. In a paper presented to London's Royal Society, entitled "Enquiry concerning the Source of Heat which is excited by Friction," Thompson suggests that heat is a form of motion, not some sort of fluid (as was commonly believed). He derived this theory from experiments begun when he supervised the boring of cannon barrels at an arsenal in Munich.

JANUARY 25

1799 The first U.S. patent for an automatic seeding machine is granted to Eliakim Spooner of Vermont for "a machine for planting." It was a commercial failure.

1812 Mathematician William Shanks is born in Corsenside, England. He spent much of his life calculating pi (the ratio of circumference to diameter in a circle), and finally reached 707 places in 1873. No one surpassed him for nearly a century. However, modern computers have now taken the calculation to over 2 billion places (and still there seems nothing special about the number), and in 1944 it was discovered that Shanks had erred at the 528th decimal place, making all of his subsequent calculations wrong.

1823 Edward Jenner, creator of vaccination, suffers the stroke that will soon kill him, in Berkeley, England.

1900 Geneticist Theodosius Dobzhansky is born in Nemirov, Ukraine, the son of a mathematics teacher. Through research on the tiny fruit fly, Dobzhansky's major contribution was to combine Darwin's theory of evolution with Mendel's theory of genetic inheritance. His *Genetics and the Origin of Species* (1937) was the first synthesis of these previously unrelated fields.

1915 Transcontinental telephone service across North America is inaugurated by the inventor of the telephone, Alexander Graham Bell. From the AT&T offices in New York City he talks to his longtime assistant Thomas Watson in San Francisco.

1945 Grand Rapids, Michigan, becomes the first town in the United States to add fluoride to its water supply.

1951 A New York City parade of persons on crutches and wheelchairs celebrates the opening of the Institute of Rehabilitation Medicine, the world's first establishment devoted to physical therapy.

1961 John F. Kennedy holds the first presidential news conference carried live on television and radio.

1959 Transcontinental jet service across North America is inaugurated; a four-engine American Airlines Boeing 707 flies from Los Angeles to New York in 4 hours 3 minutes 3 seconds.

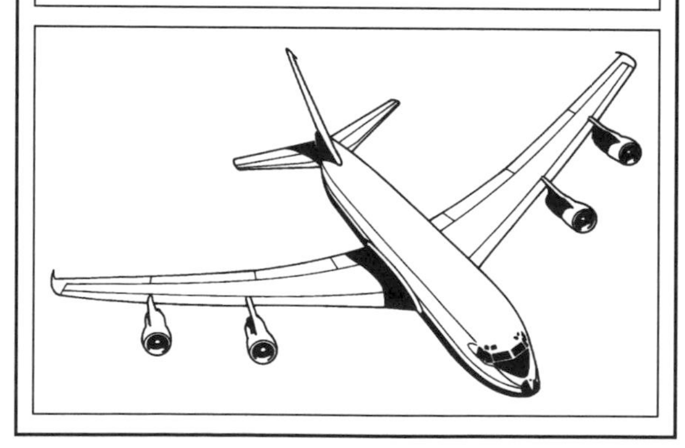

1968 In *New Scientist* magazine, experts with British Petroleum estimate that 500,000 gallons of oil is dumped annually into the oceans merely as unwanted ballast and bilge water.

1971 The Federal Trade Commission first proposes that the ingredients in washing detergents be displayed on packages, along with a warning about environmental hazards of phosphates. The move is opposed by detergent makers, advertisers, and the Nixon administration, but eventually becomes law.

1983 The first telescope in space is launched from Vandenberg Air Force Base, California. The $80 million infrared telescope can detect a speck of dust a mile away.

1994 Arby's becomes the first nationwide fast-food chain to ban smoking in all of its corporate-owned restaurants.

1788 European settlers land at present-day Sydney, Australia. Led by Captain Arthur Phillip, the party includes 700 convicts from England.

1810 Joseph Rogers Brown, inventor of famous fine-measurement tools and gauges, is born the son of a clockmaker in Warren, Rhode Island. His micrometer caliper appeared in 1867, and is still widely used.

1823 Edward Jenner, discoverer/inventor of modern vaccination, dies at 73 in the country town of Berkeley, England. His rural childhood instilled a love and curiosity about nature that stayed with him for life. Trained as a physician, he was a keen observer and experimenter. His greatest experiment was performed in 1796 when he inoculated an eight-year-old village boy with cowpox material. He endangered both his career and the boy's life when he reinoculated the lad with the killer smallpox several weeks later.

1848 Conservationist Henry David Thoreau delivers the first draft of *Civil Disobedience* to his publisher in Concord, Massachusetts. It was one of Thoreau's best-known essays (none of which sold well during his lifetime), and it was written during his famous two-year residence in a cabin on Walden Pond.

1875 An electric dental drill is first patented in the United States. George F. Green of Kalamazoo, Michigan, receives patent No. 159,028 on "electro-magnetic dental tools" for polishing, dressing, filing, and sawing.

1884 Naturalist-explorer Roy Chapman Andrews is born in Beloit, Wisconsin. He is famous for his fossil-hunting expeditions all over the globe, but especially in Asia. Perhaps his most famous discovery was fossilized dinosaur eggs. He also discovered a skull and other bones of the extinct *Baluchitherium*, the largest land mammal on record.

1915 Rocky Mountain National Park (Colorado) is established.

1932 The U.S. Patent Office receives the application of Ernest Orlando Lawrence for his cyclotron, the first "atom smasher" to move neutrons in a circular path (thereby generating unprecedented force). The device won Lawrence the 1939 Nobel Prize for Physics.

1950 The first baby-sitter's insurance policy in U.S. history is issued in St. Louis, by the American Associated Insurance Companies. The policy covers sitters hired through the Missouri State Employment Office, and bonds each for up to $2500 for fraud and dishonesty.

1962 *Ranger III* is launched in an attempt to soft-land instruments on the moon. At first all goes well, but the Atlas booster rocket does its job too well, providing too much power, and sending *Ranger III* 22,862 miles past the moon into an orbit around the sun.

1984 President Ronald Reagan announces construction of the American Orbital Space Station; it was to be completed in 1992, and would serve as an observation post, a pharmaceutical factory, and a launching pad for outer space missions. It has yet to be built.

1994 Brittany Nicole Abshire, the first human to be screened as a preembryo for Tay–Sachs disease, is born in Norfolk, Virginia. When Brittany was just eight cells old, doctors at the Jones Institute for Reproductive Medicine used a needle one-fifth the width of a human hair to withdraw individual cells from preembryos that had been fertilized in a test tube. Three preembryos that tested free of Tay-Sachs were then implanted in the mother's womb, and one of these became Brittany.

1995 A smart medicine bottle is unveiled in New York City by its manufacturer, Aprex Corp. of Silicon Valley. The bottle cap contains a tiny computer that records how often and when it is opened, and beeps when it's time to be opened again. The computer can also be linked to a telephone modem, making it possible to inform doctors, insurance companies, and your mother about how often you take your medicine.

1784 Benjamin Franklin writes to his daughter that he is displeased with the selection of the eagle as the nation's symbol. His own choice for a symbol: the turkey.

1621 Physician Thomas Willis, "the first great epidemiologist," is born the son of a manor steward in Great Bedwyn, England. Willis earned this title by being the first to accurately describe in clinical terms several diseases, including typhoid fever, myasthenia gravis, and puerperal fever, which he named. In 1664 he produced the most complete description of the human nervous system then available, in which he was the first to describe a network of brain arteries now called the circle of Willis.

1851 Artist-naturalist John James Audubon dies in New York City at 65.

1900 Engineer-naval officer Hyman G. Rickover is born in Makov, Poland. At age four he was taken to the United States, eventually graduating from the U.S. Naval Academy. Rickover was the force behind the development of the first nuclear-powered engines and the first nuclear-powered submarine, the USS *Nautilus* (launched in 1954).

1903 Sir John Carew Eccles is born in Melbourne, Australia. Eccles won the 1963 Nobel Prize for Physiology or Medicine for his analysis of the chemical signals that nerves send to each other.

JANUARY 27

1936 Physicist Samuel C(hao) C(hung) Ting, discoverer of a new class of subatomic particles, is born the son of a Chinese college professor in Ann Arbor, Michigan. Ting shared the 1976 Nobel Prize with Burton Richter, who independently and nearly simultaneously discovered the psi-particle. Also called the J-particle, the psi-particle was the first of a new class of massive, long-lived particles called "mesons"; it is thought to be composed of smaller particles called the "charmed quark" and its "antiquark."

1948 The first magnetic tape recorder goes on sale. It is the "Wireway," built by the Wire Recording Corporation of America; it is lightweight, portable, and retails for $49.50.

1950 The antibiotic tetramycin is offered by Chas. Pfizer & Co. of Brooklyn, New York. *Science* magazine announces the medicine, which is effective against penumonia, dysentery, and other infections. It was isolated from Indiana soil.

1951 A new era in nuclear testing in the Nevada desert begins. A U.S. Air Force plane drops a 1-kiloton bomb on Frenchman Flats.

1954 "Atoms for peace. Man is still the greatest miracle and the greatest problem on this earth." These are the words of Brigadier General David Sarnoff in the first message ever sent with electricity generated by nuclear power, reported in today's *New York Post*.

1964 Artificial leather shoes are introduced in stores in 20 U.S. cities. The upper sections are made with Corfam, a material that looks, feels, and wears like leather, but is made out of hydrocarbons instead of cows. It was developed by E. I. du Pont de Nemours & Company of Wilmington, Delaware.

1880 The electric incandescent lamp is patented by Thomas Alva Edison. The patent is No. 223,898.

1967 Astronauts Virgil I. "Gus" Grissom, Edward H. White, and Roger B. Chaffee perish in a fire during a test of *Apollo One* on the launch pad at Cape Kennedy.

1967 More than 60 nations sign a treaty banning the orbiting of nuclear weapons.

1994 The maiden voyage of the nation's first airlines specifically for smokers is scheduled to fly from the heart of tobacco country (Raleigh, North Carolina) to Washington, D.C., where its passengers were to lobby Congress against increased tobacco taxes. The flight of the Boeing 727 "Smokers Express" is canceled, however, because not enough tickets were sold (at $345 round trip).

1994 Reuben Mattus, founder of the Haagen-Dazs ice cream empire, dies at 81. Mattus sold his family's homemade dessert for more than 30 years to a few stores and restaurants in the Bronx, New York. Then he was struck with two marketing insights: (1) ice cream had become cheaper, and people might go for a premium, high-priced product, and (2) people might be attracted by an unusual name. He coined the name "Haagen-Dazs," which means nothing in any language.

814 The enlightened monarch Charlemagne dies at about 72 in Aachen, Germany. His importance to science lay in his realization of the dangers of an ignorant society, and in 789 he began establishing schools throughout his empire (western Europe). Charlemagne himself learned to read late in life, but he was never able to write because his hands were too feeble in maturity to form letters properly.

1608 Astronomer-physiologist Giovanni Alfonso Borelli is born in Naples, Italy. The son of a Spanish soldier stationed in Italy, Borelli was accused of political conspiracy against the occupying Spaniards while a mathematics professor in Messina, Italy. He fled to Rome, where he enjoyed protection by Christina, a former queen of Sweden who was a patron of the arts and sciences (but whose eccentric habit of being tutored before dawn caused the death of René Descartes). Borelli wrote extensively on astronomy (he was the first to suggest that comets travel in parabolic orbits), but his most famous work was the posthumous *De Motu Animalium* ("Concerning Animal Motion"; 1680–1681), in which he was the first to explain muscular movement and other body processes according to laws of physics and mechanics.

1855 William Seward Burroughs, inventor of the first recording adding machine, is born in New York.

1878 The world's first commercial switchboard is installed in New Haven, Connecticut. It has only 21 subscribers, and is not open at night. "Ahoy, ahoy" is the first greeting.

1884 Physicist-inventor Auguste Piccard is born a twin in Basel, Switzerland, the son of a chemistry professor. Early in his career he worked with Einstein in designing electrical measuring devices. Piccard's interest in cosmic rays led him to build a balloon gondola, which, in 1931, he flew higher than any one had ever gone before. It was the first time man entered the stratosphere. Piccard then built several special submarines, "bathyscaphes," which took men (including his son) deeper into the ocean than ever before.

1960 A photograph is first bounced off the moon. The image was sent by the U.S. Navy from Hawaii, and was successfully received in Washington, D.C.

1969 An oil well bursts off the coast of beautiful Santa Barbara, California, and spews oil for months. At that time the largest-ever U.S. oil spill, it kills many animals, and spurs much grass-roots environmental action, including the formation of the group GOO ("Get Oil Out").

1979 A huge storm produces 25-foot-high snowdrifts in Buffalo, New York.

1983 In Detroit Jim "The Mouth" Purol smokes 140 cigarettes—all at the same time. The stunt earns him the title of history's "Most Voracious Smoker"in the *Guinness Book of World Records*. On other occasions, Purol simultaneously smoked 38 pipes for five minutes, and 40 cigars for five minutes.

1986 Space shuttle *Challenger* explodes 73 seconds after takeoff, killing all seven astronauts aboard.

1988 The Canadian Supreme Court strikes down the nation's restrictive abortion law.

1994 Ford announces that its first-ever electric car is performing well in tests, even as Ford lobbyists are fighting a California law requiring that 2% of Fords sold in that state be free of tailpipe emissions by 1998. The electric car, so far, is the only automobile that fulfills that requirement.

1994 Eighty-two-year-old Frederic Green rests in the hospital in San Leandro, California, after an incident in which he was discovered alone, cold, and not breathing in his apartment. Presuming him dead, a coroner photographed Green, who was revived by the flash of the camera.

1611 German astronomer Johannes Hevelius, the first to make accurate drawings of the moon's surface from telescope studies, is born in Danzig (now Gdansk, Poland).

1696 Isaac Newton solves the famous *brachistochrone* (= "shortest time") mathematics problem. Proposed the previous year by Johann Bernoulli and Gottfried Wilhelm Leibniz (Newton's rival), the problem asked, "What is the shape of the quickest path a particle would take when falling from one point to another under the influence of gravity?" The problem had stumped Europe's best mathematicians (including Bernoulli and Leibniz themselves) for months. Newton hears of the problem from a friend after returning home from a full day's work as England's Warden of the Mint; he solves it (and a second Bernoulli–Leibniz problem) after dinner. The next day he deposits the answer with the Royal Society. Newton did not identify himself on the document, but when Bernoulli saw the answer he exclaimed, "Ah! I recognize the lion by his paw."

1700 Johann Bernoulli's second son, Daniel, is born in Groningen, the Netherlands. Daniel was perhaps the most distinguished in the Bernoulli dynasty of great mathematician-physicists. (See February 8, 1700, for a biographical sketch.)

JANUARY 29

1802 John James Beckley is appointed the first Librarian of Congress, a post he held until his death on April 8, 1807. He was paid $2 per day for each day worked.

1810 Mathematician Ernst Eduard Kummer, creator of the concept of "ideal numbers," is born the son of a physician in Sorau, Germany.

1845 Edgar Allan Poe's *The Raven* is published under a pseudonym in the *New York Evening Mirror*.

1850 Sir Ebenezer Howard, founder of the English garden-city movement, is born in London. He lived at a time when many were leaving the countryside to live in cities, creating urban overpopulation. Howard tried to reverse this trend by proposing the establishment of "garden cities," large self-contained communities, privately owned and surrounded by nature; each "garden city" would be limited to 30,000 people. Two such communities were eventually established, and Howard's principles influenced urban planning throughout the world.

1924 An ice-cream cone-rolling machine is first patented. Carl Rutherford Taylor of Cleveland is granted patent No. 1,481,813 for a "machine for spinning or turning a waffle."

1886 German inventor Karl-Friedrich Benz patents the "Motorwagen," history's first successful gasoline-powered car.

1926 Physicist Abdus Salam is born in Jhang Maghiana, Pakistan. He won the 1979 Nobel Prize for his mathematical theory uniting electromagnetism with the "weak forces" that hold atoms together.

1930 Clyde William Tombaugh concludes six days of photographing at the Lowell Observatory in Flagstaff, Arizona, which led to his discovery of Pluto, the ninth and farthest planet from the sun.

1958 The *Boston Herald* prints a letter from Olga Owens Huckins that attacks the pesticide DDT. This letter and a personal letter from her to Rachel Carson were the genesis of Carson's environmental classic, *Silent Spring*, the book credited with founding the modern environmental movement.

1975 Twenty-three million gallons of oil spill into the Atlantic Ocean near Portugal when the tanker *Jakob Maersk* runs around and explodes. Exactly 15 years later (1990) the former *Exxon Valdez* skipper Joseph Hazelwood goes on trial in Anchorage, Alaska, for his part in that oil spill disaster.

1993 The cellular phone industry announces in Washington plans to fund a study of cancer risks related to cellular phone use.

1993 In Los Angeles, in the first operation of its kind, surgeons transplant lung material from both a mother and father into their daughter to save her from cystic fibrosis.

1820 Antarctica is reached by Englishmen Edward Bransfield and William Smith. Expeditions from Russia and the United States also claimed priority in discovering the continent in the same year.

1820 French geologist Alexandre-Émile Beguyer de Chancourtois, the first to arrange the chemical elements in order of atomic weights (in 1862), is born the grandson of a prominent artist. Chancourtois was a mine inspector (who earned the hatred of mine owners for enforcing safety measures) and a geologist (who conducted fieldwork from Greenland to Turkey). His publication on the orderly arrangement of elements was highly significant but ignored, because: (1) it was poorly written, (2) it used the language of geology (so that chemists did not read it), and (3) editors decided to omit an all-important table that would have helped explain his system. It remained for Dmitry Mendeleyev in 1869 to publish his table and garner fame for the achievement.

1868 *Variation of Animals and Planets under Domestication* by Charles Darwin, 58, is published.

1899 Microbiologist Max Theiler is born in Pretoria, South Africa, the son of a Swiss veterinarian. Theiler won the 1951 Nobel Prize for Physiology or Medicine largely for being able to make mice sick. Theiler discovered that mice are susceptible to yellow fever, and used them to develop a vaccine, which was perfected in 1937. An estimated 59 million servicemen sent to the Tropics during World War II were protected by his vaccine. During its development, Theiler tested it by courageously injecting himself and co-workers with the disease; he survived, but six colleagues did not.

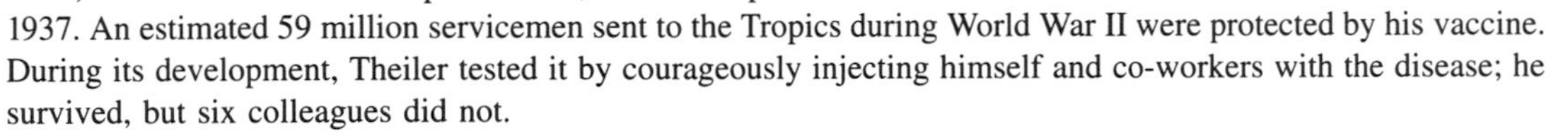

1901 The world's tallest geyser, Waimangu Geyser near Rotorua, New Zealand, is discovered by Dr. Humphrey Haines, who is investigating reports of enormous steam clouds at the time. Although it has not erupted in many years, Waimangu once shot a fountain of boiling water higher than the Empire State Building (over 1500 feet).

1933 Hitler becomes chancellor of Germany, with much implication for the development of superweapons and the flight of great scientists from Europe.

1934 The first "March of Dimes" is held. The name of the antipolio fund-raising event was the brainchild of comedian Eddie Cantor. Today is the birthday of Franklin Delano Roosevelt (born in Hyde Park, New York in 1882), history's most famous victim of polio. The March of Dimes developed from Roosevelt's Georgia Warm Springs Foundation, an organization to treat polio in the poor.

1950 President Harry Truman orders the development of a thermonuclear fusion bomb, also called the hydrogen (H) bomb. The project is code-named "Super."

1957 An artificial pacemaker is used for the first time to successfully treat a disruption of a human heartbeat. C. Walton Lillehei leads the Univesity of Minnesota team that engineered the pacemaker and connected it to the patient's heart. This original pacemaker was as big as a desk; modern pacemakers are so small that they can be inserted into the patient's body.

1964 The first U.S. spacecraft to strike the moon, *Ranger VI*, is launched from Cape Canaveral at 10:40 AM. It hit the moon on February 2 in the Sea of Tranquility. Its on-board TV cameras were not working at the time.

1994 Pierre Boulle, *Planet of the Apes* author, dies at 81 in Paris. In World War II he was a member of the French Resistance in Malaysia when he was captured by the Japanese. He served as a forced laborer until his escape in 1944. This experience was the basis for another novel that became a blockbuster movie, *Bridge on the River Kwai.*

1995 The first-even treatment for sickle-cell anemia is announced by scientists at the National Institutes of Health in Washington. In the middle of testing the drug hydroxyurea, results were proving so positive that the researchers ended the trials and notified 5000 doctors nationwide that an effective treatment had finally been found.

1917 Jazz music is first recorded. In New York City the Original Dixieland Band records *Indiana* and *The Dark Town Strutters Ball* for the Columbia label.

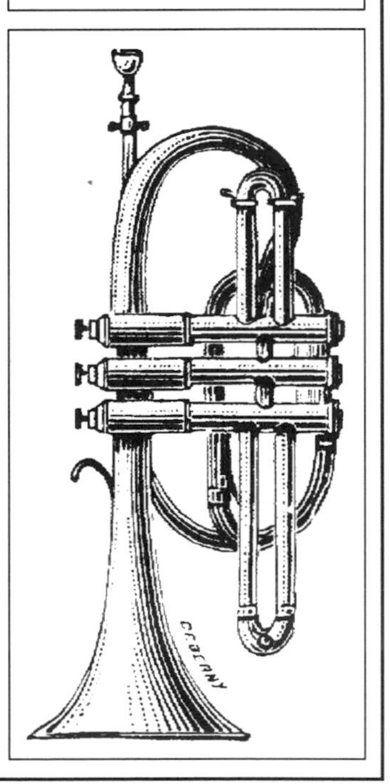

1825 The first trademark lawsuit in U.S. history (involving a newspaper) ends when Judge Nathan Sandford, chancellor of New York State, decides that the *National Advocate* of New York City will not be granted a restraining injunction against the *New York National Advocate.*

1862 Renowned lens grinder and amateur astronomer Alvan Graham Clark has just finished working with his father on a new 18-inch lens in their shop in Cambridge, Massachusetts; Clark is testing the new lens by observing the star Sirius. One of the first things Clark notices is a tiny spot of light near Sirius. Clark has just discovered an entirely new class of heavenly body, the "white dwarf" star.

1868 Chemist Theodore William Richards is born in Germantown, Pennsylvania. His father was a painter and his mother a poet. Richards achieved fame for his own obsession with detail; he won the 1914 Nobel Prize for his unprecedentedly precise determination of the atomic weights of approximately 60 elements. It was the first time the Nobel Prize went to a non-European.

JANUARY 31

1881 Chemist Irving Langmuir is born in Brooklyn, New York, the son of an insurance executive. After obtaining his Ph.D. in Germany under the supervision of Hermann Nernst, Langmuir began work at the General Electric Company in Schenectady, New York. His first and last formal assignment was to extend the life of light bulbs, which he did many times over by discovering why filaments were burning out, and by showing that different gases in the bulbs would slow down this process. Langmuir stayed at G.E. until his retirement in 1950, but with his early success, he had complete freedom to work on whatever project struck his fancy. He was awarded the 1932 Nobel Prize in Chemistry for his discovery and analysis of substances that will form layers on water that are one molecule thick.

1902 Anthropologist Julian H(aynes) Steward, founder of the "theory of cultural ecology," is born in Washington, D.C. Field studies of Indian tribes in North and South America, and insights from a range of social sciences, led Steward to develop a theory explaining how environment determines social structure among different peoples: The natural environment influences technology, which in turn influences labor patterns, which in turn influences social structure.

1929 German physicist Rudolf Ludwig Mössbauer is born in Munich. He announced discovery of the "Mössbauer effect" in 1958, the same year he received his Ph.D. from the Munich Institute of Technology. Three years later he won the Nobel Prize. In the Mössbauer effect, certain crystals are made to emit energy (called gamma rays) this is unaffected by a recoil of the source. Recoil normally occurs when individual, small atoms emit gamma rays, and it changes the energy and wavelength of the rays, thereby preventing scientists from getting an accurate picture of their nature. The first laboratory test of Einstein's theory of relativity came in 1960 through the use of the Mössbauer effect.

1953 History's most damaging storm surge sweeps the coasts of Britain and continental Europe. A recent depression produced enormous winds that moved approximately 15 billion gallons of water from the Atlantic to the North Sea, raising the general level of the North Sea by two feet. On this night the storm continues and combines with high tide to create enormous floods.

1958 The United States enters the Space Age, when it places a satellite into orbit for the first time. From Cape Canaveral, a Jupiter C Army missle launches the unmanned *Explorer I* (tubular shaped, 80 inches long, 30.8 pounds) into orbit at 10:55 A.M.

1971 Astronauts Shepard, Mitchell, and Roosa blast off aboard *Apollo 14*, the first manned flight to the moon since the harrowing *Apollo 13* expedition, in which midflight malfunctions nearly destroyed the ship and the three astronauts aboard.

1990 McDonald's opens its first restuarant in Russia (Moscow). The franchise proved an immediate success.

1995 The final resting lace of Alexander the Great has been discovered near the oasis of Siwa, in Egypt's Western Desert. Abdel-Halim Noureddin, chairman of the Egyptian Antiquities Organization, makes this announcement in Cairo after a weekend visit to the site, where a team of Greek archaeologists unearthed the tomb after four years' excavation. "I do feel that this is the tomb of Alexander," Noureddin said. "All the evidence is there." The 2300-year-old mystery of where Alexander was buried is finally solved.

FEBRUARY 1

1650 René Descartes, famous French scientist and philosopher, catches a chill during a 5 AM. trip to his patroness, Queen Christina of Sweden, who he often tutored at this hour. Complications led to pneumonia and Descartes soon died.

1844 Granville Stanley Hall, hailed as the founder of child and educational psychology, is born in Ashfield, Massachusetts.

1844 Botanist Eduard Strasburger, the first to demonstrate that sperm and egg have just half the number of chromosomes found in body cells, is born in Warsaw. He coined the terms "cytoplasm" and "nucleoplasm" to describe the watery gel that is found in the cell and in its nucleus, and in 1891 he discovered that plant sap moves upward through stems through physical processes, like capillarity, rather than through physiological activity by plant tissue.

1857 Russian psychiatrist-neurobiologist Vladimir Bekhterev is born in Sorali. A lifelong rival of the great Ivan Pavlov, Bekhterev established the first Russian laboratory of experimental psychology (1886, at the University of Kazan) and created the first Russian journal on nervous diseases (1896; the *Neurology Journal*). His most lasting work was the discovery of several brain structures and disorders, including the Bekhterev nucleus and Bekhterev's disease, or numbness of the spine.

1893 Thomas A. Edison completes work on Black Maria, the first motion-picture studio, in West Orange, New Jersey.

1905 Physicist Emilio Segrè is born the son of an Italian industrialist in Tivoli. He won the 1959 Nobel Prize for producing the first antiproton, which is a subatomic particle with all of the characteristics of a proton, except that its electric charge is opposite to that of a proton. He also created/discovered the elements technetium (1937) and astatine (1940). Technetium, meaning "artificial," was the first man-made element not found in nature.

1911 The first fingerprint conviction in U.S. history is recorded in the Criminal Court of Cook County, Illinois. Thomas Jennings is found guilty in the killing of Clarence B. Hiller. The conviction was appealed, but on December 21, 1911, the Illinois Supreme Court ruled fingerprint evidence admissible. On February 16, 1912, it was Jennings, not the jury, who was hung.

1941 The first commercial airplane flight to touch four continents over a single route begins. The 42½-ton *Dixie Clipper* departs New York City with 10 passengers and 11 crew members. It returned eight days later. This was also the first time a U.S. commercial aircraft landed in Africa.

1949 History's first 200-inch reflector telescope, the Hale, goes into operation on Mt. Palomar, California. The first constellation it sees is Coma Berenices, 6 sextillion miles away.

1951 An atomic explosion is seen on television for the first time. An NBC camera on Mount Wilson captures the blast at Frenchman Flats, Nevada, 300 miles away. Station KTLA broadcasts the event to Los Angeles.

1959 Dr. George Schaller and wife Kay depart New York City for Africa to conduct the first ever scientific study of gorillas in their natural habitat.

1964 Murray Gell-Mann publishes an eight-paragraph note in *Physics Letters* that introduces the term "quark" to science, to describe a class of subatomic particles. The future Nobel laureate stole the term from James Joyce's *Finnegan's Wake*.

1991 James MacDonald, the voice of Mickey Mouse for 30 years, dies at 84 in Glendale, California.

1993 For the first time in its 147-year history, the Smithsonian Institution will ask visitors for donations.

1788 The nation's first patent for a steamboat (and the only patent ever issued by the state of Georgia) is awarded to Isaac Briggs and William Longstreet. Their craft stayed afloat and moved, but not well enough to go into commercial production.

1046 The weather turns especially cold throughout Europe. Monks note in the *Anglo-Saxon Chronicle* that "no man alive ... could remember so severe a winter." It is the first known record of the beginning of the 200-year-long period of exceptional cold known as the "Little Ice Age."

1556 History's most lethal earthquake kills 830,000 as it rips through three Chinese provinces.

1802 Jean Boussingault, the agricultural chemist who discovered several key steps in the biological "nitrogen cycle," in which nitrogen moves out of the atmosphere and into living things, is born in Paris.

1859 Essayist-physician (known chiefly for his pioneering research into sexual behavior) Havelock Ellis is born the son of a sea captain in Croyden, England (see biographical note on July 8, 1939).

1869 Zoologist Charles Manning Child is born in Ypsilanti, Michigan. His "axial gradient theory" explains regeneration of injured body parts.

FEBRUARY 2

1870 The Cardiff Giant hoax is debunked. Supposedly a petrified human ancestor, the massive form is announced today to be carved gypsum. It was discovered on a farm in Cardiff, New York.

1893 "The Sneeze" is filmed at Edison's studio in West Orange, New Jersey. It is the first-ever motion-picture closeup. The entire footage consists of comedian Fred Ott sneezing.

1905 Psychoanthropologist Adolf Bastian dies at 78 in Port of Spain, Trinidad. His first visit to foreign lands (Africa, South America, the Caribbean, China, India, and Australia) was in 1851 as a ship's surgeon. Later trips abroad were as a scientist studying native myths, folklore, customs, and beliefs. He proposed that there is a single set of laws to explain cultural evolution, with minor variations according to different geographical locations. He further proposed the existence of a psychological unity among all humans, leading to a set of mental symbols and ideas that are common to all cultures. Psychoanalyst Carl Jung later elaborated on this idea, which has since become known as the "collective unconscious." Bastian was the founder and first curator of the world-famous Royal Museum of Ethnology in Berlin, to which he donated much of his private collection.

1907 Dmitry Mendeleyev, 73 years old and famous for developing the periodic table of the chemical elements, dies at home in St. Petersburg, Russia, while listening to a reading of Jules Verne's *Journey to the North Pole*.

1935 Accused criminals are given lie-detector tests for the first time in U.S. history. Cecil Loniello and Tony Grignano are tested in Portage, Wisconsin, by Leonarde Keeler, a detective and inventor of the Keeler Polygraph. The tests were submitted as evidence in court, and both Loniello and Grignano were convicted of assault.

1962 The USS *Diamond Head* departs for Africa from Newport News, Virginia, with staff and equipment to conduct the first field trials of the "pistola de la paz" ("pistol of peace"), the vaccination gun invented by Robert Hingson that improved the health of millions.

1964 The moon is struck by a U.S. spacecraft for the first time. The 800-pound *Ranger VI* crashes into the Sea of Tranquility at 4:24 AM EST. Its television cameras were not working at the time.

1985 O.J. Simpson and Nicole Brown wed at Simpson's mansion in Brentwood, California.

1995 For the first time in history, research links artistic talent with brain structure. At a news conference in Washington, Dr. Gottfried Schlaug of Boston's Beth Israel Hospital announces that he and colleagues have discovered that persons with perfect pitch have an abnormally large "planum temporale" region in the left hemisphere of their brains. But this artistic gift is use-it-or-lose-it: Those with enlarged planum temporales who are not exposed to music before age ten have an "extremely low" chance of developing perfect pitch.

1995 At the Parkes radio telescope outside Sydney, Australia, an international team of astronomers begins a five-month vigil of several hundred stars, in hopes of detecting unusual radio waves that might indicate the existence of alien life. It is history's most intense search for extraterrestrials in the Southern Hemisphere.

1790 Paleontologist Gideon Mantell is born the son of a shoemaker in Lewes, England. (See biographical note on November 10, 1852.)

1821 Elizabeth Blackwell, considered to be the first female doctor in modern medicine, is born in Counterslip, England. Her family moved to New York when she was 11. At age 17 she and her sister founded a boarding school for girls, to support her family after her father died. She began studying medicine on her own at this time, but her applications to medical schools were rejected. She was finally accepted in 1847 to the Geneva (New York) Medical College (now Hobart College). Despite being a social outcast there, she graduated with an M.D. at the top of her class. This was the first medical degree awarded to a woman in the United States. She established the New York Infirmary in New York City (it was entirely staffed by females) and later added a medical school to its functions as a hospital. In 1869, she returned to England where she cofounded the London School of Medicine for Women.

1857 Wilhelm Johannsen, who coined the word "gene," is born in Copenhagen. The son of a Danish army officer and unable to afford a college education, Johannsen educated himself in chemistry and biology. He rose to become a professor at Carlsberg and then Copenhagen, where he devoted his research to heredity in plants. He coined the now-standard terms "phenotype" and "genotype" to describe the external/observable characteristics of an individual (its phenotype) versus the genetic characteristics (genotype).

1862 A newspaper is printed on a train for the first time in U.S. history. The *Weekly Herald*, consisting of one sheet 7 by 8 inches, was issued on the train between Port Huron and Detroit, Michigan. It was produced by Thomas Alva Edison.

1925 *The Star* in Johannesburg, South Africa, carries the first newspaper report of Raymond Dart's discovery of the first "missing link," a fossil that demonstrates an evolutionary relationship between man and other primates. The worldwide journal *Nature* was supposed to carry the story, but it backed out at the last moment. Dart's 32nd birthday is the next day.

1958 Rachel Carson writes to E.B. White (editor of *The New Yorker* magazine and author of *Charlotte's Web*), trying to get him to write what will become the ecology classic *Silent Spring*. Unable to do it himself, White writes back and successfully urges her to write it.

1959 The most famous plane crash in music history. Rock 'n rollers Buddy Holly, Ritchie Valens, and "The Big Bopper" perish near Clear Lake, Iowa.

1960 Two men depart Chepo, Panama, in a Land Rover called "The Affectionate Cockroach," en route to driving over the last unconquered section of the Pan-American Highway (the longest roadway on Earth, stretching from northern Alaska to southern Chile). The two pioneers are Richard E. Bevir of England and Terence John Whitfield of Australia. They finished their drive on June 17 in Quibido, Columbia, after traveling over extremely harsh terrain at an average speed of 200 yards per hour.

1966 The first "soft" landing of equipment on the moon is accomplished in the Ocean of Storms by Russia's *Luna IX*.

1984 The space shuttle *Challenger* begins an erratic mission that includes the faulty deployment of several satellites. One of the satellites disappeared completely after release.

1986 President Reagan appoints a 12-person commission to investigate the disastrous *Challenger* flight/explosion of January 28 that killed the entire crew.

1994 A new era in space exploration dawns, when the space shuttle *Discovery* lifts off at sunrise from Kennedy Space Center in Cape Canaveral. It carries cosmonaut Sergei Krikalev, the first-ever Russian to ride in a shuttle. It is also the first U.S.–Russian manned mission since the 1975 *Apollo–Soyuz* docking.

1690 Paper money is issued for the first time in U.S. history. Massachusetts establishes a bank and begins production of currency, in denominations from two shillings to five pounds, to pay soldiers who had fought in the war with Quebec.

1778 Augustin Candolle, the Swiss botanist who developed the modern system of plant classification (and coined the term "taxonomy," which is the science of classification), is born in Geneva.

1790 Naturalist-Lutheran minister John Bachman is born in Rhinebeck, New York. He wrote much of the text for the books of bird and mammal paintings by John James Audubon. Bachman also published *The Unity of the Human Race* (1850), in which he theorized that all humans are of one species.

1868 Charles Darwin, 59, begins writing *The Descent of Man, and Selection in Relation to Sex* in the study of his country home in Downe, England.

1902 Charles Augustus Lindbergh, perhaps history's most famous aviator, is born the son of a Minnesota congressman in Detroit.

FEBRUARY 4

1906 Astronomer Clyde W. Tombaugh, the man who discovered the planet Pluto, is born in Streator, Illinois. His family were dirt-poor farmers, and his first telescope was built of junked agricultural equipment.

1913 Louis Henry Perlman of New York City receives patent No. 1,052,270 for the first automobile tire rim that could be removed and remounted.

1915 Dr. Joseph Goldberger begins a historic experiment with the diets of 12 inmate-volunteers at a state prison in Jackson, Mississippi. Goldberger discovered that the cause (and cure) of the killer disease pellagra is dietary.

1936 A radioactive substance is synthesized for the first time in the United States (at the University of California at Berkeley). Dr. John Jacob Livingood bombards common bismuth with deuterons to produce Radium E.

1951 The longest operation in medical history begins in Chicago on Mrs. Gertrude Levandowski. It lasts four days. With the removal of her ovarian cyst, her weight dropped from 616 to 308 pounds by the end of surgery.

1957 The first portable electric typewriters are sold in the United States. They weigh 19 pounds and are manufactured by Smith-Corona Inc. of Syracuse, New York.

1962 St. Jude's Children's Research Hospital opens in Memphis. It is the nation's first facility dedicated solely to research and treatment of catastrophic childhood diseases. Actor Danny Thomas was the founder.

1974 A historic case in "brainwashing" and cult behavior begins in Berkeley, California, when newspaper heiress Patty Hearst is kidnapped by the Symbionese Liberation Army.

1983 Singer Karen Carpenter, a woman with a haunting voice and a haunted personality, perishes at just 32 in Downey, California. The tragedy brought widespread attention to anorexia nervosa and other eating disorders.

1993 What harm can dreams do? The *New England Journal of Medicine* presents a new study by Dr. Virend K. Somers; during REM sleep (when dreaming occurs) blood pressure is high, the heart speeds up, and stress hormones are released—all conditions that strain the heart and precede a heart attack. The findings suggest an answer to one of cardiology's longstanding mysteries, namely why heart attacks are most common in the morning.

1993 A group of doctors suggest in the *New England Journal of Medicine* a new approach to the shortage of hearts available for transplantation: recycling. The idea arose after a recent case in Zurich, Switzerland, in which a transplant patient died a few days after receiving someone else's heart. The organ was removed and successfully installed in a second recipient. Within a matter of weeks, the same heart had been in three people.

1893 Paleontologist Raymond Dart is born in Brisbane, Australia. He emigrated to South Africa, where in 1924 he was brought a humanlike skull, which he hypothesized to be the long-sought missing link between human and ape evolution. Highly controversial at the time, his theory is now generally accepted.

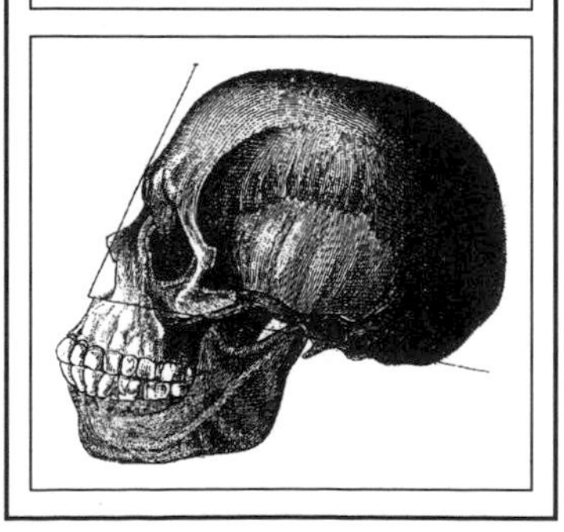

FEBRUARY 5

1770 Alexandre Brongniart, the first to describe and arrange the geological formations of the Tertiary Period, is born in Paris. He was also the first to systematically study trilobites, and he helped introduce the system of dating geological specimens by identifying the fossils in each layer of Earth's crust.

1790 Antoine Lavoisier, one of history's greatest chemists, writes from Paris to his friend Benjamin Franklin, about the French Revolution, "We look upon it as over, and well and irrevocably completed." Lavoisier was guillotined 1552 days later.

1825 The first magazine specifically for mechanics is published in the United States. *American Mechanics' Magazine* "containing useful original matter, on subjects connected with manufactures, the arts, and sciences; as well as selections from the most approved domestic and foreign journals" debuts on this date in New York City, published by J.V. Seaman. In 1826 its name changed to *The Franklin Journal and American Mechanics' Magazine*, and later to *The Journal of the Franklin Institute*.

1840 Hiram Maxim, inventor of the first fully automatic machine gun, is born the son of a farmer in Sangerville, Maine.

1850 An adding machine with depressible keys is first patented, by D.D. Parmelee of New Paltz, New York. His device was neither practical nor successful, but the name he gave it, "calculator," survived and is now the universal name for many types of machines.

1861 The "peep show" motion-picture machine is patented, by Samuel D. Goodale of Cincinnati.

1866 Sir Arthur Keith is born in Aberdeen, Scotland. He became a doctor of medicine, science, and law, but is especially remembered as an anthropologist who provided reconstructions of hominid remains from Europe, North Africa, and Israel.

1872 Lafayette Benedict Mendel, codiscoverer of vitamin A (in 1913) and the vitamin B complex (in 1915), is born in Delhi, New York. His discoveries sprang from a long-term study of rats in which he determined that pure fats, carbohydrates, and proteins are not enough to sustain life. He also determined that the nutritive value of proteins depends on their content of essential amino acids, the building blocks of proteins. His work helped establish the modern science of nutrition.

1901 Edwin Prescott of Arlington, Massachusetts, patents a much-beloved device of self-torture: the loop-the-loop centrifugal railway. This patent actually comes 2½ years after Prescott's patent for a loop-the-loop roller coaster. His device was commercially called "Boyton's Centrifugal Railway" when it was installed at Coney Island, New York, in 1900; it had a 75-foot incline and a 20-foot-wide loop.

1915 Physicist Robert Hofstadter, famous for dissecting the proton and neutron, is born in New York City. Using Stanford's linear accelerator, Hofstadter discovered that both of these nuclear particles have a positively charged core surrounded by two clouds of pi-mesons. He won the 1961 Nobel Prize.

1995 The Russian Space Agency calls off a scheduled close encounter between its *Mir* space station and the U.S. space shuttle *Discovery* which has sprung a fuel leak.

1799 British botanist John Lindley is born in Catton, Northumberland. In 1830 he organized the first-ever flower show in England (in his capacity as a secretary of the Horticultural Society). Shortly thereafter he became the first botany professor at the University of London; he held the post for the next 30 years. In 1838 he recommended that London's Kew Gardens be established as a national botanical center, and donated his outstanding orchid collection to the effort. Kew Gardens is now world-famous for its plant displays. Lindley's *Theory and Practice of Horticulture* (1842) is considered as one of the great books on plant physiology. *The Vegetable Kingdom* (1846), his most popular book, was important in advocating the now-accepted "natural system" of plant taxonomy, which stresses consideration of all plant characters in classification.

1796 John Henslow, botanist, clergyman and mentor to Charles Darwin, is born in Rochester, England.

1802 Self-taught physicist Sir Charles Wheatstone is born Gloucester, England. It is little known that he invented the concertina, the stereoscope, and an early form of the telegraph; he is primarily known for the Wheatstone bridge (a standard device for providing very accurate measurements of electrical resistance), which he did not invent.

1834 Edwin Klebs, one of the first to link bacteria to disease, is born in Königsberg, Prussia. He was able to establish this link in endocarditis, tuberculosis, and syphilis (in 1878; he was the first to transmit this disease from human to monkey, thus providing science with a valuable nonhuman subject with which to study the disease). In 1884 with Friedrich Löffler he discovered the deadly diphtheria bacillus, now known as the Klebs–Löffler bacillus.

FEBRUARY 6

1852 C. Lloyd Morgan, a founder of comparative psychology, is born in London. Morgan originally trained to be a mining engineer, but a chance meeting with biologist Thomas Huxley turned him to zoology, especially to the then-unexplored field of animal behavior. His major innovation was to focus only on what could be observed scientifically, and to eliminate interpretations of animal actions that are based on human feelings and emotions. He developed the following ethic, now known as "Morgan's Canon": "In no case may we interpret an action as the outcome of the exercise of a higher psychical faculty, if it can be interpreted as the outcome of one which stands lower in the psychological scale."

1857 The contract to produce the nation's first perforated postage stamps is signed between the federal government and printers Toppan, Carpenter and Co. of Philadelphia. The stamps were delivered to the government 18 days later.

1886 Clemens Winkler discovers the element germanium in his lab in Freiburg, Germany.

1892 Physician William Murphy is born in Stoughton, Wisconsin. He shared to the 1934 Nobel Prize for developing a liver diet treatment for the previously incurable pernicious anemia.

1913 Mary Douglas Leakey, discoverer of critically important fossils and records of human ancestors, is born in London.

1927 Physicist Gerard Kitchen O'Neill is born in Brooklyn. He invented the colliding-beam particle accelerator, and earned further notoriety by designing outer-space colonies.

1933 The history's highest officially recorded sea wave (112 feet high) is measured by the USS *Ramapo* in the Pacific Ocean during a 68-knot hurricane.

1957 The cryotron, a superconductive computer switch designed to operate at very cold temperatures, is revealed. It was developed by Dudley Allen Buck of MIT.

1959 From Cape Canaveral, the United States successfully test-fires a Titan ICBM (intercontinental ballistic missile) for the first time.

1973 Dr. Dixy Lee Ray becomes the first female chairman of the Atomic Energy Commission.

1995 The owners of Keiko (the 3½-ton star of *Free Willy*, a 1993 smash movie about a whale's quest for freedom from amusement park imprisonment) announce in Mexico City that the animal will, indeed, gain his freedom. The 15-year-old behemoth, captured as a youngster off the shores of Iceland, will be shipped to Oregon Coast Aquarium for a period of rehabilitation, then brought back to and released in the waters from which he was taken.

1995 The Associated Press reports that Ben & Jerry's has found its new boss. Saddled with red ink for the first time in its history, the ecologically minded ice-cream company ran a unique search entitled, "Yo! I Want to be CEO!" The winner is Robert Holland, Jr., a business consultant from suburban New York City. Among the 20,000 other applicants was a California man who sent in a Superman suit ("Due to recent layoffs at a major metropolitan newspaper I am looking to replace my day job ...") and a medical librarian who sent in her résumé and a picture of herself, au naturel, surrounded by strategically placed reference books.

1776 The country's first prison reform society is organized in Philadelphia. It is the Philadelphia Society for Relieving Distressed Prisoners Owing to the War of Independence. It disbanded in September 1777, but a second prison reform society was formed 10 years later by Philadelphia Quakers.

1818 *Academician*, the first successful education periodical in the United States, debuts.

1865 Gregor Mendel, discoverer of the laws of genetics, reads his first scientific paper at age 42 to the Brünn Society for the Study of Natural Science in Moravia.

1870 Psychiatrist Alfred Adler is born in Penzing, Austria. He was one of the first to join Freud in practicing psychoanalysis, and one of the first to defect from Freud's circle of pioneers. Adler theorized that power, not sex (as Freud believed), is the bottom line of human motivation. It was Adler who popularized the concept of "inferiority complex."

FEBRUARY 7

1877 Mathematician Godfrey Harold Hardy is born in Cranleigh, England. Although he solved many problems in prime number theory, his most famous discovery was the Hardy–Weinberg law in ecology. In 1908 he and Wilhelm Weinberg independently developed a simple mathematical formula showing that the proportion of dominant and recessive genes tends to remain constant in large populations. This law is a cornerstone of population genetics, and explains why even rare and harmful genes survive.

1896 One of the earliest uses of X rays—its discovery announced only weeks before—occurs in Liverpool, England, when a doctor successfully searches for a bullet in a boy's head.

1905 Physiologist Ulf von Euler is born in Stockholm. He received the 1970 Nobel Prize for his discovery of noradrenaline, a hormone in the nervous system.

1931 Historic female aviator Amelia Earhart marries publisher George P. Putnam in Connecticut.

1932 James Chadwick coins the term "neutron" and announces its discovery in *Nature*.

1934 The first electrical contract in U.S. history, in which a town contracted to purchase electric power from a federal institution, becomes effective in Tupelo, Mississippi. The 20-year agreement stipulates that Tupelo will purchase power from the Tennessee Valley Authority at approximately 5½ mills per kilowatt-hour, and sell it to citizens at a rate agreed to by the TVA.

1983 Elizabeth Dole becomes the first female Secretary of Transportation; she is sworn in by the first female Supreme Court justice, Sandra Day O'Connor.

1984 At the Texas Children's Hospital in Houston, 12-year-old "David" touches his mother for the first time, when he is removed from the germfree plastic "bubble" that has been his home since birth.

1984 Man floats free in space for the first time. *Challenger* astronauts Bruce McCandless II and Robert L. Stewart take history's first untethered "spacewalk."

1994 The Rockefeller Foundation in New York City announces a $1 million bounty on the bacteria that cause gonorrhea and chlamydia. The prize will go to anyone who devises a diagnostic test that is cheap, quick, easy, does not require refrigeration or running water (because the test is to be used in poor countries), and can detect both diseases before any symptoms occur. The contest will stay open until March 1999, or until a winner is declared.

1995 A lawsuit over a combustible Kellogg's Pop-Tart is settled in Dayton, Ohio. Kellogg agrees to pay $2400, after Thomas Nangle of Springfield claimed that one of the pastries ignited and set his kitchen on fire.

1834 Dmitry Mendeleyev, creator of the modern periodic table of the elements, is born in Siberia, the youngest of "14 to 17" children. His father was a high school principal who soon went blind. Mendeleyev's first science lessons were from a political prisoner in the northern wasteland.

1863 English chemist John Newlands arranges the known chemical elements in order of atomic weight. It is one of the earliest attempts to organize the elements in a meaningful way, but was completely overshadowed by the achievement of Mendeleyev.

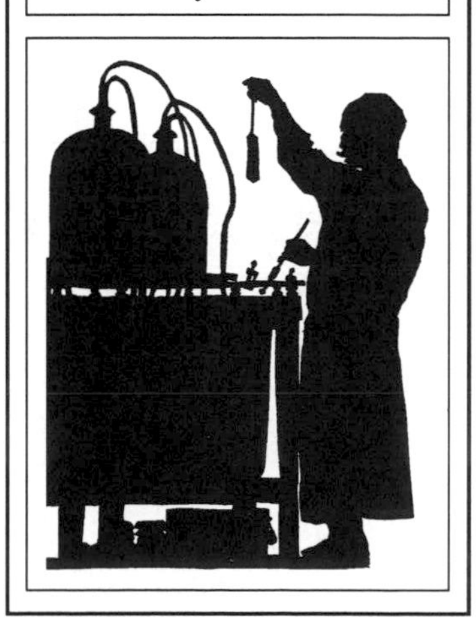

1700 Daniel Bernoulli is born into a dynasty of mathematicians-physicists in Groningen, the Netherlands. He won numerous awards and made important contributions in a variety of fields. The best known of his discoveries was Bernoulli's principle: pressure in a fluid decreases as the speed increases. He also gave the first plausible explanation of the behavior of gases in relation to pressure and temperature changes.

1777 Bernard Courtois is born in Dijon, France, the son of a gunpowder chemist. Bernard also went into chemistry, becoming the first to obtain an alkaloid in pure form (he isolated pure morphine from opium). He became famous in his own lifetime for discovering (accidentally) iodine. He was unable to capitalize on this, however (like his father, he was a bad businessman), and he died in poverty.

1866 U.S. chemist Moses Gomberg is born in the Ukraine. He developed the first practical car antifreeze, but is primarily remembered as the first to isolate a free radical (a carbon atom with three, not four, rings attached). He stumbled on this molecule during attempts to produce another.

FEBRUARY 8

1877 Charles Wilkes dies at 78 in Washington, D.C. The son of a prominent businessman, Wilkes received a solid education in a range of subjects. He was one of the nation's earliest oceanographers, and leader of its first major ocean expedition (1838–1842), which circled the globe and determined that Antarctica (named by Wilkes) is a continent.

1898 A machine that folds and gums envelopes is patented in the United States. John Ames Sherman of Worcester, Massachusetts, receives patent No. 598,716 for a "mechanism for folding and sealing envelopes." The device reduced the cost of making 1000 market-ready envelopes from 60 to 8 cents.

1825 Naturalist-explorer Henry Walter Bates is born in Leicester, England. Son of a hosiery manufacturer, the adolescent Bates worked 13-hour days in the family business, but went to school at night. His great love was entomology, the study of insects. At age 22 he and companion Alfred Russel Wallace embarked on the adventure of a lifetime: a self-financed insect-gathering expedition to unknown regions of the Amazon. The pair were able to pay for the entire trip by selling specimens to museums and private collectors in England. Wallace stayed in the Amazon for 4 years, Bates 11. Both made deep contributions to the understanding of life on Earth: Bates discovered Batesian mimicry (as well as 8000 new insect species), and Wallace formulated the theory of evolution (independently of Darwin).

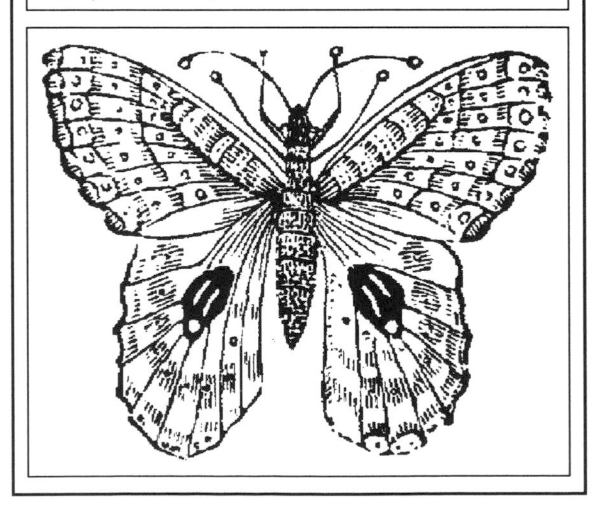

1910 "Be prepared." The Boy Scouts of America incorporates in the District of Columbia. A Congressional act granted the organization a federal charter on June 15, 1916.

1915 The first movie to gross $50 million, the epic *The Birth of a Nation*, premieres in Los Angeles.

1924 The first lethal-gas execution is conducted in Carson City, Nevada, on a Chinese immigrant.

1928 The first transoceanic TV image is received in Hartsdale, New York. The picture is of a Mrs. Mia Howe, and was sent by John Logie Baird at station 2 KZ from Purley, England.

1969 A group of meteorites falls on Chihuahua, Mexico. They date back 461 billion years, to the dawn of our solar system.

1974 *Skylab 4* splashes down, ending the longest U.S. space mission (84 days 1 hour 15 minutes 30 seconds).

1994 A second group of cleanup workers arrives at beaches near San Juan, Puerto Rico, after a month-old oil spill (750,000 gallons from the grounded barge *Morris J. Berman*) got really serious: It disrupted surfing and deterred tourism.

1994 Robert Holton, a biochemist at Florida State University, announces that he has achieved what others called "impossible": the total synthesis from common chemicals of taxol ("the most important cancer drug in 15 years"). At least a dozen other labs had been working on the same problem.

1865 Explorer-glaciologist Erich Drygalski is born in Königsberg, Prussia. His greatest discovery stemmed from near tragedy, when his research ship *Gauss* accidentally became frozen in Antarctic ice. The group was forced to spend the winter of 1902–1903 there, during which Gaussberg was discovered and named by Drygalski; it is a volcano totally free of ice, on Earth's coldest continent.

1870 Congress authorizes formation of the U.S. Weather Bureau, as a subdivision of the Signal Corps of the War Department. In 1891, the Bureau became part of the Department of Agriculture, until 1940 when it was transferred to the Department of Commerce, after which it was renamed the National Weather Service and transferred to the National Oceanic and Atmospheric Administration.

1871 The first federal fish-protection office is authorized by an act of Congress.

1871 Pathologist Howard Ricketts is born the son of a grain merchant in Findlay, Ohio. His name is forever linked to a disease that he helped conquer—*not* rickets. After his death in 1910 (from a disease he was studying), Ricketts's name was given to rickettsia, a type of disease-causing microbe that he discovered and that causes typhus and Rocky Mountain spotted fever.

1909 The Davey Tree Expert Company incorporates in Kent, Ohio. It established the nation's first school devoted to shade trees and their care. John Davey was the organization's first president. In 1914 the "Davey Institute of Tree Surgery of Kent, Ohio" became the first U.S. school to offer a correspondence course in tree surgery.

1910 Neurosurgeon Harvey Cushing is called in to consult on the condition of U.S. Army Chief of Staff Leonard Wood. Cushing, 40, diagnoses a meningeal tumor in the brain, and removes it. Cushing had never successfully removed such a tumor before. The prominence of the patient, and the fact that such tumors were virtually incurable prior to Cushing's innovations, made the incident a turning point in Cushing's fabled career and in the history of medicine.

1910 Jacques Monod is born in Paris. In 1965 he received the Nobel Prize in Physiology or Medicine for discovering mechanisms by which genes control cells. Monod shared the prize with François Jacob and André Lwoff; it was the first time in 30 years a Frenchman was so honored, and at a press conference after the ceremony, the trio lambasted the French government for inadequate support of research. Back at home, they had been trying unsuccessfully to wrest control of the Pasteur Institute from ultraconservative politicians. Three weeks after getting the Nobel Prize, they were given control of the Institute's administrative council. "We've gone from zero to the condition of movie stars," said one.

1923 Destined to become the world's largest airline, the Soviet Union's Aeroflot goes into operation.

1923 Pioneering heart surgeon Norman E. Shumway is born in Kalamazoo, Michigan.

1957 Russia, Canada, Japan, and the United States sign a seal-protection pact that establishes cooperative research, search and seizure of suspicious vessels, and a ban of seal killing at sea.

1971 *Apollo 14* returns to Earth after man's third visit to the moon.

1977 "Guas," the oldest known orangutan, dies at 59 in the Philadelphia Zoo (his home for 46 years).

1991 The worst nuclear accident in Japan's history occurs in Mihama, Japan, when a steam generator tube cracks, causing a leak of radioactive water. An emergency cooling system in the reactor's core is activated. There are no deaths, and leakage of radioactivity into the atmosphere is reported as small.

1994 The Hershey Chocolate Company celebrates its 100th birthday in Hershey, Pennsylvania. Company founder Milton S. Hershey suffered through a series of failures and bankruptcies before his belief in himself and his skill made him the most famous chocolate-maker ever.

1969 The first-ever jumbo jet, the Boeing 747, takes its maiden flight.

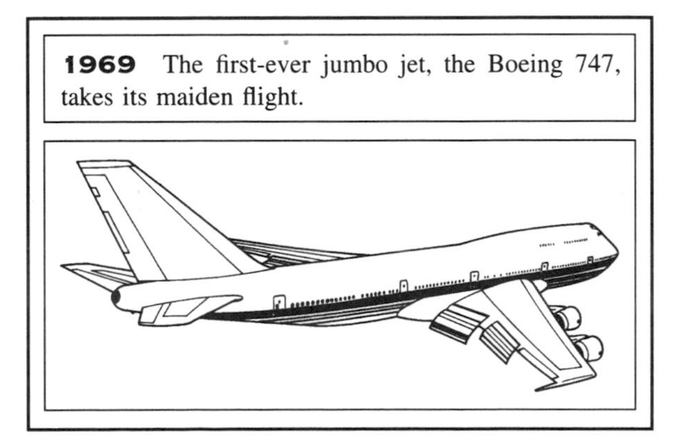

1835 Victor Hensen is born in Schleswig, Germany. He became famous as a physiologist (the "cells of Hensen" and the "Canal of Hensen" are found in the inner ear) and as an oceanographer (he coined the term "plankton" to describe tiny marine life, developed equipment for plankton study, and led history's first Plankton Expedition).

1840 Per Cleve, discoverer of elements thulium and holmium, is born a 13th child in Stockholm.

1840 Queen Victoria marries Prince Albert. She is a carrier of the genetic disease hemophilia. The couple had nine children and many descendants after that. The scattering of hemophilia through the family tree formed a famous case history in genetics.

1846 Chemist Ira Remsen, creator of the sugar substitute saccharin (in 1879), is born in New York City.

FEBRUARY 10

1853 Renowned crystallographer Victor Mordechai Goldschmidt is born in Mainz, Germany.

1863 The first fire extinguisher patent in U.S. history is No. 37,610, awarded on this date to Alanson Crane of Fortress Monroe, Virginia.

1897 John Franklin Enders is born in West Hartford, Connecticut, the son of a successful businessman. Enders abandoned a career in commerce early, then took a doctorate in English, but finally found his niche in microbiology. He won the 1954 Nobel Prize in Medicine or Physiology for developing ways to cultivate the polio virus in nonnervous tissue (which allowed others to create a vaccine).

1902 Walter Brattain, a Nobel laureate and codeveloper of the transistor, is born in Amoy, China.

1912 Joseph Lister, creator of antiseptic surgery, dies at 84 in Walmer, England.

1923 Wilhelm Roentgen, winner of the first Nobel Prize for Physics, dies at 77 in Munich. In 1895 Roentgen was already an established professor when, while experimenting with electricity in a glass tube (a cathode-ray tube), he happened to notice that a nearby piece of treated metal glowed when the tube was activated. He guessed that some form of energy was traveling from the tube to the metal. A series of experiments revealed that the energy passed through paper, wood, aluminum and other substances, and that, although it was invisible, this energy created images on photographic plates. Because of its mysterious nature, he called his discovery "X radiation" or "X rays."

1933 The singing telegram is introduced, by New York's Postal Telegram Company.

1667 An abstract of *Head of a shark dissected* by Nicolaus Steno is published by the Royal Society of London. The essay is a classic in paleontology, the study of fossils. For the first time since the ancient Greeks 20 centuries before, it correctly argued that fossils were the petrified remains of once-living creatures. This notion stirred immediate controversy at that time.

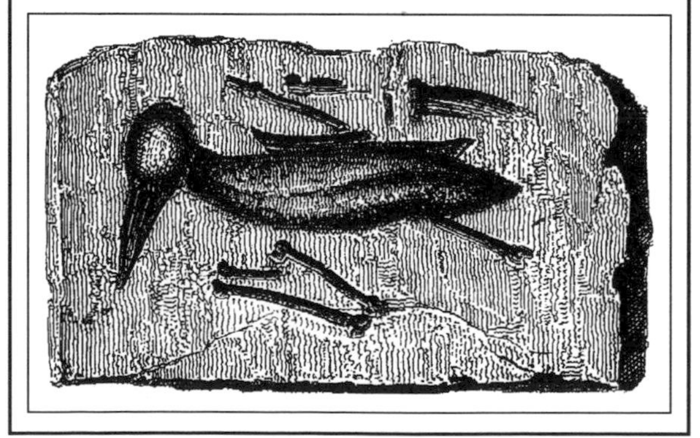

1958 "It will free man from his remaining chains, the chains of gravity which still tie him to this planet. It will open to him the gates of heaven."—Wernher von Braun, chief scientist of the U.S. Army Ballistic Missile Agency, describes the meaning of space travel in an interview in *Time*.

1967 In central Africa Dian Fossey first sees the grisly aftermath of animal poaching. After a full day of gorilla research in the jungle, a cold, hungry, and wet Fossey returns to her hut to find an illegally killed buffalo and its calf in her front yard. The incident begins her war against poachers that eventually claimed her life.

1994 The U.S. Senate approves $8.6 billion in aid to victims of the recent Northridge earthquake in southern California; it is the nation's largest-ever relief package for a natural disaster.

1994 Designer Calvin Klein announces that his company will no longer make fur garments. The action is taken partly because he supports humane treatment for animals. "We're ecstatic about his decision," said Jenny Woods, PETA spokeswoman.

1650 Famous scientist-mathematician-philosopher, René Descartes dies at 53 in Stockholm.

1657 Bernard Fontenelle, possibly the first to establish himself as a science writer for the lay reader, is born is Rouen, France. Descartes was one of his heroes.

1752 America's first true hospital opens in Philadelphia. Created mainly through the efforts of Benjamin Franklin and Dr. Thomas Bond, the "Pennsylvania Hospital" was initially part of a private house on High Street.

1800 William Henry Fox Talbot, pioneer in the development of photography, is born the son of an army officer in Dorset, England. He invented the photographic negative; his *Pencil of Nature* (1844–1846) was the first book to contain photographs

1839 Chemistry/physics theoretician J. Willard Gibbs is born is New Haven, Connecticut.

FEBRUARY 11

1858 A vision of the Virgin Mary is first seen at Lourdes, France.

1898 Leo Szilard, a father of the Atomic Age, is born in Budapest, the son of a Jewish engineer. Szilard was a physicist in Germany until the Nazis took power, after which he went to Vienna, then England, and finally the United States. At Columbia University he was part of the group that urged Franklin Roosevelt in a famous 1939 letter to build an atomic bomb. At the University of Chicago he collaborated with Enrico Fermi to build the first-ever nuclear reactor. He was also a key scientist in the Manhattan Project that built the first atomic weapons. After World War II Szilard devoted himself to curbing nuclear weapons and using nuclear power peacefully.

1908 Sir Vivian Fuchs, leader of the first land expedition across Antarctica (1957–1985), is born on the Isle of Wight. The expedition confirmed the existence of a single land mass beneath the South Polar ice.

1924 Biologist Jacques Loeb dies at 64 in Hamilton, Bermuda. Famous for his 1899 discovery of artificial parthenogenesis (this is reproduction without sex; Loeb was able to bring unfertilized eggs of urchins and frogs to maturity by manipulating their environment), Loeb is also known for studying tropisms, or reflexive responses by plants and simple animals to environmental stimuli.

1925 Virginia Johnson, collaborator in the Masters and Johnson studies of human sexuality, is born in Springfield, Missouri.

1939 *Nature* publishes a theoretical paper by Lise Meitner and her nephew, Otto Frisch, about how a nuclear chain reaction could possibly convert mass to energy. The authors coin the term "nuclear fission."

1961 Sixty-two nations sign a treaty to ban the placement of nuclear weapons on the ocean floor.

1972 A snowmobile exceeds 125 mph for the first time. In Boonville, New York, Quebec's Yvon Duhamel powers his Ski-Doo Blizzard X2R to a maximum speed of 127.3 mph at the Boonville Airport.

1994 After nine years of scrutiny by the Food and Drug Administration, a genetically engineered growth hormone for cows (called rBGH) goes on sale to dairy farmers. Genetically engineered cheese has been available for years, but this is the first time altered genes will be injected into living creatures to assist their production of milk, a food that has come to symbolize purity.

1995 Religion and science meet again. The Associated Press reports that DNA tests have begun on the 2000-year-old Dead Sea Scrolls. The genetic analysis will allow scientists to determine what individual goat or sheep was used to make which scroll fragment.

1847 Thomas Alva Edison is born in Milan, Ohio. His mother was a schoolteacher and his father a sometimes-carpenter and a sometimes-lighthouse keeper. Edison had hearing problems all of his life, and as a child had an unusual phrasing of questions. When one of his first teachers called him "addled," Mrs. Edison withdrew him from school. He eventually returned, but quit forever at age 12.

1585 Caspar Bartholin is born in Malmö, Denmark. A physician, theologian, and student of the legendary Fabricius, Bartholin wrote *Textbook of Human Anatomy* (1611); it became one of the most widely read works on the human body during the Renaissance. He also discovered Bartholin's gland, a lubricating organ in the reproductive system of female mammals.

1637 Jan Swammerdam, one of history's great microscopists, and the first to see red blood cells, is born the son of a pharmacist in Amsterdam.

1785 Chemist Pierre Dulong is born in Rouen, France. He codiscovered the Dulong–Petit law, which states that specific heat is inversely related to atomic weight.

1809 Charles Darwin is born in Shrewsbury, England. Four thousand miles away, Abraham Lincoln is born in Kentucky.

FEBRUARY 12

1813 Geologist-naturalist-explorer James D. Dana is born in Utica, New York.

1815 Edward Forbes, a pioneer in the field of biogeography (the study of how organisms are distributed on Earth), is born on the Isle of Man. He was a medical student until he happened to witness a starfish being brought up from the seabed a quarter-mile deep; this made him a lifelong naturalist. His great work was discovering and explaining the correlation between life forms and geology in Britain.

1877 A news dispatch is sent by telephone for the first time, from Salem to Boston. "This special dispatch to the Globe has been transmitted by telephone in the presence of twenty people who have thus been witnesses to a feat never before attempted—the sending of news over the space of sixteen miles by the human voice."

1878 F.W. Thayer of Waverly, Massachusetts, receives patent No. 200,358 for the baseball catcher's mask.

1893 Astronomer Marcel Minnaert is born in Bruges, Belgium. He began his professional life as a botanist, but a desire to understand light led him to physics. He used the newly invented microphotometer to analyze the intensities of light waves coming from the sun. His monumental *Photometric Atlas of the Solar Spectrum* (1940) is still a standard reference on the sun's energy. He is also known for *Light and Color in the Open Air* (1954) which analyzed how various weather conditions affect light.

1908 The first round-the-world automobile race begins in New York (to end in Paris in August).

1941 Ernst Chain and Howard Walter Florey (developers of penicillin's use as an antibiotic) supervised the first injection of penicillin into a human (see February 17 for the triumphant, then tragic, outcome of the experiment).

1918 Child prodigy Julian Seymour Schwinger is born in New York City. He was awarded the 1965 Nobel Prize for Physics for developing quantum electrodynamics.

1933 Scottish naturalist-science popularizer Sir John Arthur Thomson dies at 71 in Limpsfield, England. He spent years researching corals, but he is best known for coauthoring with Sir Patrick Geddes a series of popular books, including *Sex* (1914), *Evolution* (1911), and *The Evolution of Sex* (1889). He also attempted to reconcile biology with religion in *Science and Religion* (1925). He was knighted in 1930.

1973 Almost 200 years after its conception, the metric system finally reaches a U.S. highway. Ohio begins erection of roadside kilometer-distance markers, on Interstate 71 from Cincinnati to Cleveland.

1982 The largest known litter of mice is born in Blackpool, England. All but one of the 34 pups survived.

1633 Galileo arrives in Rome to face the Inquisition.

1741 A magazine is first published in America. In Philadelphia, Andrew Bradford and John Webbe produce *The American Magazine, or a Monthly View of the Political State of the British Colonies.*

1743 Sir Joseph Banks, explorer, naturalist, and science financier, is born in London. Banks was on the first scientific expedition to the Pacific Ocean (the 1768 voyage of James Cook), and was the first naturalist to study the unique wildlife of Australia. Shortly after this trip he journeyed to the North Atlantic, where he discovered great geysers in Iceland. In later years Banks financed research and the transplanting of crops from one part of the world to another. The infamous mutiny on the *Bounty* occurred during a Banks-financed transfer of breadfruit from Tahiti to the West Indies.

1766 Sociologist Thomas Malthus is born near Guilford, England. In 1798 he published anonymously his most famous work, *Essay on Population*, which argued that infinite human happiness was not possible because population size would always outrun food supply and other resources; only war, pestilence, and famine would limit population size. The work was highly controversial and exerted an immediate influence on English social policy. It also gave Charles Darwin and Alfred Russel Wallace a critical insight for their development of the theory of evolution.

1805 Mathematician Peter Gustav Lejeune Dirichlet is born in Düren, Germany.

1834 Industrialist-chemist Heinrich Caro is born in Posen (now in Poland). He was responsible for the explosive growth of the German dye industry in the late 1800s, and in so doing established the prototype industrial research organization that has come to dominate technology.

1851 Zoologist G. Brown Goode, ichthyologist and supervisor of the reorganization of the National Museum of Natural History in Washington, is born is New Albany, Indiana.

1875 The first well-documented birth of quintuplets in U.S. history occurs in Watertown, Wisconsin. Five sons are born to Mrs. Edna Beecham Kanouse and her husband Eddie.

1882 Jean Charcot presents results from his studies of hypnotism to the French Academy of Medicine in Paris. For the first time, a medical establishment accepts hypnotism as a valid form of therapy.

1910 William B. Shockley, Nobel laureate and a developer of the transistor, is born in London.

1923 Chuck Yeager, the first to fly faster than sound (in 1947), is born in Myra, West Virginia.

1935 Bruno Hauptman is found guilty of kidnapping the Lindbergh baby.

1960 France explodes its first atomic bomb on one of the most ecologically delicate places on Earth, the Sahara desert.

1993 The German news magazine *Der Spiegel* reports that the Soviet Union secretly sank two of its own nuclear-powered ships in the Baltic Sea. No date or motive is given for the sinking, nor is it clear whether radioactivity is leaking. *Der Spiegel* also claims that Russia dumped contaminated reactor parts and nuclear waste into the Baltic, without regard for international law or public safety.

1728 John Hunter, the founder of pathological anatomy in England, is born in Long Calderwood, Scotland. An early advocate of experimental biology, his death 65 years later was hastened by venereal disease, which he gave himself to demonstrate (incorrectly) that syphilis and gonorrhea are the same disease. As with many surgeons in the 1700s, Hunter never graduated from college; instead, he learned his craft as an apprentice and as a dissector of cadavers for lectures by his famous brother William Hunter.

1779 Captain (James) Cook is killed and probably cannibalized in Kealakekua Bay, Hawaii, in a skirmish with locals. The explorer-oceanographer is 50 years old. The son of a farmhand, Cook worked in his teens on the farm, in a haberdashery, and in a general store. In 1746 he was apprenticed to the Quaker shipowner John Walker of the English coastal town of Whitby, and spent the rest of his life at sea.

1803 Eli and John Phipps are born in Affington, Virginia. They became history's most long-lived twins; Eli died first, just 9 days after their 108th birthday.

1803 The apple parer is patented, by Moses Coats, a mechanic from Downington, Pennsylvania.

1848 James K. Polk becomes the first president to be photographed in office, by Mathew Brady.

1860 Geologist Waldemar Lindgren is born in Kalmar, Sweden. He invented methods to determine the composition of ore veins; in so doing, he established how such deposits are formed in nature.

FEBRUARY 14

1864 Sociologist Robert E. Park is born in Harleyville, Pennsylvania. He is known for researching ethnic minorities, and is credited with coining the term "human ecology" in a sociological context.

1869 C.T.R. Wilson, inventor of the cloud chamber (with which nuclear phenomena have been explored and Nobel prizes have been won), is born the son of a shepherd in Glencorse, Scotland.

1872 California becomes the first state to authorize a bird refuge, when Lake Merritt, near Oakland, is made a sanctuary.

1876 Alexander Graham Bell and Elisha Gray file separate documents with the patent office in New York City, with each man claiming priority as the inventor of the telephone. The case eventually went to the Supreme Court, which ruled in Bell's favor. Bell filed his patent application just two hours ahead of Gray, and this slight time difference decided the case.

1878 Julius Nieuwland is born in Hansbeke, Belgium. Nieuwland was first a priest, then a botanist, and finally a chemist. He discovered neoprene, the first commercially successful rubber.

1858 Joseph Thomson, the first professional naturalist in several regions of Africa, is born in Penpont, Scotland. The most common gazelle in East Africa is Thomson's gazelle (*Gazella thomsoni*), which was named in his honor.

1896 Cosmologist Edward Arthur Milne is born in Hull, England. He determined the sun's temperature at various depths and demonstrated the sun's ejection of particles at enormous speed (1000 miles per second), which introduced the notion of a solar wind.

1898 Astronomer Fritz Zwicky, known for discovering and explaining supernovae, is born in Bulgaria.

1914 Renato Dulbecco is born in Catanzaro, Italy. He won the 1975 Nobel Prize in Physiology or Medicine for studying chemical events after a virus invades a cell.

1963 Rachel Carson's ecology classic *Silent Spring* is published in England. The book contains a foreword by the eminent biologist-educator Sir Julian Huxley, who died exactly 12 years later (1975).

1984 In Pittsburgh, six-year-old Stormie Jones undergoes history's first heart-liver transplant.

1992 The British House of Commons votes down a law that would have banned fox hunting.

1994 Vegetarian actress Kim Basinger unveils a new poster of herself during a private dinner in New York City. Proceeds from poster sales will go to the antifur group People for the Ethical Treatment of Animals. Naturally, Basinger is not wearing a fur coat in the poster ... in fact, she's wearing nothing at all.

1564 Galileo is born the son of a mathematician in Pisa three days before Michelangelo dies.

1748 Social philosopher Jeremy Bentham is born in London, the son of a lawyer. Bentham founded the School of Utilitarianism on the doctrine of "the greatest happiness of the greatest number." for which he invented a "felicific calculus" for quantifying happiness. It is said that he began reading Latin at age four. As per instructions in his will, on death his body was donated to London Hospital and publicly dissected; his skeleton was reconstructed and fitted with a wax head (the real head was mummified). The body was then clothed and placed in a glass-fronted case, and for the next 92 years the body was brought to hospital board meetings, at which Bentham was always noted as "not voting."

1826 Physicist George Stoney, the first to use the term "electron," is born in Oakley Park, Ireland.

1842 The first adhesive postage stamp is introduced by a private firm in New York City.

1856 Psychiatrist Emil Kraepelin, who laid the foundation for the modern classification of mental disorders, is born in Neustrelitz, Germany.

1858 Astronomer William Henry Pickering is born in Boston. He and older brother Edward established an observatory in Peru in 1891. In 1899 William discovered Phoebe, the outermost satellite of Saturn. It was the first satellite discovered by photography; it orbits Saturn in a direction opposite to the planet's nine other satellites.

1861 Alfred North Whitehead is born in Ramsgate, England, the son of a headmaster-clergyman. Like his collaborator Bertrand Russell, Whitehead achieved fame as both a mathematician and a philosopher.

1861 Physicist Charles Guillaume is born in Fleurier, Switzerland. His father was a watchmaker, and Guillaume spent his life developing precision instruments. In particular he discovered Invar, an iron–nickel alloy that changes very little with temperature and is therefore an excellent material for precision devices in science. Guillaume was awarded the 1920 Nobel Prize for Physics for this work.

1873 Biochemist Hans Euler-Chelpin is born in Augsberg, Germany. He determined the chemical structure of the first-discovered coenzyme, for which he won the 1929 Nobel Prize for Chemistry.

1898 The U.S. battleship *Maine* blows up in Havana harbor, which starts the Spanish–American War, which leads to research that conquers yellow fever.

1951 The first atomic reactor to be used in medical therapy—the Atomic Energy Commission's unit in the Brookhaven National Laboratory in Upton, New York—is placed in operation.

1954 Man first reaches the ocean floor at a depth beyond 2000 fathoms. Georges Houot and Pierre Willm touch bottom at 13,287 feet (greater than 2½ miles) in the Atlantic Ocean off Dakar, Senegal.

1982 A huge oil-drilling rig (the *Ocean Ranger*) sinks off Newfoundland in a storm; 84 perish.

1992 Canada's Society for the Prevention of Cruelty to Animals sponsors an unusual fund-raising stunt: naked bungee jumping. Visitors to a bridge near Nanaimo on Vancouver Island pay $2 to watch people jump off the bridge wearing only a bungee cord. The Bungy Zone Company donated its facilities; volunteers are given free bungee jumps if they do it au naturel. By day's end, 106 men and 19 women took the plunge; nearly 1000 spectators attended.

1995 Kevin Mitnick, the world's "most wanted computer hacker," is arrested by federal agents in his apartment in Raleigh, North Carolina.

1768 This day in the history of indigestion: mustard is first advertised in the United States. Benjamin Jackson, who manufactured the condiment in the Globe Mills on Germantown Road in Philadelphia, bills himself in the *Philadelphia Chronicle* as "the original establisher of the mustard manufactory in America, and ... at present the only mustard manufacturer on the continent. I brought the art with me into the country."

1514 Rheticus, the first established scientist to accept the Copernican theory of a "heliocentric" universe (which correctly stated that Earth orbits the sun), is born in Feldkirch, Austria. His father was a physician beheaded for sorcery.

1698 Pierre Bouguer is born in Le Croisic, France, the son of a hydrographer (a geographer specializing in water bodies). Pierre entered the same profession, and was eminent there, but he is most famous for inventing and using devices with which to measure light intensity. He is considered a founder of photometry, the branch of science that deals with measuring light intensity.

1804 Zoologist Karl Theodor Ernst von Siebold is born into a family of eminent biologists. His 1846 *Textbook of Comparative Anatomy* was one of the first in the field, and important for focusing purely on observable facts (instead of philosophical issues common to science books at the time).

FEBRUARY 16

1822 Sir Francis Galton, explorer, anthropologist and eugenicist, is born near Sparkbrook, England, to a wealthy family. He was a child prodigy (he could read at 3 and began studying Latin at 4) and Darwin's first cousin. He studied medicine, but his father's death left him well-off in his early 20s. The first passion he indulged was travel and exploration, in Europe, Asia and most notably Africa. On settling in England, he wrote books on his travels, and then studied meteorology (he founded the modern system of weather-mapping and coined the term "anticyclone" to describe a long ridge of high pressure). During this time he also invented the dog whistle (inaudible to humans, but easily heard by canines). Galton is most remembered for his studies of human heredity. The term "eugenics" is his to describe the elimination of "undesirable" traits from a population by selective breeding. Galton was the first to emphasize the importance of applying statistics to biological data, and the first to study identical twins to help determine the role of heredity in shaping one's life. Galton also established fingerprints as unique identifiers of a person, and showed that the distribution of "intelligence" can be described by a bell curve. Much of his work remains controversial, including his attempts to statistically study the effectiveness of prayer, and his attempt to map the distribution of beauty throughout England.

1834 Ernst Haeckel is born in Potsdam, Prussia. He introduced the term "ecology."

1868 Anthropologist Wilhelm Schmidt, known for theories on how environment influences the evolution of the family, is born in Hörde, Germany. He also wrote the 12-volume *The Origin of the Idea of God* (1912–1955).

1892 Naturalist-explorer Henry W. Bates dies at 67 in London (see biographical note on February 8, 1825).

1923 The burial chamber of King Tut's tomb is unsealed by Howard Carter in the Egyptian desert.

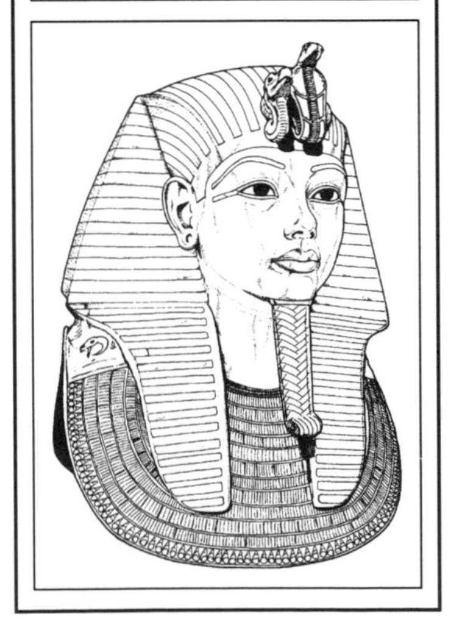

1932 James E. Markham receives the first fruit tree patent in U.S. history. It is plant patent No. 7, for a peach tree with fruit that ripens very slowly. The patent was assigned to Stark Bros. Nurseries & Orchards Co. of Louisiana, Missouri.

1937 Nylon is patented by its discoverer, the tragic Wallace H. Carothers.

1965 *Pegasus 1*, the first satellite equipped with spring-folded wings to intercept tiny meteoroids, is launched from Cape Kennedy by a *Saturn 1* rocket. Four hundred sixteen panels (each 20 by 40 inches, of various thicknesses) are used to register meteoroid impacts. Pegasus 1 is solar-powered, equipped with 25,200 solar cells and 47 nickel–cadmium batteries.

1968 The first 911 emergency telephone system begins operation in Haleyville, Alabama.

1968 Atlantic Richfield announces that oil has been struck in Alaska's Prudhoe Bay.

1980 The longest traffic jam ever reported begins in Lyon, France, and ends 109.3 miles later.

1994 "A new frontier in fetal medicine" opens. Dr. Ruben Quintero (of Wayne State University in Detroit) announces that he has performed the first successful operation of fetuses without opening the mother's body.

FEBRUARY 17

1723 Astronomer Johann Mayer, who discovered that the moon wobbles, is born in Marbach, Germany. He discovered small irregularities in the moon's orbit during his compilation of lunar tables, which were more accurate than any before.

1740 Physicist-geologist Horace Saussure is born in Geneva, Switzerland, the son of an eminent agricultural scientist. Among Saussure's contributions: the first device to measure electric potential, the first humidity-measuring device to use a hair, and introduction of the term "geology."

1795 An 18¼-pound potato is dug up in Chester, England. It remains history's grandest spud.

1817 Lights go up in Baltimore; it becomes the nation's first city to have gas streetlights. Rembrandt Peale was a founder of the Gas Lighting Company of Baltimore, the first U.S. gas company.

1850 Bacteriologist Georg Theodor August Gaffky, who isolated the organism that causes typhoid fever (in 1884 in India as an assistant to Robert Koch), is born the son of a shipping agent in Hannover, Germany.

1856 Frederic Eugene Ives, photographer and photography inventor (70 patents, including one for the modern single-objective binocular microscope), is born in Litchfield, Connecticut.

1864 A submarine sinks a man-of-war for the first time. During the American Civil War, the Confederate sub Hunley (4 feet wide, 5 feet tall, manually powered by hand crank, and having no storage of air) sneaks up to the USS *Housatonic* (1400 tons and awash off Charleston, South Carolina) and explodes a torpedo under its hull. Both ships sink.

1874 Belgian statistician Adolphe Quetelet dies at 77 in Ghent. Although highly respected as a mathematician and astronomer (he founded the Royal Observatory in Brussels in 1827 at age 31), Quetelet is best known for his pioneering, and often controversial, applications of statistics and probability theory to social phenomena, especially crime.

1888 Physicist Otto Stern is born the son of a grain merchant in Sohrau, Germany (now Poland). He won the 1943 Nobel Prize for developing the "electron beam" apparatus to study atomic properties.

1890 Sir R.A. Fisher, a founder of statistical analysis, is born this day in London.

1934 The nation's first high school driving course begins at State College High School in State College, Pennsylvania.

1941 Doctors Howard Florey and Ernst Chain run out of penicillin during its first-ever use on a human. The patient, a 43-year-old British policeman with blood poisoning (who had been responding perfectly to treatment), perishes shortly thereafter. Florey, Chain, and the discoverer of penicillin, Sir Alexander Fleming, shared the 1945 Nobel Prize for Physiology or Medicine.

1990 The largest popsicle in U.S. history is produced in Appleton, Wisconsin. It weighs 7080 pounds.

1992 "Robo-Cougar," the first cougar artificially conceived with a laparoscope, is born in the Octagon Wildlife Sanctuary in Fort Myers, Florida.

1994 Newspapers announce a machine that automatically erases commercials from taped television programs.

1781 Théophile-Hyacinthe Laënnec is born the son of a lieutenant in Quimper, France. His mother died shortly thereafter, which led to Laënnec being put in the care of his physician uncle who turned the lad to a career in medicine. Laënnec invented the stethoscope. The key insight into its structure came to Laënnec one afternoon as he watched children play on seesaw in Paris in 1816; at the time, the shy Laënnec was wrestling with the problem of how to listen to the heart of a young female patient without putting his ear on her chest.

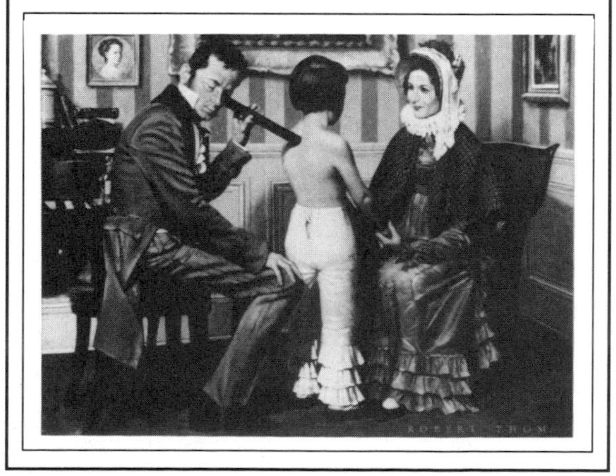

901 Renowned scholar-mathematician Thabit Ibn Qurra dies at 65 in Baghdad.

1201 Nasir ad-Din at-Tusi, Persian philosopher, scientist, mathematician, and astronomer, is born in Tus. After establishing himself as a court astrologer, he lived in the castle fortress of the terrorist Assassins, and then joined the invading Mongol army after giving them a description of the castle defenses. He wrote a number of books, and translated the works of many Greek scientists.

1404 Leon Battista Alberti is born the illegitimate son of a wealthy Florentine merchant-banker in Genoa. Leon first established himself as an artist (as a painter and as one of the premier architects of the Renaissance), then as a scientist of the arts. In 1434 he published a book on the laws of perspective, which laid the groundwork for the science of projective geometry.

1626 Poet-physician Francesco Redi is born in Arezzo, Italy. His name appears in almost every biology textbook for one experiment, in which he proved that maggots arise from flies, not from spontaneous generation.

FEBRUARY 18

1747 Alessandro Volta, inventor of the electric battery (history's first source of a continuous, reliable electric current), is born one of nine children in Como, Italy, in a noble family that had fallen on hard times. Volta did not talk until he was four, which mistakenly convinced his family that he was retarded.

1790 Marshall Hall, the first to scientifically explain bodily reflexes, is born the son of a cotton manufacturer in Basford, England. His theories won praise abroad but derision at home. He coined the term "reflex" in the biological context.

1838 Physicist-turned-philosopher Ernst Mach is born in Chirlitz-Turas, Austria. His early work centered on the study of sound, light, and waves (during which he developed the well-known "Mach number," which is speed in relation to the speed of sound), but his lifetime interest in human perception led to a series of psychology experiments, which secured him an appointment as professor of philosophy at the University of Vienna in 1895. His philosophy, not his physics, greatly influenced Einstein.

1888 Marshall McDonald begins work as the first U.S. Fish and Fisheries Commissioner.

1913 Chemist Frederick Soddy coins the term "isotope."

1930 Pluto is discovered, by astronomer Clyde William Tombaugh at Arizona's Lowell Observatory.

1930 The first flying cow, a Guernsey named "Elm Farm Ollie," boards a plane for her maiden voyage, accompanied by a horde of reporters. Her milk is sealed in paper containers and dropped by parachute over St. Louis.

1972 The death penalty in California is struck down by the state's Supreme Court.

1977 The space shuttle *Enterprise* takes its maiden voyage, piggybacking on a Boeing 747 over the Mojave Desert.

1994 Scientists at a conference in San Diego announce new hope for stroke victims. The scientists have developed a medicine called "selfotel," which seems to reduce the crippling brain damage that inevitably follows a stroke. Two ironies in the discovery are: (1) this brain damage is caused by normal, healthy bodily chemicals that are attacking the stroke-causing blood clot, and (2) selfotel mimics some psychological effects of the hallucinogen PCP, which damages the brain.

1994 U.S. Attorney General Janet Reno dedicates the Big Cypress camp in the Florida Everglades. The camp is designed to benefit "our two most precious possessions": children and the environment.

1564 Michelangelo dies in Rome at age 89. He is revered for his work as a painter, but he is also known for his extremely accurate anatomical drawings of the human body and for his works as an architect, which include several forts and St. Peter's Basilica in Rome.

1764 Gottlieb Kirchhof is born the son of a pharmacist in Teterow, Germany. He was already a prosperous industrialist-chemist when he performed his epoch-making experiment: treating starch with sulfuric acid; this produced glucose (in essence this was the discovery of glucose) and was the first-ever controlled catalytic reaction.

1792 Geologist Sir Roderick Murchison is born in Tarradale, Scotland. His work produced the names Silurian, Devonian, and Permian for geological ages, taken from the locations where he found his samples.

1855 The first-ever weather map is presented to the French Academy of Sciences by M. Le Verrier.

1859 Svante Arrhenius, winner of the third Nobel Prize for Chemistry (in 1903), is born a child prodigy (it is said that he read at age three) to a surveyor in Vik, Sweden. He determined why ionic solutions conduct electricity; many fought his theory because it forced rethinking of atomic structure.

FEBRUARY 19

1863 The first pipeline from an oil field to a refinery is completed. The two-inch line runs from the farm of James Tarr at Oil Creek, Pennsylvania, to the Humboldt refinery at Plumer, 2½ miles away. The line is plagued by leaks, and never sees commercial use.

1874 Joseph Lister, founder of antiseptic surgery, humbly acknowledges the source of his pioneering work in a letter to Louis Pasteur: "Please allow me to take this opportunity to render you my most cordial thanks for having by your brilliant researches demonstrated to me the truth of the germ theory of putrefaction, and thus furnished me with the principle on which alone the antiseptic system can be carried out."

1878 Thomas A. Edison patents the phonograph. "Mary Had A Little Lamb" was the first recording.

1909 The country's (and possibly the world's) first national mental health organization holds its inaugural meeting in the Manhattan Hotel, New York City. The event was a direct result of *A Mind That Found Itself*, Clifford Beers's exposé of his abuse as a mental patient.

1910 Neurologist William Walter is born in Kansas City. First achieving notoriety by discovering a particular brain wave associated with learning, Walter later built the most advanced robot of his day.

1916 Physicist-turned-philosopher Ernst Mach dies one day after his 78th birthday (see biographical note on February 18, 1838).

1957 The SS *Tropicana*, the world's first ship to transport fresh orange juice in stainless steel tanks, completes its maiden voyage from Port Canaveral, Florida, to Whitestone, Long Island, New York. It carried 650,000 gallons of juice, and made the trip in 56 hours.

1970 The world's largest neon advertising sign is demolished at Port Tampa, Florida.

1977 Off the Galapagos Islands in the Pacific Ocean, John B. Corliss and John M. Elmond descend to the ocean floor in the research submarine *Alvin*. There they discover a new world, a new type of ecosystem, populated by never-before-seen clams, worms, and crabs. This "oasis" of life (others have since been found) occurs around springs of hot water rising through the Earth's crust. Unlike all other known ecosystems, the ultimate energy source is not the sun, but chemical reactions in bacteria on the oasis.

1986 The U.S. Senate ratifies a treaty outlawing genocide ... 37 years after the document was submitted.

1992 A chimp named Gamma dies at 59 years 5 months, the greatest recorded age for a nonhuman primate. He spent every minute of it in captivity. He was born in the Florida branch of the Yerkes Primate Research Center, and died in the Georgia branch.

1473 Astronomer Nicolaus Copernicus is born in Torun, Poland, the son of a successful merchant. Through his research and theories Copernicus produced one of the greatest-ever revolutions in human thought: that Earth revolves around the sun, rather than being the center of the universe. Copernicus did not consent to full publication of his findings until very late in life; it is said that a finished copy of his masterwork was brought to him in 1543 on the last day of his life.

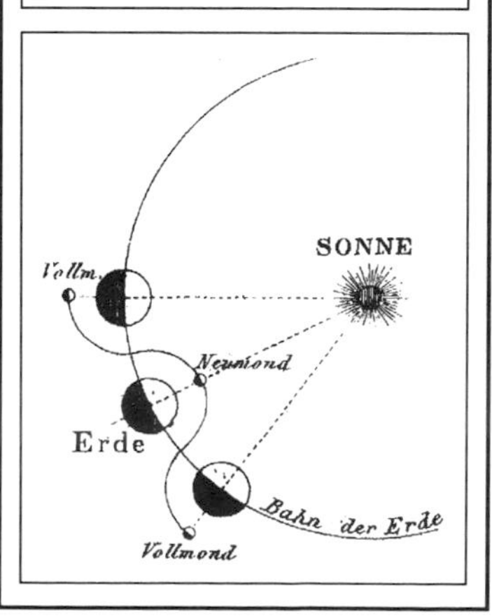

1725 A white man scalps an American Indian for the first time in recorded sociological history, outside Dover, New Hampshire.

1792 President Washington signs an act permanently establishing the U.S. postal service.

1835 At sea on the HMS *Beagle*, Charles Darwin witnesses a volcanic eruption on the Pacific island of Chloé. He is awed, as he is again several days later when his measurements reveal that the land had risen several feet. This incident showed Darwin the power of nature, and helped convince him that natural processes could produce conditions previously explained by divine intervention Presaging his theory of evolution.

1844 The gem-studded, but eventually tragic, life of physicist Ludwig Boltzmann begins in Vienna. He founded "statistical mechanics" with his mathematical study of the second law of thermodynamics, developed the kinetic theory of gases, and established the Stefan–Boltzmann law (relating radiation to temperature). He firmly defended atomism (the notion that all matter was composed of tiny particles) at a time when many thought it ludicrous; hostile opposition is thought to have precipitated his 1906 suicide.

FEBRUARY 20

1872 Silas Noble and James P. Cooley of Granville, Massachusetts, receive the nation's first patent for a toothpick-making machine. It is patent No. 123,790; the invention allows "a block of wood, with little waste, at one operation, [to] be cut up into toothpicks ready for use."

1912 Science fiction writer Pierre Boulle, author of a classic book on talking primates (*Planet of the Apes*) and one on bridges (*Bridge on the River Kwai*), is born in Avignon, France.

1934 "We felt like kicking each other's butts," summarized one physicist about the day's events, begun in the afternoon when cyclotron inventor Ernest O. Lawrence burst into his Berkeley lab with a report by Frédéric Joliot theorizing that many elements could be made radioactive by hydrogen bombardment. Lawrence's group had been doing that for years, but had never checked for radioactivity. Within minutes they were bombarding nitrogen, and discovering that they had created a new radioactive substance. No other lab, except Joliot's, had ever created artificial radioactivity.

1938 The first successful automobile–airplane combination is complete and ready for testing. The *Arrowbile*, built by the Waterman Arrowplane Corporation of Santa Monica, California, boasts a top air speed of 120 mph. Five of the vehicles were eventually delivered to the Studebaker Corporation.

1952, 1953, 1956, 1961 AND 1966 The only verified case of five single children in the same family (the Cummins of Clintwood, Virginia) being born on the same date.

1962 John Glenn becomes the first American to orbit Earth, in *Friendship VII*. A four-cent stamp is issued to mark the event, becoming the first U.S. stamp issued the day of the event it commemorates.

1963 Laser-operated TV is first demonstrated, in Bayside, New York.

1965 *Ranger VIII* transmits thousands of spectacular pictures of the moon—before crashing into it.

1982 René Dubos dies the day before his 81st birthday in New York City. He enjoyed a distinguished career, initially as a pathologist and microbiologist and later as an environmentalist and writer on human nature.

1990 The modern snail-racing record is established. A garden snail named Vern covers a foot-long course in 133 seconds in Plymouth, Michigan.

1994 Russia's whaling commissioner announces that his government lied for decades about the number of whales slaughtered by the Russian fleet. The ships were specially designed to hide deck activity from observing airplanes, and ships' logs contained entries such as "foreign aircraft have appeared" and "sink the prohibited whales." At the time of the revelation only one whale species, the minke whale, still existed in commercially exploitable numbers (500,000 to 1 million individuals), and Japan was pushing hard to break a 1987 moratorium on harvesting minkes.

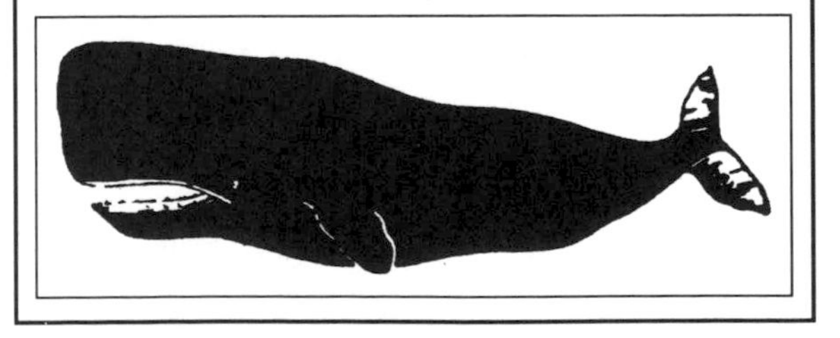

1741 Jethro Tull, agronomist and farm-implement inventor, dies at about 67 (his exact date of birth is unknown) at Prosperous Farm, Berkshire, England.

1866 August von Wassermann, who conquered syphilis by developing the diagnostic test that bears his name, the Wassermann test, is born the son of a banker in Bamberg, Bavaria.

1887 The nation's first bacteriology laboratory is incorporated. It is the Hoagland Laboratory at 335 Henry Street in Brooklyn, New York. Open for business in 1889, its first director was Dr. George Miller Sternberg, famous for finding the pneumonia organism in human saliva.

1892 Psychiatrist Harry Stack Sullivan is born in Norwich, New York. He first achieved prominence through his unique ability to communicate with schizophrenics, but he is especially known for his interpersonal theory of personality that emphasizes the role of early contact with other humans in determining personality.

FEBRUARY 21

1892 The birth of silviculture (the scientific management of woodlands) in North America: 26-year-old Gifford Pinchot (the first American to be educated at a European school of forestry) arrives at the 7000-acre Vanderbilt estate near Asheville, North Carolina, as chief forester. Within two years, Pinchot not only installed proper conservation measures, but also was able to turn a profit on the sale of excess timber. He eventually became the first head of the U.S. Forest Service.

1895 Henrik Dam is born the son of a pharmaceutical chemist in Copenhagen. Henrik achieved fame, and the 1943 Nobel Prize for Physiology or Medicine, for his own chemical work: discovery of vitamin K.

1901 Microbiologist René Dubos is born in Saint Brice, France (see biographical note on February 20, 1982).

1931 The first known flight of a liquid-fuel rocket outside the United States occurs in Breslau, Germany, when Johannes Winkler launches a small oxygen–methane missile.

1941 Sir Frederick Grant Banting dies at 49 in a plane crash on a war mission. A Nobel laureate in his early 30s, Banting (with Charles Best) discovered insulin and its role in diabetes.

1947 At an optical society meeting in New York City, Edwin H. Land demonstrates the Polaroid camera, the first device to take, develop, and print pictures on a single sheet of paper ... all in 60 seconds.

1948 The day pernicious anemia is vanquished. Dr. Karl Folkers and colleagues at Merck & Company administer a vitamin B_{12} substitute to a 66-year-old woman near death; within days she is on the road to full recovery. Production of the substance was made possible by Folkers's chance discovery of a type of bacteria on which B_{12} substitutes could be rapidly tested. This ended a 10-year search for a powerful substitute; until now, the only treatment for pernicious anemia was to have the sufferer eat a half-pound of raw liver every day, or to undergo daily, expensive, and painful injections of liver extract.

1958 The first global circumnavigation by a submarine (the *Gudgeon*) ends, in Pearl Harbor, Hawaii. The 25,000-mile journey took 228 days.

1968 The first expedition to cross the Arctic "continent" begins; a British team of 4 men and 40 dogs departs Point Barrow, Alaska. Fifteen months later they reach the other side.

1969 The most powerful rocket in history (the Russian NI booster with 5200 tons of thrust) is launched for the first time. It soars skyward for 70 seconds, then explodes.

1994 A doctor in Jerusalem announces that a 60-year-old, postmenopausal woman has delivered a healthy baby girl.

1994 The Whirlpool Corporation begins production of a freon-free refrigerator. As well as eliminating the ozone-destroying chemical freon, the appliance boasts an energy efficiency 25% greater than law demands.

1932 Another photography first: a camera light-exposure meter is patented in the United States. Technically called a "Photronic Photoelectric Cell," the device translated brightness values into aperture settings. It was invented the previous year by William Nelson Goodwin, Jr. of New Jersey, who is awarded patent No. 1,407,147 on this date.

1522 Amerigo Vespucci, the official astronomer to Ferdinand II, dies at about 58 in Seville, Spain. His name was given to the New World, America, because it was he, and not Columbus, who realized and announced that Columbus had discovered a new continent.

1630 English colonists first eat popcorn. It is brought to them in deerskin bags by the Native American Quadequina as his contribution to the first Thanksgiving dinner.

1636 Italian physician Santorio Santorio dies at 74 in Venice. He was the first to use precise medical instruments, and performed the first systematic study of human metabolism.

1796 Belgian astronomer-mathematician-sociologist Adolphe Quetelet is born in Ghent (see biographical note on February 17, 1874).

FEBRUARY 22

1824 Pierre Janssen is born the son of a musician in Paris. Although lamed in a childhood accident, Janssen established his fame as a globetrotting astronomer, mainly in studies of the sun. He invented ways to study solar prominences, noted a spectral line around the sun that led to the discovery of helium, and was the first to regularly photograph the sun.

1828 German biochemist Friedrich Wöhler announces his synthesis of urea to Jakob Berzelius, "the greatest chemical authority in the world." It was a tremendous achievement, being the first time that any man in the lab had artificially synthesized a chemical that living things normally make in their bodies (urea is produced in the excretory system of many animals; it is a major constituent of urine). Wöhler's feat destroyed the distinction between organic and inorganic chemicals, and destroyed the then-accepted classification system for chemicals, which Berzelius had created.

1857 Physicist Heinrich Hertz is born the son of a lawyer in Hamburg, Germany. In honor of his many achievements, his name was given to the international unit of frequency, the hertz.

1879 With ten days left in office, President Grover Cleveland creates 13 new forest reserves. This sets aside over 21 million wooded acres for conservation (see 1972 below).

1879 Chemist Johannes Brønsted is born in Varde, Denmark. He is known for developing the modern definition of acid and base (acids contain an excess of H^+ ions, bases an excess of OH^- ions).

1902 Fritz Strassmann is born a ninth child in Boppard, Rhineland. In 1938 he worked with Otto Hahn in Hitler's Germany to split the atom, thereby paving the way for invention of atomic weapons.

1905 The play *Damaged Goods* reopens in Paris, after being banned because it dealt with syphilis.

1804 Richard Trevithick debuts his second steam locomotive in Pennydarren, Wales. It is uncertain whether the inventor's first locomotive ever worked, making today's event the first verified trip on rails of a vehicle with a motor or engine. Today's demonstration ends when the rails collapse.

1918 The world's tallest human on record, Robert Pershing Wadlow, is born in Alton, Illinois. He was 8 feet 11 inches and still growing when he died at age 22.

1972 The Sierra Club petitions the chief of the U.S. Forest Service to reconsider the Management Plan for California's Six Rivers National Forest, which calls for clear-cutting ancient trees faster than they can grow back and for the construction of 5000 miles of roads through the million-acre wilderness. Three days later the Forest Service announces that the Plan will not change.

1993 China's legislature passes a new consumer-protection law; fraud, in some cases, will now be punished with death. The current economic boom has led to instances of unregulated, unscrupulous manufacturing. Especially harsh penalties are reserved for cases where the advertising claims are particularly false, the product is particularly harmful, or the profits are particularly outrageous.

1787 Emma Willard, champion and reformer of women's education, is born in Berlin, Connecticut.

1841 The Chemical Society of London holds its first organizational meeting. It becomes the first chemists' association to last 100 years.

1848 John Quincy Adams dies at 80 of a stroke suffered on the floor of the House of Representatives. The event leads to the nomination and election of the great educational reformer Horace Mann to fill Adams's seat in Congress.

1855 Carl Friedrich Gauss, "the Mozart of mathematics," dies peacefully at 77 in Göttingen, Germany, of dropsy, an enlarged heart, and the effects of a carriage accident he suffered during the one and only day that he ventured out of Göttingen in the last 20 years of his life. A child prodigy, Gauss achieved most of his fundamental breakthroughs by age 17; he spent the rest of his celebrated life refining them. His name lives on in a variety of forms: the gauss (a unit of magnetism), the Gaussian distribution (the "normal" or bellshaped curve), Gauss's law (also called Gauss's theorem, it explains electromagnetic phenomena), Gauss elimination (in algebra, solving a two-unknown equation by solving for one unknown first), and degaussing (removing magnetism from objects).

1868 Social scientist-political activist W.E.B. Du Bois is born in Great Barrington, Massachusetts. Harvard-educated, Du Bois's early work involved collecting and reporting data on the condition of blacks in the United States, while his sociologist colleagues were mainly theorizing about this issue. He became increasingly active in the civil rights movement, and increasingly discouraged at the impotence of scholarly findings to effect change. He developed a stance in direct opposition to the appeasement position of Booker T. Washington. Du Bois was a founder of several black organizations, including the NAACP. He abandoned his U.S. citizenship in 1962.

1884 Biochemist Casimir Funk is born the son of a physician in Warsaw. Coiner of the term "vitamine" (meaning "life-amine"), the "e" was dropped from the word when others discovered that not all such vital compounds contained the amine group ($-NH_2$).

1886 Eight months after graduating from college, Charles Martin Hall completes work on his electrolytic process for making aluminum. He patents the process, which becomes the first commercially viable means of producing the metal. The process is put in operation in November 1888 by the Pittsburgh Reduction Company, which later becomes ALCOA.

1924 Allan Cormack is born in Johannesburg, South Africa. He received the 1979 Nobel Prize for Physiology or Medicine for inventing the CAT scanner, a three-dimensional X-ray machine.

1927 The Federal Radio Commission (forerunner of the FCC) is signed into law by Calvin Coolidge.

1954 The first mass inoculation of children with the Salk polio vaccine begins in Pittsburgh.

1980 *Irenes Serenade* sings a horrible song off the Greek island of Pilos; the massive tanker explodes and sinks, sending 37,000,000 gallons of oil into the Mediterranean.

1987 The brightest supernova, or an apparent explosion of a large star, of the twentieth century is first seen on Earth; it became visible without a telescope in May.

1994 Concerned with potential health threats posed by secondhand smoke, McDonald's Corporation announces that it is banning smoking in all of its 1400 company-owned franchises. Individually owned franchises will be free to set their own policies. The debate over the exact degree of harm from "passive" smoking continues to swirl between the EPA, medical researchers, tobacco companies, and smokers.

1917 The American Society of Orthodontists, the first organization of U.S. dentists who specialize in straightening irregular teeth, is incorporated in Pennsylvania.

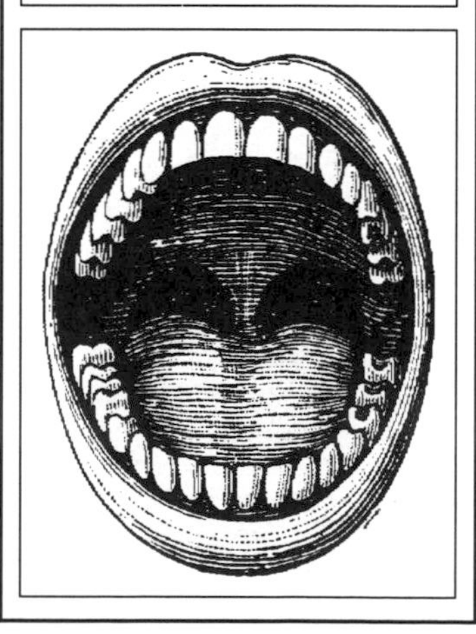

1582 Pope Gregory XIII issues an edict of calendar reforms, creating the modern Gregorian Calendar.

1663 Thomas Newcomen, harbinger of the Industrial Revolution, is born in Dartmouth, England. A blacksmith by trade, Newcomen tinkered for ten years before producing a steam engine that was the world's best for over a half-century (eventually supplanted by the engine of James Watt).

1804 Heinrich Lenz is born in Dorpat, Estonia, the son of a magistrate. Like Darwin, he studied theology and sailed on a multiyear scientific voyage before he "found himself." Lenz's calling was physics; among other contributions is Lenz's law (describing the direction of electric current).

1841 Chemist Carl Graebe is born the son of a soldier in Frankfurt am Main, Germany. In 1868 he changed the face of textiles by synthesizing the orange-red compound alizarin, one of the first synthetic dyes. Graebe also introduced the now-universal prefixes *ortho-, meta-* and *para-* to identify isomers.

FEBRUARY 24

1849 John Henry Comstock, prolific author and eminent entomologist (insect scientist), is born in Janesville, Wisconsin. With his wife-illustrator, Anna Botsford Comstock, he produced a number of very popular books, for both the layman and the scientist. His anatomical studies of butterflies and moths formed the basis for their modern classification.

1871 *The Descent of Man* by Charles Darwin is published in London by John Murray.

1874 Naturalist-Lutheran minister John Bachman dies at 84 in Columbia, South Carolina (see biographical note on February 4, 1790).

1896 Henri Becquerel tells the French Academy of Sciences in Paris about some interesting effects he's seen with uranium; he thinks he's found a new way to produce the X rays that Wilhelm Roentgen had accidentally discovered in November. Several days later, back in his lab, Becquerel accidentally discovers another totally unforeseen form of energy: radioactivity.

1917 *Livery Stable Blues* is recorded by the Original Dixieland Jazz Band in New York City. It is the first jazz record ever released.

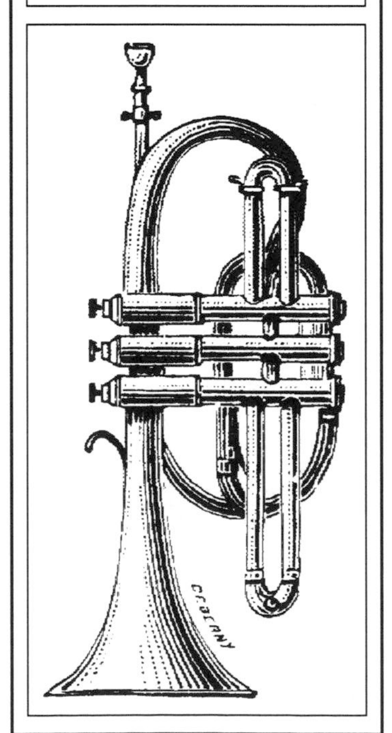

1938 Production of nylon bristles for toothbrushes begins in the factories of E.I. du Pont de Nemours & Co., at Arlington, New Jersey. This "miracle" material was invented in 1935 by Du Pont chemists under the supervision of Wallace H. Carothers, who spent many years experimenting with artificial fibers before nylon was created. Nylon became instantly famous when used in women's stockings, but Carothers did not live to see any of its great success; he committed suicide in 1937, two days after his 41st birthday, following a long depression.

1968 The discovery of pulsars (extraterrestrial pulsating radio sources) is announced from where they were first detected, the Mullard Radio Astronomy Observatory at Cambridge University.

1981 Jean Harris is found guilty in White Plains, New York, of murdering *Scarsdale Diet* author Dr. Herman Tarnower.

1992 Hundreds of construction workers at the Russian space center in Baikonur, Kazakhstan, begin several days of rioting over "inhuman treatment" by their supervisors. "Hungry and unwashed" mutineers had commandeered 17 trucks, set fire to barracks, stolen 35,000 rubles, and marched on the nearby city of Leninsk (where the supervisors live) when they were dispersed by authorities who gave them food and money and met some of their other demands. One worker summed up his ordeal at the Baikonur Cosmodrome: "When I came here I was healthy and intact. I have served for two years and now I am completely sick—a cripple."

1994 During a Congressional hearing, an attorney for the people of Bikini Island accuses the United States of deliberately and needlessly showering his clients with radioactivity during an H-bomb test in 1954. The government had consistently claimed that the atomic bath was a complete mistake resulting from a sudden change of wind; recently uncovered documents, however, reveal that officials had plenty of warning about the windshifts.

FEBRUARY 25

1616 In Rome, Cardinal Bellarmine orders Galileo "to give up altogether the said false doctrine [that Earth moves around the sun] ... and if you should refuse ... you should be imprisoned." Galileo did, indeed, renounce the "doctrine."

1751 The home of Mr. Edward Willet in New York City is the scene of the first performance by a trained monkey in America. Viewers are charged a shilling (about five cents) to see dancing, tightrope walking, and operation of a pistol by "a creature called a Japanese, of about two feet high, his body resembling a human body in all parts except the feet and tail." The firearm was presumably unloaded.

1799 Forestry legislation is first enacted by the U.S. government, which authorizes funds "not exceeding $200,000 to be laid out in the purchase of growing or other timber, or of lands on which timber was growing, suitable for the navy."

1869 Phoebus Levene is born in Sager, Russia. He discovered the D in DNA and the R in RNA.

1896 Ida Noddack, codiscoverer of the element rhenium, is born in Rhenish Prussia.

1898 William Thomas Astbury, the first to make X-ray diffraction studies of nucleic acids (1937), is born in Stoke-on-Trent, England. Although Astbury's guesses as to the complete structure of these molecules were wrong, his work opened the door to others who eventually determined the exact structure of DNA.

1904 Biologist John Bittner is born in Meadville, Pennsylvania. He discovered that genetically cancer-prone mice became cancer-resistant if nursed by cancer-resistant mothers; likewise, cancer-prone mothers' milk produced cancer-prone infants. The "Bittner milk factor" was isolated in 1949, and provided some of the earliest and strongest evidence that some cancers are virus-caused.

1909 Lev Artsimovich, whose research laid the groundwork for the modern "tokamak" nuclear-fusion devices, is born in Moscow.

1924 Molecular biologist Hugh Esmor Huxley is born in Birkenhead, England. In partnership with Jean Hanson, Huxley elucidated how muscles work; the two scientists proposed the now-accepted "sliding filament theory" (which states that muscles contract when two intertwined proteins, actin and myosin, "slide" past each other).

1952 For the first time in modern history, a third total solar eclipse is observable from the same spot on Earth—44° N, 67° E in Kazakhstan.

1956 The world's heaviest chicken egg (16 oz.) is laid in Vineland, New Jersey, by a White Leghorn.

1971 Theodor Svedberg (Nobel Prize winner in 1926 for Chemistry) dies at age 86 in Orebro, Sweden. He invented the ultracentrifuge (which generates forces that are hundreds of thousands of times greater than the force of gravity), thereby making possible extremely precise studies of proteins and other chemicals. He also helped develop the cyclotron and the process of electrophoresis, two other important advances in chemical analysis.

1993 The *New England Journal of Medicine* publishes a long-running study on "moderate exercise": it will add months to life expectancy (even when begun in one's 60s and 70s), but it is "no fountain of youth." The greatest benefits are seen in middle-aged men who both quit smoking and start exercising.

1682 Giovanni Morgagni, the founder of pathology, whose works established anatomy as an exact science, is born an only (and soon fatherless) child in Forli, Italy. After spending 56 years as a professor of anatomy at the University of Padua, Italy, his great fame came in his 80s, after a lifetime of postmortem dissections and the publication in 1761 of *On The Seats and Causes of Disease*, which convinced science that specific illness begins in specific organs and tissues (rather than arising diffusely throughout the body, as had been thought).

1786 Physicist Dominique Arago is born in Estagel, France. He discovered the solar chromosphere, proved the wave theory of light, and discovered that rotation of a conductor produces magnetism.

1799 Engineer Benoit Clapeyron is born in Paris. Interest in steam engines led him to discover a relationship between temperature, volume, and heat of vaporization of a fluid; this is often called the Clapeyron–Clausius formula, and was later developed into the second law of thermodynamics.

1876 "Princess Pauline" Musters is born in Ossendrecht, the Netherlands. At maturity, she measured only 23.2 inches, earning her the title of the world's shortest mature human. When she died at 19, she weighed 9 pounds, with measurements of 18½–19–17, indicating that she was obese!

1878 The adoption of the word "microbe" took shape in a series of letters between French scientists Sédillot and Littré; in today's letter, Littré approves the term, saying he likes it better than "microbia," even though it means "short-lived" rather than "very small" in Greek.

FEBRUARY 26

1895 The first U.S. patent for a glass-blowing machine is issued. Patent No. 534,840 goes to Michael Joseph Owens of Toledo, Ohio; his device simultaneously creates five glass pieces "without seams or roughness" by positioning five molds, surrounded by molten glass, in front of a blowing pipe.

1896 Poor weather interrupts the work of Paris physicist Henri Becquerel: "[S]ince ... the sun appeared only intermittently, I stopped all experiments [on the effects of sunlight on uranium]." Becquerel happened to leave the uranium on top of a photographic plate in a dark desk drawer until the weather got better, several days later. When he developed the plate, expecting no result whatsoever, he was amazed to find that the uranium rocks had left a very clear image. This was the critical discovery about radioactivity, that the uranium rocks themselves contained some form of invisible energy—and it happened totally by accident!

1903 Chemist Giulio Natta is born the son of a judge in Imperia, Italy. He received the 1963 Nobel Prize for Chemistry for developing polymers with tremendous commercial application.

1918 The worst-ever disaster at a modern sporting event occurs when 604 perish in the collapse/fire of the stands at the Hong Kong Jockey Club.

1866 Herbert Henry Dow, founder of the giant Dow Chemical Company, is born in Belleville, Canada. He actually created a number of chemical companies, and received over 100 patents for his innovations.

1919 Grand Canyon National Park, Arizona, is established by Congress.

1929 Grand Teton National Park, Wyoming, is established by legislation signed by Calvin Coolidge.

1952 Churchill announces that Britain has the atomic bomb.

1955 George Franklin Smith of Manhattan Beach, California, ejects at 777 mph from an F-100A Super Sabre Jet fighter, becoming the first to parachute out of a plane going faster than the speed of sound. He experiences pressure 40 times that of gravity. His clothes are shredded, and his helmet, mask and socks are torn off before he hits the ocean. He spends six months in hospital after being rescued by a passing boat off Laguna Beach, California.

1962 "Exploration and the pursuit of knowledge have always paid dividends in the long run," testifies one of the first in space, John Glenn, before a joint session of Congress.

1992 Renowned fertility specialist Dr. Cecil Jacobson takes the stand in his own defense, accused on 52 counts of fraud and perjury in a famous case in which he allegedly used his own sperm to artificially inseminate tens of unsuspecting women.

1993 After a decade of mounting controversy, the Environmental Protection Agency calls for massive research into the health effects of magnetic fields.

1824 Sir William Huggins, the man who revolutionized astronomy by wholesale application of spectroscopy, is born in London to a linen draper. Huggins was the first to make significant discoveries by studying the light emitted by heavenly bodies. One of his early triumphs was announcing that the light on these bodies had the same spectra as light on Earth, thereby ending 21 centuries of controversy begun by Aristotle's edict that the heavens were composed of stuff not found on Earth. Huggins was one of the first to photograph extraterrestrial objects.

1891 David Sarnoff, pioneer in both radio and television, is born in Uzlian, Russia. He was the first general manager of RCA and he founded NBC in 1926.

1897 Astronomer Bernard Lyot is born the son of a surgeon in Paris. His fame came through developing new ways to photograph the heavens and through his inventing the coronagraph (1930), the first device that allowed the sun's corona to be studied at leisure, without an eclipse occurring.

FEBRUARY 27

1899 Charles Herbert Best is born in West Pembroke, Maine. He became famous as part of the "Banting and Best" team that discovered, isolated, and applied insulin to fight diabetes. Banting received the Nobel Prize for this work, but Best was denied the honor (much to Banting's anger) because he was an undergraduate when the discovery was made. Best went on to a distinguished career in which he discovered several other biochemicals (choline and histaminase) and was the first to use anticoagulants to battle thrombosis (blood clots that form within blood vessels).

1905 Freud meets Jung for the first time, in Freud's Vienna office.

1919 The nation's first social organization for the deaf is formed in New York City. The American Association for the Hard of Hearing held its first official meeting the following year, and changed its name to the American Society for the Hard of Hearing in 1935 at a meeting in Cincinnati.

1933 Germany's parliament building, the Reichstag, burns to the ground in Berlin. The Nazis blame the Communists and suspend civil liberties. It is a major step in the Nazis' ascension, and precipitated what could be history's greatest "brain drain," the exodus of top scientists and intellectuals.

1934 Ralph Nader, the most famous consumer advocate in history and one of big business' most hated foes, is born in Winsted, Connecticut.

1936 Ivan Pavlov, who revolutionized psychology while studying digestion, dies at 86 in Leningrad. While directly observing the flow of saliva in dogs (Pavlov was a superb surgeon who pioneered a number of physiological techniques), he accidentally discovered that stimuli associated with food presentation, as well as the food itself, would produce saliva flow. He measured the strength of this effect by determining the rate and quantity of saliva flow. This led to his development of "Pavlovian responses" and conditioned reflexes, which provided great insights into learning, and gave science powerful tools to examine the perceptions and psychology of nonhuman animals. Pavlov received the Nobel Prize in Physiology or Medicine in 1904. He was an outspoken anti-Communist, but the government left him alone because of his prominence.

1926 Brain scientist David Hubel is born in Windsor, Canada. He shared the 1981 Nobel Prize for Physiology or Medicine with Torsten Nils Wesel and Roger Wolcott Sperry for mapping the path of nerve impulses from the eye to various centers in the brain.

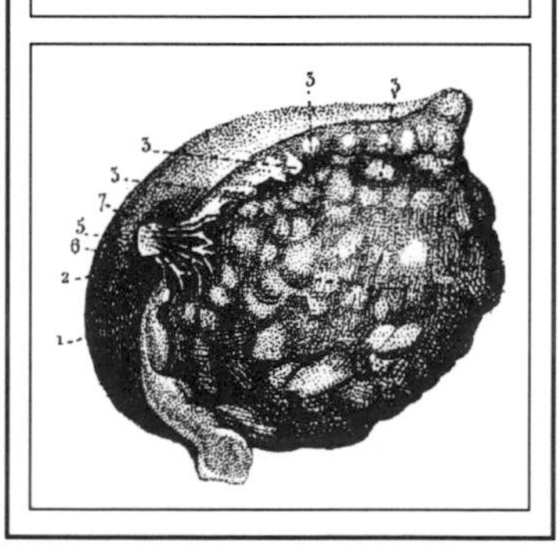

1947 The first closed-circuit broadcast of a surgical operation takes place at the Johns Hopkins Hospital in Baltimore, Maryland. On the day, five operations are broadcast, four on the heart and one on nerves along the spine. Dr. Alfred Blalock performs the first two. Ten observers in four classrooms simultaneously see each procedure.

1994 Existence of planets outside our solar system is confirmed for the first time. Wire services carry the announcement of Penn State's Alexander Wolszczan that he has verified the discovery of two planets orbiting a sun 1300 light-years from Earth. The year of one of the new planets is calculated to be 66.6 days long, the year of the other is 98.2 days.

1994 Closing arguments are delivered in the Virginia trial of world-renowned fertility doctor Cecil B. Jacobson, facing 52 counts of fraud and perjury surrounding his alleged artificial insemination of some 75 women with his own sperm.

1683 René-Antoine Réaumur is born in La Rochelle, France. The world's leading expert on insects in his day, Réaumur also invented ways to make steel and iron, devised the Réaumur temperature scale (widely used for 200 years), discovered regeneration of lost limbs in crustaceans, built the first cupola furnace (still used for melting iron), designed the first egg incubator, developed a unique formula for pottery (Réaumur porcelain), and, in 1752, became the first to collect digestive juices from the stomach of a living organism.

1743 René-Just Haüy, founder of crystallography, is born in St.-Just-en-Choisée, France. In 1781 he was handling a piece of calcite in the collection of a friend when he dropped the sample, causing the calcite to shatter. Haüy was mortified, but then he noticed that it had broken along very regular lines. This led him to investigate the structure of many other pieces of calcite and of many other minerals; these researches mark the beginning of modern crystallography (the study of the form and formation of crystals).

February 28

1814 French chemist Edmond Frémy is born in Versailles. His great struggle was to isolate pure fluorine; he never succeeded, but he did discover a variety of compounds along the way. He also created rubies in a failed attempt to produce crystals of aluminum oxide.

1849 The first gang of gold-rushers arrives in San Francisco on the ship *California*.

1896 Physician Philip Hench is born in Pittsburgh. In attempts to relieve arthritis pain, Hench found that a newly discovered hormone, which he named "cortisone," was especially effective. He won the 1950 Nobel Prize for this work.

1898 The altitude record for a single kite is set at over two miles high (12,471 feet) at the Blue Hill Weather Station in Milton, Massachusetts.

1901 Double Nobel laureate Linus Pauling is born the son of a druggist in Portland, Oregon.

1915 Biologist P.B. Medawar is born in Rio de Janeiro, Brazil. His research changed the face of immunology and transplant surgery when he found (in 1949–53) that foreign tissue would not be rejected if it were introduced when an organism was still an embryo. For this he received the 1960 Nobel Prize.

1928 John D. Rockefeller, Jr., donates $5 million toward the purchase of Tennessee land to establish the Smoky Mountain National Park. It is half the amount needed.

1930 Physicist Leon Cooper is born in New York City. He won the 1972 Nobel Prize for developing superconductivity theory, a cornerstone of which was his 1956 discovery that (normally repellent) electrons are attracted to each other in superconductors.

1936 French bacteriologist Charles-Jules-Henri Nicolle dies at age 70 in Tunis, Tunisia. He received the 1928 Nobel Prize for Physiology or Medicine for his discovery that typhus is transmitted by body lice.

1951 On his 50th birthday, Linus Pauling and collaborator Robert Corey publish in *Proceedings of the National Academy of Sciences* a theoretical (and correct) description of the structure of proteins.

1953 James Watson first suggests the base pairings of DNA. At least this is the way he remembered it. Watson's collaborator at Cambridge, Francis Crick, remembered himself as the originator of the insight.

1977 The first birth of a killer whale (*Orca*) in captivity occurs at Marineland in Los Angeles.

1994 Cardinal Joseph Bernadin is exonerated of child molestation charges by his accuser, Steven J. Cook, ("I now realize that the [repressed] memories which arose during and after hypnosis are unreliable.")

1994 The first public hearings on the mysterious "Gulf War syndrome" begin in Washington.

1913 The world's largest known seal (a bull elephant seal, 22 feet long and weighing at least 9,000 pounds) is slaughtered in Possession Bay, Antarctica.

1792 Zoologist-embryologist Karl Ernst von Baer, who discovered that women have eggs, is born in Piep, Estonia, into a family of 12. His parents were first cousins. Dissatisfied with his medical studies in his early 20s, Baer became enthralled with comparative anatomy, especially the development of embryos; his change of focus largely reflected the influence of a mentor, Ignaz Döllinger. In 1827 Baer reported his discovery of the mammalian egg. In 1828 he published the first of two monumental volumes (the second came in 1837) that surveyed all existing knowledge on the prebirth development of vertebrates. This work established embryology as a distinct field of science.

1860 Herman Hollerith, calculator-inventor, is born in Buffalo, New York. Through employment stints at the U.S. Census Bureau, the Patent Office, and the Massachusetts Institute of Technology, Hollerith became gripped by the problem of automating census calculations. By the 1890 census he had invented several punched-card machines that electrically read and sorted data. The machines were successful in both the United States and Europe. In 1896 he formed the Tabulating Machine Company in New York, which, through a series of mergers, became IBM.

1892 Britain and the United States agree to submit their dispute over seal-hunting rights in the Bering Sea to international arbitration. Eventually, Britain was granted precedence.

1904 President Theodore Roosevelt appoints a seven-man commission to speed completion of the Panama Canal.

1912 The most famous Antarctic expedition in history (the attempt of England's Robert Falcon Scott to be the first to the South Pole) is in deep trouble. "Cold night. Minimum Temp. −37.5°; −30° with northwest wind, force 4, when we got up. Frightfully cold starting.... Next camp is our depot and it is exactly 13 miles. It ought not to take more than 1½ days; we pray for another fine one. The oil will just about spin out in that event." The group of four did reach the depot, but the amount of oil was far below expectations, one of the men was found to have badly frostbitten feet, and the weather turned stormy. This expedition is remembered for its heartbreak and heroism; Scott did reach the Pole, only to find a tent and Norwegian flag that had been set there three weeks before by Roald Amundsen. On the return trip, Scott and his last two companions died just 11 miles from food and fuel.

1936 Jack Lousma, spacewalker and pilot of the *Skylab 3* mission (1973), is born in Grand Rapids, Michigan. In 1978 he became a test pilot for the space shuttle.

1936 Niels Bohr publishes his famous "bowl of balls" analogy in *Nature*; it is an attempt to explain what happens when bombarding, subatomic particles enter an atomic nucleus.

1936 Langdon, North Dakota, officially becomes the coldest place in the U.S. history; it is the 92nd consecutive day in which the temperature has stayed below freezing. The record stands to this day.

1940 Having arranged to receive his Nobel Prize in California rather than Sweden so that he would not lose a month of fund-raising for his cyclotron research, Ernest O. Lawrence delivers his acceptance speech in Berkeley; he ends it by asking for more money.

1960 Bunnies are first seen in Chicago—Playboy bunnies. Hugh Hefner opens the first Playboy Club. The last closed in 1986, but by that time the magazine and its editor had had significant societal impact.

1968 With nuclear power plants beginning to appear regularly (there were 15 U.S. plants by June 1968), Senator T. B. Morton of Kentucky introduces a resolution into Congress to reevaluate the direction and safety of the growing use of commercial nuclear power. The use of nuclear power remains controversial.

1840 John Philip Holland, father of the modern submarine, is born in Liscannor, County Clare, Ireland. A school teacher by trade, Holland emigrated to the United States in 1873. In hopes of using submarines against England, the Irish Fenian Society bankrolled the early prototypes built by Holland. In 1879 he produced the somewhat successful "Fenian Ram." He then won contracts from the U.S. Navy, and delivered the "Holland" in 1898; it was the first really useful submarine. The Navy ordered, and got, six more. His final years were spent in court, battling his financiers.

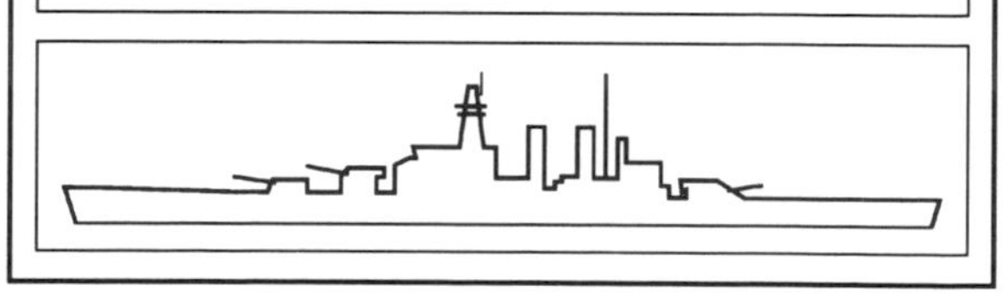

1790 Congress authorizes the first U.S. census (see August 1, 1790, for the results).

1846 Geomorphologist Vasily Vasilyevich Dokuchayev, a pioneer in soil science, is born in Milyukovo, Russia. As well as furthering new understanding of how soil is created and classified, Dokuchayev reformed Russian technical education (including establishing the first-ever university course in Quaternary geology, at the University of St. Petersburg).

1864 A black woman first receives a medical degree in the United States. Rebecca Lee is granted her "Doctress of Medicine" certificate from Boston's New England Female Medical College.

1866 Fire destroys the countryside factory (near Meaford, Canada) in which 27-year-old John Muir has been employed as a woodworker and tool inventor. A number of his inventions are destroyed, as is the stock of wood products in which he owned a 50% share of profits. It is a turning point for Muir, who soon quit industrial employment forever to resume his life as an author and wandering naturalist that brought him fame and fulfillment.

MARCH 1

1872 Another incident occurs in the feud between celebrated dinosaur-hunters Edward Drinker Cope and O.C. Marsh. Cope reads a paper to the American Philosophical Society in Philadelphia describing a huge, winged fossil that he names *Ornithochirus*. The world will forever know this same beast as *Pterodactylis*, however, because that is the name that Marsh gave it in the *American Journal of Science* (published in New Haven, Connecticut, where Marsh was a Yale professor) just five days before Cope was able to publish his paper.

1896 Becquerel discovers radioactivity—by total accident! (See February 26, 1896, for details.)

1910 British biochemist A.J.P. Martin is born in London. In 1952 he was awarded (with R.L.M. Synge) the Nobel Prize for Chemistry for his development of chromatography (a technique for separating the different components of a mixture, thereby allowing their identification and analysis). Modern DNA testing is based on Martin's work.

1927 The first paper on quantum electrodynamics ("The Quantum Theory of the Emission and Absorption of Radiation" by P.A.M. Dirac) is published in the *Proceedings of the Royal Society*. Quantum electrodynamics is the branch of physics that studies the movement of electric current according to quantum theory, first announced by Max Planck in 1900.

1872 Yellowstone National Park, the nation's first and largest national park, is authorized, as President Ulysses Grant signs "an act to set aside a certain tract of land (2,142,720 acres) lying near the headwaters of the Yellowstone River as a public park."

1932 The Lindbergh baby is kidnapped from his home near Hopewell, New Jersey.

1950 Scientist-philosopher Alfred Korzybski dies at 70 in Sharon, Connecticut. His 1933 book, *Science and Sanity*, coined, defined, and began the science of semantics (the study of the relationships between language, meaning, thought, and behavior). "We do not realize what tremendous power the structure of an habitual language has. It is not an exaggeration to say that it enslaves us through the mechanism of semantic reactions and that the structure which a language exhibits and impresses upon us unconsciously, is *automatically projected* upon the world around us."

1954 An enormous atomic bomb (a thermonuclear hydrogen bomb named *Bravo*) is dropped over the Bikini Atoll in the Pacific. The device is hundreds of times more powerful than any previous device tested, and at 18-22 megatons, it remains the most powerful bomb the United States has ever exploded. It leaves the islands radioactive for decades.

1993 Duke University scientists announce that they have discovered a gene linked to Alzheimer's disease. The finding may be helpful in identifying potential sufferers, and it represents one more step toward a cure. The picture is still muddy, however, because the gene produces a protein that occurs in several diseases of the heart and brain.

MARCH 2

1476 The battle of Grandson is fought on the shores of Lake Neuchâtel, Switzerland. It is the first encounter between mass infantry and fully armed knights on horses. Despite much simpler weaponry and lack of armor, the infantry is victorious because they formed "pike phalanxes" (squares composed of many men, armed only with long spears). The battle made pikes a mainstay of European warfare until the advent of firearms.

1730 Biologist Otto Friedrich Müller, the first to see bacteria clearly (which was made possible by other scientists' invention of achromatic, or blur-free, microscopes), is born the son of a court trumpeter in Copenhagen.

1784 From Paris, famed balloonist Jean-Pierre-François Blanchard, 30, makes his first ascent. Blanchard later became the first to cross the English Channel by air, and the first to make balloon flights in several European countries and in the United States.

1889 John B(enjamin) Murphy, a pioneer in several areas of internal medicine, performs his first preventive appendectomy, on a young laborer named Monham in Chicago. At the time it was a revolutionary procedure, in which the appendix was removed before infection had set in.

1894 Biochemist Aleksandr Ivanovich Oparin is born in the Russian village of Uglich. Because the town had no secondary school, the family moved nine years later to Moscow, where Oparin spent the rest of his life. In 1935 he cofounded the A.N. Bakh Institute [of biochemistry], at which he produced theories on the origin of life on Earth. His concepts were initially and universally ridiculed, but are now widely accepted. He conceived of a long, lifeless period on Earth during which "chemical evolution" produced a primordial "soup" that was rich in organic compounds. When they finally did arise, the first organisms fed on this soup (rather than being able to produce their own food internally, as was thought before Oparin's work).

1896 Henri Becquerel reports the discovery of radioactivity to the French Academy of Sciences.

1907 Under intense political pressure, President Theodore Roosevelt signs the Agricultural Appropriation Bill of 1907, which has an amendment that ends the presidential power to create forest preserves in Oregon, Washington, Idaho, Montana, Colorado, and Wyoming. Just before signing, however, the conservation-minded Roosevelt creates 21 new national forests in the six western states.

1913 Physicist Georgii Nikolaevich Flerov, discoverer of spontaneous fission (in which uranium spontaneously undergoes a splitting into two different elements without artificial bombardment with neutrons or other particles), is born in Russia.

1925 A joint board of state and federal highway officials, appointed by the U.S. Secretary of Agriculture, adopts the first nationwide system of numbering roads. The board eliminates confusing state-to-state differences in route symbols, and selects the familiar shield-numbered marker to indicate an interstate highway.

1949 The first nonstop flight around the world ends in Fort Worth, Texas. The 23,452-mile trip took 94 hours 1 minute, and was made in a B-50 Superfortress named *Lucky Lady 2*. It carried a crew of 14.

1994 A Houston jury orders the 3M Corporation and two other companies (Inamed and McGhan Medical Co.) to pay $12.9 million to three women claiming damages from leaky silicone breast implants. A variety of health problems allegedly resulted from the leaks. Defense attorney Joe Redden said the jury ignored the lack of supporting medical or scientific evidence. The lawyer for the women, John O'Quinn, said, "That's what cigarette companies are still saying about lung cancer."

1779 Joel R. Poinsett, the U.S. diplomat for whom the poinsettia flower is named, is born in Charleston, South Carolina. An accomplished botanist, Poinsett brought the flower from Mexico to the United States. His name was also the basis for the Mexican *poinsettismo*, which means intrusive, officious behavior. Poinsett was America's first minister to Mexico, but during his tenure (1825–1829) he became a sympathizer with revolutionaries seeking to overthrow the government; eventually he was expelled.

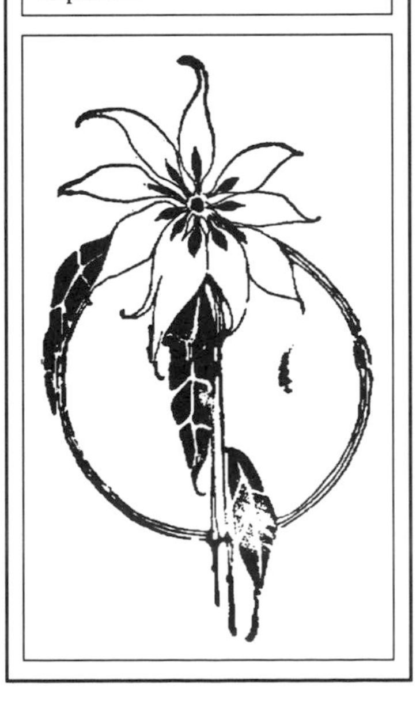

1841 Sir John Murray, founder of oceanography, is born in Cobourg, Canada.

1845 German mathematician Georg Cantor is born to a prosperous merchant and an artistic mother in St. Petersburg, Russia. He is credited with founding set theory (the basis of today's "new mathematics") and with introducing transfinite numbers. Vehement opposition to his views strained him greatly; Cantor suffered recurrent breakdowns and depressions, and eventually died in an asylum.

1847 Alexander Graham Bell is born in Edinburgh into a family of eminent speech teachers (his father, Alexander Melville Bell, wrote *Standard Elocution*, which passed through some 200 editions in English). The younger Bell first achieved notoriety by teaching the deaf to speak by adapting methods that his father had developed for the nondeaf. In his mid-20s he opened his own Boston school to train teachers of the deaf. It was actually his passion to help the hearing-impaired that led him to invent the telephone.

MARCH 3

1862 The theory behind pasteurization is put to the test, when Louis Pasteur and Claude Bernard seal blood and urine in flasks and keep them precisely heated for the next 50 days. No fermentation or decay is seen at the end of the experiment, which offered strong support for Pasteur's idea that liquid foods could be heated enough to kill disease-causing bacteria, without affecting the chemical and food properties.

1883 Sir Cyril Ludowic Burt, the first psychologist to be knighted, is born in Stratford-on-Avon, England. He made his name by developing the statistical technique of "factor analysis," and using this powerful/important tool to assess the role of heredity in intelligence. His results were remarkable: Gentiles were more intelligent than Jews, men more intelligent than women, Englishmen more intelligent than Irishmen, upper-class Englishmen more intelligent than lower-class, and so on. After he died it was discovered that he had made up much of his data to suit his own prejudices.

1887 Twenty-year-old Anne Mansfield Sullivan arrives in Tuscumbia, Alabama, to begin tutoring the blind and deaf Helen Keller. Keller would later call this "the most important day I remember in all my life." During the search for a teacher, the Kellers met with audiologist Alexander Graham Bell, who directed them to the Perkins Institute for the Blind, where Sullivan had been a trainee. Bell and Keller remained lifelong friends.

1903 Fingerprints are taken for the first time in a U.S. state or federal prison. The site is Sing Sing Prison in Ossining, New York. Fingerprints were first used to convict someone in 1911, in an Illinois murder case in which the accused was eventually condemned to death and hanged.

1918 Arthur Kornberg is born in Brooklyn, New York. He was awarded the 1959 Nobel Prize for Physiology or Medicine for discovering mechanisms by which DNA makes copies of itself.

1927 A radio probe is first sent into the upper atmosphere. French meteorologists Robert Bureau and Pierre Idrac launch a balloon from Trappes, France, which has instruments for measuring temperature, humidity, and air pressure. These instruments are linked to radio transmitters, for immediate reporting of results back to ground.

1969 *Apollo 9* blasts off from Cape Kennedy. The lunar landing module, later to carry men to the moon's surface, was given its first test flight during this mission.

1993 Chrysler engineers begin a nine-day, 2604-mile voyage from Detroit to Los Angeles, thereby breaking the world endurance record for an electric vehicle. The experimental minivan was driven to the Los Angeles "Eco Expo," which was appropriate because electric vehicles are less environmentally harmful than conventional, gasoline-burning cars.

1993 Albert Sabin, 86, dies in Washington, D.C., after a bout with congestive heart failure. He will forever be remembered as the developer of the oral vaccine for polio. Because it is so easy to take (swallowed on a sugar cube) and because it is longer-lasting, the Sabin vaccine replaced the Salk vaccine, introduced seven years before. The Sabin vaccine actually consists of living, but weakened, polio organisms.

1994 Two articles in the *New England Journal of Medicine* signal a new era in pain reduction for cancer patients. The first reports a survey of cancer patients, indicating that debilitating pain is more prevalent, and of less concern to doctors, than previously realized. The second presents physicians' guidelines and options for keeping their patients suffering-free. "The guidelines will make the burden of cancer lighter for those of us who have borne it or are bearing it," observes Dr. Fitzhugh Mullan, founder of the National Coalition for Cancer Survivorship.

1394 Prince Henry the Navigator is born in Porto, the son of Portugal's King John I (see November 13, 1460 for biographical notes).

1840 The nation's first commercial photography studio opens in New York City. It is the studio of John Johnson and Alexander S. Wolcott, who took his first photo the previous October.

1864 Marine architect David Watson Taylor is born in Louisa County, Virginia. He built the nation's first ship-model testing facility (1899), and proposed basic principles in ship design.

1881 Physicist-chemist Richard Chace Tolman is born in West Newton, Massachusetts. He demonstrated that electrons are the charge-carrying entities in the flow of electricity, and was able to measure the mass of the electron. During World War II, Tolman was the chief scientific adviser to Brigadier General Leslie R. Groves, overseer of the development of the atomic bomb.

1903 Biochemist William Clouse Boyd is born in Dearborn, Missouri. He conducted an enormous survey of the global distribution of various blood types (accomplished in the 1930s with his wife, Lyle, by taking blood samples from many people in many parts of the world).

1904 Cosmologist George Gamow, a foremost proponent of the big-bang theory, is born in Odessa, Russia, the son of a teacher and grandson of a tsarist general. Gamow's interest ranged from the very large (the formation of the universe) to the very small (the structure of DNA); in 1954 Gamow proposed the existence of a genetic code based on triplets of chemicals called nucleotides; this was proved true by 1961.

1913 The McLean Law is passed. It is the first U.S. act regulating the shooting of migratory birds.

1921 Hot Springs National Park is established in Arkansas. Though not the first area to be officially called a National Park, Hot Springs was the first area protected by federal legislation.

1933 The first father and son team to occupy the same U.S. cabinet post were both secretaries of agriculture. On this day, Henry Agard Wallace assumes his duties in the administration of Franklin Delano Roosevelt. Henry Cantwell Wallace, the father, served under Harding and Coolidge.

1950 The only person ever to be classified as both a giant and a dwarf, Adam Rainer (born in Graz, Austria, in 1899) dies at age 51. At age 21 Rainer measured just 3 feet 10.45 inches tall; then a bizarre growth spurt brought him past 7 feet in his early 30s.

1952 Nobel laureate and neurologist Sir Charles Scott Sherrington dies at 94 in Eastbourne, England. Among his contributions during 50 years of research on animal nervous systems are Sherrington's Law (when one muscle group is activated, muscles opposing this action are automatically inhibited), classification of sense receptors (as extero-, intero-, or proprioceptive), and coining of the terms "neuron" and "synapse."

1991 Dr. Maurice Buchbinder announces at the annual meeting of the American College of Cardiology that he has developed and tested a tiny drill to bore out the sludge in clogged arteries. His "Rotablator" employs a diamond head, spinning at 200,000 rpm on a small shaft that is inserted into the artery. Although successful in clearing out 95% of the arteries attacked, some problems remain; 6% of the patients suffered heart attacks after the procedure.

1992 World-famous infertility specialist Dr. Cecil Jacobson becomes instantly infamous, as a federal jury in Virginia finds him guilty of all 52 counts of fraud and perjury, in a case in which he allegedly used his own sperm to artificially inseminate tens of unsuspecting women, as well as tricking patients into thinking they were pregnant when they were not. Remarked one juror, "I think that he was a good man. It was obvious that he went wrong somewhere and mistreated a lot of women."

1948 Aldo Leopold completes the Foreword to the environmental classic *A Sand County Almanac*.

MARCH 5

1574 English amateur mathematician (and full-time minister) William Oughtred, inventor of the slide rule and the first to use the multiplication sign (×), is born in Eton. The multiplication sign was introduced in a 1631 textbook Oughtred wrote; the same book introduced the now-common abbreviations "sin," "cos," and "tan."

1616 Written by Cardinal Robert Bellarmine, a decree declaring Copernican theory "false and erroneous" is issued by the Catholic Church from Rome. No one is henceforth allowed to hold or teach the theory (that Earth moves around the sun). It was this decree that Galileo violated, which led to his famous trial and conviction, and which caused him to spend the last eight years of life under house arrest.

1623 The first antidrinking law in North America is signed by Governor Sir Francis Wyatt of Virginia. It provides that "the proclamations for swearing and drunkenness set out by the Governor and Counsell are confirmed by this assembly, and it is further ordered that the churchwardens shall be sworne to present them to the commanders of every plantation and that the forfeitures shall be collected by them to be for publique uses."

1815 Austrian physician Franz Anton Mesmer dies at age 80 in Iznang am Bodensee, Austria, after a roller-coaster career filled with controversy and tumult arising from a unique form of physio-mental therapy, called "mesmerism," that Mesmer invented and that he plied (to great profit) among wealthy and aristocratic clients in several European capitals. His techniques were the forerunner of modern hypnotism, and opened the door to the acceptance of Freud and psychotherapy.

1830 Limelight (the original incandescent form of lighting, made by heating lime to glowing) is first tested before scientists, in a trial against two other lamp designs in the Tower of London. Limelight produced much greater brilliance than any other light source available, and was soon adapted for lighthouses and theater lighting (hence the expression "in the limelight"), but constant attention by an operator was required, and this led to its demise.

1830 French physiologist Etienne-Jules Marey, inventor of the sphygmograph (a machine still used today, it records and graphically displays pulse and blood pressure variations), is born in Beaune. He is also known for developing a motion picture camera to film animals moving.

1836 The Patent Arms Manufacturing Company incorporates in Paterson, New Jersey. It was formed by Samuel Colt, who had invented the first pistol with a revolving cylinder. For his device, he was granted one of the nation's earliest patents, No. 138.

1846 Zoologist Edouard van Beneden is born in Louvain, Belgium, the son of an eminent zoologist. The younger Beneden discovered in experiments with the worm *Ascaris* that sexual fertilization involves the joining of two half-nuclei, one male (from the sperm), the other female (from the egg), each of which contains half the number of chromosomes found in other bodily cells. Beneden demonstrated that when these two half-cells unite, a new single cell is created that contains the normal number of chromosomes.

1970 A 43-nation nuclear nonproliferation treaty goes into effect.

1991 The U.S. Patent Office announces that its 5 millionth patent goes to University of Florida microbiologist Lonnie O. Ingram for a process that turns garbage into fuel, thus attacking two major eco-headaches with one stroke. At the heart of the patent are two different bacteria that Ingram genetically united to make one species.

1992 Tanzania announces that it has partially lifted the ban on elephant hunting. Tourists will now be allowed to slaughter the pachyderms.

1512 Flemish geographer-cartographer Gerardus Mercator is born in Rupelmonde (now in Belgium). Modern geography is said to begin with Mercator because of the science and mathematics he brought to map-making, especially with his innovation of the "Mercator projection" (still used on most modern world maps), which provides an accurate ratio of longitude to latitude at all points. He also introduced the name "atlas" to a collection of maps.

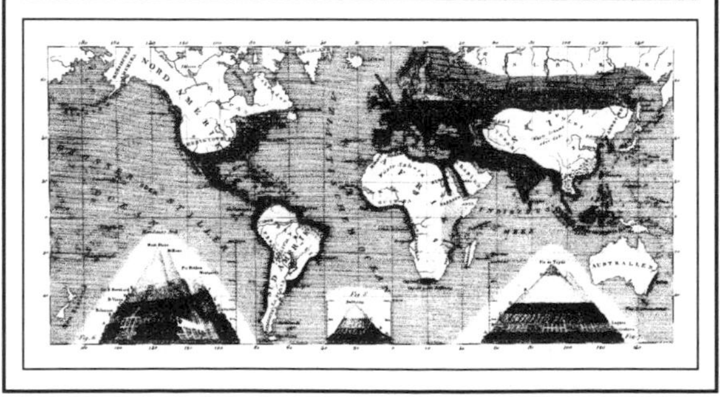

1646 The first patent in the New World for any kind of machine is issued to Joseph Jencks by the Massachusetts Bay Colony: "The Cort, considring ye necessity of raising such manufactures of engins of mils to go by water, for speedy dispatch of much worke with few hands, & being sufficiently informed of ye ability of ye petitions to pforme such workes, grant his petition, (yt no othr pson shall set up or use any such new invention or trade for fourteen yeares, without ye licence of him, ye said Joseph Jenkes)...."

1661 Sir Robert Moray is elected the first president of the Royal Society in London.

1665 *Philosophical Transactions of the Royal Society*, the world's oldest continuing periodical, is first published.

1787 German physicist-optician Joseph von Fraunhofer is born the youngest of 11 children in Straubing. His father was a glazier, and Fraunhofer became famous for his work with light and glass. While testing prisms in order to establish the properties of different types of glass, Fraunhofer discovered numerous dark bands in the solar spectrum, caused by selective absorption of these wavelengths by atmospheric elements. These are now called Fraunhofer lines, and approximately 25,000 have since been found in the solar spectrum. Fraunhofer laid the groundwork for modern spectroscopy (the study of light that is emitted, absorbed, or reflected by matter) by being the first to extensively use the diffraction grating, a device that breaks up light beams in a manner similar to a prism.

1869 Dmitry Mendeleyev publishes his first version of the periodic table of the chemical elements.

1886 The first U.S. power plant to provide alternating current (ac) electricity goes into operation in Great Barrington, Massachusetts. It starts generating power commercially two weeks later, but is later disabled by an accident and is never used again. In November of the same year, the Westinghouse company opened the first truly successful ac electricity plant in Buffalo, New York.

1913 Danish physicist Niels Bohr completes, dates, and mails (to mentor Ernest Rutherford) the first of three historic papers describing a new (and now accepted) vision of atomic structure.

1936 C. Lloyd Morgan, a founder of comparative, or animal, psychology, dies at 84 in Sussex, England (see February 6, 1852, for a biographical sketch).

1937 Valentina Tereshkova, the first woman in outer space, is born in Maslennikovo, Russia. Prior to entering her country's space program, Tereshkova had worked in tire and textile factories, but she was also an expert parachutist, which qualified her for a position as a cosmonaut. On June 16, 1963, at age 26, she made her historic flight; she was launched in the spacecraft *Vostok 6*, and completed 48 orbits in 71 hours before safely returning to Earth. She later married a cosmonaut and went into politics.

1953 Watson and Crick submit to *Nature* their historic paper describing the structure of DNA.

1978 The skull of Swedish natural philosopher Emanuel Swedenborg (1688–1772) is sold in London to the Royal Swedish Academy of Sciences for $10,505. It is the highest price ever paid for a skull.

1992 A computer virus, called "Michelangelo," is supposed to start destroying data banks, files, and programs throughout the United States, on the real Michelangelo's 517th birthday. The virus does appear, but it is not widespread. It reappeared in scattered locations on the same date the following year.

1994 Seven people from five countries walk through an airlock into the self-contained, glass-and-steel bubble, "Biosphere 2," in the desert outside Oracle, Arizona. The structure contains simulations of several ecosystems; the ultimate purpose is to solve man-made ecological problems.

1475 Michelangelo is born in Caprese, Italy.

1274 Scholar-theologian Saint Thomas Aquinas dies at about 49 in Fossanuova, Italy. His importance to science was his advocacy of reason as a means to advance human knowledge, paving the way for the reemergence of science in Christian Europe following the Dark Ages.

1644 The nation's first systematic whaling expedition (by colonists) begins from Southampton, Long Island, New York. Participants spend the day in boats, returning to shore at night. A careful plan was established to give each member of the community an equal portion of the hunt: The town was divided into four wards, with 11 persons from each ward processing the kill. Native Americans had been whaling from shore and from canoes since prehistoric times. Captain George Waymouth was the first European to record their techniques, off the coast of Maine in 1620. It was the colonists who then built large ships and invented devices of mass destruction.

1765 French inventor (Joseph-)Nicéphore Niépce, the first to make a permanent photographic image, is born the son of a wealthy lawyer in Chalon-sur-Saône.

MARCH 7

1792 English astronomer Sir John Herschel, the first to measure brightness of stars precisely, is born in Slough, the only child of William Herschel, the greatest astronomer of his time.

1849 "Do not build me a monument; plant a tree." These were the words of Luther Burbank, born on this day in Lancaster, Massachusetts, as he neared the end of a historic career in which he created more than 800 varieties of fruits, vegetables, and other plants.

1857 Psychiatrist-neurologist Julius Wagner-Jauregg is born the son of a civil servant in Wels, Austria. In the 1880s, while a psychiatrist at the University of Vienna, Wagner-Jauregg noticed that certain nervous disorders improved after a bout with a fever-producing illness. In 1887 he suggested deliberately giving fever-inducing diseases to the insane as therapy. Malaria was recommended because it could be controlled with quinine. This was history's first form of shock therapy, and it had mixed results; syphilitic meningoencephalitis responded especially well to the treatment, earning Wagner-Jauregg the 1927 Nobel Prize for Physiology or Medicine.

1854 The first U.S. patent for a sewing machine that stitches buttonholes is No. 10,609, granted on this day to Charles Miller of St. Louis. The nation's first patent for any kind of sewing machine was granted in 1842, but it was not until 1889 that sewing machines were powered by electricity.

1876 Alexander Graham Bell patents the telephone. Three days later he finally got the thing to work.

1917 A jazz record is first released. It is *Livery Stable Blues*, recorded by the Original Dixieland Jazz Band for Victor Records.

1926 The first transatlantic radio telephone conversation occurs, between London and New York.

1954 After an illustrious career in which he discovered sickle-cell anemia, Dr. James Bryan Herrick dies at 92 in Chicago. His major discovery occurred during a three-year period when he studied the anemia of a single black patient; his diligence resulted in the first description of the crumpled appearance of blood cells that characterizes the sickle-cell condition. Later workers determined that this disorder is inherited and it occurs predominantly in blacks. Herrick also defined a number of diseases of blood vessels, including the first description of coronary thrombosis (a blockage of an artery that feeds the heart muscle itself).

1993 A right-to-die legal battle ends abruptly, with the death of 12-year-old Christine Busalacchi in Barnes Hospital, St. Louis, after her life-support services were finally discontinued some days before. The discontinuation of life support followed a long and sometimes bitter court struggle between the family (who wanted life-support shut off for years) and the state and hospitals (who fought to have life support continued).

1712 Physician John Fothergill is born in Wensleydale, England. He was the first to describe coronary arteriosclerosis (hardening of the arteries that carry blood to the heart muscle itself). A very successful London physician, Fothergill first gained notoriety when he discovered the difference between diphtheria and scarlet fever, in the wake of a horrific diphtheria epidemic in 1747–1748. He also promoted the use of coffee in England and its cultivation in the West Indies.

1775 Joseph Priestley, one of the discoverers of oxygen, experiments with its effects on mice, and determines they need it to survive. The work is performed in the laboratory at his home in Calne, England.

1839 Chemist James Mason Crafts is born the son of a woolen-goods manufacturer in Boston. In 1877 he and partner Charles Friedel discovered the Friedel–Crafts reaction, a still-important method for joining together rings and chains of carbon atoms.

MARCH 8

1855 The nation's first railroad suspension bridge (in which the roadway is suspended by metal cords that are attached to structures at either end of the bridge) is traversed for the first time by a train. The bridge spans the gorge at Niagara Falls; it actually consists of two decks, the lower being a highway and the upper a railway.

1859 Kenneth Grahame (author of the animal classic *The Wind in the Willows*) is born in Edinburgh, Scotland. The enduring appeal of his book lies in Grahame's understanding and accurate portrayal of both human nature and authentic animal habits. First published in 1908, its main characters are a rat, a mole, a badger, and a toad.

1879 Otto Hahn, the first to split the atom (thereby paving the way for the atomic bomb and more peaceful uses of nuclear power), is born in Frankfurt am Main, Germany, the son of a glazier who wanted Otto to be an architect. Hahn was awarded the 1944 Nobel Prize for Chemistry; element number 105 was named hahnium in his honor.

1894 The nation's first state dog-licensing law is passed by New York "for the better protection of lost and strayed animals and for securing the rights of the owners thereof." The law specifies that the American Society for the Prevention of Cruelty to Animals is to collect a $2 license fee in cities with populations over 1,200,000. Unlicensed dogs will be destroyed if not claimed with 48 hours. The law does not apply to nonresidents or exhibitors.

1976 During a meteor shower that eventually dropped over four tons of material in the area, Jilin, China, is hit with a 3902-pound meteor fragment, the largest single piece of extraterrestrial matter ever recovered on Earth.

1993 "There is one indisputable fact that we have to understand: extinction is forever. If we don't do something to help the black bear today, there may not be any to protect in 1996, and that would be a tragic, irreversible loss," states Congressman Porter Goss at the signing of a letter to Florida Governor Lawton Chiles that urges a total ban on black bear hunting in the Sunshine State. Only one Florida House member refuses to sign (Charles Canady, R-Lakeland). Wrote Goss in the letter, "As the state population grows by 900 people a day, as more bears are killed by automobile, and as habitat is developed and fragmented, the future for the Florida black bear can hardly be considered promising."

1994 At an international conference on the manatee in Gainesville, Florida, scientists from the local university announce results of a 16-year study showing that the sea cow population has stabilized, and is unlikely to go to extinction in the near future. "This is not grounds for complacency," warned Stephen R. Humphrey, interim dean of UF's College of Natural Resources and Environment. "Collisions with boats result in about 75% of human-caused manatee deaths, and the number of boats in Florida is rising." Scientists calculate that the manatee population could become extinct with as little as a 10% increase in yearly deaths.

1886 Biochemist Edward Calvin Kendall is born in South Norwalk, Connecticut. He was the first to isolate a number of hormones, but his best-known work was on the structure and effects of hormones from the adrenal cortex (including cortisone, which he developed as a treatment for arthritis). He was a cowinner of the 1950 Nobel Prize for Physiology or Medicine.

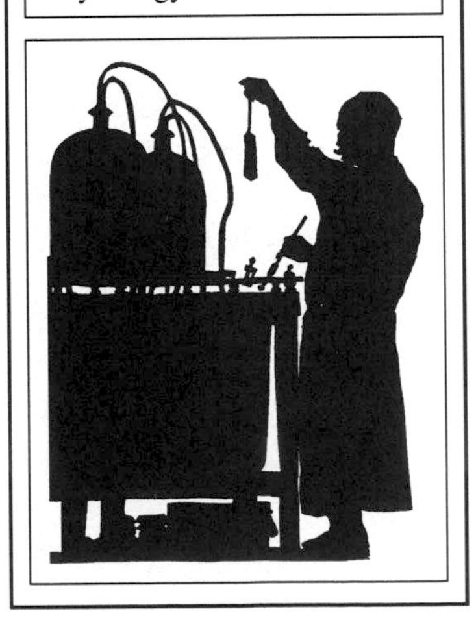

1758 Franz Joseph Gall is born in Tiefenbronn, Germany. Gall was an important brain scientist who was the first to understand that the brain's gray matter is composed of nerve cell bodies, while the white matter is composed of fibers that carry nervous impulses. He also became convinced that different functions are localized in different regions of the brain. This conviction led him and followers to develop the specious practice of "phrenology" (in which the contours on the outside of a person's skull were felt, so as to gauge his personality and mental powers); but it also presaged Paul Broca's 1861 discovery that the power of speech is controlled by one specific brain center. Since then many other functional centers have been discovered within the human brain.

1766 Robert Malthus is just three weeks old when he is visited at his family's home near Dorking, England, by two giants in the history of philosophy, Jean-Jacques Rousseau and David Hume (both are friends of Malthus's father). Years later, Malthus's *Essay on Population* gave both Charles Darwin and Alfred Russel Wallace the final insight that each needed to create the theory of evolution.

MARCH 9

1845 German botanist Wilhelm Pfeffer is born in Grebenstein. His interest in the workings of plant cells led to important discoveries about the process of dialysis (whereby molecules separate themselves by size across a membrane). In 1877, he devised an artificial membrane with the properties of biological membranes; he further devised a method for measuring the pressure that forces molecules across membranes. He showed that this pressure relates to the size of the molecules, which he was also able to measure. His findings have since been applied to all living cells, not just those of plants.

1856 Inventor Edward Goodrich Acheson, creator of carborundum (the hardest substance known on Earth, with the exception of diamonds, for half a century), is born in Washington, Pennsylvania.

1858 Wallace mails Darwin his version of the theory of evolution, from Ternate, Indonesia.

1858 The first U.S. patent for a mailbox is No. 19,578, granted on this day to Albert Potts of Philadelphia. A simple device, it consists of a box with a hole in it (through which it could be attached to a lamppost). In August, Potts's boxes were installed in the streets of Boston and New York City.

1862 The famous five-hour clash between the iron-clad ships, *Monitor* and *Merrimac*, occurs at Hampton Roads, Virginia, during the American Civil War. The event ends the age of sail-driven, wooden-hulled warships.

1934 Yuri Gagarin, the first man in space, is born on a collective farm near Gzhatsk, Russia, the son of a carpenter.

1993 Governor Lawton Chiles signs a bill making Florida the first state to guarantee breast-feeding rights. It is now illegal to apply any charges of obscenity, lewdness, or public nudity against women who publicly breast-feed their children.

1994 Another chapter unfolds in one of history's most famous swapped-at-birth cases. The Associated Press reports that 15-year-old Kimberly Mays has moved in with her biological parents, from whom she had sought, and won, "divorce" just seven months before. The case stemmed from a mysterious swap-at-birth in a rural hospital, which took Kimberly from her real parents, Ernest and Regina Twigg, and placed her in the hands of Robert Mays, who raised her in ignorance of the swap.

1564 German astronomer David Fabricius is born in Esens. Although he joined Galileo as one of the first to use a telescope, Fabricius's greatest claim to fame was his 1596 naked-eye discovery if Omicron Ceti (also called Mira), the first known "variable star" (meaning that its brightness fluctuates). This discovery put another nail in the coffin of the Aristotelian view that the heavens are perfect and unchanging. Fabricius was a Protestant minister who was murdered by one of his parishioners in 1617.

March 10

1496 Christopher Columbus departs Hispaniola for Spain, ending his second trip to the New World.

1628 Italian physician-biologist Marcello Malpighi, who founded the field of microscopic anatomy, is born is Crevalcore, Papal States. He made many discoveries that are now taken for granted. For example, he discovered taste buds, was the first to see red blood cells and to suggest that they determine the color of blood, and also the first to see the small blood vessels that connect arteries with veins (this discovery was crucial in establishing that blood does circulate). His extensive research on the fine structure of many human organs earned him the title of "the first histologist." He has also been called "the father of microscopy." But he was the object of much jealousy; in the last decade of his life, his home, library and laboratory were destroyed by unknown assailants. He spent the remainder of his days as personal physician to Pope Innocent XII.

1762 German chemist Jeremias Benjamin Richter is born in Hirschberg. In 1791 he established that acids and bases react in fixed proportions (i.e., a specific amount of acid is needed for each specific amount of base) when they neutralize each other and form salts. This was the first time a chemical reaction was shown to occur in fixed proportions; ratio constancy has since become a cornerstone in chemistry.

1849 From Springfield, Illinois, Abraham Lincoln files a patent application for a device for "bouying vessels over shoals." The design uses inflatable cylinders to float grounded craft over obstacles. The patent was granted on May 22, 1849, making Lincoln the first U.S. president to receive a patent.

1882 Scottish naturalist-oceanographer Sir C. Wyville Thomson dies five days after his 52nd birthday in Bonyside, Scotland. Trained as a physician, Thomson devoted himself from his 30s onward to the study of ocean creatures. He first achieved notoriety through two dredging expeditions north of Scotland (1868–1869) in which he discovered a host of invertebrates, many of which were believed extinct. During these studies he also discovered the presence of ocean currents deep below the surface. This work earned him the job of chief naturalist on the historic trip of the HMS *Challenger* in the 1870s. He was knighted at the end of this voyage.

1923 Physicist Val Logsdon Fitch is born in Merriman, Nebraska. He shared the 1980 Nobel Prize with partner James Watson Cronin for a 1964 experiment that disproved the longstanding belief (called "CP symmetry") that interactions of subatomic particles are indifferent to the direction of time.

1945 The deadliest conventional bombing run in history occurs during World War II; approximately 140,000 die in the air raid on Tokyo.

1975 Glasses with corrective lenses for dogs are patented in France by optician Denise Lemiere, who hit on the idea after she made sunglasses for her own dog.

1980 Dr. Herman Tarnower, renowned dietitian and weight-loss authority (he wrote *The Scarsdale Diet*, a smash bestseller), is shot repeatedly at his home in Purchase, New York. His murder is apparently the result of a lover's triangle. Jean Harris, high school principal and long-time companion of Tarnower, was later convicted and imprisoned for the crime.

1987 An end comes in Yokohama, Japan, to one of the great feats of human memory; Hideaki Tomoyori completes his recitation of pi to 40,000 places.

1992 Police and wildlife officials raid 20 restaurants in Changsa, China, and find that half of the wild animals on the menus are endangered species. Leopards, spiny anteaters, and macaque monkeys are among the dishes offered.

1993 "Prolife" activist Michael Frederick Griffin shoots physician David Gunn in the back, thus ending his life, during an antiabortion rally in Florida.

1876 The first-ever telephone call occurs. "Mr. Watson, come here! I want to see you," speaks inventor Alexander Graham Bell to assistant Thomas Watson in another room in a boardinghouse on Exeter Place in Boston, in which Bell had established his laboratory. The event comes a week after Bell's 29th birthday and three days after the device was patented.

1791 The U.S. Patent Office issues four patents to Samuel Mulliken of Philadelphia, making him the first person to receive more than one. His four devices are a "machine for threshing grain and corn," a "machine for breaking and swingling hemp," "a machine for cutting polishing marble," and a "machine for raising a nap on cloths."

1818 French chemist Henri-Étienne Sainte-Claire Deville, inventor of the first process to produce aluminum economically, is born the son of a shipowner-diplomat in St. Thomas, Virgin Islands.

1890 Electrical engineer Vannevar Bush, developer of the first analog computer, is born the son of a minister in Everett, Massachusetts.

1892 The tortured life of chemistry genius Archibald Scott Couper ends in his mother's home in the town of his birth (Kirkintilloch, Scotland). Couper was 60. He had never recovered physical or mental health after a tragic incident in 1858; it had started with such bright promise. In that year, at age 27, Couper conceived of the "tetravalence" of carbon (essentially, that carbon has four points of attachment with other atoms, and that carbon readily bonds with itself to form long chains; this is critical in understanding how living things are built). Furthermore, Couper began using a dash to indicate a chemical bond (this greatly facilitated the visualization of chemical structures, and it remains standard procedure for all chemists and chemistry teachers). He described these revolutionary changes in a single paper, which he gave to his boss, Adolphe Wurtz, to submit to the French Academy of Sciences. For whatever reason, this was not done for months, during which time August Kekulé published the same idea of tetravalence. Kekulé became famous. Couper became furious; he had a huge argument with Wurtz, who banished him from the laboratory. Within a year, Couper had spent several months in an insance asylum. Shortly thereafter he suffered a sunstroke that enfeebled him for life.

1920 Nicolaas Bloembergen, winner of the 1981 Nobel Prize for Physics for his study of how electromagnetic radiation interacts with matter, is born in Dordrecht, the Netherlands.

1927 The nation's first automobile robbery (involving an armored car) takes place. Members of the Flatheads gang plant a mine on the Bethel Road, between Pittsburgh and Coverdale; the mine is set off by an armored truck carrying a $104,250 payroll of the Pittsburgh Terminal Coal Company. Five guards are injured.

1932 The heath hen is now extinct.

1955 Nobel laureate Sir Alexander Fleming dies at 73 in London. His everlasting fame came from his ability to take advantage of chance occurrences. In 1921 he was inspecting a bacterial culture when a droplet happened to fall from his nose (he had a bad cold at the time) onto the bacteria; he subsequently observed that the liquid killed the organisms. He had discovered the natural bacteria killer "lysozyme"; work by himself and others revealed that lysozyme is present in tears, saliva, milk, and other fluids in many different animals. In 1928 a spore of airborne mold happened to land on another bacterial culture that Fleming was growing; he observed that where the mold grew, the bacteria didn't, and he later discovered that the mold produced an antibacterial liquid, which he isolated and named "penicillin." It was man's first antibiotic.

1993 Owls are apparently not the job-stealing, price-raising villains they had been accused of being. The National Association of Home Builders, the American Forest and Paper Association, and the United Brotherhood of Carpenters and Joiners recently blamed federal environmental laws and spotted owl protection for a huge leap in lumber prices over the winter. But a report released today by the Congressional Research Service (a branch of the Library of Congress) identifies the nation's economic recovery as the main cause of the price elevation; further, the report states that the increased lumber price should not have much effect on housing starts, because lumber accounts for a relatively small portion of the cost of a house, while mortgage rates are far more critical to the housing market.

1994 The environmental organization Greenpeace demonstrates at a bridge across the Bosporus, the strait connecting the Black and Mediterranean Seas, near Istanbul, Turkey. Greenpeace calls for a total ban of oil tanker traffic through the strait because of the high likelihood of a collision in the heavily traveled area. Two days later, in the Bosporus, a Greek freighter rams a tanker than is carrying 16 million gallons of Russian crude oil. Fifteen people die, twenty-nine are injured, there is a huge fire, and the tanker is finally towed out to open sea to dump the rest of its oil.

1755 A steam engine is used for the first time in America. Imported by Josiah Hornblower (America's first steam engineer) from England in 1753, the device was taken to the copper mine of Colonel John Schuyler in New Barbados Neck (now North Arlington, New Jersey), where it was used to pump water from the mine.

1784 French King Louis XVI appoints a commission of scientists (including Lavoisier, Guillotine, and Ben Franklin) to study the controversial psychotherapy methods of Franz Anton Mesmer (see March 5, 1815).

1824 Physicist Gustav Robert Kirchhoff is born the son of a law counselor in Kaliningrad, Germany. In partnership with chemist Robert Bunsen, he developed "spectrum analysis" (in which chemicals are identified by the light they emit when heated). This tool allowed the pair to discover several new elements, and allowed Kirchhoff to determine the composition of the sun.

1831 Clement Studebaker is born in Pinetown, Pennsylvania. He founded a family firm that, before it became a leading manufacturer of automobiles, was the world's largest producer of horse-drawn carriages.

MARCH 12

1832 French chemist-mineralogist Charles Friedel, codiscoverer of the Friedel–Crafts reaction for synthesizing organic molecules, is born in Strasbourg.

1835 Astronomer-mathematician Simon Newcomb is born in Wallace, Nova Scotia, the son of an itinerant country schoolteacher. Newcomb was a child prodigy, but received virtually no formal education. He is famous for producing the world's most gigantic, most accurate ephemerides (tables of computed locations of stars, planets, moons, and other bodies over time) and for producing a universal, unified system of astronomy constants. Newcomb's system was adopted by an international congress in 1896, and this decision was reaffirmed by another congress in 1950.

1838 Sir William Henry Perkin, discoverer/inventor of aniline dyes (the first synthetic material for coloring cloth that was truly successful), is born the son of a carpenter in London. His father was against William's interests in chemistry, until William discovered the aniline dyes (by accident) at age 18, and became rich and the world's expert on dyes in his early 20s.

1907 German engineer Alfred Maul receives the first patent for a rocket that can carry cameras and other scientific equipment into space, and that can return the equipment safely to Earth.

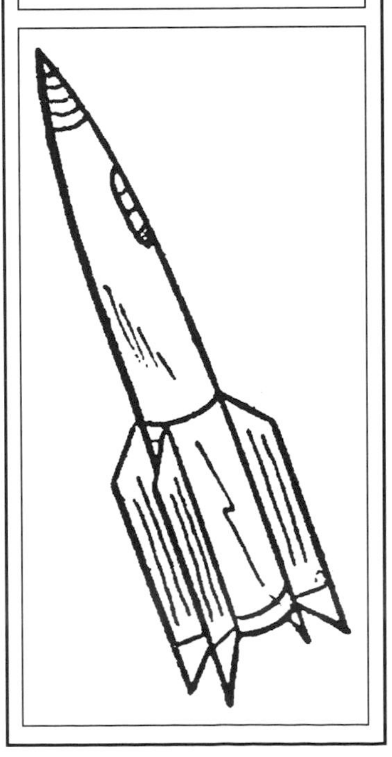

1923 The first sound-on-film motion picture is demonstrated for the press. It is called the Phonofilm by its inventor, Dr. Lee De Forest (already famous for his 1907 invention of the radio tube, which formed the basis for radio, television, radar, and the early computers). Music, but not talking, appears on the film.

1929 Coca-Cola developer Asa Griggs Candler dies at age 77 in Atlanta. Candler was a pharmacist and owner of a successful wholesale drug business when he purchased the formula for the then-obscure beverage from a colleague in 1887. He improved the manufacturing and marketing procedures, and finally sold his business for $29 million in 1919. He donated much of his wealth to the improvement of Atlanta's Emory University.

1969 The world's fastest airliner, the Anglo-French SST *Concorde*, makes its first flight.

1992 The *New England Journal of Medicine* reports a bizarre case in which a woman entered a hospital with mysterious chest pains, nausea, vomiting, and sweating, only to be told that a sewing needle was lodged in her heart! Open-heart surgery was required to remove the object because "the pumping action of the heart kind of suck[ed] that needle in," reports her surgeon, Dr. James J. Bachman. Apparently the patient fell on the needle, which she had been carrying in her hand when she tripped over her cat. She had noticed a small entry mark, but otherwise ignored it.

1993 Chrysler engineers complete a nine-day, 2604-mile drive from Detroit to Los Angeles, thereby breaking the world endurance record for an electric vehicle (see March 3 for the details).

1720 Charles Bonnet (a lawyer by trade, but a natural historian by hobby), the first to use the word "evolution" in a biological context, is born in Geneva, Switzerland. Among other contributions, Bonnet discovered parthenogenesis (reproduction without fertilization in sexual organisms).

1733 Joseph Priestley, discoverer of nitrogen, ammonia, and oxygen, is born the oldest of six children in Birdstall, England, the son of a Nonconformist preacher.

1855 Astronomer Percival Lowell is born in Boston. He is famous for predicting the discovery of the planet Pluto, and for theorizing that a series of lines on the surface of Mars were actually artificial canals that were constructed by intelligent beings.

1866 Physicist Dayton Clarence Miller is born the son of a shopkeeper in Strongsville, Ohio. He performed experiments that confirmed the famous Michelson–Morley experiments of the late 1800s, the results of which verified Einstein's theory of relativity (even though Miller himself never accepted relativity).

1877 Earmuffs are patented. Chester Greenwood of Farmington, Maine, receives patent No. 188,292 for "ear mufflers," which he invented in 1873.

1894 The starting gate for horse races is invented by J.L. Johnstone.

1925 The teaching of Darwin's evolution theory in public schools is outlawed in Tennessee.

1930 Scientists announce the discovery of Pluto. Clyde William Tombaugh had actually discovered the planet a month before at the Lowell Observatory in Flagstaff, Arizona, but the public was not informed until extensive verification could be made. Furthermore, the announcement was withheld until the anniversary of the birth of Dr. Percival Lowell (for whom the observatory was named, and who had accurately predicted Pluto's existence and location years before its discovery), which was also the date of Sir William Herschel's discovery of Uranus in 1781.

1938 Historic legal figure Clarence Darrow dies at 80 in Chicago. He is remembered as the defense lawyer in a number of famous cases, including the Leopold–Loeb case (1924), the murder trial of labor leader "Big Bill" Haywood (1907), and the 1902–1903 arbitration hearings of striking coal miners, during which he exposed horrific working conditions and child labor in the mines. He secured a place in biology history by defending John T. Scopes in the 1925 Tennessee "Monkey Trial," in which the teacher was charged and convicted of teaching Darwin's theory of evolution in public schools (see above).

1781 Sir William Herschel discovers the planet Uranus.

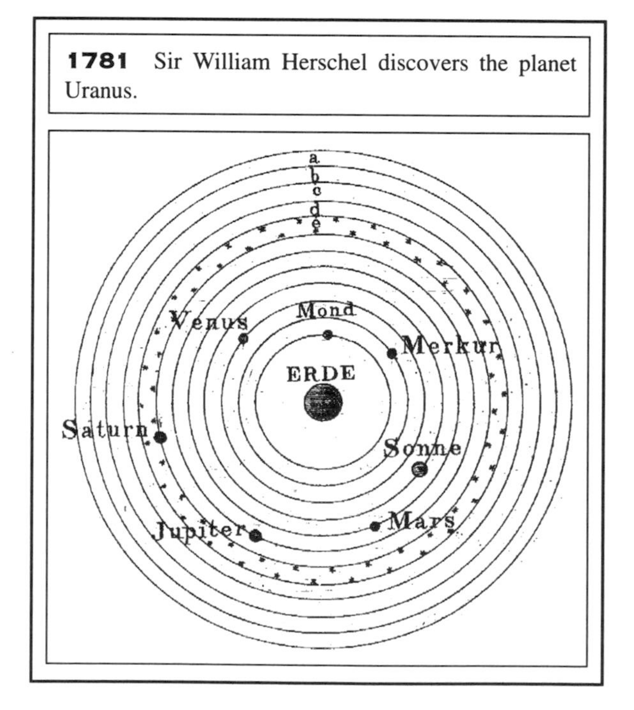

1969 After 10 days in orbit *Apollo 9* splashes down, having performed the first in-space tests of the lunar module that will eventually carry astronauts to the surface of the moon.

1988 Gallaudet College (founded in 1864 in Washington, D.C.; it is the world's oldest and only institute devoted to the deaf and to teachers of the deaf) chooses I. King Jordan to become its first deaf president.

1994 Researchers at the National Cancer Institute announce a drug that halts Kaposi's skin cancer caused by AIDS. Shuji Nakamura, first author of the study in *Science* magazine, reports that the compound works on cancers both in chicken cells and in immune-deficient mice, and it shows none of the side effects that current Kaposi-fighting drugs do. The new drug is obtained from soil bacteria.

1994 Two days after a protest against oil shipping in the Bosporus, a Greek freighter rams an oil tanker carrying 16 million gallons of Russian crude oil in the Bosporus Strait, near Istanbul.

1692 Mathematician-physicist Pieter van Musschenbroek, inventor of the Leyden jar (the first truly efficient device for storing static energy), is born in Leiden, the Netherlands.

1794 Eli Whitney patents the cotton gin. It revolutionized the textile industry, and indirectly caused the American Civil War. (By revitalizing the demand for cotton, the device stimulated farmers, businessmen, and politicians in the southern states to defend their use of slavery in the gathering of cotton.)

1835 The belief in life on Mars largely extends from the writings and research of astronomer Giovanni Virginio Schiaparelli (born on this date in Savigliano, Italy), who reported discovering groups of straight lines on the Martian surface. He called them *canali*, which means "channels" but which was incorrectly translated as "canals."

1862 Vilhelm F(riman) K(oren) Bjerknes, a founder of the science of weather forecasting, is born in Christiania (now Oslo), Norway, the son of mathematician Carl Anton Bjerknes.

MARCH 14

1879 Albert Einstein is born in Ulm, Germany, the son of a chemical engineer who failed at business several years later. Einstein showed no brilliance as a child; in fact, he was suspected of being retarded because it took him so long to learn speech.

1903 The first national bird sanctuary in established by Executive Order of President Theodore Roosevelt. His Order protects the nesting colony of pelicans and herons on Pelican Island near Sebastian, Florida.

1928 Astronaut Frank Borman is born in Gary, Indiana. As a member of *Apollo 8* in 1968 he took the first manned flight around the moon.

1934 Astronaut Eugene Andrew Cernan is born in Chicago. As a member of *Apollo 10* in 1969 he piloted the lunar module to within 9.7 miles of the lunar surface.

1936 "I've had a telegram that Priestley is dying, too, but I *think* it was an imaginary telegram," jokes John Scott Haldane on the night of his death to his son J.B.S. Haldane. The elder Haldane was referring to Joseph Priestley (d. 1804), discoverer of oxygen. Haldane's life was linked with oxygen; he discovered much about human breathing, and applied his findings to the health and safety of coal miners. In 1905, Haldane discovered what makes us breathe (the concentration of carbon dioxide in the blood triggers inspiration and expiration); he determined why carbon monoxide causes death; he developed methods for mesuring gases dissolved in the blood; he studied how the body uses these gases; and he developed "stage decompression," which allows deep-sea divers to surface safely.

1967 At a meeting in Mexico City, Mexico, the American Association of Zoological Parks and Aquariums votes unanimously not to purchase, accept, or trade any orangutans or any of a variety of other threatened species.

1994 The doctor who supplied steroids to Olympic athlete Ben Johnson (who was stripped of his 1988 gold medal when he tested positive for the muscle enhancers) is detained without bond for selling 11 pounds of steroids laced with cocaine to federal undercover agents.

1994 A magnetic grappling system is used in outer space for the first time. Several *Columbia* astronauts take turns operating the 50-foot robot arm.

1854 Paul Ehrlich, medical giant in the fields of immunology and chemotherapy, is born in Strehlen, Germany, into a Jewish family that was successful in business and industry. He became a medical student and researcher, and soon began a long string of disease-fighting innovations and discoveries. His crowning achievement was developing the "magic bullet," a shot that, for the first time, cured syphilis. In fact, Ehrlich's development of artificial chemicals to combat disease is acknowledged as the birth of chemotherapy. He was awarded the 1908 Nobel Prize for Physiology or Medicine.

1493 Christopher Columbus docks in Cabo de Palos, Spain, concluding his first voyage to the New World. A variety of new life forms are aboard his ships, including syphilis (according to some theorists).

1614 Physician-chemist-physiologist Franciscus Sylvius is born in Hanau, Germany. As the founder of the seventeenth-century "iatrochemical school of medicine" (which held that all living processes are based on chemical reactions), Sylvius was instrumental in shifting the focus of medicine from mysticism to physics and chemistry. He was among the first to incorporate ward training in a doctor's education, established what was possibly the first university chemistry laboratory, and, in 1641, discovered the Sylvian fissure, which separates sensory from motor areas in the human brain.

1801 Conservationist George Perkins Marsh is born in Woodstock, Vermont. His 1864 *Man and Nature, or Physical Geography as Modified by Human Action* is now considered a classic in ecology, geography, and resource management.

MARCH 15

1806 A chondrite meteor (laden with carbon and organiclike chemicals) is unequivocally identified for the first time, when it crashes into Earth outside Alais, France, at 5:30 in the afternoon. The chemical makeup of the meteor suggests it came from some planet or other body in outer space that contains life.

1854 Bacteriologist Emil von Behring, winner of the first Nobel Prize for Physiology or Medicine, is born the son of a schoolmaster in Deutsch-Eylau, Prussia.

1887 Salary and expenses for the nation's first game warden are approved by the legislature of Michigan.

1897 "I really love my subject." These were the words that James Joseph Sylvester (who died at 83 on this day in London, following a paralytic stroke) used to summarize his life in mathematics. Ironically, Sylvester had once quit mathematics, at age 27, after a three-month stint as a professor at the University of Virginia. He tried work as an insurance actuary, then became a lawyer and tutored mathematics privately (one of his pupils was Florence Nightingale). In 1846 he happened to meet Arthur Cayley, another disappointed-mathematician-turned-lawyer. The relationship rejuvenated both; they went on to make mathematics history by jointly developing a number of areas in algebra.

1959 George Schaller finally gets very close to gorillas, during the first-ever scientific study of the ape in its own habitat (in the Virunga Mountains, the Congo). Today is coincidentally the 17th anniversary of the 1942 death of "M'bongo" in the San Diego Zoo; at 660 pounds he was the heaviest gorilla in captivity at the time (until the Zoo acquired "N'gagi," who tipped the scales at 683 pounds and also died behind bars in San Diego).

1960 President Dwight D. Eisenhower creates the first national underwater park. It is a coral reef, 21 miles by 3½ miles, lying on the floor of the Atlantic Ocean off Key Largo, Florida. It had previously been the John Pennekamp State Park, but will now be under federal protection. It is given the name "Key Largo Coral Reef Preserve." It contains 40 of the 52 known species of coral.

1989 A 23-pound gall bladder is removed from a 69-year-old woman in Bethesda, Maryland. The abnormal organ is three times the weight of a newborn baby, and is the heaviest gallbladder in history.

1992 Seventy members of the youth group Eclaireurs de France (which means "Those Who Show the Way") spend this Sunday in public service, steel-brushing graffiti off the walls of the Mayrieres cave, near the village of Brunquiel in southwestern France. In their zeal, they destroy parts of Paleolithic bison paintings, estimated to have been created around 13,000 BC (see March 21, 1992).

1992 Two poachers climb a power pole on U.S. Route 1 in the Florida Keys, and start stealing chicks from a nest of an osprey (a species of fish-eating hawk that is protected by federal law). Passing motorists Will Gilbert (a natural science teacher) and his daughter Nancy (an animal trainer) are appalled at the sight. As they pull to the side of the road to confront the poachers, Nancy leaps out of the car while it is still moving; "I had to restrain my daughter, pull her back," said Will. The criminals flee, but their license plate number is recorded by someone in the gathering crowd, and they are eventually caught.

1521 The explorer Magellan reaches the Philippines, where he soon dies in a skirmish with natives.

1787 German physicist Georg Simon Ohm is born in Erlangen, the son of a master mechanic. After achieving his Ph.D., Ohm was only able to secure a high school teaching position. Needing impressive research publications to get a university post, he began experiments in the new field of electric current. Lack of funds forced him to make his own wires, which led to experimentation on wires of different thicknesses, and his discovery of Ohm's law (electric current is inversely proportional to resistance and directly proportional to voltage).

1802 An act of Congress establishes the U.S. Army Corps of Engineers.

1819 The first clinical description of an allergy is delivered in London before a meeting of the Royal Medical and Chirurgical Society by Dr. John Bostock. His paper described hay fever.

1867 Lister publishes the first of a series of articles in *Lancet* on his discovery of antiseptic surgery.

1914 Sir John Murray, a founder of oceanography, dies at 73 near Kirkliston, Scotland. In his late 20s, Murray established his reputation as an explorer and naturalist during a trip to the Arctic islands of Jan Mayen and Spitsbergen. He helped organize the historic expedition of the HMS *Challenger* (1872–1876), during which he was the chief curator of biological specimens. On the 1882 death of the *Challenger*'s scientific leader, Sir Wyville Thomson, Murray took over the writing and publication of the 50-volume report of the *Challenger*'s findings.

1926 Robert Goddard of Clark University conducts man's first liquid-fuel rocket flight. The missile is powered by gasoline and liquid oxygen, ignited by a blowtorch; it travels 184 feet in about 2.5 seconds, in a field in Auburn, Massachusetts. All modern space shuttles and rockets now use liquid fuel.

1927 Cosmonaut Vladimir Mikhaylovich Komarov is born in Moscow. In 1964 he piloted *Voskhod 1* on the first journey to carry more than one person into space, and on April 23, 1967 (in *Soyuz 1*) he became the first to die during a space mission.

1935 The esteemed physiologist-educator J.J.R. Macleod dies at 58 in Aberdeen, Scotland. His research centered on how the body processes sugar and on diabetes. It was Macleod who coined the name "insulin" for the hormone that is responsible for proper sugar metabolism. He shared the 1923 Nobel Prize with Sir Frederick Banting "for the discovery of insulin." The award, and Macleod's true role in the discovery, were always a point of outrage for Banting, who claimed that Macleod had originally scoffed at, and refused to support, the experiments that led to the discovery.

1978 The horrific *Amoco Cadiz* oil spill occurs off Portsall, France, when the tanker drifts onto shore, breaks up, and dumps 246,400 tons of crude oil into the ocean. The oil slick was 100 miles long.

1993 A jury in Newark, New Jersey, finds four men guilty in the sexual assault of a handicapped teenager. The defense argued that the victim had instigated the vicious attack, but the jury believed that the girl was unable to give informed consent because of her undeveloped IQ (64) and social skills (those of an eight-year-old). The case is viewed as a landmark in the rights of the handicapped.

1993 On the day before St. Patrick's Day, amateur botanist John Piergross announces that he is marketing a genetically engineered four-leaf clover.

1662 The first public transportation coach goes into service in Paris.

March 17

1834 Automobile giant Gottlieb Daimler is born in Schorndorf, Germany. An engineer by training, Daimler was the technical director in the firm of Nikolaus A. Otto (inventor of the four-stroke internal-combustion engine) until starting an independent engine-building shop with Wilhelm Maybach in 1882. They produced one of the first high-speed internal-combustion engines, and invented a carburetor that made it possible to use gasoline as fuel. They attached their engines to an assortment of vehicles, including a bicycle (1885), a horse-drawn carriage (1886), and a boat (1887). Then they concentrated on making a vehicle that was designed as a car from the start. They founded the Daimler-Motoren-Gesellschaft firm in 1890, and produced the first Mercedes in 1899; this car was named after the daughter of Emil Jellinek, a diplomat and major investor in Daimler's enterprises.

1854 The first land to be used for a park in a U.S. city is purchased by Worcester, Massachusetts, from Levi Lincoln and John Hammond. It is 27 acres, and becomes Elm Park.

1881 Ophthalmologist-turned-physiologist Walter Rudolf Hess is born in Frauenfeld, Switzerland. He shared the 1949 Nobel Prize for Physiology or Medicine for determining the role of several brain areas (the medulla and hypothalamus, especially) in the coordination and stimulation of behavior and organ function. Perhaps his most famous experiment was making a cat react as if it had just seen a threatening dog, when no dog was actually present; Hess produced the cat's behavior and stress physiology simply by applying current through precisely located electrodes in the cat's brain.

1884 The first flight of a glider in the United States is made by John Joseph Montgomery (the father of gliding) from a hill outside Otay, California. Montgomery weighs about 130 pounds, his glider only 30. The flight goes 600 feet and creates no air pollution.

1885 A medical description of the deformities of Joseph Carey Merrick (a.k.a. the "Elephant Man," now world famous) is reported to the Pathological Society of London by Dr. Frederick Treves.

1897 Bob Fitzsimmons knocks out Jim Corbett in the 14th round to win the world heavyweight boxing championship in Carson City, Nevada. It is the first boxing match filmed with a motion-picture camera.

1912 "I am just going outside and may be some time," says army officer Lawrence "Titus" Oates, as he leaves his Antarctic tent to die. He is one of the hand-picked men in the ill-fated attempt by England's Robert Falcon Scott to be the first to reach the South Pole (the group did reach the Pole, only to find a tent and flag planted there three weeks before by Norwegian Roald Amundsen). Oates has terribly frostbitten feet, and, rather than endanger his companions, decides to end his own life. Scott remarked in his journal, "Oates' last thoughts were of his mother.... We can testify to his bravery."

1958 *Vanguard 1* is launched from Cape Canaveral. A three-pound satellite (containing only a single radio transmitter) is put into orbit. It is the first U.S. object in space, but many felt it to be a disappointing anticlimax because the Soviet Union has already launched a 184-pound object (*Sputnik I*) and a 7000-pound object (*Sputnik II*) into orbit.

1937 Famed aviator Amelia Earhart, 39, departs Oakland, California, on the last journey of her life (an attempt to be the first to fly around Earth at the equator). She and copilot Fred Noonan mysteriously disappeared during the trip.

1992 Soviet officials report a leak of radioactive iodine from a nuclear power plant in Sosnovy Bor—five days after the leak began. Estonia and Finland seem to have taken the brunt of the airborne material, which is the same material that was released during the 1986 Chernobyl disaster.

1994 Tree-free paper, made of fiber from the kenaf plant, is displayed in Washington D.C., on National Agricultural Day by Kenaf International of McAllen, Texas. Also shown are kenaf-based substitutes for: foam "peanuts," cardboard boxes, wood shavings, peat moss, and fiberglass. Kenaf is related to cotton and okra; it grows very quickly, and its paper is stronger than tree-based paper. It has been cultivated since prehistoric times in Asia, but as yet there are no kenaf mills to produce packaging and paper substitutes.

1690 Mathematician Christian Goldbach is born the son of a minister in Königsberg, Prussia. He is famous for Goldbach's conjecture (that every even number over 2 can be expressed as the sum of two prime numbers). Since it was proposed in 1742 (in a letter to Leonhard Euler, another famous mathematician), it has never been disproved ... nor has it ever been proved.

1845 Johnny Appleseed dies of exposure at age 70. Born as John Chapman, he was actually a professional nurseryman in Massachusetts who began at about age 25 trekking through the U.S. wilderness west of the Allegheny Mountains, during which he planted a series of apple nurseries. He also sold or gave away thousands of apple seedlings to western-moving pioneers. His peaceful nature, affinity for the outdoors, friendship with Indians, knowledge of herbs, bedraggled appearance, flowing white hair, bare feet, and gentleness with animals all were the inspiration for pioneers' stories that developed into folktales. He perished in the countryside near Fort Wayne, Indiana, and was buried in an unmarked grave.

MARCH 18

1858 Engineer Rudolf Diesel, inventor of the internal "diesel" combustion engine that now bears his name, is born of German parents in Paris.

1886 Kurt Koffka, explorer of the mind, is born in Berlin. In collaboration with Wolfgang Köhler and Max Wertheimer, Koffka founded Gestalt psychology in the early 1900s. Their doctrine stressed wholeness; they maintained that people perceive their environment as a whole and that mental phenomena, such as thoughts and feelings, occur as wholes. The Gestalt outlook was a total departure from "structuralism" (the first major school of psychology, it tried to understand the mind by identifying and studying the minute, individual components that comprise mental activity) and "behaviorism" (which emphasized studying only external, observable actions).

1909 Einar Dessau of Denmark sends a shortwave transmittance to a government radio post six miles away. This is thought to be history's first "ham" radio broadcast ("ham" radio refers to broadcasts by amateurs).

1911 Biochemist David Shemin is born in New York City. One of the first to use the newly discovered carbon-14 for physiology research, Shemin worked out the pathway by which the body manufactures heme (a major component in red blood cells).

1922 Mohandas K. Gandhi, one of history's great motivators, is sentenced to six years' imprisonment in India for civil disobedience.

1931 The first electric razor is marketed by Schick, Inc., of Stamford, Connecticut. It is manufactured according to the 1928 patent of Colonel Jacob Schick for a "shaving implement."

1965 Aleksei Leonov takes history's first spacewalk, for 20 minutes outside the orbiting *Voskhod 2*.

1967 The giant oil tanker *Torrey Canyon* breaks apart on Seven Stones Reef off the picturesque Cornwall coast of England and 119,000 tons of oil enter the ocean.

1991 Dr. Michael Gilbert, former expert witness on the insanity defense, fails in his own attempt to use the insanity defense, in a case in which he bribed a public official. The prominent psychiatrist, 75, who had testified on behalf of some 5000 defendants since 1957, is found guilty by a six-person jury of trying to get a police officer to frame a public defender. "Look at him. The truth is he's crazy. Demented. Deranged. Nuts," offered Gilbert's attorney in his defense. The jury was not swayed by this, nor by the testimonies of three other experts who claimed him to be criminally insane. "Don't be fooled by this man. He knows every nook and cranny of the criminal justice system," warned prosecutor Russell Killinger.

1993 Singer Jimmy Buffet signs a peace agreement with the Audubon Society at Florida's Wekiwa Springs State Park, ending a bitter dispute over the protection of the endangered manatee. In 1981 Buffet cofounded the Save the Manatee club, which operated under the auspices of the Audubon Society. But disagreements over money and administration led to firings, lawsuits, and picket lines. Today's agreement creates a joint, nonprofit organization called TEAM (Toward Education and Advocacy for the Manatee).

1687 French explorer La Salle (the first European to navigate the whole Mississippi River) is murdered by his own crew in present-day Texas.

1782 Austrian astronomer Wilhelm von Biela is born into Bohemian aristocracy in Rossla. He did not discover the amazing Biela's Comet, but he did calculate its orbit/periodicity, and therefore his name was given to it. In 1846 the comet was observed to break into two, and was seen again as two comets in 1852. It was never observed again, for, in its anticipated returns in 1872 and 1885, it came back as bright meteor showers (called the Andromedids or Bielids). This was the first time a heavenly body had been observed to die before scientists' eyes, and it provided the first direct link between comets and meteors.

1800 Legendary explorer-scientist Alexander von Humboldt and his partner Aimé Bonpland capture electric eels in the eels' natural habitat, a jungle river in South America. Several horses drown during the expedition, and both men receive massive electric shocks during behavioral experiments.

1827 Eighteen-year-old Charles Darwin obtains some specimens of the marine polyzoan *Flustra* from Scotland's Furth of Forth, and dissects them. The incident, which begins his love affair with barnaclelike creatures, provides his first original scientific discoveries, and is instrumental in turning him toward a life in natural history.

1883 English chemist Sir (Walter) Norman Haworth is born the son of a successful businessman in Chorley. He is famous for determining the structure of various carbohydrates (depicting them with figures now called Haworth formulas) and for synthesizing vitamin C in 1934 (this was the first artificial preparation of any vitamin). Haworth won the 1937 Nobel Prize for Chemistry.

1900 Physicist Jean-Frédéric Joliot is born the son of a merchant in Paris. Marrying Marie Curie's daughter Irène in 1926, he added her last name to his (so that one of the most illustrious names in all of science history would not be wiped out by his own). Frédéric Joliot-Curie shared the 1935 Nobel Prize for Physics with his bride for creating new radioactive elements in the lab.

1942 Clinton Hart Merriam, cofounder of both the National Geographic Society and the U.S. Fish and Wildlife Service, dies at age 86 in Berkeley, California. In his early 20s, Merriam began traveling throughout the western United States as a naturalist (eventually he explored and collected specimens in 48 states and Bermuda), after which he was placed in charge of the U.S. Biological Survey, which eventually became the Fish and Wildlife Service. Merriam was also an anthropologist, collecting data on the distribution, mythology, and language of 157 tribes of Pacific Coast Indians.

1949 The nation's first museum devoted exclusively to atomic energy opens to the public in Oak Ridge, Tennessee, where, during WWII, uranium-235 was processed for the Hiroshima bomb.

1954 A sled fitted with rocket engines (built by the U.S. Air Force to simulate the effects of bailing out of supersonic airplanes at high altitudes) is tested for the first time, at Holloman Air Force Base in Alamogordo, New Mexico, by Lieutenant Colonel John Paul Stapp, chief of the base's Aero Medical Field Center. On this day, six rockets send Stapp down the tracks at 421 mph. Later designs sent the "abrupt deceleration vehicle" over 3000 mph.

1991 In a special ceremony, U.S. Commerce Secretary Robert A. Mosbacher presents the nation's 5 millionth patent to University of Florida President John Lombardi; the patent is actually issued to University of Florida microbiologist Lonnie O. Ingram, who genetically joined two species of bacteria to form a new wonder bug that converts garbage into fuel. "This organism gives mankind its first economical, totally renewable, environmentally benign liquid fuel," observed an agent of the company that markets Ingram's invention.

1992 Dow Corning quits production of silicone breast implants. In announcing its withdrawal from the business that it pioneered and led for 30 years, Dow denies that science has proved the devices harmful, and denies that it was "hounded out" of the field. Even though it is no longer in the business, Dow pledges $10 million to researching harmful effects of the implants, and says it will donate up to $1200 per patient to have implants removed if a doctor deems it necessary. This seems magnanimous, but a lawyer for women who are suing the company pointed out that removal costs $9000.

1800 Alessandro Volta reports his invention of the electric battery to Sir Joseph Banks, president of the Royal Society.

1846 Pathologist-university administrator Giulio Bizzozero is born in Varese, Italy. In the late 1800s he made the University of Turin a world leader in medical scholarship.

1856 Frederick W(inslow) Taylor, the father of scientific management, is born the son of a lawyer in Philadelphia (see March 21, 1915, for biographical notes).

1899 Martha M. Place becomes the first woman in the United States to be electrocuted for murder. In February 1898 she killed her stepdaughter, Ida, in Brooklyn. The scene of today's historic event is Sing Sing Prison in Ossining, New York.

1904 Physicist Walter M(aurice) Elsasser, known for his "dynamo model" that explains the origin and properties of Earth's magnetism, is born in Mannheim, Germany.

1904 Pioneering and controversial behaviorist B(urrhus) F(rederic) Skinner is born in Susquehanna, Pennsylvania. His name will forever be attached to the "Skinner box" (a highly simplified environment containing some means of delivering a reward or punishment and some means of detecting voluntary behavior, such as a lever for an animal to press), which Skinner used to develop original theories of learning. His experimental subjects were mainly rats and pigeons, but in his books *Walden Two* (1948) and *Beyond Freedom and Dignity* (1971) he applied his theories to the behavior of humans, arguing that man is simply another animal whose free will is largely controlled by the consequences of his previous actions and experiences. Critics argued vigorously against this view, but Skinner's scientific findings remain bulwarks in the fields of learning and behavior.

1947 The heaviest individual animal on record is a 209-ton female blue whale, killed today by a whaling fleet in the southern Atlantic Ocean. Her heart weighed 1540 pounds and her tongue, 4.72 tons.

1948 The first television broadcast of a symphony occurs on this date. Two concerts are shown: At 5 PM CBS shows Eugene Ormandy conducting the Philadelphia Symphony Orchestra. At 6:30 Arturo Toscanini conducts the NBC Symphony Orchestra in an all-Wagner program.

1987 The FDA approves the sale of AZT, a drug proven to extend the lives of AIDS victims.

1991 The U.S. Supreme Court strikes down "fetal protection" employment policies that barred women from hazardous (and top-paying) jobs in many major companies. Employers had justified such policies as protection of a woman's childbearing ability, but the Court saw the policies as illegal sex discrimination. "Decisions about the welfare of future children must be left to the parents who conceive, bear, support and raise them, rather than to the employers who hire those parents," said the majority opinion, written by Justice Harry A. Blackmun.

1991 The *Journal of the American Medical Association* reports results from the most extensive study of fatalities in nuclear power-plant workers: After 20 years' employment, nuclear workers show a slight elevation in death rate from all causes, but they are actually 21% *less* likely to get a fatal cancer than people not working in nuclear plants. Researchers suggest that power plant employees are generally in better health.

1931 Prolific author and eminent entomologist John Henry Comstock dies at 82 in Ithaca, New York. With his wife-illustrator, Anna Botsford Comstock, he produced a number of very popular books for both the layman and the scientist. His anatomical studies of butterflies and moths formed the basis for their modern classification.

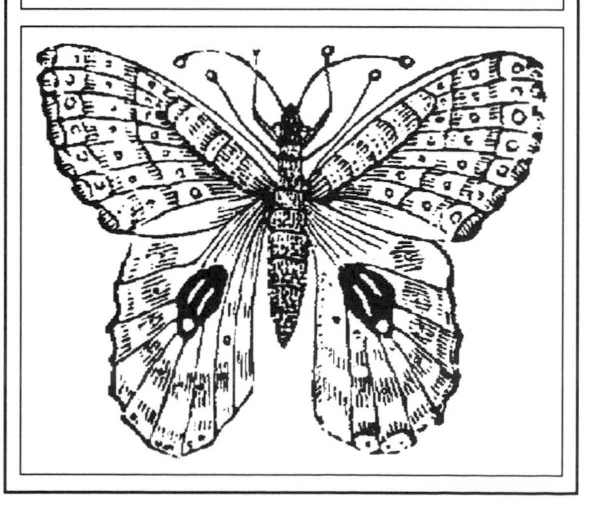

1768 French mathematician-physicist-Egyptologist Baron Joseph Fourier is born the son of a tailor in Auxerre. Orphaned at eight, Fourier went to live in a military academy run by Benedictine monks, where he exhibited behavioral problems until he discovered mathematics, at which he immediately excelled. His greatest accomplishment was *The Analytic Theory of Heat* (begun in 1807, published 1822), which described how the conduction of heat can be analyzed mathematically, using an operation now called "the Fourier series." His equations have since been applied to many natural phenomena, including tides, sunspots, and the weather.

1788 Swiss geologist Ignatz Venetz, one of the first to realize that glaciers once covered huge expanses of Earth, is born the son of a poor carpenter in Visperterminen.

1877 Virulent anthrax bacteria are first brought into Pasteur's laboratory at Lille. He and his assistants culture the organism for years, and use it to create the first vaccine for the fatal disease.

MARCH 21

1884 George David Birkoff, the foremost U.S. mathematician of the early 1900s, is born in Overisel, Michigan. Among his achievements is a mathematical model of gravity and a mathematical theory of aesthetics (art, music, and poetry).

1898 Biologist Paul Alfred Weiss is born in Vienna. He is known for discoveries in nerve regeneration, wound healing, and the mechanics of nerve development.

1915 Frederick W. Taylor, the father of scientific management, dies the day after his 59th birthday in Philadelphia. He originally wanted to follow his father as a lawyer, but abandoned this plan when excessive night studying at Harvard temporarily ruined his eyesight. He then signed on as a laborer in Philadelphia's Midvale Steel Company where he rapidly rose through a series of positions to chief engineer. In 1881 he and a partner won the National Doubles Championship in tennis. In that same year he introduced scientific time study (now known as time-and-motion study) at Midvale. His idea was that factory efficiency could be improved by carefully observing workers and eliminating waste in time and actions. His methods were successful, made him famous, and revolutionized factory work.

1925 Wolfgang Pauli, 24, publishes his "exclusion principle" (that two nearby electrons cannot be in exactly the same state at the same time) in *Zeitschrift für Physik*. The principle is a cornerstone in both quantum mechanics and chemistry, and won Pauli the 1945 Nobel Prize for Physics.

1932 Molecular biologist Walter Gilbert is born in Boston. He developed a procedure that is used to determine the exact sequence of molecules in DNA and RNA, for which he was awarded the 1980 Nobel Prize for Chemistry.

1933 The first cross-country test of all-blind flying equipment is conducted by James Kinney, as he flies from College Park, Maryland, to Newark, New Jersey. He is accompanied by radio technician William La Violette and by Henry Diamond, a scientist with the U.S. Bureau of Standards who helped develop the instrument landing system.

1991 Nationwide newspapers report that Monroe County (home of the Florida Keys) has abandoned plans to build a highway parallel to U.S. Route 1. Opposition to the new road was unusual in that it united landowners with environmentalists; the former feared that they would not be able to erect more houses, and the latter feared an increase in roadkills of the endangered Key deer (only 250–300 are still alive, and 40 were killed by cars in 1990).

1887 Biochemist David Keilin, discoverer of cytochrome (an iron-containing enzyme critical to the cell's use of oxygen), is born of Polish parents in Moscow.

1992 Officials of the French youth group Eclaireurs de France (meaning "Those Who Show the Way") publicly acknowledge that some of their members had, indeed, made a mistake when they destroyed portions of 15,000-year-old cave paintings during their previous weekend's graffiti-cleaning trip to Mayrieres cave in southwestern France. The officials, however, are "indignant that the actions of well-meaning youths should be called into question." "Absolutely stupid!" was the way one cultural official had summed up his take on the group's error. The paintings were of bison, and created during the Paleolithic era. Unfortunately, the graffiti cleaners could not see more than the writing on the wall.

1394 Mongol astronomer Ulugh Beg is born in Soltaniyeh, Persia, a grandson of the last great barbarian conqueror, Tamerlane (see October 27, 1449, for biographical notes).

1785 British geologist Adam Sedgwick is born in Dent, England. He was a mathematician who knew very little about rocks when he was appointed (because of his overall scientific prowess) professor of geology at Trinity College of Cambridge University in 1818. He then devoted himself to researching and writing about geology. He conducted many expeditions throughout the British Isles. He was the first to apply the names "Cambrian" (taken from the ancient name for Wales) and "Devonian" (from the area of Devon in England) to geologic eras. Cambridge's Sedgwick Museum was built in his honor. Interestingly, he attacked Darwin's theory of natural selection when it was first published.

1799 Astronomer Friedrich Wilhelm August Argelander is born into wealth in Memel, Germany. He established the study of "variable stars" (i.e. those stars whose brightness is variable) as a distinct branch of astronomy, and is renowned for *Bonner Durchmusterung* (1859–1862), an enormous catalog that lists the positions and magnitudes of 324,188 stars. It took him 25 years to compile the information.

1832 Versatile genius Johann Wolfgang von Goethe dies at 82 in Weimar, Germany. A prolific author and thinker right to the end, Goethe published the second part of his great drama, *Faust*, in the last year of his life. Although best known as a poet and dramatist, Goethe also wrote on biology and botany, and advanced a theory on the nature of color. The abundant mineral goethite is named for him.

1841 The first cornstarch patent in U.S. history is No. 2,000, issued to Orlando Jones of City Road, England. His innovation is a process for producing the flour.

1868 Physician Robert Andrews Millikan, the first to precisely measure the electronic charge on single electrons (with his famous oil-drop experiment of 1911), is born in Morrison, Illinois. This experiment won him the 1923 Nobel Prize. Millikan later took an interest in radiation from outer space, which he named "cosmic rays."

1882 The United States outlaws polygamy.

1895 The Lumière brothers (Auguste and Louis) present their film *La Sortie des ouvriers de l'usine Lumière* to an invited audience at 44 Rue de Rennes in Paris. It is the first motion picture to be shown on a screen.

1935 The first low-definition (180 line) television service begins broadcasting, in Berlin, Germany.

1941 The Grand Coulee Dam on the Columbia River in Washington State begins operation. It is not only the world's largest concrete dam, it is also the largest concrete structure of any kind.

1960 The first laser patent is No. 2,929,922, issued to Arthur L. Schawlow of Madison, New Jersey, and to Charles Hard Townes of New York City. The patent is assigned to the Bell Telephone Laboratories.

1980 A compact disk system is first introduced commercially. Created by RCA, it soon failed because of the superiority of laser systems that followed shortly.

1987 The sanitation barge "Mobro" departs from Islip, New York, in search of a place to dump its load of 3200 tons of New York City garbage. After 155 days at sea, in which the barge was rejected by several states and three foreign countries, the garbage is finally dumped back in New York City, in Brooklyn.

1993 The first "Great Sex-Out Day" is held throughout York County, Pennsylvania. Declared by the county's Teen Pregnancy Coalition, the aim of the event is to raise consciousness about sex-related issues, such as AIDS and teen pregnancy, and to celebrate abstinence. Among the sex substitutes recommended are moonlight walks, long workouts, and baking cookies.

1906 Baron Schröder pays £1207.50 (about $6000) for a single orchid at an auction by Protheroe & Morris of Bow Lane, London. This remains the highest price ever paid for an orchid.

1699 John Bartram, the father of U.S. botany, is born near Darby, Pennsylvania. A self-educated naturalist and explorer, Bartram was the chief botanist for the American colonies for King George III. The first North American to hybridize flowering plants, he established America's first botanical gardens (at what is now 43rd and Eastwick Streets in Philadelphia), an institution that became internationally famous. The legendary biologist Linnaeus called him "the greatest natural botanist in the world."

1749 French mathematician/physicist/astronomer Pierre-Simon Laplace is born in Beaumont-en-Auge, Normandy. In his famous *Mechanics of the Heavens* (a five-volume set published from 1798 to 1827) he expressed his life's vision, "Given for one instant an intelligence which could comprehend all the forces by which nature is animated and ... sufficiently vast to submit these data to analysis—it would embrace in the same formula the movements of the greatest bodies of the universe and those of the lightest atom: for it, nothing would be uncertain and the future as the past, would be present to its eyes." Although he ultimately failed in producing a single mathematical formula that explained/predicted all physical activity in the universe, his accomplishments were considerable. He produced many original calculations and theories on the movement of planets, he made great advances in probability and statistics, with Lavoisier in 1780 he showed that respiration is a form of combustion, and he was a key force in developing the metric system. One of his important accomplishments was avoiding the imprisonment and/or beheading that befell many of his scientific colleagues, including the great Lavoisier, during the French Revolution. He accomplished this partly because he was apolitical and partly because his mathematics were useful in warfare, especially in calculating the flight of cannonballs.

MARCH 23

1754 Thomas Chippendale, 35, publishes *Gentleman and Cabinet-Maker's Directory*, one of history's most influential books on furniture design and construction.

1769 Geologist William Smith, the father of stratigraphy, is born in Churchill, England. (Stratigraphy is the branch of geology that studies the layers, or strata, of the outer Earth).

1867 Lister publishes the second in a series of *Lancet* articles announcing the discovery of antiseptic surgery.

1907 Pharmacologist Daniel Bovet is born in Neuchâtel, Switzerland, the son of a professor of pedagogy. In 1944 he discovered pyrilamine, the first antihistamine. For this work and work on curare (he discovered the first curare substitute, and adapted curare as a muscle relaxant during surgery), Bovet received the 1957 Nobel Prize for Physiology or Medicine.

1912 Rocket scientist Wernher von Braun is born into a prosperous aristocratic family in Wirsitz, Germany.

1858 The nation's first patent for a cable streetcar is No. 19,736, issued this day to Eleazar A. Gardner of Philadelphia for an "improvement in tracks for city railways." This early mass-transit device incorporates an underground tunnel housing the pulleys and cables that move the cars.

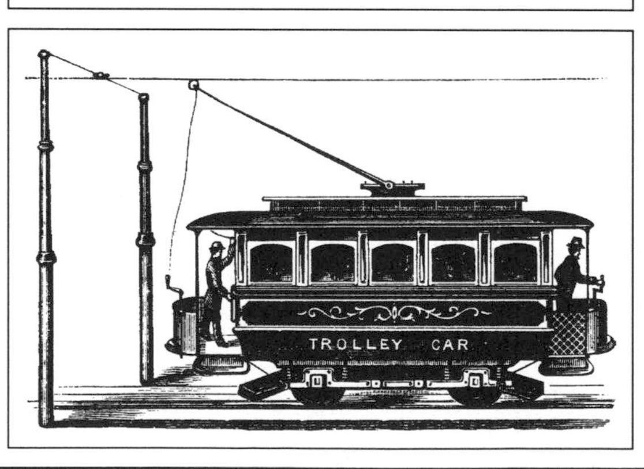

1983 The "Star Wars" missile-interception defense system is first proposed by President Reagan.

1989 "Gentlemen, it's not a question of 'what if,' but 'when,' " speaks Riki Ott at a town meeting in Valdez, Alaska, concerning the possibility of an oil spill off the shores of their community. The tanker *Exxon Valdez* is just leaving the Alyeska Terminal with a load of 53,094,510 gallons of crude oil (see March 24, 1989).

1989 A 1000-foot-diameter meteor roars past Earth, missing a collision by just six hours. The near miss is not detected by scientists until after the meteor passed.

1992 The 12-member European Community bans production and consumption of ozone-destroying chemicals such as chlorofluorocarbons (CFCs) by 1995. The action steps up an existing timetable for phasing out the environmental villains that was established at a Montreal conference in 1987. A hole in Earth's protective layer of ozone was first discovered in 1985 over Antarctica.

1494 Georgius Agricola, the father of mineralogy, is born the son of a draper in Glauchau, Germany.

1629 Virginia passes the first game law in the colonies; it provides that "no ... hides or skins whatever be sent or carried out of this colony upon forfeiture of thrice the value, whereof the half to the informer and the other half to public use."

1834 Geologist-ethnologist-conservationist John Wesley Powell is born in Mount Morris, New York.

1873 Psychologist Edouard Claparède is born in Geneva. A leader in the functionalist school of psychology, Claparède's theories on sleep anticipated Freud, and his theories on thinking and child development anticipated Piaget.

1882 Robert Koch, 38, announces in Berlin the discovery of the tuberculosis bacteria.

MARCH 24

1884 Chemist Peter Debye is born in Maastricht, the Netherlands. He won the 1936 Nobel Prize for Chemistry for theories relating to molecules that have a negative charge at one end and a positive charge at another.

1903 Biochemist Adolf Butenandt is born is Bremerhaven-Lehe, Germany. In 1939 he was a corecipient of the Nobel Prize for Chemistry for his work on sex hormones: In 1929 he isolated estrone (responsible for sexual development and function in human females); two years later he isolated androsterone (a male counterpart to estrone); and in 1934 he isolated and identified another female hormone, progesterone, which has a strong influence over the female menstrual cycle.

1955 The United States's first offshore oil-drilling rig is placed in service. Built by Bethlehem Steel for the C.G. Glasscock Drilling Company, the rig can drive piles with a force of 827 tons, and yank them out with a force of 942 tons.

1962 Auguste Piccard, famous for exploring the ocean depths and the upper atmosphere in machines he built himself, dies of a heart attack at 78 in Lausanne, Switzerland. In 1932 he built a balloon for studying cosmic rays at great heights (10.4 miles); his major innovation was constructing an airtight passenger cabin fed with compressed air. In the 1940s and 1950s Piccard adapted this cabin to undersea vessels, and he and his son Jacques built several "bathyscaphes" which they took to record-setting depths. In 1954 they descended to a depth of 2 miles, and in 1960 Jacques was the pilot on a dive of about 7 miles in the Challenger Deep of the Marianas Trench.

1989 The *Exxon Valdez* strikes a reef in Alaska's Prince William Sound. The hull rips open; almost 11,000,000 gallons of oil spew into the ocean, and 1300 miles of coastline are covered with toxic, black goo.

1992 The shuttle *Atlantis* carries seven astronauts into a 184-mile-high orbit, on man's first space mission devoted to studying Earth and its atmosphere. It is also the first flight for a Belgian astronaut, Dirk Frimout. NASA plans 45 future environmental missions; of special interest is Earth's protective stratospheric layer of ozone, which has been compromised by industrial chemicals.

1993 NASA's John D. Rather and David Morrison testify before Congress's space subcommittee that the chance of asteroids hitting Earth is "a real problem," but one that could be handled. Rather proposes diverting the course of such invaders with nuclear explosions in outer space.

1693 Instrument maker-horologist John Harrison, inventor of the first marine chronometer that allowed navigators to accurately determine their longitude at sea, is born the son of a carpenter-instrument maker in Foulby, England. He died in London exactly 83 years later (1776). He began inventing chronometers after several disasters at sea prompted the British government to offer several prizes for new navigational tools.

1639 Colonists in Dedham, Massachusetts, begin digging America's first canal to provide water power for industrial use. The canal, dug at Mill Creek, will take water from the Charles River to the Neponset River.

1712 Nehemiah Grew, one of the earliest plant physiologists, dies at 70 in London. His training in medicine and animal physiology led him to explore similarities with plant structures, at a time when improved microscopes were opening up science's eyes to the fine structure of living things. He noted the existence of cells in plants, and described and named a number of other plant structures, including the radicle, plumule, and parenchyma. Grew also advanced the idea that the stamen is a plant's male reproductive organ and the pistil is the female organ.

1825 Cell biologist Max Schultze is born in Freiburg, Germany. He was the first to define the living cell as a combination of protoplasm and a nucleus (1861), and the first to see protoplasm as a basic constituent of both plant and animal cells. Technically, he made a number of innovations in preparing material for viewing with microscopes.

MARCH 25

1865 Physicist Pierre-Ernest Weiss, famous for theories/investigations of magnetism, is born the son of a haberdasher in Mulhouse, France.

1911 The Triangle Shirtwaist Company burns to the ground in New York City, killing 146 immigrant laborers. Public outrage spurred workplace safety reforms.

1923 Astronomer Kenneth Linn Franklin, who discovered that Jupiter emits radio waves, is born in Alameda, California.

1928 Astronaut James A(rthur) Lovell, Jr. is born in Cleveland, Ohio. Most famous as the heroic commander of the near-disastrous *Apollo 13* flight, Lovell has been aboard several flights that have established "firsts" or records in outer space.

1937 The first newspaper advertisement containing perfume is run in the *Daily News* in Washington, D.C.

1945 British Prime Minister Winston Churchill urges in a memo to President Franklin Roosevelt that neither France nor Russia be given any secrets about nuclear technology. Such secrecy was a major cause of the East–West "Cold War" in the years following World War II.

1954 RCA announces production of color TV sets.

1993 One of the key mysteries about AIDS has been its roller-coaster patterning, in which many patients show an initial bloom of virus particles in the blood, followed by a lengthy disappearance of the virus (up to 10 years), followed by a second bloom. Two reports in today's *Nature* solve this mystery: during the latent period when no virus is detectable in the blood, the virus is hiding out in lymph tissue where it is reproducing and attacking disease-fighting cells. Eventually billions and billions of virus-carrying cells burst from the lymph tissue throughout the whole body. This finding shows that any effective vaccine must block the virus before it gets a foothold in lymph material.

1993 *Nature* reports "one of the most extraordinary fossil finds of recent times": a population of woolly mammoths that survived 6000 years after they were thought extinct. Russian paleontologists discovered their remains on Wrangel Island, northwest of Alaska.

1786 Astronomer-optician-physicist Giovanni Battista Amici is born the son of a government official in Modena, Italy. He made various improvements of telescopes and microscopes, including the oil-immersion microscope (invented in 1840 and now standard microbiology equipment, this microscope eliminates some sources of visual imperfections in very high magnification work).

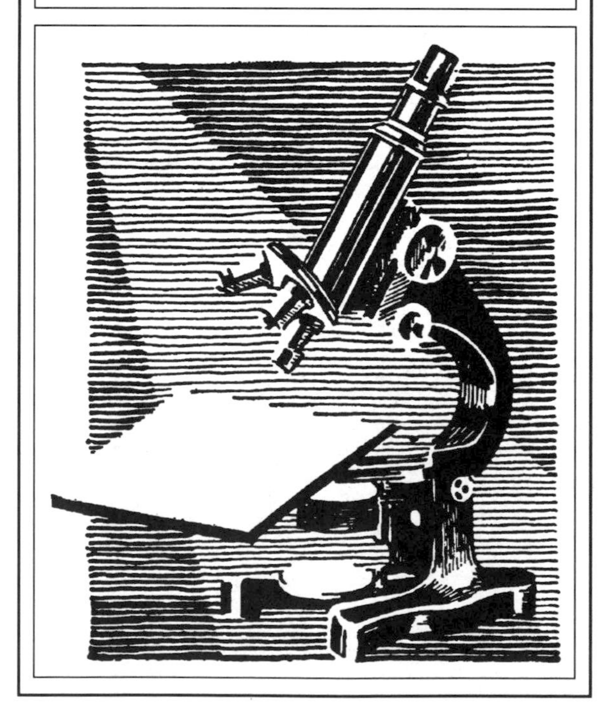

March 26

1753 Physicist Sir Benjamin Thompson (also known as Count Rumford) is born the son of a farmer in Woburn, Massachusetts. He is famous for historic investigations of heat (which led to the modern realization that heat is a form of motion), but he was also a distinguished public administrator and inventor (among his creations were a double boiler, a kitchen stove, and a drip coffeepot). During the American Revolution he was an English spy and Redcoat officer, which eventually led to his exile to Britain, where he was knighted and where he performed much of his research. He held no grudges, and later established the Rumford professorship at Harvard and the Rumford medal of the American Academy of Arts and Sciences.

1789 Meteorologist William C. Redfield is born the son of a seafarer in Middletown, Connecticut (see September 3, 1821, for Redfield's major discovery).

1835 Darwin, 26, is repeatedly bitten by *Triatoma infestans* (the "great black bug of the Pampas") at Luxan, Argentina, during the historic voyage of the *Beagle*. This attack probably inflicted trypanosome parasites on Darwin, thereby giving him Chagas' disease. Although it has never been proven, theorists suggest that this disease caused the heart, digestive, and nervous troubles that plagued Darwin for the rest of his days, and that finally killed him.

1843 Education pioneer Horace Mann becomes engaged to Nathaniel Hawthorne's sister. The couple spent their honeymoon researching social institutions in England.

1845 "Adhesive and medicated plaster,"forerunner of the Band-Aid, is first patented in the United States on this day by Dr. Horace Harrel of Jersey City, New Jersey, and Dr. William H. Shecut. They receive patent No. 3,965 for a process that involves dissolving rubber in a solvent and spreading the mixture on a fabric. They sold their process to Dr. Thomas Allcock, who turned it into Allcock's Porous Plaster.

1885 Motion-picture film (commercial) is manufactured for the first time, by the Eastman Dry Plate & Film Company of Rochester, New York. They were also the first to produce film in continuous strips on reels.

1916 Robert Stroud, the "Birdman of Alcatraz," stabs a prison guard to death in Leavenworth, Kansas. This act ensures that Stroud will spend the rest of his life behind bars, where he conducted and published important research on bird diseases.

1916 Biochemist Christian B(oehmer) Anfinsen is born in Monessen, Pennsylvania. He shared the 1972 Nobel Prize for Chemistry for discovering relationships between structure and function in enzymes and other proteins.

1985 The first test-tube quintuplets are born to Linda and Bruce Jacobssen in London.

1993 Richard Guimond (acting assistant administrator of the Superfund of the Environmental Protection Agency) writes a memo that summarizes the failure of the EPA to recover costs of cleaning up polluted sites from the companies that created the pollution. The agency simply lacks sufficient resources to make the polluters pay. Only $843 million of a possible $4.3 billion (or about one-fifth) has been reclaimed. Some of the difference is tied up in court cases, some has been written off as unrecoverable, and the remainder has yet to be pursued. "The situation is likely to become more difficult in the near future," writes Guimond grimly.

1516 Natural historian-encyclopedist Conrad Gesner is born is Zurich. He published a variety of monumental works during his lifetime: a Greek–Latin dictionary (1537), a "Universal Bibliography" (1545; it listed and evaluated works by some 1800 authors), a 21-volume encyclopedia (19 books were published in 1548; Gesner's aim was to survey all of the world's recorded knowledge), the elaborately illustrated *Historiae animalium* (the first of five volumes appeared in 1551; each book covered a portion of the animal kingdom), and a book describing some 130 languages (1551).

1794 Congress passes an act to establish the U.S. Navy. The bill authorizes the president to obtain six ships (four to carry 44 guns and two to carry 36 guns) to protect U.S. shipping against Algerian pirates. Construction began in a Virginia shipping yard (lent by the state) later in the year under the supervision of Captain Richard Dale. A peace treaty in 1796 halted production until 1797, when it resumed under the supervision of Commodore Samuel Barron. The *Chesapeake* was finally completed at the navy yard at Gosport, Virginia, and was launched on December 2, 1799. It was the first ship constructed by the U.S. government.

1827 Charles Darwin, 18, submits his first report of an original scientific discovery to the Plinian Society in Edinburgh, Scotland. Darwin had discovered several things about the biology of tiny marine organisms from the Scottish coast.

1845 Wilhelm Roentgen, is born the son of a textile merchant in Lennep, Germany. He was the recipient of the first Nobel Prize for Physics.

MARCH 27

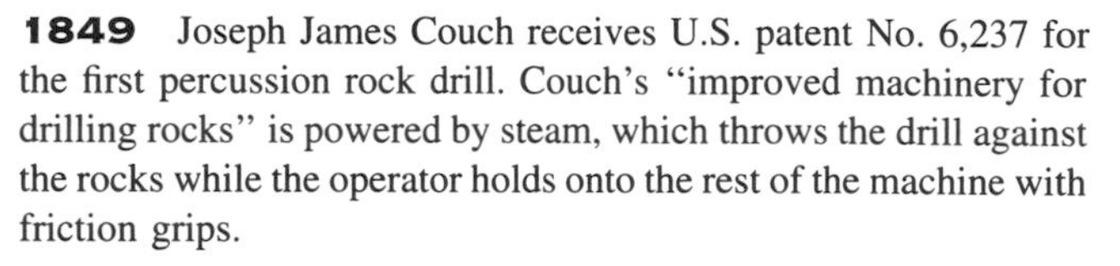

1849 Joseph James Couch receives U.S. patent No. 6,237 for the first percussion rock drill. Couch's "improved machinery for drilling rocks" is powered by steam, which throws the drill against the rocks while the operator holds onto the rest of the machine with friction grips.

1884 The first-ever long-distance telephone call is made, between branch managers of the American Bell Telephone Company in Boston and New York. The *Boston Journal* reported the triumph; "The words were heard as perfectly as though the speakers were standing close by, while no extra effort was needed at the other end of the line to accomplish the result."

1910 Eminent zoologist-mining engineer Alexander Agassiz dies at sea at age 74. He was born the son of a giant in natural science, Louis Agassiz, and followed him to a career in academia. Among his most notable achievements were a reorganization of the classification of starfish and the discovery of a coral reef 3000 feet above sea level (this discovery called for a revision of Darwin's theory of reef formation). Agassiz also distinguished himself by taking over an unsuccessful copper mine in Calumet, Michigan, in 1866, and by 1869 turning it into one of the world's foremost mines. He installed modern machinery and the latest safety devices, as well as establishing pension and accident funds for the miners; he also took measures to protect the environment and community surrounding his mine. He became wealthy, and donated heavily to biology programs at Harvard and other institutions.

1968 Yuri Gagarin, the first man in orbit, dies at 34 in a plane crash near Kirzhach, Russia.

1991 Wildlife biologists at a conference in Alberta, Canada, are told of a strange twist of nature: The fate of the Florida panther is in the hands of its prey, the whitetail deer. The panther is on the brink of extinction, with only a few dozen left on Earth, all of which live on a tiny plot of land in southwest Florida. Without a plentiful supply of deer, and without sufficient territory in which to conduct normal activities, the big cat is in big trouble. The problem is exacerbated because much of the panther's remaining habitat is "marginal" in its suitability, and much is privately owned. Further, recent studies have found high levels of mercury both in the local water and in dead panthers. On top of all this, signs of inbreeding are beginning to emerge because there are just so few panthers in such a small area of land.

1968 The windsurfer is patented.

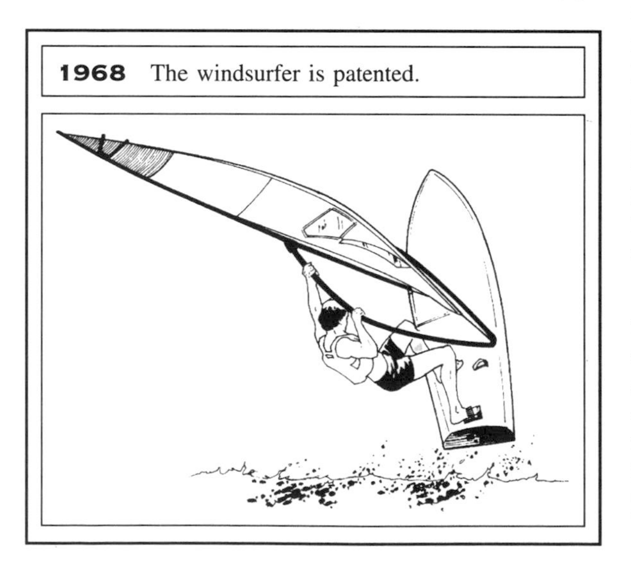

1994 Acknowledging that cigarettes are killers and that they are addictive, Surgeon General Joycelyn Elders argues against a total ban on them. "To immediately ban cigarettes would not necessarily solve the problem because you have all those millions of Americans out there that are addicted to this drug. We just can't immediately ban cigarettes without making sure we treat our American people," she says during a television interview.

MARCH 28

1797 The nation's first washing machine patent is granted on this day to Nathaniel Briggs of New Hampshire for an "improvement in washing cloaths."

1837 Physiologist Wilhelm Friedrich Kühne, who coined the term "enzyme" (meaning "in yeast"), is born the son of a prosperous merchant in Hamburg, Germany.

1849 Botanist Stephan Endlicher dies at 44 by his own hand in Vienna, after a life devoted to collecting and classifying plants. From 1836 to 1840 he published an innovative classification scheme that was widely used for more than 50 years, and some of which remains influential. He gathered a huge collection of over 30,000 specimens, which he donated to the Vienna Museum of Natural History when he became its curator. He further used his own money to buy specimens and to publish writings by himself and others while he was a professor at the University of Vienna from 1840 until his suicide.

1865 Outdoor advertising is first regulated by state law. New York amends an 1853 law with "an act for the more effectual prevention of wanton and malicious mischief and to prevent the defacement of natural scenery." Stones, rocks, trees, and other natural scenery cannot be painted, printed on, or otherwise marred; offenses are classified as misdemeanors, punishable by a $250 fine and/or six months in jail.

1892 Belgian physiologist Corneille Heymans is born in Ghent. In his late 20s he started working with his father at the University of Ghent in researching how blood pressure and blood content affect breathing and heart activity. He succeeded his father there as professor of pharmacology in 1930, after which he made some dramatic discoveries. Working on anesthetized dogs, he discovered tiny sense organs (called pressoreceptors) in neck arteries that are sensitive to blood pressure, and that control breathing and heart rate. Near these receptors, in the aorta, he discovered other sense organs that respond to the oxygen content in the blood, and that also influence breathing by sending messages to the brain's medulla. He won the 1938 Nobel Prize for Physiology or Medicine.

1895 Ronald Ross, 37, leaves his wife and children in England, to return to a desolate, primitive lab in the outback of India, where, in 1897 he discovered that malaria is transmitted by mosquitoes. Ross was the second recipient of the Nobel Prize for Physiology or Medicine (in 1902).

1979 The infamous Three Mile Island nuclear accident occurs outside Harrisburg and Middletown, Pennsylvania. Human and mechanical errors compound each other, causing cooling system malfunction; this produces partial meltdown of the reactor's core, and radioactivity leaks into the atmosphere. To date, it is the nation's worst nuclear mishap.

1983 Giovanni Vigliotto is sentenced in Phoenix, Arizona, to a heavy fine and 34 years in prison for fraud and bigamy. Vigliotto (whose real name was either Nikolai Peuskov or Fred Jipp) was married to 104 women at the same time. To date, he is history's biggest bigamist.

1984 The first baby born from a once-frozen embryo is delivered by Cesarean section in Melbourne, Australia.

1994 The Supreme Court decides that individual states can abolish the insanity defense. The Court announces that it will not hear the appeal of a Montana man who beat a U.S. Forest Service employee senseless "while [allegedly] in the throes of psychotic delusion" in 1990. The Montana Supreme Court, and now the U.S. Supreme Court, effectively upheld that state's abolition of the insanity defense, which means that the man must serve his 60-year prison term.

1899 A wireless telegraph message is first sent between England and the Continent, by the great radio pioneer Guglielmo Marconi.

1561 Groundbreaking physician-researcher Santorio Santorio (also known as Santorius and Sanctorius) is born in Capodistria, Italy. He was the first to use medical instruments of precision, performed the first systematic study of basal metabolism, and introduced quantitative experimentation to medical research. He frequently corresponded with Galileo, some of whose inventions he adapted for medical research; he produced a clinical thermometer in 1612.

1629 The first forestry legislation in America is passed by Plymouth Colony. Under the new act, approval by the governor and council are required for the sale or transport of lumber out of the colony.

1807 Amateur astronomer Heinrich Wilhelm Olbers discovers *Vesta 4*, the brightest asteroid on record and the only one visible with the naked eye.

1819 Edwin Laurentine Drake is born in Greenville, New York. He drilled the world's first successful oil well (at Titusville, Pennsylvania, in 1859).

MARCH 29

1853 Inventor Elihu Thomson is born in Manchester, England. A cofounder of the General Electric Company (in 1892, in a merger with an Edison company), Thomson received approximately 700 patents during his lifetime. A number of these involved alternating current discoveries, which laid the foundation for the development of modern ac motors.

1890 English astronomer Sir Harold Spencer Jones is born the son of an accountant in London. He was known for his precise determinations of the distance between Earth and the sun

1900 Biololgist-naturalist Charles (Sutherland) Elton is born in Liverpool, England. He was credited with framing the principles of animal ecology (which he described as "the sociology and economy of animals").

1903 Meat king Gustavus Swift dies in Chicago at age 63. Starting as a butcher's helper at 14, Swift had worked his way up to cattle buyer and partner in a Boston meat company when he moved his office to Chicago, which had become the center of the nation's cattle industry. There he conceived a plan of slaughtering cattle before they were shipped to the East (to replace the common procedure of shipping live cattle, to be slaughtered once they got to the East). He hired engineers to design and build the first refrigerated railroad cars. His scheme was a huge financial success. He broke with his old partner and formed a new venture, Swift & Company, with his brother. He also formed an alliance with two other prominent meat executives, J.O. Armour and Edward Morris, but this "Beef Trust" was dissolved by the Supreme Court in 1905. Swift was also a leader in using previously discarded cattle parts to make other products, like glue, soap, and margarine. His company was worth $25 million at his death.

1912 Robert Falcon Scott, 43, writes for the last time in his diary during his heartbreaking attempt to be the first to the South Pole. "We shall stick it out to the end, but we are getting weaker, of course, and the end cannot be far. It seems a pity, but I do not think I can write more."

1919 A total eclipse of the sun is observed by two expeditions (one in Brazil, one off the coast of western Africa), which establish that light from distant stars is bent as it passes our sun. This is dramatic support of Einstein's theory of relativity, which predicted that energy, just like matter, is affected by gravity.

1980 The first transplant of a human fingernail takes place in Strasbourg, France. Dr. Guy Foucher transplants a toenail to the thumb of 12-year-old Christopher Kempf, who lost the thumbnail because of a poorly treated hangnail.

1993 A number of archaeological artifacts are returned to Egypt by Israel at a ceremony in Jerusalem. They had been dug up during Israel's 15-year occupation of the Sinai Peninsula; they include ten Byzantine tombstones and hundreds of ceramic pots. Three days later, 39 more boxes of Greek, Roman, and Islamic artifacts are returned to Egyptian authorities in the border town of Rafah in northern Sinai.

1135 Philosopher-physician Maimonides is born into a distinguished family in Córdoba, Spain (see December 13, 1204, for biographical notes).

1842 "For etherizing and incising of a tumor, $2.25." This is the modest record in the ledger of Dr. Crawford W. Long in Jefferson, Georgia, for his removal of a half-inch cyst on the neck of James Venable. The patient felt no pain because he was the first in history to be given a true anesthetic during surgery.

1858 The pencil-with-eraser combination is patented (No. 19,783) by Hyman L. Lipman of Philadelphia. To a regular pencil he "secured a piece of prepared rubber, glued in at one end."

1867 The English medical journal *Lancet* publishes the third of four articles by Joseph Lister, in which he introduced the concept/practice of antiseptic surgery.

1876 Clifford Beers is born in New Haven, Connecticut, the home of Yale University. As a student there, Beers began suffering mental illness, which developed into a full-blown breakdown after his graduation. He spent three years in abusive asylums. When finally released, he wrote *A Mind that Found Itself* (1908), and began a lifelong campaign for more understanding and better treatment for the insane. Also in 1908 he formed the Connecticut Society for Mental Hygiene, the nation's first such organization. He was a cofounder of the first national societies for the insane in both the United States and Canada, and a cofounder of the International Committee for Mental Hygiene in 1920.

1879 Optical engineer Bernhard Voldemar Schmidt, inventor of the Schmidt telescope (an instrument widely used in photographic surveys of the cosmos because it provides distortion-free viewing over an exceptionally large field of view), is born on Naissaar Island, Estonia. Schmidt was a tragic alcoholic who literally drank himself to death at age 56; he spent his last year in an insane asylum.

1895 The heaviest object ever removed from a human stomach was a 5-pound 3-ounce hairball, taken on this day from the gut of a 20-year-old woman in England's South Devon and East Cornwall Hospital.

1902 The German giant Constantine dies in Mons, Belgium. His thigh bone was 29.9 inches long, still the longest human bone on record.

1910 Theodor Wulf, Jesuit priest and amateur physicist, takes an electroscope to the top of the Eiffel Tower (the electroscope measures electric charge and indicates the presence of radioactivity; Wulf had invented an extremely sensitive electroscope in 1909). Over the next four days' observation, Wulf discovers the first indications that Earth is constantly bombarded by high-energy rays from outer space.

1950 Invention of the phototransistor (a transistor activated by light instead of electricity) is announced by the Bell Telephone Laboratories of Murray Hill, New Jersey. Its inventor was Dr. John Northrup Shive.

1981 President Ronald Reagan is shot in Washington, D.C., by John Hinckley Jr., who is later found not guilty by reason of insanity. The case focused international attention on the validity of the insanity defense.

1993 Scientists at a San Diego meeting of the American Cancer Society announce history's first anti-cancer gene therapy. Independent groups from UCLA and the University of Pennsylvania both claim success with the same strategy: chemically blocking an "oncogene" (a cancer-causing gene) from producing a protein that causes normal cells to grow and reproduce out of control. Although oncogenes were discovered a decade ago by the Pennsylvania team of Dr. Mark I. Greene, developing the agents to block them took a long time.

1994 Protesting the end of funding for pollution studies of the St. Lawrence River, members of the environmentalist group "SVP" send fish by mail to the Prime Minister of Canada, the Premier of Quebec, and 18 other Canadian bureaucrats. "If [the fish] didn't arrive before the long weekend, there's going to be an awful smell," remarked SVP copresident Daniel Green.

1852 The first electric whale-killing machine is patented by Dr. Albert Sonnenberg and Philip Rechten of Bremen, Germany.

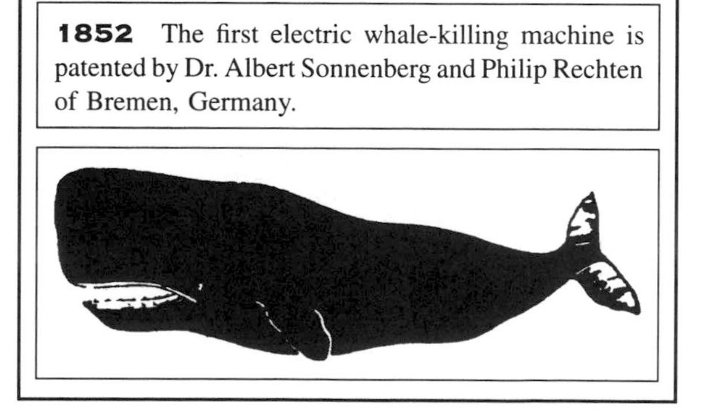

MARCH 31

1596 René Descartes, variously called the "founder of psychology," the "father of modern philosophy," and the "father of analytic geometry," is born with chronically poor health in La Haye, France.

1727 Scientific colossus Sir Isaac Newton dies at 84 in his home in Kensington, outside London. He had lapsed into a coma after days of severe pain from an old bladder infection. He became the first commoner to be buried in Westminster Abbey.

1732 The nation's first circulating library places its initial order of books. The Library Company of Philadelphia, organized by Benjamin Franklin in 1731 through his Junto Society, purchased the books with a donation of 40 shillings from each of 50 people. Louis Timothee was the first librarian.

1811 German chemist Robert Wilhelm Bunsen is born the son of a philology professor in Göttingen. Developer of the famous Bunsen burner, he also discovered the elements cesium and rubidium, using the technique of "spectrum analysis" (or spectroscopy, it is the study of light emitted, reflected, or absorbed by matter, and is now an extremely important tool in many sciences), which Bunsen invented in 1860 with Gustav Kirchhoff.

1831 The tortured chemistry genius Archibald Scott Couper is born in Kirkintilloch, Scotland (see March 11, 1892, for a biographical sketch).

1880 Wabash, Indiana, becomes history's first town with electric streetlights. Four lights (each with over 4000 candlepower) are attached to a pole outside the courthouse.

1889 The Eiffel Tower is completed. Designer Alexandre Gustave Eiffel, 56, unfurls the French flag at its top. Construction took 2 years, 2 months, 2 days. It was the world's tallest tower before the advent of enormous television towers.

1903 History's second motorized flight of any appreciable distance occurs in South Canterbury, New Zealand, when Richard William Pearce (1877–1953) flies his homemade gas-engined monoplane for at least 50 yards along the Main Waitohi Road. Nine months later the Wright brothers made their historic flight in North Carolina.

1976 The New Jersey Supreme Court rules that the comatose Karen Anne Quinlan can be removed from her artificial respirator, in accordance with her family's wishes. It is a historic precedent in right-to-die cases. Although she never regained consciousness, Karen Anne lived for years after being detached from the machine.

1993 "The mother of all spectrographs" is finished and ready to be attached to a telescope, announces astronomer Steven Vogt of Lick Observatory and the University of California, Santa Cruz, at a press conference. The $3.6 million device is a six-ton prism that splits entering light into thousands of colors, the patterning of which is analyzed by a huge computer. Extremely faint amounts of light from the edges of the universe will be observable. Because each chemical has a unique, fingerprintlike pattern of light emission, scientists will be able to tell what chemicals are present in far-distant stars and galaxies. Findings are expected to shed light on many questions, including the birth of the universe. The spectrograph is to be attached to the Keck Telescope, which is under construction atop the extinct Mauna Kea volcano in Hawaii.

1993 Some patients with inoperable brain tumors show improvement after taking the illegal abortion pill RU-486, announces Dr. Steven Grunberg of USC to an American Cancer Society meeting in San Diego. In his preliminary study of 28 men and women, 6 showed measurable tumor reduction, 2 had improved vision, and 5 exhibited a variety of other improvements, including fewer headaches and less double vision. "This paper is exciting," said brain surgeon Howard Tung. "We have run out of things to do for these patients."

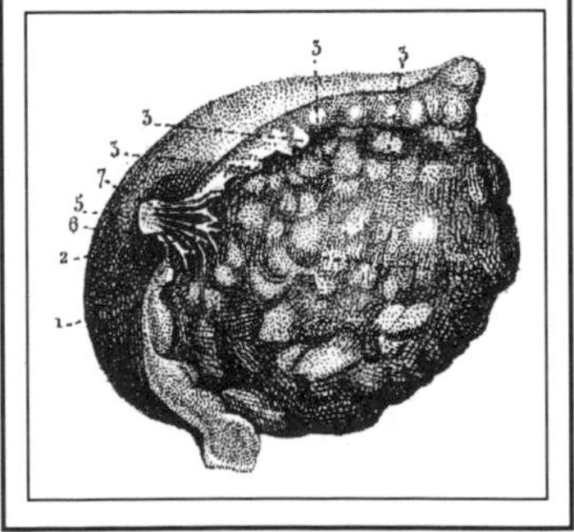

1867 Using the antiseptic methods he invented, Dr. Joseph Lister treats the last in a series of 11 compound fractures that formed the basis of a series of reports in *Lancet* that changed surgery forever.

1868 Gregor Mendel, 46, is elected abbot of the monastery in Brünn (now Brno, Czechoslovakia). Ironically, the honor was to deprive the world of one of its greatest-ever researchers (who single-handedly, without research funding, discovered the basic laws of genetics). The abbot's heavy administrative duties, coupled with some disappointing experiments and total indifference to his few publications, caused Mendel to abandon his biology research while he was still relatively young and in good physical health.

1869 A westerner sees an adult giant panda for the first time, when local hunters bring the body of a panda to French missionary-naturalist, Peré Armand David in Moupin, China. Although he considered himself a follower of the animal-loving St. Francis, Peré David paid high prices for the bodies of unusual animals, thus encouraging their slaughter.

APRIL 1

1938 Su-Lin, the first giant panda to reach the West alive, chokes to death on a piece of wood in Chicago's Brookfield Zoo. On the same day, Ruth Harkness, who brought Su-Lin to the United States, departs the country for China in search of more live pandas.

1948 Tsavo National Park is established in Kenya.

1948 The famous "Alpher, Bethe, Gamow" paper appears in *The Physical Review*. Physicists Ralph Alpher and George Gamow actually wrote the paper (which was a mathematical analysis of atomic events during the creation of the Universe), and later persuaded the esteemed Hans Bethe to lend his name as coauthor for poetic/comic effect ("Alpher, Bethe, Gamow" sounds like the first three letters of the Greek alphabet, alpha, beta, gamma). Bethe actually did make important contributions to subsequent discussions of the theory, but, according to Gamow, when the "theory went temporarily on the rocks, Dr. Bethe seriously considered changing his name to Zacharias."

1950 Dr. Charles Richard Drew, a world-famous authority on preservation of blood, is seriously injured in a car accident on route to a scientific meeting. In desperate need of a blood transfusion, he is taken to the nearest hospital, but he is turned away because he is black. He dies near Burlington, North Carolina, on the way to another hospital.

1960 History's first weather-observation satellite, *Tiros 1*, is launched from Cape Kennedy.

1972 Free contraceptives are offered for the first time in Great Britain, by the Hounslow Bureau Council.

1972 As one of his first actions as the new head of the Environmental Protection Agency, Russell Train lifts a ban on the use of the indiscriminate killer DDT on 650,000 acres of Pacific Northwest forest.

1992 Earth's oldest living creature turns out to be a fungus. Weighing 100 tons, it spreads over 38 acres, in a Michigan forest and is estimated to be as much as 10,000 years old. News of the discovery is made public today, setting off a flood of inquiries to the scientists and to the state's Department of Natural Resources, which had been given no advance warning of the announcement.

1994 The Russian news agency ITAR-Tass announces that food scientists have produced vodka candy bars. The announcement is an April Fools' Day prank, as are reports of spray-on tights and an attack on New York City by Siberian deer hunters.

1578 "I began to think whether there might not be a movement, as it were, in a circle." These were the thoughts of William Harvey (born this day in Folkestone, England, the eldest of nine children of a prosperous businessman) as he was achieving his great insight/discovery: the circulation of blood through the human body.

1513 Spanish explorer Ponce de Leon lands in Florida, searching for a "fountain of youth,"

1618 Italian physicist Francesco Maria Grimaldi is born in Bologna. He discovered the diffraction of light (the bending of light as it goes through an opening or past a solid object); this effect is used in the phase-contrast microscope to allow viewing of tiny organisms without having to stain or preserve them.

1819 John Stuart Skinner founds the nation's first important agricultural journal, *American Farmer*, in Baltimore. It is an eight-page, quarto-sized weekly magazine, and was published under a variety of names until 1867.

1875 Automobile baron Walter P. Chrysler is born in Wamego, Kansas.

1879 Toll-line commercial telephone service is first instituted, between Springfield and Holyoke, Massachusetts. The line is operated by the District Telephone Company of New Haven, Connecticut.

APRIL 2

1902 The nation's first motion-picture theater opens in Los Angeles, at 262 South Main Street. Called the "Electric Theater," the structure is a tent at a circus. Among the first movies seen: *New York in a Blizzard* and *The Capture of the Biddle Brothers*. The first building devoted entirely to motion pictures was the Nickelodeon, which opened in Pittsburgh three years later.

1931 Dr. Jacques Miller, the first to demonstrate a function of the thymus gland, is born in Nice, France. The gland is critical in the formation of the immune system, which he showed by removing the thymus in very young animals.

1935 The first practical radar system is patented by England's Robert Watson-Watt.

1957 An insert containing aluminum foil is issued in a newspaper for the first time. Milwaukee's *Sentinel* runs the feature as part of an advertising campaign of the Aluminum Corporation of America.

1958 The greatest speed ever recorded in a tornado is 280 mph, measured on this date at Wichita Falls, Texas.

1969 An FDA scientist (frustrated that his superiors have apparently ignored research reports showing that cyclamates—food sweeteners—cause cancer) brings two horribly deformed chicken embryos into the office of FDA Commissioner Herbert Vey in protest. Both embryos had been exposed to cyclamates, which eventually were banned from the U.S. food supply. The government imposed a very slow withdrawal from supermarket shelves, during which both government and industry played down the chemicals' dangers.

1981 John Hinckley, Jr., is ordered by a federal judge to be committed to a psychiatric facility for evaluation, following Hinckley's shooting of President Ronald Reagan.

1992 *Nature* reports the discovery of Earth's oldest and largest known living organism: a huge fungus! Weighing 100 tons, spreading over 38 acres, and estimated to be as old as 10,000 years, the organism was found in a Michigan forest. Most of it is underground with little mushrooms sprouting out through the surface every September. The discovery was announced on April 1, leading some to wrongly believe that the announcement was an April Fools' Day prank.

1994 Actress-turned-consumer advocate Betty Furness dies at 78 of stomach cancer at Sloan-Kettering Memorial Hospital. Following a series of B movies in the 1930s and 1940s, she became an advertising spokesperson and then a consumer affairs reporter for the *Today* show, specializing in car safety, product liability, and fetal alcohol syndrome. "She pioneered consumer TV news reporting, and she pursued it with intelligence, inquisitiveness and irrepressibility," observed fellow advocate Ralph Nader.

1889 Renowned chemist Charles Martin Hall receives patent No. 400,766 for the first practical process for producing aluminum. Hall's method involved passing electricity through a liquid solution containing dissolved aluminum. His process was put into commercial use by the Pittsburgh Reduction Company, which later became the Aluminum Corporation of America.

1778 Phyisician Pierre-Fidèle Bretonneau, who performed the first successful tracheotomy (1825), is born in Tours, France. An epidemiologist by training, Bretonneau was the first to distinguish typhoid from typhus and the first to clinically describe diphtheria, which he named.

1794 The first military air force is created. The French revolutionary "Public Safety Committee" authorizes a fleet of balloons, to be used mainly as tethered observation posts.

1798 Charles Wilkes is born in New York City, the son of a prominent businessman. This affluence helped him gain a solid education in many subjects. Wilkes was one of the nation's earliest oceanographers, and leader of its first major ocean expedition (1838–1842), which circled the globe and determined that Antarctica (named by Wilkes) is a continent.

1837 Legendary naturalist John Burroughs is born on a farm in the Catskill Mountains, near Roxbury, New York.

1841 Astronomer Hermann Vogel is born the son of a high school principal in Leipzig. He became a pioneer in the study of light emitted by distant stars, and introduced the use of photography to this field in 1887. He is best known for his discovery of spectroscopic binaries (pairs of stars revolving so close around each other that they seem to be one body; only spectroscopic analysis shows them to be two stars).

1860 The illustrious history of the Pony Express mail service begins; one rider heads west from St. Joseph, Missouri, while another rider heads east from Sacramento, California. The fee for mail transport is $5 per half ounce. The service was put out of business after 18 months by the transcontinental telegraph.

1940 Isle Royale (on Lake Superior) is established as a National Park by an act of Congress. An island wilderness without roads and accessible only by boat and seaplane, the Park forms a "natural laboratory" that was later used in David Mech's important studies of wolves and their interactions with moose. These studies showed wolves to be "prudent predators" (killing only what they need for food) and excellent parents, thus helping to improve public opinion about them.

1965 The "SNAP 10A," the first nuclear reactor in space, is launched from Vandenberg Air Force Base, California, at 1:25 PM. It starts generating power 220 minutes later, upon an electric signal from scientists on Earth. The reactor stayed operable for 43 days, generating 500,000 watt-hours of electricity.

1973 The nation's first "twilight zoo" (in which the level of illumination is held constant night and day) is dedicated in Pittsburgh.

1991 An ugly hoax in the food industry begins. Handbills appear in black and Hispanic sections of New York City, proclaiming that several brands of soft drink are manufactured by the Ku Klux Klan and will sterilize black men. As the pamphlets spread to other parts of the city over the coming days, soda sales plummet. The reason and the source of the handbills remain unknown, but sales rebounded when (1) chemical analysis by the Food and Drug Administration found no known sterility agent present in the beverages, (2) the KKK denied involvement, and (3) soda makers mounted a public relations campaign that included David Dinkins, the city's black mayor, drinking one of the slandered sodas on television.

1994 The *Orlando Sentinel* exposes a toxic paint hoax in which public-housing authorities have spent thousands of dollars removing perfectly safe paint from housing projects. At the center of the scam are two brothers, David and Jonathan Mingus; one runs a testing firm and the other a cleanup firm. The testing firm reported dangerous levels of lead where none existed, and the cleanup firm was recommended for, and won, lucrative contracts for the unnecessary follow-up work.

1934 Primatologist Jane Goodall is born in London. In 1960 she began observing chimpanzees at Tanzania's Gombe Stream National Park; the project is now history's longest-running study of a nonhuman primate in its natural habitat. Among the multitude of discoveries by Goodall and associates: chimpanzees occasionally group-hunt and eat meat, they have a rich repertoire of social customs (including intergroup warfare), and they manufacture and use tools (and teach these skills to their offspring). The findings have forced a reevaluation of man's place in the universe. In recent years Dr. Goodall founded the Jane Goodall Institute and has devoted herself to protecting chimps in captivity.

636 Spanish scholar Isidore of Seville dies at about 76 in Seville. A prolific author, his most influential contribution to science was the encyclopedia *Etymologies* (published between 622 and 633), which incorporated all of the work of the ancient Greeks that he could collect. *Etymologies* was very influential during the Dark and Middle Ages, and thousands of copies of the manuscript still exist.

1597 Galileo, 33, writes to another famed astronomer, Johannes Kepler, that he is reluctant to profess his belief that Earth and other plants move around the sun because he fears public ridicule.

1617 John Napier, inventor of the decimal point and logarithms, dies at 67 near Edinburgh, Scotland.

1688 Joseph Delisle is born the ninth child of a historian-geographer in Paris. He was the first astronomer to use the transit of Venus across the face of the sun to calculate the distance between Earth and the sun.

APRIL 4

1726 Stephen Hales, the first to measure human blood pressure, today becomes the first to measure the pressure that drives sap movement in plants.

1802 Dorothea Lynde Dix, "godmother of the insane," is born in poverty to an alcoholic, abusive preacher in Hampden, Maine. She ran away from home for the last time at age 12, and opened her first school for girls at age 19. She crusaded all her life for improved conditions and state support for mental institutions, prisons, and poorhouses.

1823 Inventor Sir William Siemens is born into a dynasty of inventors in Lenthe, Germany.

1826 Zénobe Gramme, who revolutionized the use of electricity by creating the first practical continuous-current generator, is born the son of a Belgian bureaucrat in Jehay-Bodegneé.

1850 The nation's first school devoted to the needs of the mentally retarded is incorporated in Boston. The Massachusetts School for the Idiotic and Feeble-Minded Youth now functions as the Walter E. Fernald State School, renamed in honor of its first resident superintendent.

1902 Rhodes Scholarships are established by a $10 million bequest in the will of entrepreneur-colonizer Cecil Rhodes, 10 days after his death.

1914 Man's first prehistoric remains from sub-Saharan Africa are announced by their discoverer, Hans Reck, in the *Illustrated London News*. The report introduces science to Olduvai Gorge, Tanzania, in which many prehuman fossils were to be found by others.

1964 L.S.B. Leakey, John Napier, and Phillip Tobias report in *Nature* the discovery of a new hominid species, *Homo habilis*, in Olduvai Gorge.

1821 Lock inventor Linus Yale is born in Salisbury, New York.

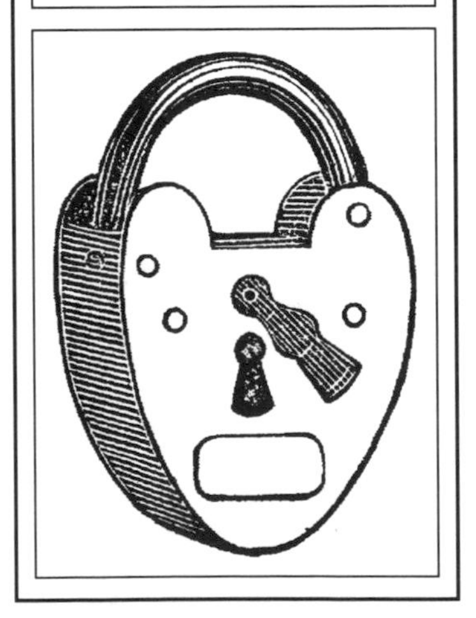

1972 The burning of municipal garbage is used for the first time in the United States to generate electrical power. St. Louis's Meramec Plant uses shredded rubbish mixed with coal to operate an industrial boiler.

1979 For the first time in over 20 years, naturalists discover that a free-living peregrine falcon (once plentiful but now endangered) has laid an egg east of the Mississippi. The nest was found on the 33rd floor of the U.S. Fidelity and Guarantee Building in downtown Baltimore.

1994 Biosphere 2 (the three-acre steel-and-glass environmental dome in the desert outside Oracle, Arizona) is broken into and vandalized. Equipment is sabotaged and outside air is let in. The main suspects are two members of the crew of Biosphere 1 who had recently been barred from the property by the new management.

1622 Italian mathematician Vincenzo Viviani is born in Florence. Perhaps the leading geometer of his time, Viviani is most remembered for founding the Accademia del Cimento, one of the first important scientific societies and the progenitor of England's Royal Society.

1753 The British Museum is founded by an Act of Parliament.

1804 Matthias Schleiden is born the son of a prosperous physician in Hamburg, Germany. Schleiden suffered through a law career and a failed suicide attempt before devoting himself to his hobby, botany. At age 34 he produced one of the major ideas in biology: that all plants are composed of cells.

1827 Joseph Lister, inventor of antiseptic surgery, is born the son of J.J. Lister (a wine merchant whose development of the achromatic, or blur-free, microscope was very important to the work of his son and others on the "germ theory of disease," which states that many diseases are caused by tiny organisms).

1870 Geneticist Clarence E. McClung is born in Clayton, California. His skill and creativity as a microscopist enabled him to discover that sperm exists in two forms, each with a different chromosome configuration. From this he proposed that chromosomes determine an individual's sex. Others later confirmed this deduction, making it one of the earliest demonstrations that chromosomes carry hereditary information. McClung also cofounded *Biological Abstracts*, a major international research periodical.

1887 The blind and deaf Helen Keller, age 6, suddenly achieves her dramatic insight into the meaning of language. She and her teacher, the "miracle worker" Anne Sullivan, are at the well of the family home in Tuscumbia, Alabama. Keller later described the moment in her autobiography: "As the cool stream gushed over one hand she [Sullivan] spelled into the other the word *water*, first slowly, then rapidly. I stood still, my whole attention fixed upon the motions of her fingers. Suddenly I felt a misty consciousness ... and somehow the mystery of language was revealed to me. I knew then that 'w-a-t-e-r' meant the wonderful cool something that was flowing over my hand."

1899 Alfred Blalock, coinventor of the first surgery to successfully treat blue baby syndrome, is born in Culloden, Georgia.

1924 Victor Hensen dies at 89 in Kiel, Germany. Famed as both a physiologist (the cells of Hensen and the Canal of Hensen are in the inner ear) and an oceanographer (he coined the term "plankton" to describe tiny, floating sea life), Hensen developed equipment and methods for plankton study, and led the world's first "Plankton Expedition."

1942 Racked with malaria, blood poisoning, and severe weight loss, Lt. Colonel Arthur F. Fischer furiously strips and grinds bark from cinchona trees that he has been growing for 20 years on Mindanao, the Philippines. This one group of trees is the world's only source of the antimalaria drug quinine that is not already in the hands of the fast-approaching Japanese army. When he learns on this day that Bataan is about to fall and that he will not be able to ship the bark back to his malaria-ridden compatriots there, Fischer races to extract what quinine he can. When Bataan does fall several days later, and Mindanao is about to fall next, Fischer takes two tins of seeds and flies to Australia on one of the last planes out of the Philippines. Trees from these seeds are still growing today.

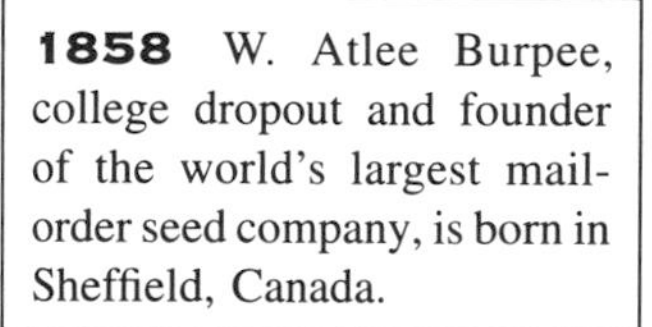
1858 W. Atlee Burpee, college dropout and founder of the world's largest mail-order seed company, is born in Sheffield, Canada.

1994 The Miccosukee Indians announce plans to sue the governments of Florida and the United States because the proposed cleanup of the Everglades does not sufficiently stop pollution by sugar and vegetable industries. Billy Cypress, leader of the Indians living in the Everglades, laments at a news conference, "We love the Everglades. The land and the water is part of our lives, our heritage. And yet, the special interests continue to do away with whatever they need because I guess they have more power in the political sense."

1994 Harry A. Blackman, author of the historic *Roe v. Wade* decision, announces his retirement from the Supreme Court at age 85. Since that decision in 1973 he has received more than 60,000 pieces of hate mail, all of which he insisted on reading himself.

1748 Excavation begins on Pompeii, the Greco-Roman city that has been buried for nearly 17 centuries under lava from a volcanic eruption of Mount Vesuvius.

1801 William Hallowes Miller, creator of the Miller indices (still in use, this is the system in which the form of any crystal face can be described by a set of three whole numbers), is born in Llandovery, Wales.

1829 Mathematical genius Niels Henrik Abel dies in poverty at 26 in Froland, Norway. Two days later word arrives that the brilliance of his work has been recognized and he is wanted as a professor in Berlin.

1839 Russian explorer Nikolai Przhevalsky in born in Smolensk. In his explorations of central Asia he discovered a wild camel and the primitive horse, which were subsequently named for him.

1857 The nation's first important veterinary college (the New York College of Veterinary Surgeons) incorporates.

APRIL 6

1859 Massachusetts enacts the nation's first law establishing an Inspector of Milk post.

1903 Charles R. Jackson, author of *The Lost Weekend*, is born in Summit, New Jersey. The book described five days in the life of an alcoholic. It was Jackson's first and best-known work. A bestseller and the basis of a successful movie (Ray Milland played the alcoholic), the book was, in the words of critic Philip Wylie, "the most compelling gift to the literature of addiction since DeQuincy."

1909 After several failed attempts, man finally reaches the North Pole. First there are two Eskimos and one African American, Matthew A. Henson (carpenter, blacksmith, linguist, cook, dog trainer, and explorer). Robert E. Peary, the leader of the expedition, arrives 45 minutes later.

1810 Philip Henry Gosse is born the son of an impoverished miniaturist. This artistic background served Gosse well in his career as a self-taught naturalist; his illustrated writings elevated popular appreciation of nature, especially marine biology. Gosse established the world's first public aquarium in Regent's Park, London.

1911 Biochemist Feodor Lynen is born in Munich. He shared the 1964 Nobel Prize for Physiology or Medicine for elucidating the pathway by which the body makes cholesterol.

1928 Biochemist James D. Watson is born a child prodigy in Chicago. In 1953 he and Francis Crick described DNA's structure. This was the key to understanding how traits are passed from parent to offspring.

1985 William J. Schroeder is moved to an apartment in Lexington, Kentucky, becoming the first artificial heart recipient to be discharged from hospital.

1988 Resolution finally comes in the famous "Baby M" surrogate parenting case. The mother, Mary Beth Whitehead Gould, was fertilized with the sperm of William Stern, under a contract in which she agreed, for $10,000, to relinquish all rights to the child once it was born. She reneged on the deal. A lengthy and bitter series of court battles ensued. In today's court ruling, Stern and his wife retain custody of the child, but Gould is granted visitation rights.

1993 A huge radioactive cloud is launched and starts drifting over the Russian wilderness, after a tank of radioactive waste explodes at the secret military plant of Tomsk 7, 1700 miles east of Moscow. Government officials call it the worst, but not the only, Russian nuclear accident since the April 1986 incident at Chernobyl. The blast prompts delegates from the world's richest nations, meeting in Tokyo a week later, to agree that Russia needs urgent help with her nuclear waste, obsolete nuclear warheads, and crumbling nuclear reactors.

1794 Chemist Joseph Priestley leaves England for the United States. His laboratory, library, and home were recently destroyed by a mob in Birmingham, England, that did not like his support of the French Revolution.

1795 The metric system is adopted by France, the country that created it.

1815 Mount Tombora erupts on the Indonesian island of Sumbawa, producing a volume of discharge greater than that of any known volcano. The eruption creates a crater 5 miles in diameter, and the height of the entire island is lowered by about 4000 feet. Over 92,000 people die, and the subsequent dust cloud creates a cold spell over the entire world.

1825 Explorer-naturalist David Douglas lands in the Pacific Northwest on a plant-collecting expedition for England's Royal Horticultural Society.

APRIL 7

1890 Marjory Stoneman Douglas (author, conservationist, and leader in the fight to save the fragile Florida Everglades from further degradation by developers) is born in Minneapolis. She joined the struggle at age 78; her first battle was blocking the construction of a huge airport in the Everglades and an oil refinery on its outskirts. "The environmental movement in Florida owes a great debt to Mr. Ludwig (the refinery's developer). His idea was so ridiculous and it stimulated such widespread opposition that many people who'd otherwise been sitting back were enlisted into the environmental movement right then."

1927 The first public demonstration of television transmitted over a long distance is made. An audience in New York is treated to a speech by Secretary of Commerce Herbert Hoover in Washington, D.C.

1933 The nation's decade-long Prohibition against alcoholic beverages ends at 12:01 AM, when "3.2" beer goes on sale in 19 states. To protect against rowdyism during "New Beer's Eve," no immediate deliveries are made from breweries in the nation's capital, except to the White House and to the National Press Club.

1948 The World Health Organization is founded.

1959 NASA announces the selection of its first group of astronauts.

1959 Electricity is first produced by a nuclear reactor. In the experimental model built by New Mexico's Los Alamos Scientific Laboratory, a "plasma thermocouple" is inserted into the reactor's core, thus bypassing the need for the costly and bulky turbine apparatus. The first power output is just enough to illuminate a common light bulb.

1966 The United States recovers a hydrogen bomb it had lost off the coast of Spain.

1983 Astronauts Story Musgrave and Don Peterson take the first U.S. spacewalk in nearly a decade.

1993 A mystery plaguing beachgoers is finally solved, reports today's *Journal of the American Medical Association*. For years, tiny organisms (nicknamed "sea lice") had been causing fevers, welts, and horrible itching on swimmers, without anyone really knowing what they were. The creatures were not lice at all, but baby jellyfish.

1988 Paleontologists at the Museum of Natural History in Cleveland announce that a new dinosaur genus, a "pygmy tyrannosaurus," has been discovered—in the museum's own exhibits. Originally misidentified over 40 years before as part of a known species, the animal's skull had been mysteriously fitted with fake, plaster eyebrows that disguised its real identity. "We thought for a while we might have a Piltdown dinosaur [referring to a famous hoax involving a doctored skull of a bogus human ancestor] on our hands," said the controversial Dr. Robert T. Bakker, who discovered the error while on a visit to the museum.

1779 Johann Schweigger is born the son of a theology professor in Erlangen, Germany. He invented the galvanometer, a device that measures the strength of an electric current.

1805 Botanist-cell biologist Hugo von Mohl is born in Stuttgart, Germany. He was the first to propose that new cells arise from the division of previously existing cells.

1818 Chemist August Wilhelm von Hofmann is born the son of an architect in Giessen, Germany. He founded the German Chemical Society, and his work on aniline (among the many substances he researched) helped found the aniline dye industry. He also discovered formaldehyde.

1858 Herbert Spencer Jennings is born in Tonica, Illinois. He was one of the first ethologists (a specialist in animal behavior).

APRIL 8

1860 French biologist Félix Dujardin dies three days after his 59th birthday in Rennes, France. Largely self-educated, his research focused on tiny animals he found in "infusoria" (mixtures of water and decaying matter). He was the first to propose that one-celled animals should be classified in a group by themselves (now called "protozoans"). His studies of flatworms formed the basis for later development of parasitology. 1835 was a huge year for him: He successfully disproved Ehrenberg's theory that tiny animals have the same organs as big ones, and he became the first to describe the jellylike substance of life that is now called protoplasm.

1862 An aerosol dispenser is first patented in the United States, by John D. Lynde of Philadelphia.

1873 Alfred Paraf of New York City is granted patent No. 137,564 for a process to manufacture margarine. It is not the first patent for margarine making, but it is the first commercially successful process; it is so successful, and butter sales are so threatened, that a series of state and federal taxes were subsequently levied on margarine.

1898 The nation's first college of forestry, established at Cornell University and headed by the German forester Dr. Bernhard Fernow, is created today by legislation signed by New York Governor Frank Swett Black.

1911 Melvin Calvin is born to Russian immigrants in St. Paul, Minnesota. In 1961 he received the Nobel Prize in Chemistry for determining the Calvin cycle (which is the part of photosynthesis that occurs in plants during the night, the so-called "dark reactions," in which plants turn carbon dioxide into sugar).

1918 Betty Ford, founder of the famous addiction-treatment center that bears her name, is born in Chicago.

1947 The largest sunspot ever seen is discovered on the sun's southern hemisphere. It is estimated to cover an area of 7 billion square miles.

1869 Legendary neurosurgeon Harvey Cushing is born into a long line of family doctors in Cleveland.

1953 The first 3-D motion picture produced and released by a major company makes its debut at the Globe Theater in New York City. It is *Man in the Dark*. Two days later *The House of Wax* premieres at the Paramount Theatre, New York City; it is the first 3-D feature motion picture in color.

1991 Two professional health organizations air "beefs" with current U.S. eating habits. The National Cholesterol Education Program announces support of a low-cholesterol, low-fat diet for children over the age of two and the Physicians Committee for Responsible Medicine asks the U.S. Department of Agriculture to replace the traditional four food groups with four new groups: grains, fruits, vegetables, and legumes.

1992 Lifesaving pigs have arrived, announce worldwide news services. Virginia Tech scientists, funded by the Red Cross, have genetically engineered pigs to produce "protein C," a chemical that is normally found in humans and that hinders excessive bloodclotting.

1770 Physicist Thomas Johann Seebeck is born the son of a prosperous merchant in Revel, Estonia. Early in life Seebeck abandoned his medical training to pursue research. He discovered (but neither understood nor exploited) the Seebeck effect, in which an electric current is produced if two different metals are joined at two places, and the junctures kept at different temperatures.

1830 Eadweard Muybridge is born in Kingston on Thames, England. He first achieved fame for his spectacular photographs of Yosemite Valley, California, but he is best remembered for his photographic studies of human and animal movement. These latter studies suffered two notable interruptions: once when he had to invent a superfast camera shutter, and once when he was tried (and acquitted) for the murder of his wife's lover.

1865 Charles Proteus Steinmetz is born hunchback and Jewish in an anti-Semitic society, in Breslau, Germany (now Wrocław, Poland). He became an extraordinary inventor and electrical engineer; his theoretical/mathematical analysis of alternating current established it as the preferred form of electrical energy throughout the world.

1869 The American Museum of Natural History incorporates in New York City. "Boss" Tweed was enlisted by the Board of Trustees to ensure that Governor Hoffman signed the act of incorporation.

1889 French chemist Michel-Eugène Chevreul dies at 102 in Paris, a longevity that allowed him to witness both the French Revolution and the Eiffel Tower. Chevreul broke ground in several fields. He was the first to chemically analyze fats (he discovered fatty acids), the first to find sugar in urine, and, in his 90s, became one of the first to study the psychology of the elderly.

1901 Howard A. Rusk is born in Brookfield, Missouri. He was a founder of the modern science of rehabilitation medicine.

1919 John Presper Eckert, Jr. is born in Philadelphia. He codesigned the first important computers (ENIAC in 1946 and UNIVAC in 1951).

1947 The U.S. Senate approves President Truman's appointees to the first Atomic Energy Commission (AEC), a body of five civilians devoted to developing atomic energy to improve public welfare, to enhance man's standard of living, and to further world peace. The AEC was abolished in 1974 when its duties were transferred to two new organizations, the Nuclear Regulatory Commission (NRC) and the Energy Research and Development Administration (ERDA).

1970 The National Industrial Pollution Control Council begins advisory work to President Richard Nixon; it consists of 53 industrial leaders. Remarked Colman McCarthy of *The Washington Post*, "Until the President appoints an advisory pollution council composed of independent citizens with nothing at stake but their lungs and the balance of nature, the goats will continue to guard the cabbage patch."

1981 The longest scientific name in history is published in *Nature*. It is the systematic name for human mitochondrial DNA; the substance contains 16,569 nucleotides, and the word contains 207,000 letters.

1992 Science finally explains the beer belly. Swiss researchers in today's *New England Journal of Medicine* report that it's not the number of calories in alcohol that is critical; it's the fact that alcohol disrupts the body's normal burning of fat (from any and all dietary sources) that accounts for drink-induced corpulence. Less fat is burned, more is retained on the legs, stomach, face, and elsewhere.

1993 Investigators announce that they have solved the mystery of Milwaukee's water supply, in which thousands of people have become ill over a two-week period. The disease has been caused by a tiny animal, *Cryptosporidium*, which lives in the intestines of cattle. Water runoff, contaminated with animal waste from outlying farms, brought the culprit into the Milwaukee River and from there to a water purification plant that had been experimenting with a new water-filtering chemical that was not effective.

1903 Endocrinologist Gregory Pincus is born in Woodbine, New Jersey. His research into the antifertility effects of steroid hormones led to the development of the first oral contraceptives. The "sexual revolution" of modern times was given another boost exactly 23 years (1926) later when publisher Hugh Hefner was born in Chicago.

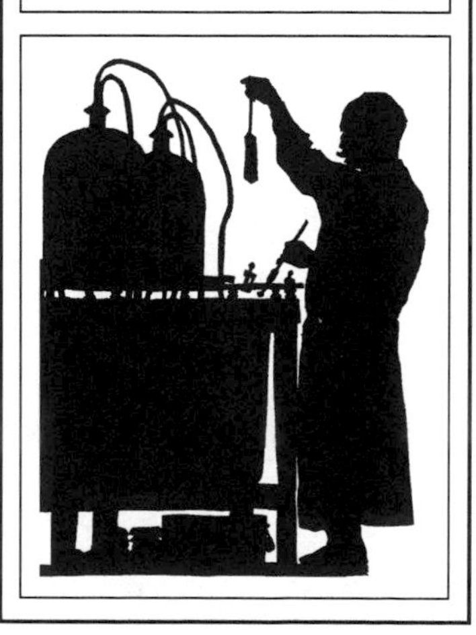

1727 Educator Samuel Heinicke is born in Nautschütz, Germany. Despite only an elementary school education, he founded the first German school for the deaf, and was responsible for elevating lipreading from a novelty to a major tool.

1755 Samuel Hahnemann, the founder of homeopathic medicine, is born in Meissen, Germany. While a doctor, he made the observation that quinine given to healthy persons produces symptoms of malaria, the very disease that quinine cures. This led him to propose (1796) the "law of similars": treat a disease with agents that produce the disease's symptoms. This is the main precept of homeopathy. Modern vaccination is based on a similar principle: prevent a disease by giving a person that disease (in a very weak form). Hahnemann's insight/proposal enjoyed widespread popularity in the early 1800s, and now is making a comeback, along with other forms of "alternative medicine."

APRIL 10

1790 The nation's first patent act becomes law "to promote the progress of useful arts." Patent applications originally were reviewed by the attorney general, the secretary of war, and the secretary of state (the first of whom was Thomas Jefferson, an inventor himself).

1835 The nation's first school for homeopathic medicine is founded in Allentown, Pennsylvania.

1866 The American Society for the Prevention of Cruelty to Animals incorporates in Albany, New York.

1887 Physiologist Bernardo Houssay is born in Buenos Aires, Argentina, to French parents. He was awarded the 1947 Nobel Prize for Physiology or Medicine for discovering that pituitary hormones regulate blood sugar.

1912 The *Titanic* departs Southampton, England, on its maiden voyage.

1927 Biochemist Marshall Warren Nirenberg is born in New York City. In 1968 he received the Nobel Prize for Physiology or Medicine for determining the code by which DNA manufactures proteins.

1944 The antimalarial chemical quinine is produced in a laboratory for the first time, by Drs. Robert Burns Woodward and W.E. Doering, working for the Polaroid Corporation at Harvard's Converse Memorial Laboratory. It is Woodward's 27th birthday.

1872 The first Arbor Day is held in Lincoln, Nebraska. Nearly 1 million trees are planted. In 1885 the date of Arbor Day was changed to the birthday of J. Sterling Morton (April 22) who founded the event.

1955 On this Easter Sunday, social philosopher-paleontologist (he codiscovered Peking man) Pierre Teilhard de Chardin dies at 73 of a heart attack in his New York City apartment. He had spent the day leading a holy mass, attending another, going to a concert, and then walking in Central Park.

1963 The nuclear-powered submarine *Thresher* fails to surface off Cape Cod; 129 perish.

1972 The United States, Soviet Union, and 70 other nations sign a treaty banning biological warfare.

1993 Ron Webeck, who achieved international notoriety in the late 1980s by being "cured" of AIDS, dies at 45 in St. Petersburg, Florida. He had lymphoma and had suffered a stroke two weeks earlier. In 1985 doctors gave him a month to live, but by 1989 he had recovered so strongly, and for so long, that he declared himself cured. His story hit the newspapers, and he received letters and phone calls from all over the world. By the end of 1989, however, he had developed heart inflammation and pneumonia, and his health went downhill from there.

1994 Science has known for some time that African Americans are much more likely (50% more likely) to get lung cancer from smoking than are whites. But it was unclear whether the difference was caused by behavior, genes, body chemistry, diet, or some other unknown factor. Today, at the annual meeting of the American Association for Cancer Research in San Francisco, John Richie (of the American Health Foundation) announces evidence pointing to a metabolic mechanism: Blacks have a poorer ability to detoxify NNK, one of the deadliest poisons in cigarette smoke.

1755 Amateur paleontologist-physician James Parkinson is born in England. He was the first to recognize a burst appendix as a cause of death, wrote the first scientific article on appendicitis (1812), and was the first to describe the neuromuscular disorder Parkinson's disease (1817).

1798 Physicist Macedonio Melloni, the first to extensively analyze infrared radiation, is born in Parma, Italy.

1810 Archaeologist Sir Henry Rawlinson is born in Chadlington, England. In 1833 he was sent as an army officer to help reorganize the Persian army, and became interested in antiquities. His great feat was translating a multilingual cuneiform inscription of Darius I on a hillside at Bisitun, Iran. The translation effectively provided a dictionary of the early languages of Mesopotamia.

1854 Conservationist John Burroughs begins his teaching career in a small country school in Tongore, New York, eight days after his 17th birthday.

APRIL 11

1890 "The Elephant Man" Joseph Carey Merrick, dies at 27 in London.

1900 Oceanography takes a step forward, as the U.S. Navy accepts the 54-foot "Holland," the first submarine to be accepted by any navy. It was the sixth submarine built by inventor John P. Holland over a 22-year period, but the only one he was able to sell.

1901 Astronomer Donald Menzel is born in Florence, Colorado. He is best known for arguments against the existence of UFOs and for supervising the assignment of names to newly discovered lunar features.

1926 Horticultural wizard Luther Burbank dies at 77 in Santa Rosa, California, after suffering a heart attack and nervous exhaustion.

1952 Parkinson's disease is successfully treated with surgery for the first time, when the team of Irving Cooper operates on the brain of Raymond Walker in Islip, New York.

1953 The Department of Health, Education and Welfare is established. Mrs. Oveta Culp Hobby is sworn in as its first Secretary, and the second woman cabinet member.

1962 The first amendment to the "International Convention for Prevention of Pollution of the Sea by Oil" is signed in London. Fifty-one nations agree to ban intentional discharge of oil at sea by any ship of 20,000 tons or more; prohibitions on dumping by smaller oil tankers are also adopted.

1970 The ill-fated *Apollo 13* mission begins.

1991 The tanker *Haven* is rocked by an explosion, and begins sinking off the Italian Riviera with 41 million gallons of oil on board.

1991 The theory of mimicry is thrown on its ear. Until today, the accepted theory behind mimicry said that viceroy butterflies look like monarch butterflies because monarchs taste bad to their predators. An article in today's *Nature* challenges that notion. University of Florida researchers report that both species are unpalatable to predators. This "dramatic" result forces a new interpretation of why the two species look alike.

1994 A vaccine to prevent cancer is announced at the annual meeting of the American Association of Cancer Research in San Francisco. A team at Thomas Jefferson University in Philadelphia made up the vaccine with the patients' own cancerous cells and with another chemical that stimulates the immune system. Three years after their tumors were surgically removed, 70% of patients treated with the vaccine remained cancer-free; only 20% of patients treated with surgery alone were cancer-free. To date, the vaccine has been used to battle skin cancer only, but "there is no reason why it's not applicable to other cancers," says team spokesperson Dr. David Berd.

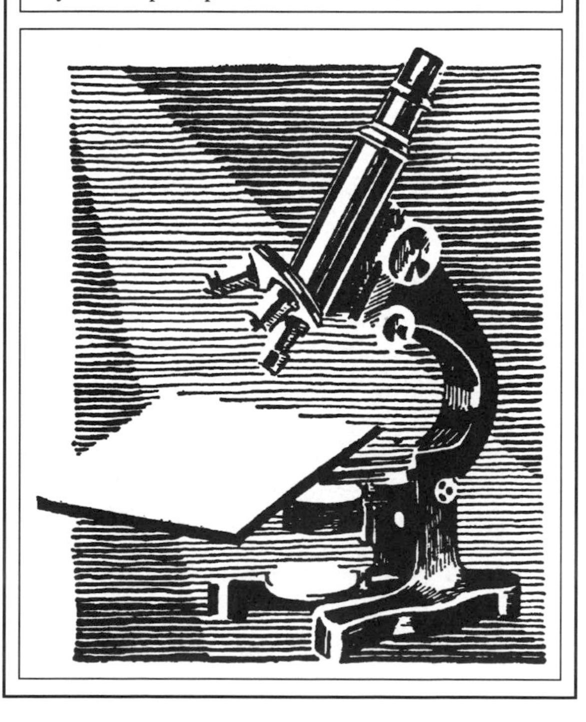

1829 Famed explorer-naturalist Alexander von Humboldt, 59, begins a scientific expedition to uncharted regions of Siberia.

1838 John Shaw Billings is born in Switzerland County, Indiana. He became both a physician and a librarian, and in 1879 with Dr. Robert Fletcher began publishing *Index Medicus*, a monthly review of medical publications that still remains a major research tool.

1877 The catcher's mask is first used in an actual baseball game, in Lynn, Massachusetts.

1884 Otto Meyerhof is born in Hannover, Germany. He is best known for discovering the role of lactic acid in muscle contraction (for which he received the 1922 Nobel Prize for Physiology or Medicine). He also determined (with Karl Lohmann) how ATP drives muscle contraction, plotted the Embden–Meyerhof pathway (whereby glucose is converted to lactic acid), and was a master teacher (Nobel laureates Krebs, Lipmann, and Ochoa were among his students).

1888 Ludwig Nobel, brother of dynamite inventor Alfred Nobel, dies at 56 of heart trouble in Cannes, France. A French newspaper mistakenly runs an obituary of Alfred, calling him a "a merchant of death." Alfred is shocked and saddened that he would be thus remembered; his desire to change public opinion leads to his decision to establish the Nobel Prizes.

1898 Marie Curie is sitting high in the gallery of the French Academy of Sciences when her discovery of substances much more radioactive than uranium is announced by one of her teachers, Professor Gabriel Lippmann.

1912 Nursing pioneer Clara Barton dies at 90 in Glen Echo, Maryland.

1938 New York becomes the first state to require marriage license applicants to undergo medical tests.

1955 "The vaccine works! It is safe, effective and potent." So announces Dr. Thomas Francis, Jr., one-time professor of Jonas Salk, at the conclusion of the year-long field trial of the Salk vaccine against polio. The announcement is made at a press conference in Ann Arbor, Michigan. It is the tenth anniversary of the death of Franklin Roosevelt (1945), a famous victim of the disease.

1961 Yury Gagarin becomes the first man to orbit Earth.

1968 A "mysterious epidemic" of sheep deaths in Skull Valley, Utah, is explained. The National Communicable Disease Center in Atlanta reports finding a nerve gas in the sheep and in their environment that was identical to gas sprayed a month before by the Army on nearby Dugway Proving Ground. The Army had previously denied any involvement, had not reported any misspraying, had not found any of the gas when it checked the sheep, and was slow to provide adequate gas samples for independent investigations by other agencies. The entire incident resulted in an overhaul of procedures relating to the development of chemical weapons at Dugway.

1971 *Water Wasteland*, a study of the nation's aquatic pollution, is released by Ralph Nader's Study Group. Two days later, EPA director Ruckleshaus announces a conference on pollution in the Houston Ship Channel.

1981 The first reusable spacecraft, the shuttle *Columbia*, is given its maiden flight in outer space.

1988 A patent is granted on an animal life form for the first time. Harvard scientists receive a patent on a genetically engineered mouse.

1991 Medical researchers in Covington, Louisiana, are granted permission by the Supreme Court to euthanize and complete nerve-damage experiments on two monkeys, Titus and Allen, thus ending a ten-year custody battle between scientists and PETA (People for the Ethical Treatment of Animals), who charged that the animals were being abused under the auspices of the National Institutes of Health.

1992 Euro Disneyland opens in Marne-La-Vallee, France.

1993 Doctors in St. Petersburg, Florida, announce the death of Ron Webeck, the man who achieved international notoriety for having been "cured" of AIDS (see April 10, 1993, for further information).

1625 The word "microscope" is coined. Writing from Bamberg, Germany, Johannes Faber suggests the term in a letter to Federigo Cesi, Duke of Aquasparta and founder of Italy's "Accademia dei Lincei" (Academy of the Lynx). This Academy is arguably the world's first scientific society; it appropriately named itself after an animal with exceptional vision.

1771 Mechanical engineer Richard Trevithick, the first to harness high-pressure steam, is born the son of a coal mine manager in Illogan, England. In 1803 he produced the world's first steam railway locomotive, and shortly thereafter adapted his engine to driving an iron mill and propelling a barge. The headmaster of his grade school described him as "disobedient, slow and obstinate."

1796 An elephant is first imported into the United States. Just a youngster at two years of age, the elephant came from India and is to be exhibited on Broadway by showman Jacob Crowninshield.

1808 The nation's first temperance society, the Union Temperate Society of Moreau and Northumberland, is established in Saratoga Springs, New York. Members pledge not to drink, except at public dinners.

1892 Radar inventor Sir Robert Watson-Watt is born in Brechin, Scotland.

1905 Physicist Bruno Rossi is born the son of an electrical engineer in Venice, Italy. A pioneer in the study of cosmic radiation, he proved that cosmic rays are protons, and that they have enormous energy (penetrating deep into lead shields). He was one of the first to use rockets to study cosmic rays before they contact Earth's atmosphere; he detected X rays from deep space, which gave rise to the discovery of stars that emit X rays.

1913 Robert Hingson, son of a mule trader, is born in his grandmother's home in Anniston, Alabama. Hingson is known for two inventions: the "pistola de la paz" (the fast, efficient inoculation tool shaped like a pistol) and "continuous caudal analgesia," the first procedure to provide painless childbirth.

1924 Architecture prophet Louis Sullivan is given the first bound copy of *Autobiography of an Idea*, his masterwork on architecture theory. He dies the next day at 67 in Chicago.

1929 Ted and Kermit Roosevelt (sons of ex-President Teddy Roosevelt) simultaneously shoot a giant panda near Yehli, China. They are the first westerners to destroy a panda in this fashion, and the news of their "success" spurs others to hunt this rare and gentle animal.

1970 A liquid oxygen tank aboard *Apollo 13* explodes, causing the crew to report, "We've got a problem here."

1992 At a conference on heart disease in Dallas, four different research teams present studies on why women receive less aggressive treatment than men. The results are split right down the middle: Two studies conclude that the difference is related to women being generally older and in poorer health when heart troubles are reported and doctors don't give them the aggressive procedures; two other studies conclude that a definite sex bias exists in which women are given inferior care just because they are women.

1994 The interconnected water system of the Everglades, Florida Bay, and the nearby offshore coral reefs is brought one step closer to being saved, as Florida's Senate okays state purchase of thousands of acres of farmlands that currently both pollute and block the flow of natural waters.

1743 Thomas Jefferson is born on a plantation in Albemarle County, Virginia. A lifelong scholar and natural scientist, Jefferson collected fossils and tried to create new breeds of plants. As author of the Declaration of Independence, Jefferson had a much greater impact on science and technology.

1775 Benjamin Franklin and medical innovator Benjamin Rush organize the nation's first society to abolish slavery.

1813 The nation's first private psychiatric hospital is founded near Frankford, Pennsylvania, by the Religious Society of Friends. The 52-acre grounds of "The Asylum for the Relief of Persons Deprived of the Use of Their Reason" are five miles from Philadelphia and have no iron shackles, handcuffs, gates, or bars. In 1914 the name was changed to Friends Hospital.

1828 Noah Webster's *American Dictionary of the English Language* is first published.

1866 The great speech teacher Anne Sullivan is born in Feeding Hills, Massachusetts, and the great architect Louis Sullivan dies in Chicago 58 years later (1924).

1886 Edward C. Tolman, a founder of the psychology school of behaviorism, is born in West Newton, Massachusetts.

1894 The first "peep show" (involving film viewed through a vending machine) is put on exhibit at 1155 Broadway in New York City. The machine was the collaboration of two geniuses: George Eastman created the film, and Thomas Alva Edison produced the viewer.

1894 Ecologist Kenneth Alexander Reid is born in Connellsville, Pennsylvania. He became a driving force in the conservation of North American resources, especially water, in the 1900s. A cofounder of the Izaak Walton League, Reid fought for the first federal water pollution legislation, and was behind the creation of the Jackson Hole National Monument (now part of Grand Teton National Park), in which Mount Reid is found.

1912 The *Titanic* hits an iceberg and begins sinking.

1921 Frederick G. Banting, 29, begins a series of experiments on dog pancreases that would eventually bring him the Nobel Prize, and the cure for diabetes.

1629 Physicist-astronomer Christiaan Huygens is born the son of an important bureaucrat in The Hague, the Netherlands. At age 28 he published history's first formal book on probability, but his most famous accomplishments were a series of astronomy discoveries (enabled by his improvements to existing telescopes) and his invention of the first highly-accurate timepiece (1656).

1931 Television is first demonstrated in France.

1956 The first magnetic audiovisual tape recorder (commercial) is demonstrated in Chicago and Redwood, California, by the Ampex Corporation.

1958 *Sputnik II*, the first rocket to carry an animal into orbit around Earth, disintegrates in space. Its passenger, Laika the dog, had died months before; scientists had left her with only a ten-day's supply of air when the ship was launched the previous November, five months earlier.

1964 Rachel Carson, author of *Silent Spring*, dies of cancer at 56 in Silver Spring, Maryland.

1970 Man ventures farther from Earth than ever before (or since). The crew of *Apollo 13* reaches an altitude of 248,655 miles during their harrowing mission around the moon.

1981 The maiden orbital flight of the shuttle *Columbia* (the first reusable spacecraft) ends successfully.

1992 Executives from the seven largest U.S. tobacco companies testify before Congress, depicting smoking as a pleasurable habit, like dessert or a morning cup of coffee.

1452 Leonardo da Vinci is born illegitimately in Vinci, Italy. (For details of his life, see May 2, 1519.)

1707 Mathematician Leonhard Euler, "the greatest man of science that Switzerland has produced," is born the son of a Calvinist minister (who dabbled in mathematics) in Basel. At age 28 he blinded himself in one eye by staring at the sun in an attempt to invent a new way of measuring time.

1741 Charles Willson Peale, artist and founder of America's first major museum (the Peale Museum in Philadelphia), is born in Maryland.

1854 A state legislature earmarks funds ($1000) for the study of insects for the first time in U.S. history. The next month Asa Finch is appointed by the New York State Agricultural Society as the country's first publicly funded entomologist. His inaugural report concerned insects that harm fruit.

1895 Swiss high school teacher Johann Balmer reports in *Annalen der Physik* that light frequencies emitted by hydrogen are all mathematically related to each other. Although curious, the note was largely ignored by science—until it shook the brain of Nobel-winning physicist Niels Bohr, allowing him to understand how electrons are organized in an atom. "As soon as I saw Balmer's formula the whole thing was immediately clear to me."

1912 The *Titanic* sinks.

1920 President Woodrow Wilson commutes the death sentence of Robert Stroud, the "Birdman of Alcatraz," thus allowing him to continue his valuable behind-bars research into bird diseases.

1922 Insulin first becomes available for general use as a treatment for diabetes.

1955 The first modern McDonald's restaurant opens in Des Plaines, Illinois.

1963 History's heaviest accurately weighed snake (a 320-pound python named "Colossus") dies in the Highland Park Zoo in Mifflin, Pennsylvania. Exactly 14 years later (1977) the oldest accurately aged snake (a 40-year-old boa named "Popeye") dies in the Philadelphia Zoo. Pennsylvania apparently is not a good place for eminent reptiles.

1984 The first international congress of robot makers concludes in Albuquerque, New Mexico.

1985 South Africa announces it will repeal laws banning interracial sex and marriage.

1993 The largest-ever fine for dumping garbage at sea has been agreed on, announce federal prosecutors in Miami. Princess Cruise Lines promises to pay $500,000 for a 1991 incident in which a crew member on the passenger ship *Regal Princess* threw at least 20 plastic bags of garbage into the Florida Strait, just a few miles from the Florida Keys. This is also the first U.S. criminal penalty for dumping plastics at sea. The dumping happened in the dead of night, and was luckily seen by a passenger who videotaped it. Not only does the garbage constitute a pollutant, but plastic in the ocean also kills thousands of animals annually.

1994 Pessimism can be fatal and optimism can prolong life, report researchers at a Boston meeting of the Society of Behavioral Medicine. Dr. Daniel Mark of Duke studied heart disease patients, and reports today that patients who felt they would not recover were much more likely to perish than those who had a brighter outlook.

1877 The first flight of a helicopter occurs on this date in Italy. A machine built by Enrico Forlanini is airborne for about 20 seconds, with a maximum height of 42.6 feet (about 13 meters).

1495 Astronomer Peter Apian, the first to describe the shape of comets, is born in Leisnig, Germany.

1660 Naturalist Sir Hans Sloane is born in Killyleagh, Ireland. At death, his enormous collection of books, manuscripts, and artifacts was bequeathed to Britain, and became the collection opened to the public as the British Museum in 1759.

1728 Chemist Joseph Black is born the son of a Scotch-Irish wine merchant in Bordeaux, France. He is best known for rediscovering carbon dioxide and examining its properties.

1786 Sir John Franklin is born in Spilsby, England. A naval officer by training, he participated in the battles of Trafalgar (1805) and New Orleans (1814). His real fame came as an explorer and geographer in trips to Australia and northern Canada. Although he ultimately died on the expedition, his 1845 Arctic exploration proved the existence of the Northwest Passage, a waterway across North America connecting the Pacific and Atlantic Oceans.

APRIL 16

1838 Belgian chemist Ernest Solvay is born in Rebecq-Rognon. Because he was a sickly child, he had little formal education; instead, he indulged his curiosity in chemistry experiments (his father was a salt refiner and his uncle managed a gasworks). At age 23 he created and patented a process to produce sodium bicarbonate, and soon was manufacturing virtually the world's entire supply. He used his vast wealth to endow schools to provide others the education that he lacked.

1850 Marie Tussaud, founder of the world-famous Tussaud's Museum in London, dies at 88 in London.

1867 Aviation inventor Wilbur Wright is born near Millville, Indiana.

1912 Harriet Quimby becomes the first woman to fly across the English Channel.

1943 The hallucinogen LSD affects a human for the first time. The chemist Albert Hoffman of Sandoz Laboratories in Basel, Switzerland, had synthesized the drug five years before in hopes of treating respiratory problems. It didn't show promise for this use, and samples were shelved. On this day Hoffman accidentally gets some on his hands after touching a container; it is absorbed through his skin, into his nervous system. He later recounted the results, "I was forced to stop my work in the laboratory in the middle of the afternoon and to go home, as I was seized by a peculiar restlessness associated with a mild dizziness ... characterized by extreme activity of imagination." Three days later he purposely swallowed some.

1947 The zoom lens (for television cameras) is demonstrated for the first time, by NBC in New York City.

1958 Rosalind Franklin dies at age 37 in London. Although she received relatively few accolades during her brief life, it was her X-ray diffraction studies of DNA that gave Watson and Crick important clues about its structure, the description of which won them Nobel Prizes. Franklin died of cancer shortly after spending a period of convalescence in Crick's home.

1867 Aviation inventor Wilbur Wright is born near Millville, Indiana.

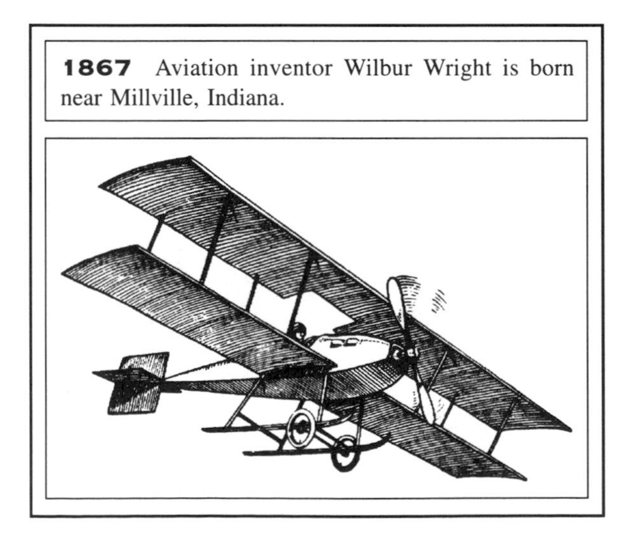

1972 *Apollo 16* departs Florida for the moon.

1976 A rocket makes the closest-ever approach to the sun; the U.S.–German *Helios B* gets within 27 million miles.

1976 India announces a dramatic population-control measure: Citizens will be paid to be sterilized. The policy is very effective; by September 3.7 million people have accepted the offer.

1992 The *New England Journal of Medicine* reports that thalidomide (the drug that caused horrific birth defects in thousands of babies in the 1950s and 1960s) effectively fights graft-versus-host disease, a major complication in bone marrow transplants.

1524 New York harbor is discovered by Giovanni da Verrazano.

1598 Italian astronomer Giovanni Riccioli, the first to observe a double star (two stars so close together that they appear as one), is born in Ferrara. He made his discovery in 1650, looking at the middle star in the handle of the Big Dipper. He also studied the moon extensively and was the first to maintain that it contained no water. The names he gave to various lunar craters survive today. As a Jesuit, Riccioli never accepted Copernicus's heliocentric theories, and his crater names honored anti-Copernicans.

1629 America's first commercial fishing enterprise is established in Medford, Massachusetts.

1790 Benjamin Franklin, author, inventor, scientist, diplomat, and oceanographer (among other occupations), dies in his Philadelphia home at age 84. His last will was very specific about his son William, whose pro-British activities during the Revolution were despised by Franklin; to William he left "all my books and papers which he has in his possession, and all debts standing against him on my account books."

1810 Pineapple cheese is patented by L.M. Norton of Troy, Pennsylvania.

1861 History's first oil well fire breaks out today when the Little and Merrick well (at Oil Creek near Rouseville, Pennsylvania) ignites shortly after gushing. It burned for three days, and left 19 dead.

1863 Geophysicist Augustus Edward Hough Love, discoverer of "Love waves" (deformations of Earth's crust during earthquakes), is born the son of a surgeon in Weston-super-Mare, England.

1866 Human physiologist Ernest Henry Starling is born in London. Among his prolific contributions are Starling's hypothesis (which describes the forces that move fluids in and out of blood vessels), discovering how hormones and nerves control digestion, coining of the term "hormone," and proposing the "law of the heart" (the strength of heart contraction is proportional to the stretching of the muscle).

1880 Archaeologist Sir Charles Woolley is born in London. His excavation of the ancient Sumerian city of Ur in Iraq created a sensation in the 1920s because it involved the earliest great civilization and Biblical events.

1906 The great San Francisco earthquake strikes.

1933 Nuclear physicist Harriet Brooks dies at 54 in Montreal. She was probably the first to observe radioactive recoil (in which a radioactive molecule is thrust backwards as it emits atomic particles).

1958 Toronto's "soiling index" (a measure of air cleanliness recently adopted to warn housewives about the risks they'd be taking if they hung out wash at a particular time) hits a whopping 7.8 at 9:38 AM. This means that the air is nearly eight times dirtier than it would be on a clear, breezy day.

1964 Ford Motor Company unveils the Mustang.

1970 *Apollo 13* splashes down safely in the Pacific, four days after an explosion crippled the craft.

1973 A benchmark in nice weather is set. The Brazilian offshore island of Fernando de Noronha hits a temperature of 90.0°F, the highest temperature there between 1911 and 1990. During the same period, the lowest temperature was 63.9°%. The total range of just 26.1°F is the smallest in history, giving the island the distinction of "most equable temperature."

1975 History's greatest genocide (if defined as percentage of a nation's total population) begins. The Khmer Rouge capture Phnom Penh; over the next four years under the rule of Pol Pot, one-third of the Cambodian citizenry perish.

1492 Columbus receives his commission from Spain's Ferdinand and Isabella to seek a westward passage to Asia.

1675 "Turnip Townshend" is born in England. Although best known for his political achievement (he directed British foreign policy from 1721 to 1730), Lord Charles Townshend earned his nickname by advocating the use of turnips in the newly developed technique of crop rotation.

1781 Joseph Priestley ignites a mixture of hydrogen and common air at his lab in Birmingham, England. He is unimpressed with the explosive power ("Little is to be expected from the firing of inflammable air in comparison with the effects of gunpowder"), and fails to see the great significance of the experiment: that water is produced, showing that water is composed of the combination of gases. It remained for Henry Cavendish several years later to perform the same experiment and earn eternal fame by realizing what was happening.

1838 French chemist Lecoq de Boisbaudran is born in Cognac. He discovered of the elements gallium (1874), samarium (1879), and dysprosium (1886).

APRIL 18

1846 The first telegraph to print letters is patented by R.E. House of New York City.

1857 Historic legal figure Clarence Darrow is born near Kinsman, Ohio (see March 13, 1938, for a biographical sketch).

1865 Monk Gregor Mendel, 42, sends the results of his seven-year study of peas to the eminent biologist Karl Wilhelm von Nägeli in Munich. Within these results are the basic laws of genetics, which the humble Mendel discovered single-handedly in a small garden at the Brünn monastery (now in Brno, Czechoslovakia). Nägeli failed to see the importance of the work; he suggested that Mendel try new experiments with different plants. Nägeli's cool reception, coupled with the failure of the new experiments, were instrumental in Mendel's abandoning serious research.

1911 Physicist Maurice Goldhaber is born in Lemberg, Austria. Hitler's rise caused him to flee to England, then to the United States, where he discovered (1944) that the element beryllium can function as a "moderator," or speed controller, of nuclear chain reactions. This was a critical tool in making the atomic bomb work.

1947 The largest "conventional" explosion during wartime occurs. The Allies explode a charge of 4253 tons to destroy fortifications and U-boat pens in Helgoland.

1955 Albert Einstein dies at 78 in Princeton, New Jersey. Befitting his modest nature, he was cremated without ceremony, and his ashes scattered at an undisclosed location.

1956 An umpire wears eyeglasses in a major league baseball game for the first time. Edwin ("Eddie") Americus Rommel wears them to officiate today's game between the New York Yankees and the Washington Senators, in Washington, D.C. Rommel had been an umpire since 1938; it is unknown how long his eyesight has been defective.

1934 The nation's first laundry opens for business in Fort Worth, Texas. J.F. Cantrell is the proprietor of the "washateria"; it contains four electric machines, which patrons rent for an hour. As in contemporary facilities, hot water and electricity, but not detergent, are provided.

1994 "You don't know how wonderful it feels," says 87-year-old Floridian George Weick, about a successful operation that relieved his hydrocephalus, a rare swelling of the brain that had robbed him of his memory for the last 15 years. He remembers nothing that happened between 1979 and 1993, but he is very grateful to be functioning again. At the time of the operation he was demented, unable to walk or talk or care for himself.

1994 The Australian state of Queensland returns a huge chunk of land to the aborigines and to nature. The property was part of an old pastorage in the state's north, and was purchased from a developer the previous year. About 425,000 acres are to be a national park that will be managed by aborigines; the remainder will be an aboriginal homeland. Elder spokesman Goombra Jacko who remembers state troopers taking his parents from the land in chains, is pleased, "My grandchildren can now look forward to taking care of the home of their ancestors."

1601 Dutch surgeon Henry De Voogt receives a permit to sail solo across the Atlantic. It is unknown if the voyage was attempted. The first documented long-distance solo voyages occurred 300 years later.

1739 John Winthrop of Cambridge, Massachusetts, begins a series of observations on sunspots. He was the first astronomer of note in colonial America. His reports are held in Harvard University Library, but have never been published.

1775 The battles of Lexington and Concord begin the formation of the United States.

1795 Christian Gottfried Ehrenberg, scientific explorer and founder of micropaleontology (which is the study of microscopic life-forms from previous geological eras), is born in Delitzsch, the son of a German magistrate. His career (once he abandoned plans to become a minister) got off to a fast start when he proved in his doctoral research that fungi arise from spores and molds and mushrooms reproduce sexually. Shortly thereafter he took part in a five-year expedition to northern Africa, during which he collected 34,000 animal specimens and 46,000 plant specimens; he was the expedition's only survivor.

1801 Psychophysiologist Gustav Fechner is born the son of a minister in Grossächen, Germany.

1882 Charles Darwin, 73, dies in his country home outside Down, England, at about 4:00 PM. He suffered a heart attack the previous midnight. He is buried next to Newton in Westminster Abbey.

1887 The nation's first kindergarten for blind children, established within the Perkins Institution and Massachusetts School for the Blind, is dedicated.

1892 The country's first automobile that was regularly made for sale is completed and taken on its maiden test-drive by its manufacturer, Charles E. Duryea of Springfield, Massachusetts.

1906 Pierre Curie dies in a tragic carriage accident.

1910 Halley's Comet becomes visible to the naked eye in Curaçao.

1912 Physicist Glenn Seaborg is born in Ishpeming, Michigan. He began college as a literature major, but changed to science under the influence of a great teacher. In 1951 he shared the Nobel Prize for Chemistry for discovering a host of transuranium elements.

1937 The first letter to encircle the world by commercial airmail is dispatched in New York City.

1943 The first deliberate ingestion of the hallucinogen LSD occurs. The drug's synthesizer, Albert Hoffman of Sandoz Laboratories in Basel, Switzerland (who had made it five years before in hopes that it would alleviate respiratory ailments), had accidentally absorbed some through his skin three days before by touching a container of it. This first exposure caused "restlessness," "dizziness," and "extreme activity of imagination." Impressed with the experience, Hoffman decides on this day to swallow some.

1975 India announces that it has launched its first satellite.

1979 The first-ever 18-theater Cineplex opens in Toronto.

1982 The United States announces its first black astronaut will be Guion S. Bluford, Jr., and its first woman astronaut, Sally K. Ride.

1991 *Science* reports a breakthrough genetic treatment in the battles against cystic fibrosis and emphysema, "the two most common lethal hereditary diseases in the United States." Researchers at the National Heart, Lung and Blood Institute in Bethesda, Maryland, transferred a human gene into rat lung cells. These cells then began producing a protein that fights the diseases.

1991 A controversial federal rule banning the feeding of dolphins in the wild goes into effect. The rule pits the welfare of dolphins against the interests of charter boat operators.

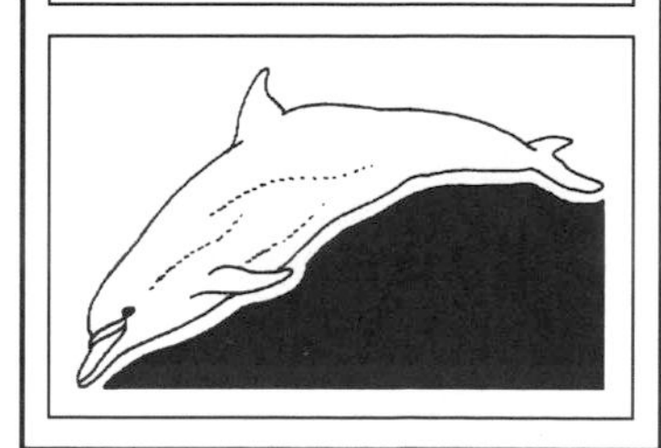

1775 A sad day in ecology history. The day after the American Revolution began, Daniel Boone leads the first group of colonists through the Cumberland Gap to the site of Boonesborough, Kentucky, ready to "tame the west." The Raid on Resources had begun.

1832 A good day in ecology history. Congress establishes the first national nature preserve. The area is 911 acres in Arkansas, named Hot Springs National Park in 1921.

1861 The first oil well fire in U.S. history is finally extinguished. It started three days earlier on the Little and Merrick well on the Buchanan farm at Oil Creek near Rouseville, Pennsylvania.

1862 Louis Pasteur and Claude Bernard complete the first test of pasteurization. The two had sealed dog's blood and urine in glass jars the previous March, and kept them heated at 30°C. On this day, they boldly open the jars for the first time at a meeting of the French Academy of Sciences. It is another great triumph for Pasteur; there has been no rotting or putrefaction of the liquids.

APRIL 20

1888 History's deadliest hailstorm kills 246 in Moradabad, Uttar Pradesh, India.

1889 Hitler, one of history's great psychological mysteries, is born in Braunau, Austria.

1902 Marie and Pierre Curie isolate the radioactive element radium in their Paris laboratory. They processed eight tons of pitchblende rock to extract just one gram of radium.

1906 The oldest marsupial on record, a 22-year-old wombat, dies in the London Zoo.

1923 The first of 33 operations for cancer is performed on Freud's throat and mouth.

1745 Philippe Pinel, the man who "unchained the insane," is born the oldest of seven children of a poor country doctor in Saint-André, France. Prior to his work, insanity was commonly thought to result from demoniac possession; those afflicted were often locked up, fettered, purged, bled, steamed, and blistered. When Pinel became the chief physician of Bicêtre (the Paris asylum for men) in 1792, his first bold reform was to remove the chains from inmates (some of whom had been chained for 40 years) and instituted therapies involving counseling on personal matters, friendly contact and meaningful activity. In 1794 he became the director of the female facility at Salpêtrière and instituted the same changes.

1928 Gerald Stanley Hawkins is born in Norfolk, England. His *Stonehenge Decoded* (1965) established Stonehenge as a prehistoric astronomy observatory, in which the placement of the enormous stones enables the tracking of sun and moon and the prediction of solstices and lunar eclipses.

1940 The electron microscope is first demonstrated, at a meeting of the American Philosophical Society in Philadelphia. The device is 10 feet high, weighs a half-ton, and magnifies objects 100,000 times. It was invented by Dr. Vladimir Kosma Zworykin of the RCA Laboratory in Camden, New Jersey.

1972 Astronauts from *Apollo 16* land on the moon.

1984 The largest permanently installed movie screen (70.5 by 90 feet) opens to the public in the Keong Emas Imax Theatre in Jakarta, Indonesia.

1993 Circuit Judge John Parnham refuses to delay the Pensacola murder trial of Michael Griffin, who stands accused of shooting abortion doctor David Gunn. Griffin had requested the trial be delayed until his "material witnesses" (unborn fetuses) could be born.

1993 The Pentagon reports to Congress that hundreds of once-military, now-civilian sites throughout the nation contain buried weapons (including TNT, live grenades, artillery shells, and chemical agents). Some of the weapons have been buried for 75 years. Officials estimate that simply finding all of the sites could take 15 years and cost $4 billion.

1142 French scholar Peter Abelard dies at about 63 in Chalon-sur-Saône. He is known for founding the University of Paris, for being an early and eloquent advocate of the use of reason, and for his romance with Heloise (a gentle and beautiful woman who was his student; her uncle Fulbert, the powerful canon of Notre Dame, hired thugs to castrate him in 1121).

1782 Friedrich Froebel, the developer of the kindergarten, is born the fifth child of a devout clergyman in Oberweissbach, Germany. Ironically, his own childhood was a disaster. His mother died when he was just nine months old; his father was absorbed with his own career, and his stepmother was cold and indifferent. The young Froebel grew up moody, introspective, and weak in his studies. After a series of short-term jobs, he came on the great turning point in his life—a job at age 24 as the tutor of three sons of a German baron. From then on he devoted himself to the education of young people. He wrote extensively and founded a number of schools. In the mid-1830s, in Blankenburg, he started the Child Nurture and Activity Institute, which he later renamed "Kindergarten."

1838 Naturalist John Muir is born in Dunbar, Scotland. He was largely responsible for the establishment of Sequoia and Yosemite National Parks and for the founding of the Sierra Club.

1843 Anatomist Walther Flemming is born in Sachsenberg, Germany. He was the first to study and describe the process of cell division (and coined the terms "mitosis," "chromatin," and "aster").

1847 The nation's first health insurance company is organized in Boston. It is the Massachusetts Health Insurance Company, and it went out of business shortly after organizing.

1889 The *New York Times* reports that the Panama Canal will never be completed—at least not by engineer Ferdinand de Lesseps and the French company that had begun digging the Canal in 1882 (the Canal was taken over by the U.S. military, and eventually opened in 1914).

1915 Ecologist-essayist Garrett Harden is born in Dallas, Texas.

1925 The Albert National Park is established in the Congo.

1933 The *Vineyard Gazette* (published on Martha's Vineyard, Massachusetts) carries an obituary for the newly extinct heath hen. The last of the species perished on March 11, 1932.

1960 Brasilia, the city cut out of the Brazilian jungle, is inaugurated as the nation's capital.

1972 *Apollo 16* astronauts John Young and Charles Duke roam the surface of the moon.

1976 Full-scale testing of swine-flu vaccine begins in Washington, D.C.

1980 Biological theorist Aleksandr Oparin dies at 86 in Moscow. Prior to his work, it was thought that the first living creatures on Earth were autotrophs (like plants, able to produce their own food internally). Oparin conceived of a long period of "chemical evolution" on Earth prior to the appearance of any living thing. Once life did arise, the environment was plentiful in organic molecules, which the organisms were able to use. His ideas were initially resisted, but are now widely accepted. Laboratory experimenters have proved them workable.

1993 The "best-designed study yet conducted to investigate the link between toxic chemicals and a major disease in women" is published in today's *Journal of the National Cancer Institute*. It shows that high exposure to DDT increases a woman's chance of getting breast cancer by 400%. Although DDT was eliminated as a pesticide in the United States in 1972, the environment and many human bodies still contain it. DDT is broken down very, very slowly by animal bodies, so it tends to accumulate in meat, in dairy products, and in people. Babies are exposed to it through mother's milk, and many countries, including Mexico, still use it widely.

1994 "This is it. We finally have solid, irrefutable evidence that there are planets outside of our solar system," announces Penn State astronomer Alexander Wolszozan, who discovered a cluster of three planets orbiting a star about the size of Philadelphia, in the constellation Virgo.

753 BC Rome is founded.

1451 Queen Isabella I of Spain, Columbus's sponsor, is born in Madrigal.

1724 German philosopher Immanuel Kant is born the son of a saddlemaker in Kaliningrad, a town he never traveled more than 60 miles from during his 80 years. He was trained as a mathematician-physicist, and his *General History of Nature and Theory of the Heavens* (1755) was an important work on astronomy in that it anticipated several cosmic features that were confirmed centuries later; among these was his description of the Milky Way and his realization that it is just one of many "island universes."

1832 J(ulius) Sterling Morton, founder of Arbor Day, is born in Adams, New York.

1838 The first transatlantic crossing under continuous stream power ends in Sandy Hook, New Jersey. The 787-son *Sirius* had departed Queenstown (now Cobh), Ireland, 18 days before.

1846 Under direction of archaeologist Sir Austen Henry Layard, several rafts [loaded with the famous winged bull and many other art treasurers from the ancient Assyrian capital of Calah (now Nimrud)] begin a journey down the Tigris River, from Nimrud, Iraq, to the British Museum. Several sheep are sacrificed to appease local gods.

1889 The Raid on Resources of North America continues. At noon, the Oklahoma Land Rush begins.

1892 The nation's first anatomy research establishment (the Wistar Institute of Anatomy and Biology in Philadelphia) is incorporated. Funding came from a $20,000 bequest by General Isaac James Wistar in memory of his granduncle, Caspar Wistar, physician and author of the first anatomy text by an American.

1904 Physicist J. Robert Oppenheimer is born in New York City. He was scientific head of the Manhatten Project (which designed and constructed history's first two atomic bombs, during World War II).

1909 Neuroembryologist Rita Levi-Montalcini is born in Turin, Italy. She won the 1986 Nobel Prize for Physiology or Medicine for discovering NGF (nerve growth factor), an important chemical in the normal and abnormal development of nerves.

1915 German troops bring several metal cylinders of chlorine gas up to the front lines at Ypres, Belgium, and open the valves. A yellow-green cloud arises, and is wind-blown over French soldiers. 5000 died horribly. It is the first use of modern chemical weapons.

1952 The first atomic blast shown over a network of television stations occurs on this day. The explosion takes place in the Nevada desert, at "News Nob," and is filmed by KTLA of Los Angeles. The picture is relayed to an Atomic Energy Commission station a quarter-mile away, which then transmits it to four other stations.

1970 Earth Day is first observed.

1975 "Critical habitat" is defined in the *Federal Register*, as part of the Endangered Species Act.

1977 An offshore well blows out, sending 8,200,000 gallons of oil into the North Sea.

1980 Nuclear chemist Fritz Strassmann dies at 78 in Mainz, Germany. Early in his career he established a place in history as a codeveloper of rubidium-strontium dating of geological samples. But it was his partnership with Otto Hahn and Lise Meitner, begun in 1934, for which he is most famous. The group was the first to produce energy from nuclear reactions, when they bombarded, and split, uranium with neutrons in 1938. They had split the atom; it was nuclear fission; it opened the door to tapping nuclear energy.

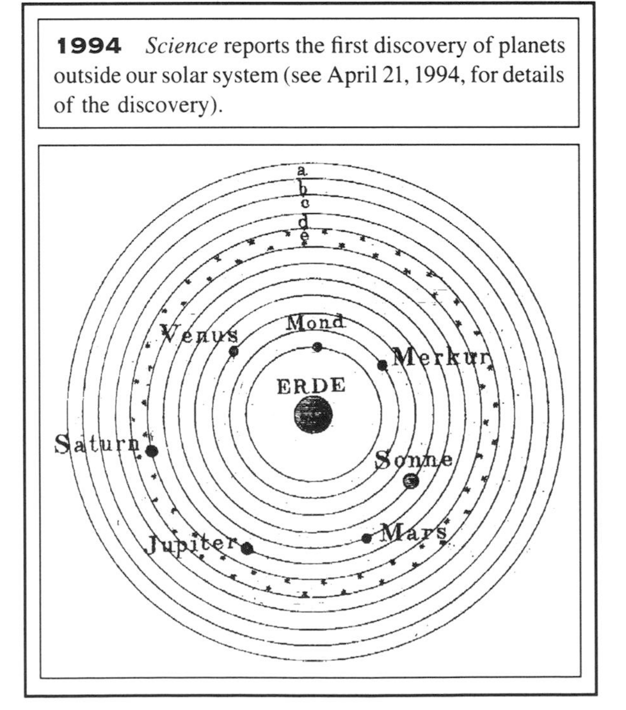

1994 *Science* reports the first discovery of planets outside our solar system (see April 21, 1994, for details of the discovery).

1827 One of the great works on optics, "Theory of Systems of Rays," is presented to the Royal Irish Academy in Dublin by William Rowan Hamilton (just 21 and still an undergraduate). It presented a single function that unified mechanics, optics, and mathematics, and it helped establish the wave theory of light (which states that light is an energy form that travels in waves).

1843 Disheartened at the nation's growing urbanization, and at the destruction of nature and natural living that this entails, Frederick Law Olmsted sets sail from New York City as a deckhand on the *Ronaldson.* It is three days before his 21st birthday. During his four years abroad, he was especially impressed with landscaping in England. Eventually Olmsted became one of history's greatest designers of city parks. Perhaps his greatest triumph was New York City's Central Park, but he also designed the grounds around the Capitol in Washington, D.C., the grounds of Stanford University, and the park around Niagara Falls.

1858 Max Planck, the founder of quantum physics (he even coined the term "quantum" in physics), is born the son of a civil law professor in Kiel, Germany. His Nobel Prize came in 1918. His greatest insights came in 1900; physics before this date is now known as "classical physics," while that after this date is now called "modern physics."

1867 The Zoetrope, the first device to show "moving pictures," is awarded patent No. 64,117 on this day. This "Wheel of Life" was invented by William E. Lincoln of Providence, Rhode Island, who assigned his patent to Milton Bradley & Co. of Springfield, Massachusetts. The device consisted of a cylinder with a sequence of figures drawn on the inside wall; these figures appeared to move when viewed through a slit as the cylinder was spun around.

1901 Edmund Brisco Ford, one of the first to join genetics and ecology, is born in Papcastle, Britain.

1969 A female gray seal is shot in the Shetland Islands, Scotland. Dental rings indicate she is at least 46, making her the oldest pinniped on record.

1970 Invented by Professor Aime Limoge, electrical anesthesia is first used in obstetrics, at the Rothschild Hospital in Paris.

1971 The Reserve Mining Company, one of the greatest polluters of the Great Lakes, receives notice that it must eventually halt dumping waste into Lake Superior.

1976 The drug/chemical company F. Hoffmann-La Roche of Basel, Switzerland, sets a record for highest value of stock. On this date, one share is worth $38,486.

1985 After 99 years of unrivaled success, the Coca-Cola company announces that it is changing the secret formula of the world's best-selling soft drink. Poor sales and poor public reaction later convinced the company to bring back the original formula.

1992 Doctors announce in Rome that a 62-year-old woman has become pregnant, through a test-tube fertilization that united her husband's sperm with an egg of a 30-year-old donor. The fertilized egg was then successfully implanted in the 62-year-old Concetta Ditessa. The birth will call for a rewriting of history; the *Guinness Book of World Records* lists the previous oldest age of birth at 57.

1994 Dirt is ceremonially tossed into "the big ditch" by state and federal officials, signaling the start of a $372 million effort to restore Florida's Kissimmee River to the condition it was in before the U.S. Army Corps of Engineers "improved" it for agriculture in the 1960s. It is the nation's biggest-ever river-restoration project. Observes Senator Bob Graham at the ceremony, "When we try to manipulate nature, we often end up doing more harm than good."

1896 The first motion picture screening in a theater (also the first successful screening of a movie for a paying audience) occurs today at Koster & Bial's Music Hall on 34th Street in New York City. It is presented by Thomas Armat in association with inventor Thomas Alva Edison. Shown are several ballet scenes, a burlesque boxing match, surf breaking on a shore, and a comic allegory entitled *The Monroe Doctrine*. Edison's device, the Vitascope, presents the films. Edison is in attendance, and the audience calls for him to take a bow, but he does not.

1284 King Alfonso X, astronomer and science patron, dies at 62 in Seville, Spain.

1833 The soda fountain is patented by Jacob Ebert of Cadiz, Ohio.

1873 The nation's first free lunch is offered. The New York Diet Association Kitchen opens this day on 23rd Street in New York City, for the benefit of sick poor people.

1895 The first ship to circumnavigate the globe with a one-man crew sets sail on this day from Boston. *The Spray* is a sloop, 34 feet 9 inches long, weighs 9 tons, and cost $553.62. Captain Joshua Slocum completed the 46,000-mile trip on July 3, 1898.

1905 Conservationist-prize-winning author Robert Porter Allen is born in South Williamsport, Pennsylvania. He is credited with saving the whooping crane from extinction by discovering the nesting grounds of the last remaining flock near the Arctic Circle in 1955.

APRIL 24

1915 The second use of modern chemical weapons begins. Germans float chlorine gas over Canadians in Belgium.

1925 This is the day in Dayton, Tennessee, that John Scopes "unlawfully did willfully teach in the public schools ... a certain theory and theories [i.e. Darwin's theory of evolution] that deny the story of the divine creation of man." Scopes is later prosecuted in the famous "Monkey Trial," which became the subject of the play and movie *Inherit The Wind*. In fact, Scopes was a physics teacher who was teaching biology as a favor to a sick friend, and he was not even in school the day evolution was taught.

1962 The first satellite relay of a television signal occurs between California and Massachusetts. MIT scientists bounce signals off the two-year-old orbiting balloon *Echo I*. The picture is poor but recognizable.

1964 Gerhard Domagk dies at 68 in Burgberg, Germany. In 1939 he was awarded the Nobel Prize for Physiology or Medicine for discovering the antibiotic properties of the first "miracle drug," a sulfa compound called Prontosil. Great drama surrounded his award. Because World War II had already started and because Domagk was German, there was considerable opposition to his nomination within the Nobel awarding committee at the Caroline Institute in Sweden; eventually a majority voted for him because of the importance of his discovery (it saved many lives during the war). The Nazi government, however, informed the committee that the award was not wanted. The Nobel committee persisted, and awarded the Prize to Domagk anyway. He was caught in the middle. He wrote to the committee, thanking them and saying he would have to wait for further government action before he could accept. His government acted: He was arrested twice at gunpoint, interrogated, and manhandled in prison. He decided not to accept the Prize. It was the first time in history that a Nobel Prize was refused.

1967 The first undisputed astronaut fatality. Colonel Vladimir Komarov, 40, perishes in *Soyuz 1*.

1990 *Discovery* blasts off from Cape Canaveral with the Hubble Space Telescope.

1990 Joe Junior Cowan, a paranoid schizophrenic with an extensive history of mental treatment, violently assaults U.S. Forest Service employee Maggie Doherty, at the remote Lolo Work Center in Missoula County, Montana. Cowan's lawyers tried to use the insanity defense by challenging Montana's abolition of this defense. In 1994 the case reached the Supreme Court, which upheld Montana's position and ruled that individual states are free to abolish the insanity defense.

1766 Robert Bailey Thomas, founder and first editor of *The Old Farmer's Almanac* (1792), is born in Grafton, Massachusetts.

1800 The Library of Congress is established with the appropriation of $5000 "for the purchase of such books as may be necessary."

1661 Alcoholic beverage merchants are first licensed in England.

1792 The guillotine is first used on a living human (a highwayman named Nicolas Jacques Pelletier) in Paris. The device was invented in 1788 by the eminent surgeon Antoine Louis, and tested on corpses and sheep. It is named for Joseph Ignace Guillotin, a politician who lobbied for its use because he felt it would alleviate suffering. It "alleviated the suffering" of 2498 during the French Revolution.

1853 U.S. army surgeon William Beaumont dies at 67 in St. Louis. His place in science history began on June 6, 1822, when he was summoned to Michilimackinac (now in Michigan) to treat a 19-year-old trapper, Alexis St. Martin, who was near death after an accidental shotgun blast ripped a gaping wound in his abdomen. The wound never completely closed; it developed a flap of skin that, when depressed with a finger, allowed Beaumont to directly observe and experiment on activity within the stomach.

1859 Ground is first broken for the Suez Canal.

APRIL 25

1865 Colonel Edward A.L. Roberts of New York City receives patent No. 47,458 for a process that drills oil wells with torpedoes. His method had been tried experimentally the previous January at the Ladies' Well on Watson Flats near Titusville, Pennsylvania.

1898 The United States declares war on Spain. A side effect of the conflict was a cure for yellow fever (see February 15, 1898, for the connection).

1900 Wolfgang Pauli is born the prodigy son of a chemistry professor in Vienna. Winner of the 1945 Nobel Prize for Physics, Pauli "was impossibly clumsy with his hands and was a poor and stumbling lecturer; but it was his brain that was nonpareil" (Isaac Asimov).

1901 License plates are required by law for the first time. New York State passes legislation that requires car owners to register with the secretary of state within 30 days of the law's enactment. The license plates bore the owner's initials, and had to be over three inches high.

1953 *Nature* publishes the structure of DNA. The one-page article by James Watson and Francis Crick begins modestly, "We wish to suggest a structure ... for the salt of DNA." They were awarded the Nobel Prize nine years later. Once its structure was known, the process that DNA uses to pass hereditary information from cell to cell and from generation to generation became obvious.

1954 The first solar-powered battery is announced in New York City by Bell Telephone Laboratories.

1959 The earliest known case of AIDS meets medical science. David Carr walks into the Royal Manchester Hospital (England), and dies mysteriously 4½ months later. He was diagnosed posthumously in the 1980s through preserved tissue samples.

1959 The St. Lawrence Seaway opens to shipping, on the 100th anniversary of the Suez Canal start.

1983 *Pioneer 10* crosses Pluto's orbit, on its endless voyage through the cosmos.

1992 Wire services report an incident in which a turkey shoots a hunter. Larry Lands of Potosi, Missouri, assumed he'd killed the turkey, and stuffed the bird into the trunk of his car. When he went to show off his trophy, the animal started thrashing and triggered a loaded shotgun. Lands was wounded in the leg. On top of that, Lands and his son were hunting out of season, and will be fined.

1994 Nabisco, the nation's largest cookie and cracker company, introduces low-fat Oreos. Nutritionist Jayne Hurley is not impressed. "Reduced-fat versions of fatty foods don't make them healthy. These cookies are still fatty cookies. This doesn't give you license to eat a box."

1874 Radio inventor Guglielmo Marconi is born in Bologna, Italy.

1558 Jean François Fernel dies at 61 in Paris after a historic career in medicine and physiology. He coined the terms "physiology" and "pathology," introduced dissection to clinical practice, and was the first to describe appendicitis and peristalsis.

1607 Captain John Smith goes ashore at Cape Henry, Virginia, to found the first permanent English settlement in the New World.

1822 The great landscape architect Frederick Law Olmsted is born in Hartford, Connecticut.

1826 On his 41st birthday, John James Audubon leaves his wife and son in Bayou Sara, Louisiana, and sets sail for England in hopes of finding a publisher for his paintings of birds and mammals.

1826 The nation's first engineering college, the Rensselaer School in Troy, New York, graduates its first ten students.

APRIL 26

1829 Theodor Billroth, founder of modern abdominal surgery (and the first in continental Europe to use antiseptic surgery), is born in Bergen, Germany.

1848 The adventure of a lifetime begins. Alfred Russel Wallace and Henry Walter Bates, both in their early 20s, depart Liverpool "in a small trading vessel" for the unexplored Amazon, planning to finance the trip by collecting and selling animal specimens to museums and private collectors in England. Both men later made everlasting contributions to biology theory.

1879 Physicist Sir Owen Richardson is born in Dewsbury, England. His exploration of electron-emission by heated substances earned him the 1928 Nobel Prize, and enabled others to develop radio and television tubes.

1882 Charles Darwin is buried next to Newton in Westminster Abbey.

1900 Charles Richter, creator of the Richter earthquake scale, is born in Hamilton, Ohio.

1908 The first International Psychoanalytic Congress convenes in Salzburg, Austria.

1921 The first U.S. radio broadcast of the weather occurs on this day from St. Louis. The broadcast is made for the federal government by station WEW.

1922 At noon a horrific earthquake rips through Tokyo. The only large building that remains safe and intact is the Imperial Hotel, proving architect Frank Lloyd Wright's design to be truly revolutionary.

1933 Astrophysicist Arno Allan Penzias, explorer of radio waves from outer space (and Nobel laureate for Physics in 1978), is born in Munich, into a Jewish family ten weeks after Hitler seized power in Germany.

1785 John James Audubon is born illegitimately on a tropical island (Haiti).

1942 History's worst mining disaster occurs in Benxihu, China. A coal dust explosion kills 1549.

1954 Mass testing of the Salk polio vaccine begins. About 1,830,000 children will be involved.

1986 The Chernobyl nuclear disaster begins at 1:23 AM. in Pripet, Russia.

1989 History's deadliest tornado kills some 1300 in Shaturia, Bangladesh.

1993 The world's largest animal preserve is under development, according to an official Chinese announcement today. Set aside will be 80,000 square miles in the Qangtang grassland in northern Tibet, currently home to 80 species of rare animals, including blue sheep, yaks, and snow leopards.

1521 Ferdinand Magellan, 41, is killed by natives on Mactan Island in the Philippines during the first circumnavigation of Earth.

1791 Artist-telegraph inventor Samuel Morse is born the son of a minister in Charlestown, Massachusetts.

1820 English sociologist-philosopher Herbert Spencer is born the son of a schoolteacher in Derby. His magnum opus *The Synthetic Philosophy* (1896) combined biology, psychology, morality, and sociology. Spencer was one of the first to embrace Darwin's theory (indeed, Spencer was uniquely responsible for popularizing the term "evolution," which Darwin seldom used), but Spencer's misapplication of its principles to political/social phenomena led him to create the controversial doctrine of "social Darwinism."

1867 *Lancet* publishes the last of four articles by Joseph Lister that revolutionized surgery by introducing antiseptic methods.

1880 The first electric hearing aid is awarded patent No. 226,902 on this day to Francis D. Clarke and M.G. Foster for a "device for aiding the deaf to hear." The machine operates by bone conduction, in which sound is passed through the skull by bone vibration. The first commercially produced hearing aid was the Acousticon, which appeared in 1901.

1887 George Thomas Morton (son of William Thomas Green Morton, who gave the first successful public demonstration of surgical anesthesia in 1846) saves the life of a 26-year-old man with acute appendicitis by performing the first-ever appendectomy, in Philadelphia.

1896 Chemist Wallace H. Carothers, inentor of nylon, is born the son of a college administrator in Burlington, Iowa.

1900 Walter Lantz, creator of Woody Woodpecker (1941) and the penguin Chilly Willy (1954), is born in New Rochelle, New York.

1913 Nuclear chemist Philip Hauge Abelson is born in Tacoma, Washington. A codiscoverer of the element neptunium, Abelson conceived the "thermal diffusion" principle for creating the enriched uranium needed in atomic bombs.

1968 Abortion becomes legal in Great Britain, through the Abortion Act (passed the previous October). The bill was introduced by the liberal MP David Steele, and had been defeated on seven previous occasions. The first country to legalize abortion was Russia, in 1920.

1970 The nation's first automated "tellerless" bank opens on this day in Los Angeles. It is the Civic Center branch of the Surety National Bank. It contains a money-dispensing machine and six stations (viewed by closed-circuit televisions, which a teller does monitor) that validate checks by code rather than signature.

1982 The trial of John Hinckley, Jr. (would-be assassin of President Reagan) begins in Washington. His eventual acquittal by reason of insanity made legal and psychiatric history.

1991 Newspaper services introduce Americans to the latest craze in Japanese cuisine: the eating of live fish and eels. "The food moves around a lot—that's the whole idea," claims Sunao Uehara, chef of a well-known Tokyo seafood restaurant. Shrimp, flounder, and lobster are among the live offerings.

1871 The American Museum of Natural History first opens its doors to the public, in New York City.

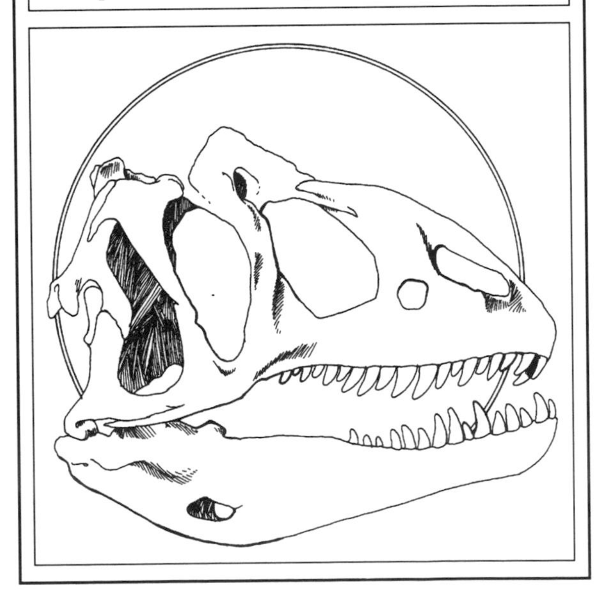

1774 Francis Baily, discoverer of Baily's beads (brilliances on the moon's surface at the start and finish of total solar eclipses), is born in Newbury, England. Despite only an elementary school education, he became a prosperous stockbroker; he retired at 51 to devote himself to his real love, astronomy.

1855 The first U.S. college of veterinary medicine, the Boston Veterinary Institute, incorporates. Its first president is medical doctor Daniel Denison. Among the offered courses are chemistry, anatomy, physiology, surgery, and the theory and practice of medicine.

1858 Johannes Peter Müller dies at 56 in Berlin, Germany, after a historic career in animal physiology and anatomy. Because he suffered from manic depression through his life, some scholars speculate that he died by his own hand. One of his earliest triumphs was an enormous study of human and animal vision, during which he realized that each sense organ responds in its own characteristic manner, with its own specific energy, to specific external stimuli. This finding was important to philosophy as well as science. Müller also did important work on nervous systems, reproductive organs (he discovered the Müllerian duct), blood, and cancer (his 1838 microscope study of cancerous cells was the foundation for the modern discipline of pathological histology).

1882 Pasteur okays the final design of the experimental test of his anthrax vaccine.

1900 Dutch astronomer Jan Hendrick Oort is born the son of a physician in Franeker. In 1927 he convinced science that our galaxy rotates around its center, and in 1930 he presented the now-accepted estimate of its size. In the 1950s radio telescopy allowed his team to map galaxy movements.

1906 Mathematician Kurt Gödel ("Gödel's proof" states that all mathematical systems contain unresolvable paradoxes) is born in Brno, Czechoslovakia, and astronomer Bart Bok ("Bok globules" are thought to be stars forming) is born in Hoorn, Holland.

1926 Nuclear physicist Erwin Schrödinger coins the term "wave mechanics" in a letter to Einstein, to describe the newly emerging branch of physics that interprets the behavior of subatomic particles according to the mathematics of wave motion.

1932 The first yellow fever vaccine for humans is publicly announced at a Philadelphia meeting of the American Societies for Experimental Biology. It was developed by Drs. Stuart Kitchen, Wray Devere Marr Lloyd, and Wilbur Augustus Sawyer.

1947 The balsa raft *Kon Tiki* departs Peru with a crew of six. They reach Polynesia 101 days later, supporting the theory of the raft's captain, anthropologist Thor Heyerdahl, that ancient Polynesians may have originated in South America.

1962 History's greatest oil flare (burning for five months with flames 450 feet high in the Algerian Sahara) is finally extinguished, by "Red" Adair with 550 pounds of dynamite.

1986 Russia announces the Chernobyl nuclear disaster.

1991 Space shuttle *Discovery* departs Earth with a crew of seven to perform "Star Wars" defense research.

1673 Although their countries are at war, Reinier de Graaf writes from Delft, the Netherlands, to England's Henry Oldenburg (Secretary of the Royal Society) about discoveries that Antonie van Leeuwenhoek has been making with his homemade microscope. Oldenburg then writes directly to Leeuwenhoek; it is the start of a 50-year correspondence between Leeuwenhoek and the Royal Society. During his lifetime of discovery, Leeuwenhoek was the first to see one-celled animals and human sperm.

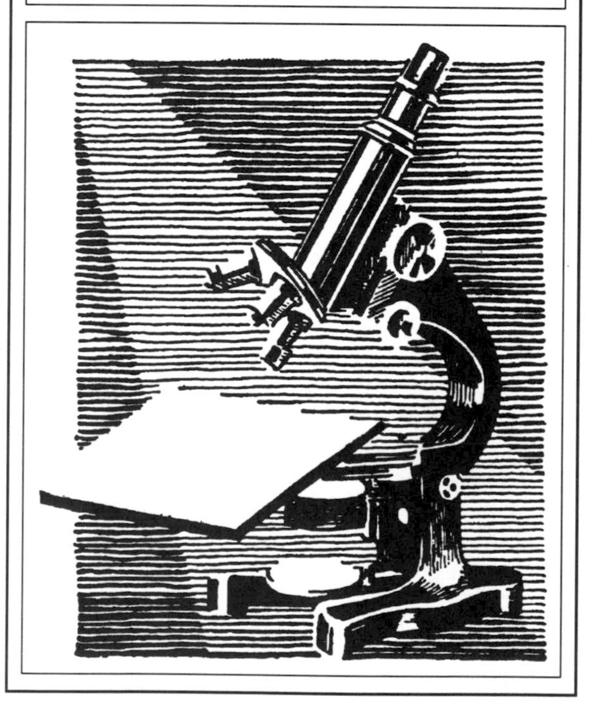

1699 The French Academy of Sciences holds its first public meeting, in the Louvre.

1813 The first U.S. patent on a rubber product is granted on this day to Jacob Frederick Hummel of Philadelphia for a "varnish of elastic gum to render water-proof" shoes and other things.

1854 Jules-Henri Poincaré, "the last of the universal mathematicians," is born in Nancy, France. As well as making important contributions to most branches of math, Poincaré produced respected theories in astronomy, Earth science, and creativity.

1872 Astronomer Forest Ray Moulton is born in Le Roy, Michigan. He is best known for developing the planetesimal theory of how our solar system formed (this theory hypothesizes that another star came very close to our sun sometime in the distant past, which caused matter to be ripped free from both stars; this matter first coalesced into small fragments called "planetesimals," which further coalesced into our present-day planets).

1893 Chemist Harold Urey is born the son of a schoolteacher/lay minister in Walkerton, Indiana. He is famous for theories on the origin of life on Earth, and for producing "heavy hydrogen" (which won him the 1934 Nobel Prize and which has the potential to destroy much of life on Earth as a key ingredient in hydrogen bombs).

1898 Funds are appropriated in New York State for the nation's first cancer laboratory.

1921 David Sarnoff is made general manager of RCA. It begins a communications revolution.

1925 Dr. Florence Rena Sabin, histology professor at Johns Hopkins University, becomes the first woman elected to the U.S. National Academy of Sciences.

1937 Chemist Wallace H. Carothers, the creator of nylon, commits suicide in Philadelphia two days after his 41st birthday. Carothers led a team of Du Pont chemists that created nylon in 1938, after years of experimenting with man-made fibers. It revolutionized the synthetics industry once it hit the market. Unfortunately, Carothers never saw the success; he was already gone by the time researchers had created a process to commercially produce nylon in 1938.

1945 The "Peace" rose is presented at the Conference of American Rose Growers in Pasadena, California. It becomes a huge success in the United States and Europe. It was the first flower ever patented in France, and was created by Antoine Meilland.

1949 John Talbot Robinson discovers the first fragments of the hominid *Telanthropus capensis* in a limestone crevasse in Swartkrans, South Africa. Resembling *Homo erectus*, the species was another of the semihuman forms that roamed Earth prior to man.

1976 In a groundbreaking right-to-die case, after a series of court battles won by her family, Karen Ann Quinlan is taken off life-support systems at a hospital in Denville, New Jersey. She survived, in a coma, for ten more years.

1993 Flooded with huge payments for earthquakes, hurricanes, pollution, and asbestos, Lloyd's of London announces that it is changing its liability policy for the first time in its more than 300-year history. Investors had previously been liable for everything they owned; but a number of recent natural and man-made disasters (including Hurricane Andrew and the *Exxon Valdez*) had bankrupted so many of Lloyd's backers that the company's directors felt a change was necessary, to a limited-liability structure.

1994 Scientists at Northwestern University report in today's *Science* that they have discovered a gene in a mammal (the mouse) that regulates the body's internal "biological clock."

1913 The zipper is patented by Gideon Sundback of Hoboken, New Jersey.

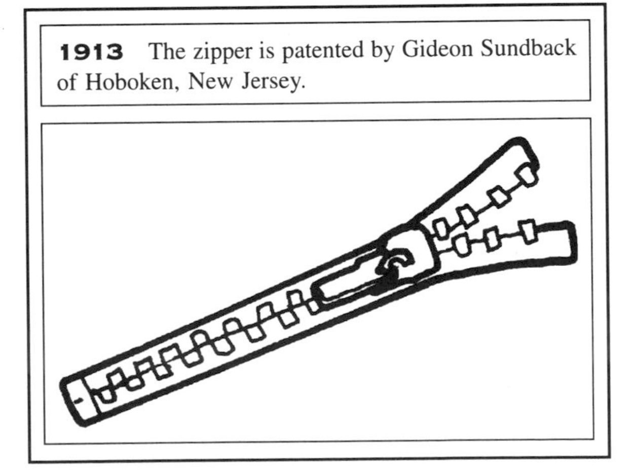

1770 David Thompson, the first European to explore the Columbia River from source to mouth, is born in London. A trapper and geographer, his maps of the western United States formed the basis for all subsequent ones.

1777 Carl Friedrich Gauss, the Mozart of mathematics, is born in Brunswick, Germany (see February 23, 1855, for a biographical sketch).

1796 The nation's first patent for a pill of any kind is granted on this day to Connecticut's Samuel Lee, Jr., for a "composition of bilious pills." Commercially his products were called Lee's Windham Pills and Lee's New London Bilious Pills.

1878 Still under attack by some surgeons, Pasteur deliverys a passionate, celebrated lecture to the French Academy of Medicine on the germ theory of diseases (which states that many diseases are caused by tiny organisms). Denouncing "those [opposing] opinions fatal to medical progress," Pasteur describes methods to prevent infection, and presents a simple experiment that skeptics can use to prove his theory to themselves.

1894 The ship *Dochra* sights an Antarctic iceberg fragment at a latitude equal to that of Rio de Janeiro. No iceberg has ever been sighted so close to the equator before or since.

1897 J. J. Thomson announces discovery of the electron. He calls it a "corpuscle" at today's meeting of the Royal Society of London.

1916 Mathematician Claude Elwood Shannon is born in Gaylord, Michigan. In 1941 he went to work at Bell Telephone Laboratories to increase the efficiency of information transmission. He developed an analysis (published in 1949) in which the "bit" is the fundamental unit of information, and all aspects of communication (including noise and redundancy) can be analyzed mathematically. It was the birth of information theory.

1932 The Seeing Eye, the nation's first organization to train dogs to assist the blind, is incorporated in Morristown, New Jersey. The first U.S. Seeing Eye dog was "Buddy," a shepherd imported from Vevey, Switzerland, by Morris S. Frank.

1939 Cosmic rays are harnessed for the first time to produce electricity. The rays are trapped by a Geiger-Mueller counter at the Hayden Planetarium in New York City, turned into an electric current, and sent to the World's Fair Grounds in Flushing Meadows to operate switches on colored lights at the Lagoon of Nations. Also at today's opening of the Fair, Franklin Roosevelt becomes the first president to be seen on television.

1960 The oldest bat on record is found in a cave on Mount Aeolis, Vermont. It is a female little brown bat (*Myotis lucifugus*), at least 24 years old, and had been banded on June 22, 1937, in Mashpee, Massachusetts.

1877 The first known design for a phonograph is deposited in sealed papers with the French Academy of Sciences by poet-scientist Charles Cross.

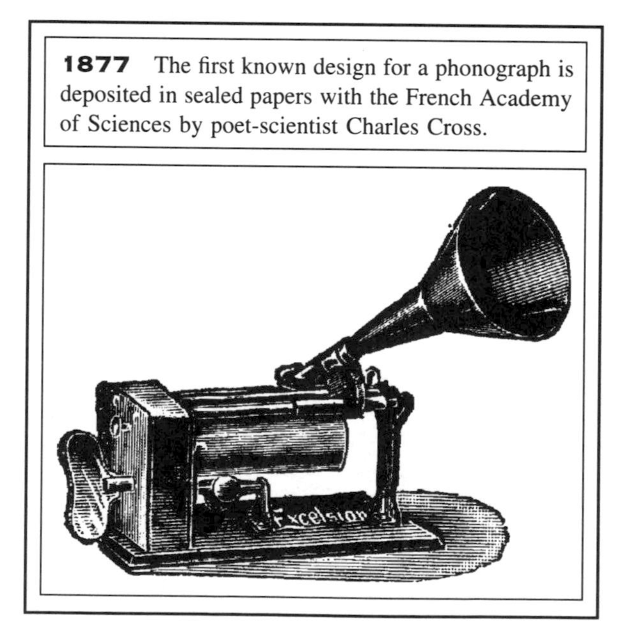

1993 German physicist Hans Schlegel becomes the first human to be infused in outer space. The "landmark experiment" takes place aboard the shuttle *Columbia*; it involves pumping saline solution into the astronaut's body through a needle, to relieve the common space problems of dehydration, puffy faces, and skinny legs. "The infusion of saline itself isn't a very big problem because the liquid is heated up to body temperature prior to infusion," reported Schlegel. "The only thing you feel after one or two hours is that you have the urgent need to see the toilet."

1993 The world's largest animal, the blue whale, is making a comeback (after nearly being hunted into extinction by commercial whalers), according to wire service reports of a new federal study. The International Whaling Commission banned the killing of both the blue and humpback whales in the north Pacific in 1966; since then the number of individuals has more than quadrupled.

1839 French chemist Louis-Marie Chardonnet, inventor of rayon (the first commonly used artificial fiber), is born in Besançon. The new material created a sensation at the Paris Exposition of 1891, where it was called "Chardonnet silk."

1840 The first state school for the blind was the Ohio Institution for the Blind, established and opened in 1837. On this day William Chapin begins his job as the Institution's first superintendent. He held the post for six years. In 1903, the facility changed its name to the Ohio State School for the Blind.

1906 Marie Curie becomes the first female professor at the Sorbonne in Paris in its 650-year history.

1913 The Henry Phipps Psychiatric Clinic admits its first patient. Operating under the auspices of the Johns Hopkins Hospital in Baltimore, the Clinic became one of the world's foremost research and teaching facilities in psychiatry. Its first director was the legendary Adolf Meyer.

1925 M. Scott Carpenter is born in Boulder, Colorado. Not only was he one of the first humans in space (1962, as one of the original seven Mercury astronauts), but he was also one of first humans to live under the ocean surface for an extended period of time (1965, as one of the aquanauts in *Sealab II* off the California coast).

1928 Sir Ebenezer Howard, founder of the English garden-city movement, dies at 78 in Welwyn Garden City, England. Famed entomologist Leland Ossian Howard dies at 92 in Bronxville, New York, exactly 22 years later.

1931 The 102-story Empire State Building is dedicated in New York City by President Herbert Hoover. It is the first building taller than 1250 feet.

1958 Discovery of the Van Allen radiation belts around Earth is published in the *Washington Evening Star*, following a report by James A. Van Allen to a joint symposium of the National Academy of Sciences and the American Physical Society in Washington, D.C.

1961 The first commercial U.S. airplane to be hijacked is a National Airlines twin-engine Convair CV 440, taken on this day mid-flight on its way from Miami to Key West. The hijacker is Antillo Ortiz, using the alias Elpirata Cofresi ("The Pirate Cofresi"); he is armed with a pistol and knife. Cofresi, the three crew members, and the eight passengers fly to Havana. All hostages are then freed, and the plane returns to Key West 4½ hours later.

1988 *Newsweek* magazine reports that astrology was used to make scheduling decisions for President Ronald Reagan. The claim was made by former chief of staff Donald Regan.

1991 Charles Osborne (1894–1991) dies in Anthon, Iowa, having just ended a hiccuping fit that started in 1922.

1993 The Florida Audubon Society releases its 100th rehabilitated bald eagle back into the wild. The released animal, "Helen," had suffered internal injuries and a broken collarbone during a territorial fight the previous October. She was nursed back to health at the Audubon Center for Birds of Prey in Maitland, and released near Lake Dalousie.

1994 "Tempest in a B cup." Europe's Wonderbra goes on sale in the United States. The undergarment accentuates the female figure by forcing the bust upward and inward simultaneously.

1852 Histologist Santiago Ramón y Cajal is born the son of a doctor in Petilla de Aragón, Spain. Blessed with exceptional creativity, great powers of observation, and the sensitivity of an artist, he developed new stains and microscope techniques that allowed major advances in our understanding of how the nervous system is built. He was awarded a Nobel Prize in 1906. Born on May 1, 1852, Ramón y Cajal was a poor student as a youth, and served apprenticeships with a barber and a shoemaker before turning to science.

1519 Leonardo da Vinci dies at age 67 in Cloux, France. A leading figure of the Renaissance, he combined great talents as an engineer and an artist in producing unprecedented studies of the structure and function of animals and plants. Leonardo's numerous dissections of humans and animals were severely criticized by the Church, which may be one reason he kept all of his notebooks in code, written backwards and in Latin. This code prevented his scientific discoveries from having any impact on others until long after his death.

1601 German scholar Athanasius Kircher is born the youngest of six sons in Fulda. A number of minor inventions and discoveries are attributed to Kircher (he was once lowered into Mount Vesuvius shortly after it erupted in order to study its features), but his great contribution was as a disseminator/reporter of knowledge. He wrote over 44 books, compiling knowledge from Europe and the far-flung areas visited by Jesuit missionaries (Kircher was a Jesuit himself). He has been called "the last Renaissance man."

MAY 2

1775 Benjamin Franklin completes the first scientific study of the Gulf Stream. He began the study in 1769 as deputy postmaster of the British colonies trying to explain why ships took two weeks longer to bring mail from England than was required in the opposite direction. Franklin became the first to chart the Gulf Stream.

1800 English chemist William Nicholson builds one of the first batteries. He is the first to attach wires to the two poles of the battery and place these wires in water. Bubbles of oxygen and hydrogen are released as the current flows through the liquid. He has created electrolysis; it is the first time that electricity has been shown to produce a chemical reaction.

1855 The first U.S. company to produce zinc commercially incorporates on this day in Bethlehem, Pennsylvania. The Pennsylvania and Lehigh Zinc Company mill was built in 1853 by Samuel Wetherill; it produced the metal from calamine ore. Wetherill received a number of patents for his innovations in zinc processing.

1860 Physiologist Sir William Maddock Bayliss is born in Wolverhampton, England. He is most famous for his collaboration with Ernest Starling, who died exactly 67 years after Bayliss was born at 61 in Kingston Harbor, Jamaica. Among his prolific contributions are Starling's hypothesis (which describes the forces that move fluids in and out of blood vessels), discovering how hormones and nerves control digestion, coining of the term "hormone," and discovering the "law of the heart" (the strength of heart contraction is proportional to the stretching of the muscle). He and Bayliss were a famous research team, credited with discovering hormones. Bayliss married Starling's sister in 1893.

1866 Dr. Jesse William Lazear, a tragic martyr in the battle against yellow fever (see September 26, 1900), born in Baltimore County, Maryland.

1939 Lou Gehrig (namesake of a now-famous disease) does not play in today's game between the New York Yankees and the Detroit Tigers, which Gehrig's Yankees win 22–2. Gehrig's consecutive games-played streak ends at 2130. He never plays again.

1993 A robot in space (in the shuttle *Columbia*) catches a floating object (an inch-long aluminum die) for the first time in history. The 2½-foot-long robot arm, called "Rotex," missed the die on its first try, but snagged it successfully on its second. Through much of the mission, astronauts have been working in near darkness, and have turned off computers and other appliances not in use, to save energy. Their strategy worked, as NASA also announces today that the flight can be extended for one day. The astronauts are thrilled.

1993 One of history's most bizarre, most tragic cases of group mind-control nears its final chapter; government officials announce that they have identified the body of cult leader David Koresh among the ashes of the Branch Davidian compound that burned to the ground after a 51-day siege by federal agents outside Waco, Texas. Koresh had apparently been shot in the forehead.

1903 Dr. Benjamin Spock, the first to be fully trained in both psychiatry and pediatrics, is born in New Haven, Connecticut. His *The Commonsense Book of Baby and Child Care* (1946) revolutionized child-rearing; it has now sold more copies than any book except the Bible.

1649 The first law in colonial America to regulate medical practice is enacted today in Massachusetts. "Physicians, chirurgians [i.e. surgeons], midwives or others [are forbidden] to exercise or put forth any act contrary to the known rules of art, nor exercise any force, violence, or cruelty upon or towards the bodies of any, whether young or old."

1654 The General Court of Massachusetts establishes a fee structure for using the colonies' first toll bridge. The bridge was built by Richard Thurlow over the Newbury River at Rowley, Massachusetts. People will henceforth be allowed free passage over the bridge, but a charge may be levied for the passage of animals.

1743 The first automaton (a robotlike machine) in the United States arrives on this day from England. According to advertisements by its exhibitor, a "Mr. Pacheco" of New York City, it performed "several strange and diverting motions to the admiration of the spectators" who paid one shilling, one pence to see it.

1860 Physiologist-philosopher John Scott Haldane is born in Edinburgh (see March 14, 1936, for a biographical sketch).

1892 Phsysicist Sir George Paget Thomson, destined for the 1937 Nobel Prize for discovering diffraction in electrons (which showed that electrons behave like waves as well as particles), is born the only son of the famous physicist Sir J.J. Thomson (who discovered electrons).

1902 German physicist Alfred Kastler is born in Guebwiller, Alsace. His 1966 Nobel Prize-winning research led directly to the creation of lasers and masers.

1933 Steven Weinberg, a cowinner of the 1979 Nobel Prize for Physics for developing a theory that united several of the known forces in the universe, is born in New York City.

1969 Late on a Saturday night, consumer advocate Ralph Nader meets secretly with Joseph A. ("Jock") Yablonski, Yablonski's brother Leon, and Leon's son Steven. They are in Steven's law offices in Washington. Nader urges Jock Yablonski to run for the presidency of the United Mine Workers against the incumbent Tony Boyle. Under Boyle's allegedly corrupt leadership, little had been done for the health and safety of the coal miners, especially regarding black lung disease. Although Nader is successful in convincing Yablonski to run, the episode ended in profound tragedy; amid a score of improprieties, Yablonski lost the election, and later was assassinated with his wife and daughter in his home.

1974 The first photographic evidence for the existence of charmed quarks (a type of subatomic particle) is discovered, by scientists at the Brookhaven National Accelerator Laboratory on Long Island, New York.

1981 The world's largest sundial (a 25-foot column with a readable shadow of 125 feet) is installed at the Science Museum of Virginia, in Richmond. The sun's shadow travels 7 inches per minute at equinox.

1992 The Greek islands are struck with tragedy today when a Russian oil tanker rams a cargo ship, sending 12,000 barrels of crude oil into the Aegean Sea, 17 miles south of the island of Skiros.

1994 The Clinton administration announces that a compromise has been reached in the revision of the 1980 Superfund law, which ordered cleanups of hundreds of toxic waste sites. Relatively few sites have been worked on so far, and the outlook for improving the record is poor, largely because of opposition by industrial interest groups. The compromise specifies that some, but not all, of the needy sites will receive attention; the agreement was reached after a closed meeting of bankers, insurers, industrialists, and environmental officials.

1830 The first regular train service for passengers is inaugurated in Kent, England. The engine *Invicta* will carry passengers over the 6¼ miles of track between Whitstable and Canterbury, although in today's maiden run only 1 mile of track is traveled.

1626 Dutch explorer Peter Minuit lands at present-day Manhattan Island.

1776 A letter arrives in France from America's revolutionary Committee of Secret Correspondence. It is the first time the American government used invisible ink in a diplomatic correspondence. The letter tells American officials abroad to purchase military supplies on credit. The writing is done with a solution of tannic acid, which turns dark when touched with ferrous sulfate. Sir James Jay invented the ink in 1776.

1777 French chemist Louis-Jacques Thénard is born in La Louptière. Struggling through an impoverished youth of hardship and deprivation, Thénard rose to be a researcher and teacher at the Sorbonne and then chancellor of the University of Paris. He ensured his own financial prosperity in 1799 by discovering "Thénard's blue," a dye adopted by the porcelain industry. He performed other researches, both on his own and in collaboration with Gay-Lussac, including studies of esters and organophosphorous compounds. He discovered hydrogen peroxide in 1818.

MAY 4

1780 The nation's first scientific society, the American Academy of Arts and Sciences, is chartered on this day in Boston. Its purpose is "to cultivate every art and science which may tend to advance the interest, dignity, honor and happiness of a free, independent and virtuous people." Its first president is James Bowdoin, who served from 1780 to 1790.

1796 Education reformer Horace Mann is born into a home of poverty, hardship, and self-denial.

1825 Biologist T(homas) H(enry) Huxley, "Darwin's bulldog," is born the seventh of eight children of a schoolmaster in Ealing, England. Huxley's nickname came from his fierce defense of Darwin's theory of evolution in its earliest, most controversial days.

1856 Louis Thuillier, perhaps the greatest of Pasteur's assistants, is born in Amiens, France.

1869 Thomas F. Rowland of Greenpoint, New York, patents a "submarine drilling apparatus." It is the first offshore oil-drilling rig.

1886 The first practical phonograph is patented. Chichester Bell and Charles Sumner Tainter receive three separate patents today, one in collaboration with Alexander Graham Bell, to cover the innovations in their Graphophone.

1855 Credited with helping to establish gynecology as a distinct branch of science, James Marion Sims began practicing medicine in rural Alabama. He invented operations for previously incurable conditions, performed the first artificial impregnation of a woman in the United States, and was a cofounder of the nation's first women's hospital, which opened in New York City on May 4, 1855.

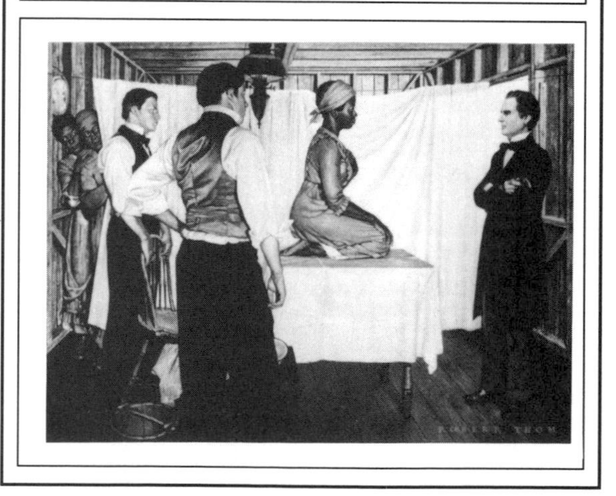

1892 The process for producing acetylene on a commercial scale is discovered—by happenstance. In Spray, North Carolina, chemist Thomas Leopold Wilson has been trying to produce metallic calcium by combining lime and coal tar in a furnace. Wilson discards his failure in a nearby stream. A gas is liberated, which Wilson discovers to be acetylene.

1927 The Academy of Motion Picture Arts and Sciences is founded.

1972 Chemist Edward Kendall dies at 86 in Princeton, New Jersey. In 1950 he shared the Nobel Prize in Physiology or Medicine for isolating the steroid hormone cortisone from the adrenal cortex, and for treating rheumatoid arthritis with this hormone. He also isolated the active component, thyroxine, from thyroid hormone.

1973 The world's tallest office building, the Sears Tower in Chicago, is "topped out."

1994 The Associated Press today reports a new trend in the U.S. diet: an increase in vegetarian offerings in college cafeterias. An estimated 15% of college students eat at least one vegetarian meal per day, and nearly 90% of college cafeterias offer vegetarian dishes at every meal.

1494 Crew members on the second voyage of Columbus to the New World sight Jamaica.

1809 The first woman to receive a U.S. patent is Mary Kies of South Killingly, Connecticut; on this day she is granted her patent for "a new and useful improvement in weaving straw with silk or thread."

1811 Chemist John William Draper, a pioneer in photographing very big things and very tiny things, is born the son of a minister in Saint Helens, England. In 1840 he made the first photograph of the moon; this was the very first astronomical photograph. Draper was also one of the first to photograph objects through a microscope. Draper also recognized that all substances glow at 525°C, a temperature now called the Draper point. His son Henry Draper was the first to photograph a nebula and the spectrum of a star.

1818 Karl Marx is born; Napoleon dies three years later in 1821.

1876 Archaeologist John Garstang, best known for excavating ancient Jericho and various Asia Minor sites, is born in Blackburn, England.

1881 Several cows, an ox, and 25 sheep are inoculated against anthrax under the direction of Louis Pasteur at Pouilly le Fort, France. The experiment was a huge success and another Pasteur milestone in medical history.

1912 *Pravda* is first published.

1925 John T. Scopes is arrested for teaching Darwin's theory of evolution in a Tennessee public school. In the famous "Monkey Trial" the next month in Dayton, Scopes was found guilty and fined $100. The verdict was reversed on appeal. Ironically, Scopes was not a biology teacher (he was a physics teacher substituting for a sick biology teacher), and he was not even in school the day when evolution was taught.

1945 In the only fatal attack of its kind during World War II, a Japanese balloon bomb kills five children and the pregnant wife of a minister when it explodes in Oregon.

1961 Alan B. Shepard, Jr. becomes the nation's first man in space, in a 15-minute 22-second suborbital flight from Cape Canaveral, Florida, in the capsule *Freedom 7*. "O-OK full go" is his statement during takeoff. The flight remains history's shortest-ever manned spaceflight.

1963 Dr. Thomas E. Starlz performs the world's first liver transplant in Denver on a 48-year-old patient who survived for 22 days. Four months before, Starlz had performed the world's first spleen transplant.

1991 Researchers at a Seattle meeting of the American Society for Clinical Investigation announce a breakthrough treatment for rheumatoid arthritis: They have genetically combined a human protein with diphtheria poison to create a substance that attacks and kills the malfunctioning white blood cells that cause arthritis (by breaking down the lining of joints). "Interleukin-2 fusion toxin," the technical name for the treatment, has also been used to battle some forms of blood cancer.

1994 A large, meat-eating dinosaur roamed the South Pole 200 million years ago, when its climate resembled that of today's Seattle, according to William Hammer of Augustana College, Illinois, speaking at a press conference about his discovery of fossilized bones from the previously unknown beast.

1847 The American Medical Association is organized in Philadelphia. Dr. Jonathan Knight is the first president. This organization is the first national doctors' organization of any longevity; it supersedes the National Medical Association, which was organized exactly one year before. Today's meeting, the first national medical convention in U.S. history, is attended by 250 delegates from 22 states, 28 medical schools, and 40 local medical societies.

1635 "The chemists are a strange class of mortals, impelled by an almost insane impulse to seek their pleasure among smoke and vapor, soot and flame, poisons and poverty, yet among all these evils I seem to live so sweetly, that may I die if I would change places with the Persian King." This is a quote attributed to chemist-physician-adventurer Johann Joachim Becher (born on this day, the son of a Lutheran minister in Speyer, Germany).

1840 An adhesive postage stamp is first sold/used. The "penny black" and the "twopenny blue" go on sale in Great Britain on this date. Queen Victoria appears on the stamps.

1851 The Yale lock is patented. Linus Yale of Newport, New York, receives patent No. 8,071 for his lock and key device, sold as the "Yale Infallible Bank Lock."

MAY 6

1851 The first patent for a refrigerator is granted on this day to Dr. John Gorrie of Apalachicola, Florida, for an "improvement in the process for the artificial production of ice." He had presented blocks of ice at a formal dinner the previous July at Mansion House in Apalachicola, but his real inspiration for inventing the refrigerator was treating yellow fever victims. He originally wanted to create cool hospital rooms to lower patients' temperatures. He made no money from his invention, but the principle of refrigeration that he created is still popular.

1856 Freud is born at 6:30 PM in Freiberg, Moravia (now Pribor, Czechoslovakia).

1856 Polar explorer Robert E(dwin) Peary is born in Cresson, Pennsylvania.

1862 Author-naturalist Henry David Thoreau dies of tuberculosis at age 45 in Concord, Massachusetts. It is thought that Thoreau's death was hastened by the arrest and hanging of slavery abolitionist John Brown, who Thoreau revered. Thoreau wrote many essays and several books, which were largely ignored and sold poorly during his lifetime. His *Walden; or, Life in the Woods* has become a classic ecological analysis of man in harmony with nature in an industrial society.

1871 French chemist Victor Grignard is born the son of a sailmaker in Cherbourg. He shared the 1912 Nobel Prize for Chemistry for developing the Grignard reagents and Grignard reaction, in which magnesium compounds are used to catalyze the synthesis of organic molecules. His methods soon became popular worldwide.

1872 Willem de Sitter, astronomer-cosmologist and one of the first to arouse public interest in Einstein's relativity, is born the son of a judge in Sneek, the Netherlands.

1937 The Hindenburg explodes in Lakehurst, New Jersey, having recently set the all-time record for most people (117) carried across the Atlantic in a dirigible.

1829 Although its ancestry goes back to prehistory, the first true accordion is patented on this date by C. Demian of Austria.

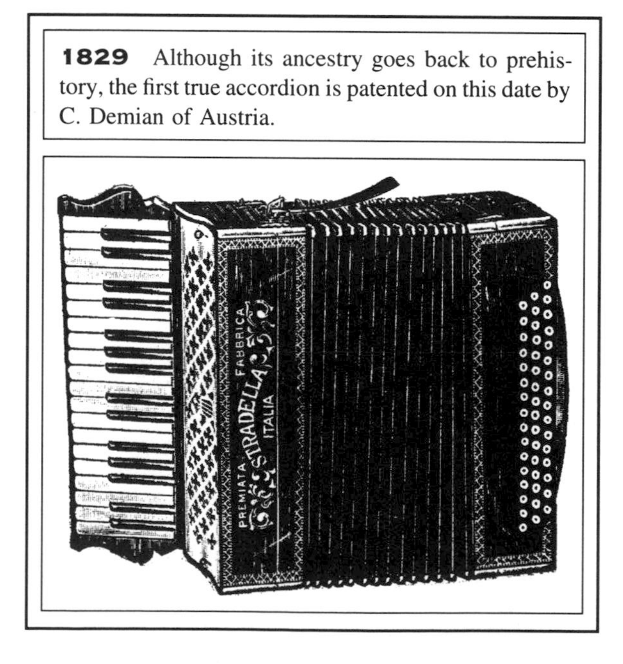

1992 A Johns Hopkins researcher announces one more reason for breast-feeding a child: It stops diarrhea. Dr. Robert Yolken reports to a combined annual meeting of three different pediatric societies in Baltimore that mother's milk contains mucin, a protein that inhibits reproduction of a virus that causes the disorder. Some 500 U.S. infants die annually from diarrhea, and many more perish from it in underdeveloped countries.

1993 *Science* publishes the discovery of a gene defect that can be used to predict with great accuracy a person's likelihood of contracting colon cancer, the world's second most deadly cancer (lung cancer is the most common). "We have proven beyond a shadow of a doubt that a genetic predisposition for colon cancer exists," said Johns Hopkins researcher Bert Vogelstein during his announcement of the discovery. "We can now tell who will get it [the inherited cancer] and who will not."

1713 Mathematician-astronomer Alexis Claude Clairaut is born a child prodigy (he was studying calculus at age 10, and wrote his first mathematics book at 18), the son of a mathematician in Paris. He was the first with a close estimate of Venus's mass.

1909 Inventor Edwin Herbert Land is born in Bridgeport, Connecticut. While a Harvard undergraduate, he began experimentation on a new form of light polarizer (which aligns all rays in the same plane); in 1932, during a leave of absence from college (he never went back, but later was awarded honorary degrees from many schools) he succeeded in creating Polaroid (a plastic sheet embedded with certain crystals), which has found many applications. Land then created the Polaroid Land Camera (first demonstrated in 1947), which revolutionized photography by providing fully developed prints in 60 seconds.

1912 The first U.S. airplane outfitted with a machine gun is given its maiden flight today over College Park, Maryland. It is a Wright biplane, flown by Lieutenant Thomas deWitt Milling. Charles De Forest Chandler of the Army Signal Corps worked the Lewis Machine Gun.

1932 The first all-blind solo flight by a member of the U.S. Army is accomplished today by Captain Albert Francis Hegenberger at Patterson Field, Dayton, Ohio. He flew a standard Douglas BT-2 plane with a completely hooded cockpit.

1941 Anthropologist Sir James George Frazier dies at 87 in Cambridge, England. His most famous work was *The Golden Bough; a Study in Magic and Religion*, which suggested a historical progression in man's major modes of thought, from magic to religion to science. Although the evolutionary sequence he suggested and the general psychological theory he applied are no longer widely accepted, *The Golden Bough* has a secure place in history because it synthesized a huge quantity of anthropological data, and because it made intelligible many primitive cultures and beliefs.

1952 The concept for the integrated circuit (the microminiature chip so important in today's computers) is first published, by Geoffrey W.A. Dummer in Washington, D.C.

1955 Surgeon General Leonard Scheele suspends all polio vaccinations after several children had developed the fatal disease from their vaccinations. "I got the credit for other people's ineptitudes," complained vaccination inventor Jonas Salk after the trouble was traced to poor workmanship from one laboratory.

1963 The communications satellite *Telstar 2* is launched from Cape Canaveral. It is history's second privately owned satellite (like the first, *Telstar 1*, it was built by Bell Telephone Laboratories for AT&T, who paid NASA to put it in orbit). Unlike *Telstar 1*, it is built to withstand radiation from nuclear test explosions (*Telstar 1* ceased operating in 1962 after radiation from a high-altitude nuclear test damaged some of its transistors). On its tenth orbit, *Telstar 2* transmitted the first transatlantic color-television program.

1984 A $180 million out-of-court settlement is announced in a class-action lawsuit brought by Vietnam War veterans damaged by the defoliant Agent Orange. Today is the 30th anniversary (1954) of the end of the 55-day battle of Dien Bien Phu, in which the French were driven from Vietnam. And today is the 9th anniversary (1975) of President Gerald Ford's declaring an end to the "Vietnam era."

1985 Baseball Commissioner Peter Ueberroth announces mandatory drug testing for all baseball personnel except players.

1992 Space shuttle *Endeavor* departs Earth on its maiden voyage.

1698 Naturalist Henry Baker is born the son of a law clerk in London. He introduced microscopy and its growing menagerie of discoveries to the general lay audience, and made a series of pioneering observations on crystals. He also developed new methods of teaching the speech-impaired.

1790 The French Revolutionary Assembly adopts the proposal by Talleyrand that scientists in the French Academy of Sciences be involved in reforming the country's system of weights and measures. By the time the metric system was created four years later, various members of the Academy had been exiled, tortured, and decapitated (see 1794 below).

1794 The immortal chemist Antoine Lavoisier, 51, is sentenced to death by France's Revolutionary Tribunal. In the afternoon he is taken by cart through the Paris streets of his childhood, past the French Academy of Sciences where he made history, to the Place de la Révolution. He is forced to watch the guillotining of his beloved father-in-law, Jacques Paulze, and then he is guillotined with the same blade. "Only a moment to cut off that head, and a hundred years may not give us another like it," mourned mathematician-physicist Joseph-Louis Lagrange the next day.

1821 The 44-ton sloop *Hero* and its six-man crew return to Stonington, Connecticut, from a ten-month voyage in which they discovered Antarctica. The party's leader is Captain Nathaniel B. Palmer. They first sighted the continent in November 1820. Russian and English explorers also claimed to have been the first to see Antarctica in 1820.

1842 Botanist Emile Christian Hansen, who revolutionized beer-making by developing new ways to culture yeast, is born in Ribe, Denmark.

1873 English chemist Nevil Vincent Sidgwick is born in Oxford. He helped explain how atoms combine by advancing the notions of the hydrogen bond and "coordinate links" via electron donation. His *Organic Chemistry of Nitrogen* (1910) is now a classic.

1879 George Baldwin Selden of Rochester, New York, files the first automobile patent in U.S. history. It was granted in 1898; and covered his unique combination of an internal-combustion engine with a carriage.

1886 Coca-Cola is invented by pharmacist John Styth Pemberton in Atlanta, Georgia.

1902 French biologist André Lwoff is born in Ainay-le-Château. He shared the 1965 Nobel Prize for Physiology or Medicine for exploring what happens when viruses invade (bacterial) cells.

1926 Naturalist-filmmaker Sir David Attenborough is born in London.

1928 Red Cross founder Henri Dunant is born in Geneva, Switzerland.

1978 Reinhold Messner (Italy) and Peter Habeler (Austria) complete the first oxygenless ascent of Mount Everest.

1992 Scientists at a conference on the ecology of the Gulf of Mexico announce that they may have found the source of the "red tide" (huge blooms of reddish algae that kill fish and cause severe respiratory problems in humans). The source may be a "green river" that starts some 60 miles off the Florida coast; nutrients at the floor of the Gulf are occasionally brought to the surface by winds and water currents. The population of the red algae then explodes when it contacts the additional nutrients at the surface.

1980 The first nonstop balloon flight across North America begins in San Francisco at 12:30 AM, local time. The *Kitty Hawk* (named for the flight's intended destination in North Carolina) is 75 feet tall, filled with helium, and manned by Maxie Leroy Anderson and his son, Kris, of Albuquerque, New Mexico. The voyagers landed in Canada's Gaspé Peninsula three days later.

MAY 9

1502 Christopher Columbus departs Cádiz, Spain, on his fourth (and last) trip to the New World.

1754 The *Pennsylvania Gazette* publishes the first cartoon in U.S. newspaper history. A snake is shown cut into sections, each of which represents a colony. "Join or Die" reads the caption. Benjamin Franklin owns the newspaper and designed the cartoon.

1794 "Only a moment to cut off that head, and a hundred years may not give us another like it," laments mathematician-physicist Joseph-Louis Lagrange over the previous day's guillotining of Antoine Lavoisier, one of history's greatest chemists. He fell victim to the Reign of Terror in France at age 51.

1893 Thomas Edison presents the first motion picture film exhibition in U.S. history. A blacksmith and two helpers are shown passing a bottle and forging a piece of iron. The images are presented through Edison's Kinetograph, before 400 persons at the Department of Physics, Brooklyn Institute, Brooklyn, New York.

1926 The first airplane flight over the North Pole is made today by Lieutenant Commander Richard E. Byrd and Floyd Bennet in a triple-engine Fokker monoplane, the *Josephine Ford.* The two fly nonstop to and from King's Bay, Spitsbergen, in 15 hours 30 minutes; the journey covered 1545 miles.

1927 German physicist Manfred Eigen is born the son of a musician in Bochum. He shared the 1967 Nobel Prize for Chemistry for studying very rapid chemical reactions.

1931 Astronaut Vance Brand is born in Longmont, Colorado. In 1975 he piloted *Apollo* in the *Apollo–Soyuz* Test Project (the first international manned space mission), and in 1990 he set the mark as the oldest person in space (at 59) in a *Columbia* space shuttle flight.

1944 The nation's first eye bank is opened at New York Hospital in New York City. Dr. Richard Townley Paton and Dr. John McLean collaborated to create the facility, in which a total of 21 hospitals participated in providing eyes.

1961 Newton N. Minow (chairman of the Federal Communications Commission) attacks television programming as "a vast wasteland," in a speech to the National Association of Broadcasters.

1962 Light from Earth is shown on another celestial body for the first time. Scientists at MIT in Massachusetts use a 48-inch telescope to shine a laser beam on the moon. The spot of light is estimated to be four miles in diameter on the lunar surface.

1986 Tenzing Norgay, the man who conquered Earth's highest peak (Mount Everest, in Nepal) with Sir Edmund Hillary, dies at age 71 in Darjeeling, India. In Nepalese his name meant "Wealthy-fortunate Follower of Religion." Indeed, Tenzing was a devout Buddhist throughout his life; he and Hillary spent just 15 minutes on Everest's peak on May 29, 1953, "taking photographs and eating mint cake." Before leaving the peak, Tenzing left an offering of food, as befitted his religious beliefs.

1994 At an Orlando conference on gynecology, a team of Northwestern University researchers report that teens who are good students and regular churchgoers are just as likely to have sex as those with poor grades, divorced parents, and low self-esteem. Adolescents who have reached puberty are more likely to be sexually active, and, paradoxically, students who have considered suicide are also more sexually active than those who have not.

1994 In today's *Physical Review Letters* UCLA physics professor Alfred Y. Wong publishes a unique plan to halt the destruction of Earth's ozone layer: Wong suggests zapping the atmosphere with huge electric charges. It worked in the lab. Wong found that electricity prevented chlorine (which arises from the ozone-destroying chemicals, chlorofluorocarbons) from reacting with ozone, thus keeping ozone levels stable in laboratory chambers.

1936 The *Hindenburg* inaugurates regular, commercial airship (i.e., dirigible) service across the Atlantic, when she docks in Lakehurst, New Jersey, after a 61-hour 38-minute flight from Germany. On board were 51 passengers and 56 officers and crew.

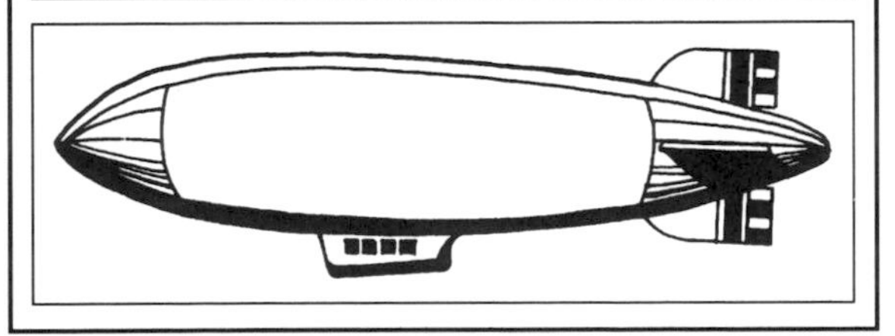

1788 Physicist Augustin-Jean Fresnel, a pioneer in optics, who established the wave theory of light mathematically, is born the son of an architect in Broglie, France. Fresnel was a slow child who could not read until he was eight. His name lives on in the Fresnel lens, which is especially useful in searchlights and lighthouses because it provides tremendous concentration of light beams using little material and weight.

1805 Botanist Alexander Braun, the first to define the cell as protoplasm encased in a flexible membrane, is born in Regensburg, Bavaria.

1852 English chemist Edward Frankland, 27, announces the theory of valence (which states that any atom can combine only with a certain, limited number of other atoms). This theory is now a cornerstone of modern structural chemistry. Frankland was knighted in 1897 at the age of 72.

1860 Discovery of the element cesium (by Robert Bunsen and Gustav Robert Kirchhoff) is announced.

MAY 10

1746 Gaspard Monge, one of the creators of the metric system, is born in Beaune, France, the son of a merchant. His skill as a draftsman earned him employment in the Mézières military school while still a teenager. There he developed a geometric method for determining gun placements; his technique replaced a tedious method based on a long series of arithmetic calculations. His method was so rapid that the school's commandant refused to believe it at first. It was soon validated, and then protected as a military secret for many years. This technique and others of Monge's applications of geometry to construction problems formed the foundation of descriptive geometry. Monge became a confidant of Napoleon's, and accompanied him in Egypt from 1798 to 1801, helping to establish the Institut d' Egypt. In chemistry, he was the first to liquefy a substance normally occurring as a gas (sulfur dioxide, liquefied in 1784).

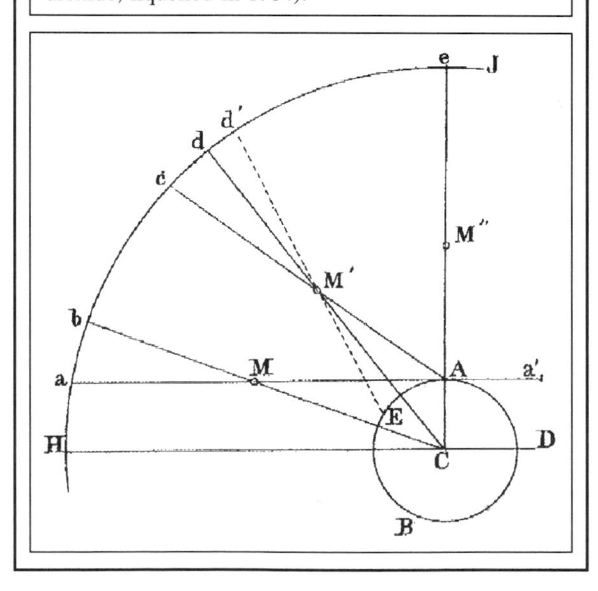

1869 A golden spike is driven at Promontory, Utah, by Senator Leland Stanford for the Union Pacific Railroad. The event marks the completion of the first coast-to-coast train line in North America.

1898 Biochemist Rudolf Schoenheimer is born in Berlin, Germany. Although he did important research into cholesterol and its relation to heart disease, his most famous work was the invention of radioactive "tagging" of molecules. A host of Nobel Prizes have been won using the procedure that Schoenheimer created. He took his own life at age 43.

1919 Sociologist Daniel Bell is born in New York City. He is known for analyzing how institutions shape personality and how science, technology, and politics interact.

1930 Chicago's Adler Planetarium and Astronomical Museum becomes the first planetarium to open its doors to the U.S. public. The $1 million facility was donated to the city by Max Adler, and is under the direction of Professor Philip Fox. The facility dramatically presents a view of our solar system and the 5400 stars visible from Earth.

1933 The Nazis conduct massive public book burnings throughout Germany.

1977 Oklahoma becomes the first state to legalize execution by injection of lethal drugs.

1993 Wisconsin Pharmacal Co. announces that the federal Food and Drug Administration has approved its female condom. It is the first such device in history, and will be sold under the trade name "Reality." Although its failure rate in preventing pregnancy during tests was a whopping 26%, the condom appealed to the FDA because it affords some protection against AIDS and other sexually transmitted diseases.

1994 During a raid for counterfeit money in the German state of Baden-Württemberg, authorities discover a quantity of weapons-grade plutonium in the possession of criminals. It is the first known case of fissionable material falling into unauthorized hands.

MAY 11

1751 Encouraged by Benjamin Franklin and Thomas Bond, the governor of Pennsylvania approves a law for more humane treatment of "lunatiks."

1752 Johann Friedrich Blumenbach, the father of physical anthropology, is born in Gotha, Germany.

1752 The first American company chartered to provide fire insurance holds the first meeting of its board of directors. The company is the Philadelphia Contributionship for the Insurance of Houses from Loss by Fire. At today's meeting, an official seal is agreed on; the seal is to be affixed to each insured house. The company's charter was approved by England's King George III in February 1768. The first name of a private citizen on the charter is that of Benjamin Franklin.

1811 Chang and Eng Bunker are born in Maklong, Siam (now Thailand); they are joined by cartilage at the chest, and remained that way for life. Their father died in a cholera epidemic when they were eight, and they spent the next few years supporting their family of five as fishermen and merchants. Later they were exhibited around the world as the original "Siamese twins." They married sisters and fathered 22 children. They died within three hours of each other in 1874.

1820 HMS *Beagle* is launched at Woolwich, England. From 1831 to 1836 the 90-foot sloop brig was Charles Darwin's home on his famous expedition around the world, during which he conceived much of the theory of evolution.

1910 Glacier National Park is established in Montana.

1918 Richard Feynman, renowned for his skill as a lecturer, for his ability to play bongo drums at parties, and for his reworking the field of quantum electrodynamics (which deals with interactions between subatomic particles and energy radiations, for which he shared the 1965 Nobel Prize for Physics), is born in New York City. Feynman diagrams are simple, visual expressions of mathematically complicated particle phenomena.

1924 Astrophysicist Antony Hewish is born in Fowey, England. He shared the 1974 Nobel Prize for Physics for his part in the discovery of pulsars.

1947 The B.F. Goodrich Company of Akron, Ohio announces manufacture of tubeless, self-sealing automobile tires.

1963 Nerve physiologist Herbert Spencer Gasser dies at 74 in New York City. In 1944 he and partner Joseph Erlanger were awarded the Nobel Prize for Physiology or Medicine for fundamental discoveries about nerve functioning. In 1924 the two had a tremendous breakthrough by adapting the oscilloscope to detect, and visually display, the movement of impulses through isolated nerves. They discovered that different nerves transmit different sensations, like cold, pain, and heat. Their work was used by others to develop machines to diagnose and treat brain/nerve disorders.

1987 Doctors in Baltimore transplant the heart and lungs from an auto fatality into Clinton House, whose own heart was removed and installed in someone else. House survived 14 months as the nation's first living heart donor.

1992 Scientists at the California Institute of Technology announce that they have found "little biological bar magnets" in the human brain. "This is really an exciting discovery," proclaims geobiologist Joseph L. Kirschvink. It puts man in a select group of other organisms that contain such magnets: homing pigeons, whales, salmon, bacteria, honeybees, and some shellfish. The magnets seem to come in two sizes: one-millionth and ten-millionths of an inch wide. The finding may be an important clue as to why electromagnetic fields have been linked to brain cancer.

1993 A British scientist announces that smoking shortens the time it takes HIV infection to become AIDS. "Cigarettes and HIV together double the insult on the immune system," explains researcher Richard Nieman of London's National Heart and Lung Institute. Nonsmokers developed AIDS in about 14.5 months, whereas smokers got AIDS in just 8.2 months.

1928 Television programs are first broadcast on a regular schedule. The picture consists of 24 scanning lines repeated 20 times a second, three times a week. The broadcast station is WGY, the General Electric affiliate in Schenectady, New York.

1003 French scholar-education reformer Gerbert of Aurillac (a.k.a. Pope Sylvester II; he was the first French pope) dies at about 58 in Rome. The rebirth of European learning following the Dark Ages is dated from Gerbert's life.

1732 Botanist Carolus Linnaeus, 24, departs his native Sweden for an extended plant-gathering expedition to Lapland. "How I wished that I had never undertaken my journey," he wrote of the trip, which involved near starvation, treacherous bogs, a boat wreck, and working in arctic conditions. Publication of his findings, however, established his reputation as a first-class scientist. In 1735 his *Systema Naturae* introduced a new system for classifying living things; it is his system that science uses today.

1803 Justus von Liebig, a giant in the development of organic chemistry and biochemistry, is born the son of an amateur chemist in Darmstadt, Germany. Among Liebig's triumphs were the creation of a simple method for the analytic determination of carbon and hydrogen, the development of the hydrogen theory of acids, and an elucidation of how plants obtain nutrition. Liebig was also a master teacher. At the University of Giessen he established the first real laboratory course in chemistry, and he trained scores of chemists. Largely through his influence, Germany became the preeminent force in nineteenth-century chemistry, a field that France dominated in the eighteenth-century.

1820 Florence Nightingale, founder of modern nursing, is born in Florence, Italy.

1895 Chemist William Francis Giauque is born in Niagara Falls, Ontario, Canada. He won the 1949 Nobel Prize for Chemistry for studying matter at temperatures near absolute zero. Giauque also discovered that oxygen is actually a mixture of three isotopes, which led to a reevaluation of international atomic weight standards.

1896 New York City's Department of Health passes the nation's first law prohibiting spitting on sidewalks and in other public places.

1896 New York State passes the first law creating a U.S. institute for psychiatric research and training. The Pathological Institute opened later in the year; Dr. Ira Van Gieson was its first director. He was succeeded by Dr. Adolf Meyer in 1902. In 1909 its name was changed to the New York State Psychiatric Institute and Hospital, by decision of the State Commission on Lunacy. In 1927 it became the Psychiatric Department of the Columbia University–Presbyterian Hospital Medical Center.

1910 English biochemist Dorothy Crowfoot Hodgkin is born the daughter of an archaeologist in Cairo, Egypt. Hodgkin's determination of the structure of vitamin B_{12} brought her the 1964 Nobel Prize for Chemistry.

1932 The body of the kidnapped son of Anne and Charles Lindbergh is found in a wooded area near Hopewell, New Jersey.

1945 Streptomycin is first used successfully on a human. This is not the first antibiotic, but the first one against the group of bacteria called "gram-negative." Streptomycin was developed by Nobel laureate Selman Waksman, from a "pet" fungus he had studied since college days.

1970 Harry A. Blackmun (author of the *Roe v. Wade* abortion decision) is unanimously confirmed by the Senate as a Supreme Court justice.

1978 The Commerce Department announces that hurricanes will no longer be named for women only.

1993 In the first ruling of its kind, a circuit judge holds cigarette makers liable when their product causes death and cancer, regardless of whether or not the smoker was aware of the danger. "Cigarettes are, as a matter of law, defective and unreasonably dangerous for human consumption. Cigarettes are defective because when used as intended, they cause cancer, emphysema, heart disease and other illnesses," rules Mississippi circuit court judge Eugene Bogen, in a $17 million lawsuit by two children of Anderson Smith, Jr., who died of emphysema and lung cancer in 1986.

1994 The U.S. Senate passes legislation that outlaws the blockading of abortion clinics by protesters. The law prevents blocking access to clinics and using force or threats against patrons of the clinics. It imposes prohibitions and penalties equal to those in the blockading of places of worship. The law comes after years of increasing violence at clinics, including arsons, bombings, death threats, and the murder of Dr. David Gunn in Pensacola.

1607 The first permanent English settlement in America is founded in Jamestown, Virginia, when settlers arrive in three ships, the *Susan Constant*, the *Godspeed*, and the *Discovery*. This date also marks the inaugural meeting of the first colonial council.

1857 Sir Ronald Ross, winner of the second Nobel Prize for Physiology or Medicine, is born in Almora, India. English-educated, Ross was an officer in the Indian Medical Service when he proved that malaria is transmitted by mosquitoes. His research was done alone, under arduous and primitive conditions in the Indian outback, starting in 1892, but he was not able to find malaria parasites in mosquito intestines until 1898. His Prize was awarded in 1902.

1885 Friedrich Henle, one of history's outstanding anatomists, dies at 75 in Göttingen, Germany. While still a student he published the first description of the location and structure of "epithelial" (covering and lining) tissue in humans. In 1840 he firmly defended the little-known, little-accepted microorganism theory of diseases (that disease is caused by tiny parasites); this theory has since been proved correct over and over again. In 1841 he published the first systematic account of histology (the study of tissues). His *Handbook of Rational Pathology* (1846–1853) is recognized as the birth of modern pathology (the study of disease).

1906 Marie Curie, 38, is appointed to the physics professorship at the Sorbonne vacated when her husband Pierre was killed the previous April in a carriage accident. Marie was the first woman to teach at the Sorbonne.

1908 President Theordore Roosevelt convenes the Governors' Conference on the Conservation of Natural Resources in the East Room of the White House.

1917 Three peasant children report seeing an image of the Virgin Mary near Fátima, Portugal.

1949 The Holland Tunnel (a twin tube structure under the Hudson River connecting New York City with the state of New Jersey) is given a spectacular test: A truck loaded with toxic and flammable carbon disulfide catches fire and explodes in the tunnel. Five hundred feet of ceiling and twenty-three trucks are destroyed, but no one dies and the tunnel is still in use today.

1953 The first U.S. city to have two educational television stations was Pittsburgh. The Metropolitan Pittsburgh Educational Station is granted a permit today to operate WQED, Channel 13; the second channel, WQEX, received its permit in 1958.

1960 A McDonnell Douglas Delta rocket (a mainstay of the U.S. space program) flies for the first time.

1992 The longest spacewalk in history (8 hours 29 minutes) is conducted today outside space shuttle *Endeavor* by Pierre Thuot, Rick Hieb, and Tom Akers.

1992 A report in today's *Nature* indicates that science has found one of the keys to unlocking the operations of DNA. Within a cell's nucleus strands of DNA copy themselves, and these copies are then passed on from one generation to the next. But a mystery has been: what triggers the copying process? According to today's report, a complex of proteins (called the "origin recognition complex") starts the replication process by attaching to DNA.

1994 The Coalition on Smoking or Health reveals that tobacco companies had developed and patented safer ("medically acceptable") cigarettes as far back as the 1960s, but never sold them because it was feared that they would make other tobacco products look bad.

1873 The sewing machine lamp holder is patented by Ludwig Martin Nicolaus Wolf of Avon, Connecticut. It was introduced by the Singer Sewing Machine Company in 1876, who claimed it "quite obviated the difficulty experienced by operators when sewing at night."

1852 Antioch College in Yellow Springs, Ohio, was the nation's first nonsectarian college to grant equal rights to women and men; it is chartered on this day, and opened October 5, 1853. Its first graduating class had three women; Horace Mann was its first president.

1856 Charles Darwin, 47, sits down in the study of his country home in Down, England, and starts to write On *The Origin of Species*, one of history's most influential books. It revolutionized the way man views himself and his place in the universe.

1872 Russian botanist Mikhail Semyonovich Tsvet, inventor of chromatography (now a very important, widely used technique that separates the constituents of complicated compounds from each other), is born in Asti, Italy. Tsvet is known for researching plant pigments, especially the chlorophylls (of which he discovered several new forms) and the carotenoids (a term he coined). It seems fitting that he made his mark by studying plant coloration, because Tsvet means "color" in Russian, and chromatography means "written in color."

MAY 14

1884 Aircraft pioneer Claudius Dornier is born in Kempten, Germany. He designed history's first all-metal airplane (1911), and founded the Dornier aircraft works at Friedrichshafen, where he made seaplanes and bombers for World War II. The Dornier Do X was the world's largest airplane at the time (1929), but it was abandoned after several were built because of the great cost. In the 1950s Dornier constructed Starfighter planes for the U.S. government.

1906 Marie Curie, the first two-time winner of the Nobel Prize, writes a letter in her diary to her deceased husband. "I want to tell you that they have nominated me to your chair of Physics at the Sorbonne, and that there are people imbecile enough to congratulate me on it." Her nomination was successful, making her the first woman to teach at the renowned French institute.

1796 Dairymaid Sarah Nelmes walks into Dr. Edward Jenner's country medical office in Berkeley, England, with a mild case of cowpox. Outside in the street eight-year-old James Phipps happens to be playing with friends. Jenner calls the boy into his office and inoculates him with material taken from the sores of Nelmes's infection. It is history's most famous vaccination. Jenner turns 47 in three days, but a horrible decision overshadows the event: whether or not to inject Phipps with the incurable killer smallpox (which had taken the life of Jenner's father years earlier), to see if the boy has been protected by the previous dose of cowpox.

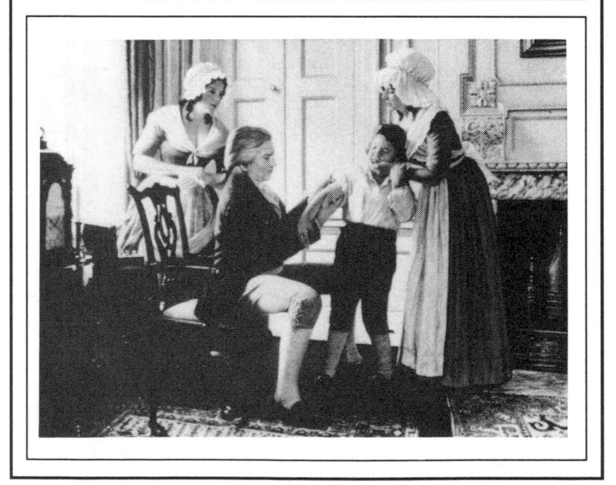

1907 Einstein publishes $E = mc^2$ in *Annalen der Physik.*

1930 Carlsbad Caverns National Park is established in New Mexico.

1932 Ernest O. Lawrence, 30, is married in New Haven, Connecticut, just as final touches are being put on the cyclotron he designed in California. The device was the largest atom smasher of its day, and it won Lawrence the 1939 Nobel Prize for Physics.

1973 *Skylab 1* is launched. It is the first U.S. manned orbiting laboratory. Over the next nine months it was inhabited by three successive crews of astronauts. It set a record as the largest payload sent into space, and then another record as the largest object to fall back to Earth. In July 1979, it came back as a fiery mass spreading debris (some pieces weighing several tons) over the Indian Ocean and Australia.

1991 A Florida State University chemist patents a process that will both fight cancer and save trees. Robert Holton has developed a method for artificially creating taxol, an anticancer drug that has previously been available only from the slow-growing yew tree of the Pacific Northwest. The bark of 12,000 trees is currently required to produce just 2.5 pounds of taxol. Many of these trees are in areas occupied by threatened species. Once the bark is taken the tree dies. "The bark is not the only source of taxol any longer," reports Holton. "There is now a way, at least on a small scale, to make taxol from other sources." Ironically, Holton's process eliminates one reason environmentalists have used to preserve trees from logging.

1048 Astronomer-mathematician-poet Omar Khayyám is born the son of a tentmaker in Nishapur, Persia (see December 4, 1131, for a brief biography).

1513 Ponce de Leon "discovers" the Florida Keys, and exactly 89 years later (1602) Bartholomew Gosnold "discovers" Cape Cod.

1672 America's first copyright law is enacted. Boston's General Court of Massachusets grants book merchant John Usher sole rights to his publication *The General Laws and Liberties of the Massachusetts Colony* "for at least seven years ... there shall be no other or further impression made by any person thereof in this jurisdiction." The fine for violating this copyright was treble the cost of printing and paper.

1845 Russian zoologist-microbiologist Elié Metchnikoff is born near Kharkov, Ukraine, the son of an officer in the Tsar's Imperial Guard. While studying digestion in starfish he made the curious discovery of free-moving cells (unconnected with digestion) that went to, surrounded, and ate debris in the animals' bodies. The discovery was unrelated to the topic of his research, but it made him forever famous. He called the cells "phagocytes" (meaning "devouring cells"); in man, white blood cells perform this role and their action is a major defense against disease/infections. Metchnikoff shared the 1908 Nobel Prize for Physiology or Medicine.

1854 The world's first hospital for alcoholics is organized in Binghamton, New York, by Dr. James Edward Turner. The United States Inebriate Asylum "for the reformation of the poor and destitute inebriates" was incorporated exactly a month before. In 1857 its name was changed to the New York State Inebriate Asylum; the cornerstone was laid in September 1858. Turner was the corporation's first treasurer, and John D. Wright was its first president.

1914 Tenzing Norgay, the man who conquered Mount Everest (Earth's highest peak) with Sir Edmund Hillary, is born in Solokhumbu, Nepal (see May 9, 1986, for details).

1930 Flight attendant service is launched. Ellen Church (a registered nurse from Cressbill, Iowa) is history's first stewardess, flying today on a Boeing Air Transport (a forerunner of United Air Lines) between San Francisco and Cheyenne, Wyoming. Her salary was $125 per month for 100 hours of flying.

1940 Nylon stockings first go on sale, in New York City (where four million pairs are sold in several hours).

1958 *Sputnik 2* is launched.

1963 The last Project Mercury spaceship is launched (it is *Faith 7*, with astronaut L. Gordon Cooper aboard), and ten years later (1973) the nation's first manned space station (*Skylab I*) is launched.

1992 "Don't have a cow, man." This is the catch phrase of the rascal Bart Simpson, cartoon star of TV's *The Simpsons*. Today eight-year-old Chris Bencze of Auburn, Washington, takes Bart's advice when he calmly saves his brother's life with the Heimlich maneuver, which he had seen on the show several months before. The specific episode, "Homer at the Bat," involved Homer's workmates standing in front of a poster about the maneuver, oblivious to the fact that Homer is choking on a donut next to them. A network publicist later said that this is the second case in which someone learned, and used, the maneuver after seeing the show.

1993 A Paris woman is surgically given two new lungs, both of which were cut from a single lung of a large man. This is the first time such a procedure has been performed on a human, having been successfully developed in animal trials. It holds great promise for surgery with children, because getting two lungs of the right size has been a big problem in the past.

1981 The 20 millionth Volkswagen "Beetle" is produced in Mexico. Originally designed by Ferdinand Porsche and first made in 1938, the Beetle has been in production for more years than any other auto in history.

1631 A newspaper is first published in France. Théophraste Renaudot (1586–1653) issues the first edition of his *Gazette*, under the patronage of Cardinal Richelieu. It has four pages and a print run of 1200 copies.

1763 French chemist Nicolas-Louis Vauquelin, discoverer of the elements chromium and beryllium, is born in St. André. The son of a peasant, he was a field laborer himself until a village priest noticed his fondness for books, and got him a lowly position in an apothecary's shop.

1830 French mathematician-physicist-Egyptologist Baron Joseph Fourier dies at 63 in Auxerre, France.

1856 Massachusetts authorizes the nation's first fish commission (on a state or federal level) "to ascertain, and report to the next General Court, such facts respecting the artificial propagation of fish, as may tend to show the practicability and expediency of introducing the same into this Commonwealth." In other words, the state wanted to build a fish farm. The commission consisted of three men, and disbanded when their report was completed.

MAY 16

1947 Sir Frederick Gowland Hopkins, codiscoverer of vitamins, dies at 85 in Cambridge, England. Hopkins found that rats failed to grow on a diet of artificial milk, but thrived when a little cow's milk was added to their diet; he went on to show that something other than pure protein, fat, and carbohydrate was needed in the animals' diet. He called this missing something "accessory substances"; these are now called vitamins. Revealing their existence won Hopkins the 1929 Nobel Prize for Physiology or Medicine. Among his other discoveries were "essential amino acids" (certain protein-forming molecules that must be consumed whole because the body cannot synthesize them from smaller molecules).

1960 Laser light is first produced, by 32-year-old Theodore H. Maiman at the Hughes Research Laboratories in Malibu, California. Maiman's device was about as big as a flashlight battery. "And then I set this up with my instrumentation," he later remembered, "and I fired it, and at a certain point it went."

1973 The nation's first solar-powered balloon flight starts today at 8:30 AM from the soccer field of the University of North Carolina in Charlotte. It lasts ten minutes. The *Solar FireFly*, a huge inverted pyramid, is piloted by Tracy Barnes, president of the Balloon Works in Statesville, North Carolina. No lifting gas is used in the flight; instead the sun's energy is absorbed by the balloon's dark skin, which heats the air inside, thus causing the balloon to rise.

1975 Japan's Junko Tabei, 35, becomes the first woman to reach the peak of Mount Everest.

1988 Surgeon General C. Everett Koop declares that nicotine is addictive in ways similar to heroin and cocaine.

1991 Since its discovery 122 years ago, Lou Gehrig's disease has remained a complete mystery—until today. Northwestern University researchers report in the *New England Journal of Medicine* that they have found the gene that causes the disease (technically called amyotrophic lateral sclerosis, or ALS; it is in the family of muscular dystrophies). "This is a difficult disease," said team director Dr. Teepu Siddique. "A hundred years of research has not yielded a clue as to what causes the disease or how to treat it. This is the first time that we have a handle on the disease."

1993 According to a 234-page report released today by the environmental group Friends of the Earth, the U.S. petroleum industry annually deposits 11 billion gallons of oil into the environment through leaks, spills, ventings and evaporation. This is 1000 times the amount of oil lost in 1989's Exxon Valdez disaster. The data came from state and federal governments and from the American Petroleum Institute, which pulled no punches in its assessment of the study. "Utter nonsense" were the exact words of Institute Vice President William O'Keefe.

1770 The worst fireworks disaster in history claims the lives of at least 800 at the dauphin's wedding on the River Seine in Paris.

1749 Edward Jenner, discoverer/inventor of modern vaccination, is born in the country town of Berkeley, England, the son of a vicar who died five years later. Jenner's rural background instilled a love and curiosity about nature that stayed with him for life. Trained as a physician, he was a keen observer and experimenter. In his greatest experiment (1796) he inoculated an eight-year-old village boy with cowpox material. He endangered both his career and the boy's life when he reinoculated the lad with the killer smallpox several weeks later. The experiment was impromptu and totally unethical, but it worked and has since saved untold numbers of lives.

1796 The nation's first major publication on pediatrics is presented today to the University of Pennsylvania Medical School by its author, Charles Caldwell, in partial fulfillment of his degree requirements. The study was later published in Philadelphia by Thomas Dobson. It was an attempt to determine if *Hydrocephalus Internus, Cynanche Trachealis, and Diarrhoea Infantum* were all the same disease.

1836 British astronomer Sir Joseph Norman Lockyer is born the son of an apothecary-surgeon in Rugby, England. In 1868 he discovered a previously unknown element in the sun's atmosphere; Lockyer called it "helium." He also coined the term "chromosphere" to describe the sun's outer layer, and he established the world-famous journal *Nature* in 1869.

1883 In a flash during a fit of sleeplessness, Swedish chemistry student Svante Arrhenius, 24, is hit with his "dissociation theory" (which states that a substance like salt dissolves into electrically charged ions when added to water) to explain numerous experiments/data he was contemplating. At the time the theory was highly controversial, and earned Arrhenius the lowest possible passing grade for his doctoral thesis. This theory is now accepted as fact, and its creator eventually won the Nobel Prize.

1897 Norwegian physical chemist Odd Hassel, a founder of conformational analysis (the study of the three-dimensional shape of molecules, and the relation of this shape to chemical properties), is born in Oslo. He shared the 1969 Nobel Prize for Chemistry with Britain's Derek H.R. Barton.

1955 Nathan Kline (credited with founding the field of psychopharmacology with his discovery and use of reserpine to treat the mental disorder psychosis) makes his first appearance before Congress about his work. "I discovered a drug that had been in use for 2000 years."

1977 Physicist Erwin Wilhelm Müller dies at 65 in Washington, D.C. Through the 1930s and 1940s he was a researcher at several German universities; at age 40, in 1951, he emigrated to the United States and took a post at Penn State University. Five years later he invented the field ion microscope, which produced the first ever photographs of individual atoms.

1979 The highest-ever voltage produced in the lab (32 1.5 million volts) is generated on this date by the National Electrostatistics Corporation at Oak Ridge, Tennessee.

1994 Women who gain weight at around age 30 are hit with a double whammy, says Dr. Noreen Aziz to a cancer conference in Tampa. Not only do the commonly known cosmetic and health difficulties arise with the extra weight, but Dr. Aziz and her colleagues now report that just ten extra pounds can dramatically elevate the chances of getting breast cancer. Although the cause of the link between weight and breast cancer is unknown, the message is clear. "Of all the decades in which you should lose that extra weight, it's the third," says Aziz.

1994 The Phil Donahue Show will not televise a live execution, rules North Carolina's Supreme Court. Speaking for the Court, Justice Sarah Parker says that neither the state nor federal Constitution provides either Donahue or the condemned man (David Lawson, scheduled to die in June for a 1980 offense) the right to videotape the execution. Lawson and Donahue fought to air the event to draw attention to the problem of depression and its consequences, which Lawson claims was instrumental in his crime.

1933 Louis Leakey, 29, exhibits the infamous "Kanam mandible" before England's Royal Society. It was named for the site in Kenya where Leakey dug it up in 1932. The jawbone is definitely from a human ancestor, but Leakey's overestimation of its age, and some careless field documentation on his part, brought him much criticism. The whole episode was a harsh lesson in Leakey's fabled career.

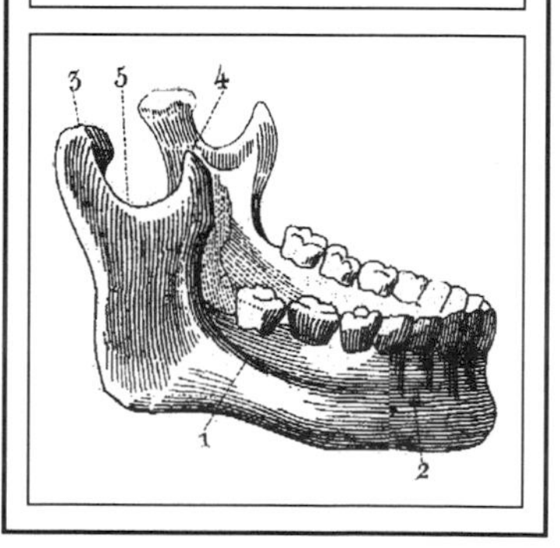

1850 English physicist Oliver Heaviside is born the son of an artist in London. Like Edison, he had little formal education, became a telegrapher, and was plagued by deafness. Deafness forced Heaviside to abandon telegraphy in 1874, after which he devoted himself to researching electricity. His three-volume *Electromagnetic Theory* predicted the existence of an electrically charged layer in the atmosphere. This layer was subsequently discovered and named the ionosphere or the Kennelly–Heaviside layer.

1872 Mathematician-philosopher Bertrand Russell is born into a lonely, unhappy childhood in Trelleck, England, the grandson of John Russell (the English prime minister from 1846 to 1852 and 1865 to 1866).

1901 Biochemist Vincent Du Vigneaud is born in Chicago. He was awarded the 1955 Nobel Prize for Chemistry for isolating and synthesizing two hormones that are critical to human existence: vasopressin (which regulates blood pressure by controlling the muscles around blood vessels) and oxytocin (which causes the uterus to contract and the breasts to secrete milk).

MAY 18

1910 Halley's Comet is seen from Earth to move across the face of the sun.

1922 Alphonse Laveran dies at 77 in Paris. He was a major force in the development of tropical medicine (he established the Laboratory of Tropical Disease within Paris's Pasteur Institute in 1907, and founded the Societé de Pathologie Exotique in 1908), and pioneered the study of human diseases caused by protozoa (one-celled animals). His discovery of the organism that causes malaria won him the 1907 Nobel Prize for Physiology or Medicine.

1923 The first patent application for a rotary-dial telephone is submitted in France by engineer Antoine Barnay.

1934 Capital punishment is first authorized by federal law in the United States.

1967 Governor Dewey Follett Bartlett of Oklahoma signs the nations's first law legalizing artificial insemination in humans. The state already was operating birth control centers, available to anyone regardless of financial condition.

1969 *Apollo 10* is launched with astronauts Cernan, Stafford, and Young.

1980 Mount St. Helens in Washington State erupts with the force of 27,000 large atomic bombs. About 13,000 feet is gone from the top of the mountain, 57 people are dead or missing, and the resulting ash cloud circles the Earth.

1991 Helen Sharman becomes the first Briton in space, aboard a *Soyuz* spacecraft with two cosmonauts.

1992 Paris's "pooper-scooper" law goes into effect. It is an attempt to reduce the 11 tons of animal waste that are deposited on the city's streets *each day*. "It's a fantastic idea, but I don't think it will ever work," said one dogless citizen, who guesses that authorities "just won't enforce it." Some 10,000 posters have been displayed in city streets and sanitation workers have been warning dog owners for weeks about the possible $250 fine for offenses. Some 50 plainclothes officers are now on the streets to catch violators.

1994 The Massachusetts Institute of Technology's *Tech Talk* publishes an article about a student who has built a tiny, remote-controlled robot designed to crawl up into the human intestine. The event caught the attention of humorist Dave Barry who titled his syndicated column for the week, "This might hurt just a little bit ..." Barry went on to observe, "This will be a great boom to the medical profession, which, as you know, is always looking for new things to stick as far as possible into our various bodily orifices."

1960 Three Dublin priests see a huge Loch Ness-like monster in Ireland's Lough Ree during a clear, sunny afternoon's fishing trip. All three are reportedly sober. Risking ridicule, they submit a written account to the Inland Fisheries Trust and are questioned for several hours by Colonel Harry Rice, an authority on Irish rivers. The incident is one of the best-documented sightings of a giant serpent in European history.

MAY 19

804 Alcuin, the foremost scholar of the Carolingian Renaissance, dies at about 72 in Tours, France.

1780 Darkness covers New England and part of Canada in the middle of the day. The cause of this cosmic blackout has never been explained.

1857 Biochemist John Jacob Abel is born the son of a prosperous farmer in Cleveland. Abel invented the first artificial kidney (1912), but he is best known for his work on glandular secretions. In 1897 he was the first to isolate an active molecule from the adrenal medulla; he named this molecule "epinephrine," but it is more popularly known as "adrenaline." In 1925 Abel became the first to crystallize insulin.

1857 The first U.S. patent for an electric fire alarm is No. 17,355, granted today to William Francis Channing of Boston and Moses Gerrish Farmer of Salem, Massachusetts, for "a magnetic electric fire-alarm." In June 1851 Boston allocated $10,000 to test the alarm.

1864 Inventor-naturalist-explorer Carl Akeley is born in Clarendon, New York. His innovations in taxidermy made possible museum displays of unprecedented realism, especially at Chicago's Field Museum of Natural History and at New York City's American Museum of Natural History (which houses the Akeley African Hall). He invented the Akeley cement gun (used in mounting animals) and the Akeley camera (with which he made the first movies of gorillas in the wild). He made five hunting and specimen-gathering expeditions to Africa; he died during the last and is buried on Mt. Mikeno in Zaire's Albert National Park (now Virunga National Park), which was central Africa's first animal preserve. Akeley headed the effort to establish it.

1868 Engineer John Fillmore Hayford, a founder of the modern science of geodesy (the precise determination of Earth's shape), is born in Rouses Point, New York. Hayford established the theory of isostasy (which proposed that the pressure exerted by Earth's crust is approximately the same over the entire globe, regardless of whether a particular area is covered with, say, lowlands or with huge mountains). This theory has since been accepted, with modification, and has helped explain phenomena within the crust.

1910 Earth passes through the tail of Halley's Comet. It is the most intimate contact between Earth and any comet in recorded history.

1914 Because of the research of Max Ferdinand Perutz (born on this day in Vienna), we now know what hemoglobin looks like, how it is built, and how it carries oxygen to the body's cells. His X-ray diffraction studies revealed that hemoglobin (a pigment is found in red blood cells) is constructed of four protein chains wound together, and that the molecule's shape changes when oxygen is added. He received the 1962 Nobel Prize for Chemistry. He also studied glaciers, providing the first-ever measurements of different rates of flow in different parts of the same glacier.

1936 The Akeley African Hall is dedicated in the American Museum of Natural History in New York City. Akeley (see 1864 above) hoped that "African Hall will tell the story of jungle peace."

1992 A group of 14 live whales is found beached on Chilean shores, 310 miles south of Santiago. Unfortunately, it is fishermen who first find the whales, and they start harpooning the helpless beasts. A tour of students encounters the scene, and the group's teacher is able to persuade the hunters to stop the slaughter. Three whales are killed; the other 11 are returned to the ocean. Unlike most industrialized nations, Chile has banned whale hunting.

1993 An Orlando meeting of the American Association for Cancer Research hears a group of reports on cancer-fighting properties of common foods. Science has long known that people who consume lots of fruits and vegetables have reduced cancer risks; scientists are now trying to isolate the key chemicals in these foods. Among the recent findings: soybean extracts have been effective in battling mouth cancers and orange peel extracts have both prevented and shrunk breast cancers in lab animals.

1796 The first federal game law in U.S. history is approved. Intended "to regulate the trade and intercourse with the Indian tribes and to preserve peace on the frontiers," the law established a fine of $100 and six months in jail for anyone who went into Indian territory to hunt.

1506 Christopher Columbus dies at 55 in Valladolid, Spain.

1537 Hieronymus Fabricius ab Aquapendente, one of the founders of embryology, is born in Aquapendente, Italy. An outstanding anatomist, in 1603 he provided the first complete description of valves in blood vessels (William Harvey was one of Fabricius's pupils, and used his teacher's work in discovering the circulation of blood). Fabricius also gave the first description of the human placenta, the first full description of the larynx as the organ of speech, and the first demonstration that the eye's pupil changes its size.

1825 Eminent astronomer George P(hillips) Bond is born the son of eminent astronomer William Cranch Bond in Dorchester (now part of Boston), Massachusetts. The pair discovered Hyperion (the eighth satellite of Saturn), and George further distinguished himself by discovering a number of comets, and by making the first photograph of a double star.

MAY 20

1830 D. Hyde of Reading, Pennsylvania, receives the first patent for a fountain pen. However, it was not until 1884 that the first truly practical fountain pen was invented, by Lewis Edson Waterman, who made some 200 pens by hand the first year he was in business. Waterman later developed machinery for the job.

1851 Inventor Emil Berliner, developer of the telephone (the modern mouthpiece is based on his design) and phonograph record (the modern "platter" is his design), is born in Hannover, Germany.

1860 German biochemist Eduard Buchner is born in Munich. His older brother by ten years, the great bacteriologist Hans Buchner, first interested Eduard in chemistry. Eduard won the 1907 Nobel Prize for Chemistry for demonstrating that the fermentation of carbohydrates is caused by enzymes in yeast, not by the yeast cells themselves. This conclusion was fought by many scientists who still clung to the doctrine of vitalism (that life contains special properties beyond the laws of chemistry and physics). Eduard died young in World War I.

1899 The nation's first arrest for speeding in an automobile is made. Jacob German, a cab driver, is arrested for traveling at the "breakneck speed" of 12 mph on Lexington Avenue in New York City. The arresting officer took German to the East 22nd Street police station, where he was booked and jailed.

1747 British ship's surgeon James Lind begins an experiment on HMS *Salisbury*, in which he strictly controlled the diets of 12 sailors with scurvy; those receiving lemons and oranges showed "sudden and visible good effects." The study is now a classic; it produced a cure for the disease that had killed millions, led to reforms in naval health practices, and shaped the fates of nations.

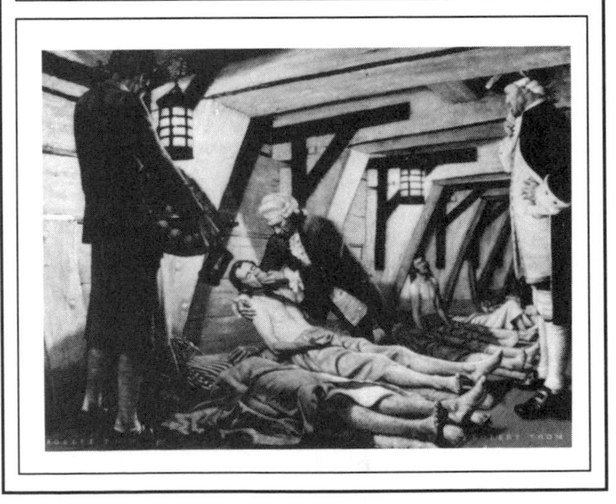

1927 Lindbergh departs Roosevelt Field (on Long Island, New York) in the *Spirit of St. Louis* on the first solo flight across the Atlantic.

1932 Amelia Earhart departs Harbor Grace (Newfoundland) at 5:50 PM; the following day she arrives in Londonderry, Ireland, thereby becoming the first woman to fly *solo* across the Atlantic.

1939 Regular transatlantic airplane service begins, as the *Yankee Clipper* (a Pan American Airways airplane) departs Port Washington, New York, bound for Lisbon, Portugal. The event also inaugurates regular transatlantic airmail service.

1986 Helen Brooke Taussig dies four days before her 87th birthday in Kennett Square, Pennsylvania. She pioneered the use of X rays and fluoroscopy to pinpoint heart defects in newborns. In the 1940s she and surgeon Alfred Blalock invented a surgical procedure for treating blue baby syndrome. In the early 1960s Taussig was a major figure in exposing and ending the harm to newborns caused by thalidomide, taken by pregnant women as a sedative.

1990 The Hubble Space Telescope sends back its first photos of the cosmos.

MAY 21

1471 Renowned artist Albrecht Dürer is born the son of a goldsmith in Nuremberg, Germany. He is famous for developing mathematical rules for creating very precise/lifelike renditions of nature. His 1525 treatise on geometric constructions is considered the oldest surviving text on applied mathematics.

1619 Hieronymus Fabricius ab Aquapendente, one of the founders of embryology, dies the day after his 82nd birthday in Padua, Italy (see May 20, 1537, for a biographical sketch).

1792 French engineer-mathematician Gustave-Gaspard Coriolis is born in Paris. In 1835 he became the first to describe the inertial process now called "Coriolis force" (which is a special type of pressure pushing away from the equator of a spinning object, like Earth, caused by different rotation speeds of different latitudes). This force creates ocean currents and the whirling motions of hurricanes, tornadoes, and other weather systems, and influences artillery fire and rocket launches.

1845 Botanist Charles E(dwin) Bessey is born near Milton, Ohio.

1860 Willem Einthoven is born in Semarang, Java, the son of a doctor. Einthoven combined this medical background with a lifelong interest in physics, to become the first to tap small electrical signals in the body for the diagnosis of normal and abnormal function. In 1903 he invented the string galvanometer, which detects small electric currents under the skin. Einthoven applied the device to heart function, to create a graphic record of heart activity, which he called an "electrocardiogram," a term still in use today. His work provided an important tool to cardiology, and was adapted by others to study electrical activity in other body systems, especially the nervous system. He won the 1924 Nobel Prize for Physiology or Medicine.

1881 The American Red Cross is organized in Washington, D.C., by its first president, Clarissa Harlowe Barton. On this day, a constitution is also adopted. The society incorporated the following July 1. The International Red Cross was organized in 1863 by Switzerland's Jean Henri Dunant; the United States joined this organization on March 16, 1882, when the Senate ratified the governing treaty.

1894 Gifford Pinchot and Teddy Roosevelt (two legendary conservationists who transformed the nation's wilderness policy) meet for the first time, in Washington, D.C.

1927 Lindbergh completes the first solo transatlantic airplane trip, landing the *Spirit of St. Louis* outside Paris.

1956 The first airborne hydrogen bomb is exploded over the Bikini atoll in the Pacific.

1968 The nuclear-powered submarine Scorpion is last heard from. Ninety-nine sailors were aboard. Its remains were later found on the ocean floor 400 miles southwest of the Azores.

1991 The English government bans the importing of pit bull dogs. "It is clear that such dogs have no place in our homes," explains Prime Minister John Major, following a series of pit bull attacks over the last few days. The most recent involved the half-hour mauling of a six-year-old child, Rucksana Khan, by a neighbor's dog. Newspaper and television pictures of her battered and bloody face were the most potent trigger for today's ban. The following day, the government announces plans for legislation to outlaw pit bulls completely.

1994 "It's easy for [space aliens] to get you anywhere," reports author Karla Turner in describing her alien abduction to attendees at a weekend-long convention on "UFOlogy" in Tampa, Florida. Turner reports that she and her family were taken from their Texas home to a world where nonearthlings use humans for brain surgery, cloning, and cross-breeding. Another conventioneer reports her sexual encounters of a third kind. Other lectures (for which listeners paid $12 each to hear, or $245 for eight lectures and seven workshops) included "The Watchers," "The Secret, The Answer," and "Symptoms of Hidden UFO Abductions." Organizers are hoping that this will be the first in an annual series of UFO conventions.

1819 Bicycles are used for the first time in the United States. The imported "velocipedes," also known as "swift walkers," first appear in New York City. In August 1819 the city's Common Council passed legislation "to prevent the use of velocipedes in the public places and on the sidewalks of the city of New York."

1783 Inventor-physicist William Sturgeon is born in Whittington, England. He was a shoemaker's apprentice early in life, but after observing a severe thunderstorm he became obsessed with electricity. He founded the first English journal devoted to electricity, invented devices to measure electric current, and (most importantly) invented the first electromagnet capable of supporting more than its own weight. This led to the invention of the telegraph, the electric motor, and numerous other modern devices.

1819 The 350-ton, wooden *Savannah* departs Savannah, Georgia, bound for Europe. On the following June 20, it reaches Liverpool, England, to become the first steam-powered ship to cross the Atlantic. The ship had 32 staterooms, but no passengers dared make the maiden transatlantic voyage. See 1958 below for another first by a ship named *Savannah*.

1841 The reclining chair is first patented, by Henry Peres Kennedy, a cabinetmaker and upholsterer from Philadelphia.

MAY 22

1849 Abraham Lincoln receives patent No. 6,469 on a device for "buoying vessels over shoals." In his design, inflated cylinders would float grounded vessels over shallows. Lincoln was the first U.S. president to receive a patent.

1900 The Associated Press incorporates in New York, as a nonprofit news cooperative.

1902 Crater Lake National Park is established in Oregon.

1906 The Wright brothers, Orville and Wilbur, receive the nation's first airplane patent. It is No. 821,393 for "new and useful improvements in Flying-Machines."

1912 Chemist Herbert Charles Brown is born in London. His Ukrainian-born parents soon emigrated to Chicago, where he attended college and received his Ph.D. (from the University of Chicago) in 1938. His research into the element boron, especially its combination with hydrogen and its use in the synthesis of organic molecules, won him the 1979 Nobel Prize for Chemistry.

1920 Astronomer-cosmologist Thomas Gold is born in Vienna, which he fled as the Nazis came to power. Gold was the chief architect of the steady state theory of the universe (i.e., as galaxies move away from each other in the expanding universe, new matter and galaxies are created in the vacated space, so that the density and positioning of material in the universe stays approximately the same through time).

1931 Rattlesnake meat in cans goes on sale for the first time, in Florida. It is produced by George Kenneth End, president of the Florida Products Corporation.

1969 Astronauts conduct a dress rehearsal for man's first walk on the moon. A module is released from *Apollo 10* and flown to within nine miles of the lunar surface.

1958 The first nuclear-powered merchant ship, the 21,000-ton NS *Savannah*, has its keel laid in a shipyard in Camden, New Jersey.

1991 The British government says it will introduce legislation to ban American pit bulls. This would instantly imperil an estimated 10,000 of the breed already in Britain. (See May 21, 1991)

1994 UFOlogists conclude a weekend-long convention in Tampa, Florida, on space visitors. "The evidence is overwhelming that Earth is being visited by intelligence-controlled aliens," observed nuclear physicist Stanton Friedman, who spoke on "Cracking UFOlogy's Watergate." Sold at the convention were caps, paperweights, ceramic angels, and hundreds of titles of books.

1707 Famed botanist-explorer Carolus Linnaeus is born in Råshult, Sweden, the son of a curate. His fascination with flowers blossomed early in life; his childhood nickname was "the little botanist." His earliest professional recognition on an international level came from several expeditions to Lapland (see May 12, 1732).

1718 Physician William Hunter is born in Long Calderwood, Scotland. He is credited with establishing obstetrics as an accepted branch of medicine.

1734 Austrian physician Franz Anton Mesmer is born in Iznang am Bodensee, the son of a forester. His life was a roller-coaster filled with controversy and tumult arising from a unique form of physio-mental therapy, called "mesmerism," that Mesmer invented and plied (to great profit) among wealthy and aristocratic clients in several European capitals. His techniques were the forerunner of modern hypnotism, and opened the door to the acceptance of Freud and psychotherapy.

1829 The expedition of HMS *Victory*, eventually to discover the magnetic North Pole, departs England under the leadership of Sir James Clark Ross and his uncle Sir John Ross.

1848 Aeronautical engineer Otto Lilienthal is born in Anklam, Prussia. His work laid the foundation for the Wright brothers. He specialized in designing, building, and flying gliders. After 2000 flights he died in 1896 in a crash.

1903 Dr. Horatio Nelson begins the nation's first coast-to-coast automobile trip in San Francisco. He is accompanied by a mechanic, Sewell K. Crocker of Seattle. The pair reached New York City on July 26. Of the 63 days required for the journey, 44 were spent traveling and 19 were passed waiting for parts and supplies.

1908 Physicist John Bardeen is born in Madison, Wisconsin. He shared two Nobel Prizes for Physics, the first (in 1956) with Shockley and Brattain for inventing the transistor, and the second (in 1972) with Cooper and Schrieffer for developing the theory of superconductivity.

1925 Joshua Lederberg, who discovered sex in bacteria, is born in Montclair, New Jersey. In 1946 he and Edward Tatum reported that genes from two strains of bacteria can recombine (hence, the discovery of a sexual interaction in this life form) to form a different strain. He, Tatum, and George Beadle shared the 1958 Nobel Prize for Physiology or Medicine. While others were still confirming and expanding this discovery, Lederberg and student Norton Zinder reported (in 1952) that bacteriophage viruses could transport genes from one bacterium to another, a process they named "transduction."

1962 The first human limb transplant takes place in Massachusetts General Hospital in Boston. Drs. Donald A. Malt and J. McKhann replace the entire right arm of a 12-year-old boy.

1994 The Indian State of Rajasthan outlaws amniocentesis and other tests that reveal a child's sex before birth. The law is an attempt to stop parents from murdering female offspring, either before or after birth. Because Indian women are required to present a dowry when they get married, female children are less desirable than male children. And indeed, India does have a lopsided birth rate in favor of males. Rajasthan is the third state to take such a step, after Punjab and Maharashtra.

1994 A "deadly flesh-eating bug" (which has been attacking Englishmen in Gloucestershire, about 100 miles west of London) is explained by British doctors. The outbreak has been caused by the common *Streptococcus* bacterium, normally responsible for sore throats and fever. What has been unusual about the Gloucestershire cases is that a cluster of infections have occurred together, and the infections have been so intense that the bacteria produced a toxin that dissolves flesh and fat.

1785 From Passy, France, a pleased Benjamin Franklin writes to friend George Whatley, describing the success of the new bifocal eyeglasses he has just invented. "I have only to move my eyes up and down as I want to see distinctly far or near." Franklin invented the glasses because he was tired of having to carry around several pairs of glasses to see objects at different distances.

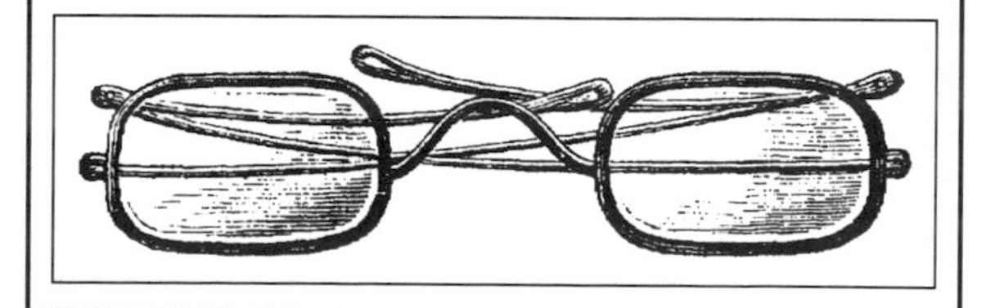

1544 Physicist-physician William Gilbert, England's most distinguished scientist during the reign of Elizabeth I (known especially for researching magnets), is born in Colchester. It was Gilbert who first suggested that Earth itself is a large magnet. Today, "gilberts" are units of magnetomotive force.

1640 Physiologist-chemist John Mayow, an early investigator into gases, is born in London. Long before Priestley and Lavoisier, he saw that breathing and combustion are related, and he identified *spiritus nitroaereus* (oxygen) as a distinct atmospheric substance.

1686 Physicist Gabriel Fahrenheit, inventor of the mercury thermometer, is born in Danzig, Germany.

1743 Jean-Paul Marat, one of the great villains in science history, is born in Boudry, Switzerland. Marat failed as a scientist, but used his considerable writing and political skills to have Lavoisier condemned and finally guillotined during France's Reign of Terror.

MAY 24

1844 Telegraph service in the United States begins with the message "What hath God wrought," sent on this day by inventor Samuel Morse, from the U.S. Supreme Court room in Washington, D.C., to his partner Alfred Lewis Vail in the Baltimore offices of the Baltimore and Ohio Railroad Company. Vail then retransmits the same message back to Morse.

1876 The expedition that began modern oceanography, the four-year circumnavigation by HMS *Challenger*, ends in Spithead, England.

1898 Helen Brooke Taussig, renowned investigator of heart problems in babies, is born in Cambridge, Massachusetts (see May 20, 1986, for a biographical sketch).

1899 The nation's first public car-repair shop is established. W.T. McCullough opens the "Back Bay Cycle and Motor Company" in Boston. He advertised it as a "stable for renting, sale, storage and repair of motor vehicles."

1543 The astronomer Copernicus dies at 70 in Frauenberg, Prussia (now Frombork, Poland). Through his research and theories, Copernicus was responsible for one of the greatest ever revolutions in human thought: that Earth revolves around the sun, rather than being the center of the universe. Copernicus did not consent to full publication of his findings until very late in life; it is said that a finished copy of his masterwork was brought to him on the last day of his life.

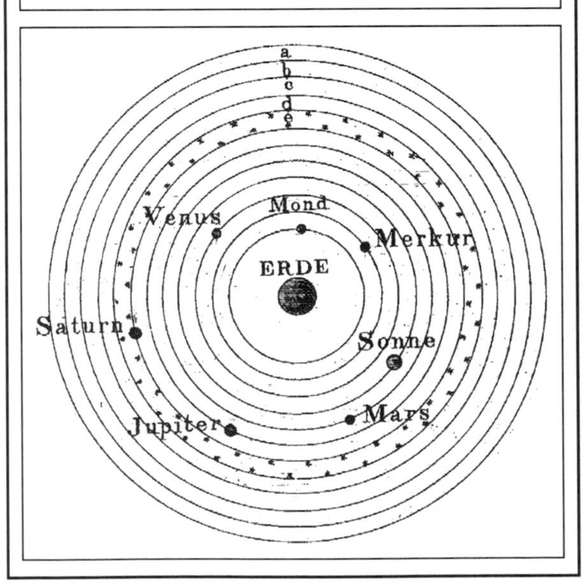

1959 The first home in the United States with its own built-in nuclear bomb shelter is exhibited. It was built at Hi-Tor Woods, Pleasant Hills, Pennsylvania, by Pittsburgh's Obie Construction company. The shelter sleeps four, has a fully functional kitchen, a Geiger counter and a fire extinguisher. The walls are ten inches thick and insulated with radiation-blocking lead.

1976 Britain and France begin flights to Washington of the SST *Concorde*.

1993 Louisiana-Pacific Corp., one of the world's largest lumber companies, is fined $11 million for violating clean-air laws at 14 facilities in 11 states. It is the nation's second largest civil fine for an environmental offense. The company (which had falsified documents and failed to get proper permits for gaseous discharge) is further ordered to install $70 million worth of antipollution equipment.

1994 Before a single ounce of soil is moved, the struggle to cleanup Florida's Everglades has been complicated and costly. To date, just deciding who will pay for the cleanup has cost federal taxpayers $9 million and Florida taxpayers $7 million in legal fees in trying to get sugar growers to pay their share. The Florida Sugar Cane League has put $11.5 million in battles against the state since 1988. The actual cleanup is estimated to cost $685 million.

1721 Apparently the first insurance advertisement in America's history appears today in Philadelphia's *American Weekly Mercury*. John Copson (said to be the country's first fire insurance agent) announces that he is opening an office to insure "vessels, goods and merchandise."

1775 In a letter to England's Royal Society, Joseph Priestley summarizes a series of experiments comparing the combustion of nitrogen and oxygen. He interprets the results to support the now-defunct phlogiston theory of matter. Priestley's procedure for isolating oxygen was used by Lavoisier to discredit the phlogiston theory forever.

1804 The preface is dated on the first U.S. book on distilling. Writing about this august craft is Michael August Krafft. His book *American Distiller, or The Theory and Practice of Distilling, according to the latest discoveries and improvements, including its most important methods of constructing stills and of rectification* was published in Philadelphia by Thomas Dobson. It had 219 pages and was dedicated to Thomas Jefferson.

1832 Twenty-year-old mathematical genius (famous for contributions to the branch of higher algebra called "group theory") Evariste Galois writes from prison (he is serving a six-month sentence for antigovernment agitating) to his friend Auguste Chevalier. "On re-reading your letter, I note a phrase in which you accuse me of being inebriated by the putrefied slime of a rotten world which has defiled my heart, my head, and my hands.... Inebriation! I am disillusioned of everything, even love and fame. How can a world which I detest defile me?" Four days later he was released from prison. The following day he was mortally wounded in a pistol duel with an unknown assailant for unknown reasons.

1844 History's first news dispatch telegram is sent from Washington, D.C. to the Baltimore *Patriot.*

1860 Geologist Daniel Moreau Barringer is born in Raleigh, North Carolina. In 1905 he suggested that an enormous crater in Arizona had been created by a meteor strike, rather than by a volcano. Although laughed at originally, the theory is now accepted, and the Great Barringer Meteor Crater is world famous.

1865 Dutch physicist Pieter Zeeman is born the son of a Lutheran minister in Zonnemaire. For discovering the Zeeman effect (in which spectral lines from a light source are split into two or more components when that light source is placed in a strong magnet; the effect established the relationship between magnetism and light, and has been used to explore the structure of atoms and stars), he shared the second Nobel Prize for Physics (in 1902) with one of his professors, Hendrik A. Lorentz.

1913 Stung by Carl Jung's criticisms and recent defection from the mainstream psychoanalytic movement, Freud convenes the first meeting of the Psychoanalysis Committee in Vienna.

1961 President John Kennedy delivers his famous speech about getting a man to the moon before the decade ends. "I believe this nation should commit itself to achieving the goal, before the decade is out, of landing a man on the moon and returning him safely to Earth." His vision was achieved on July 20, 1969.

1973 The first crew of *Skylab* is launched from Cape Kennedy, over a week after the space station itself was put in orbit. *Skylab* was the first manned U.S. space laboratory, and was in space for over six years. One of the astronauts in today's launch, Dr. Joseph P. Kerwin, became the first U.S. doctor to practice in space.

1994 Who says you can't take it with you? Wire services report that George Swanson and his white Corvette were recently buried together in Irwin, Pennsylvania. The 1984 car only had 27,000 miles on it. Following his wishes, Swanson's ashes were placed in the driver's seat, two caps were put in the back, and red roses were draped on the hood. An Engelbert Humperdinck tape was in the cassette player, with the song *Release Me* ready to play.

1994 New York's Suffolk County opens carpool lanes on the fabled Long Island Expressway in an effort to reduce the horrendous pollution caused by automobile commuters. Only multioccupant vehicles are allowed in the reserved lanes from 8 AM to 6 PM weekdays. It didn't take long for the first cheat to be caught—one day. Amelian Woff, 58, is caught the next day with a dummy beside her.

1990 "Rodney" the oldest caged rat on record, dies in Tulsa, Oklahoma, at the age of 7 years 4 months.

735 English scholar Bede (who established the tradition of dating events from the birth of Christ, as AD or BC) dies at about 62 in Jarrow.

1667 French mathematician Abraham de Moivre, founder of analytic geometry and the first to develop fundamental formulas in probability, is born the son of a surgeon in Vitry. Religious intolerance drove him from his homeland, after a period of imprisonment, in his 20s.

1676 Delft, Holland, is having a torrential rainstorm. Antonie van Leeuwenhoek (43-year-old haberdasher with a hobby of making microscopes with his own handmade lenses) seizes the opportunity. He takes some roof runoff water and studies it with one of his microscopes. The water is alive! It contains "very little animalcules." He sees none of this live activity in pure rainwater. In satisfying his own curiosity, he has made a fundamental discovery about bacteria and one-celled animals: They do not fall from the sky.

MAY 26

1815 Glassblower-inventor Heinrich Geissler is born the son of a German burgomaister in Igelshieb. Geissler developed techniques to produce vacuums of unprecedented purity. His products were called Geissler tubes, and they enabled physicists to study electricity and the structure of matter in more detail than previously possible.

1848 Alfred Russel Wallace, 25, and Henry Walter Bates, 23, reach South America. Their lives and biology history change forever (see May 28, 1848, for more on the expedition).

1876 Robert Mearns Yerkes, one of the founders of comparative (i.e., animal) psychology in the United States, is born the son of a farmer in Breadysville, Pennsylvania. His 1907 book, *The Dancing Mouse* helped establish mice and rats as standard subjects in psychology experimentation. His 1929 book *The Great Apes* was science's standard work on primates for generations.

1826 English astronomer Richard Christopher Carrington is born the son of a brewer in London. Carrington entered Cambridge as a divinity student, but lectures there brought him from the spiritual contemplation of the heavens to scientific study of the heavens. He was the first to map the movements of sunspots, and from this deduced that the sun rotated more rapidly at the equator than at the poles ("equatorial acceleration") and that the spots were not fixed to any solid object. He was the first to observe/record a solar flare (1859).

1906 An act of Congress incorporates the country's first national archaeological society. The Archeological Institute of America was founded in Boston in 1879, with its first annual meeting held in 1880. Charles Eliot Norton was its first president.

1908 The first family automobile trip across the United States is over. Mr. and Mrs. Jacob Murdock, their three children, and a mechanic reach New York City in a four-cylinder, 30-horsepower Packard, having left Los Angeles on April 24. They did not travel at night or on Sundays, and suffered just one flat tire.

1938 Biochemist John Jacob Abel dies at 81 in Baltimore. He pioneered the study of human endocrine glands, and isolated adrenaline (1897), and he crystallized insulin (1926). He also invented the first artificial kidney.

1951 Sally Kirsten Ride, the third woman in space, is born in Encino, California.

1969 *Apollo 10* returns to Earth after a successful dress rehearsal for the first assault on the moon. During reentry, the capsule and its three astronauts (Stafford, Cernan, and Young) traveled at 24,971 mph, the fastest any humans have ever moved.

1994 President Clinton signs legislation that prevents the blockading of abortion clinics. After today, those blocking access to clinics will face jail and stiff fines. The first conviction under the Freedom of Access to Clinic Entrances Act was that of ex-minister Paul Hill the following October.

MAY 27

1647 America's first known execution for witchcraft occurs today. Achsah Young of Massachusetts is hanged.

1818 Ophthalmologist Frans Cornelis Donders, who discovered the causes of farsightedness and astigmatism, is born in Tilburg, the Netherlands. Left fatherless in infancy, Donders was raised by his mother and eight sisters. His 1864 masterpiece *On the Anomalies of Accommodation and Refraction* was the first important book on ophthalmology, the study of the structure, function, and diseases of the eye.

1887 Chemist-nuclear physicist Kasimir Fajans is born in Warsaw, Poland. He discovered several elements that are created through nuclear disintegration, and also discovered (simultaneously with Frederick Soddy of Great Britain) the all-encompassing "radioactive displacement law" (during radioactive decay, emission of an alpha particle reduces the atomic number by two, while emission of a beta particle increases the atomic number by one).

1897 Sir John Douglas Cockcroft is born the son of a textile manufacturer in Todmorden, Yorkshire, England. He was the first to build an atom smasher with artificially accelerated particles (the Cockcroft–Walton generator, completed and bombarding lithium in 1932). In 1951 he and partner Ernest T.S. Walton shared the Nobel Prize for Physics. During World War II, Cockcroft helped develop radar and the atomic bomb.

1907 Rachel Carson is born in a home surrounded by trees, in Springdale, Pennsylvania. Her lifelong love of nature enabled her to write a series of award-winning books: *Under the Sea-Wind* (1941), *The Sea Around Us* (1951), *The Edge of the Sea* (1955), and *Silent Spring* (1962). This last book told a naive public about the hazards of pesticides, produced important antipesticide legislation, and is credited with starting the modern environmental movement.

1910 Robert Koch, cofounder of the science of bacteriology, dies at 66 in Baden-Baden, Germany. His scientific contributions were enormous, foremost of which were discovering the organisms that cause cholera and tuberculosis. He won the 1905 Nobel Prize for Physiology or Medicine, but the award resulted in the greatest fiasco in Nobel history. The great scientist was awarded the prize while his antituberculosis serum was still in development; the drug started killing people six months after the award was made. The Nobel judges were denounced as idiots.

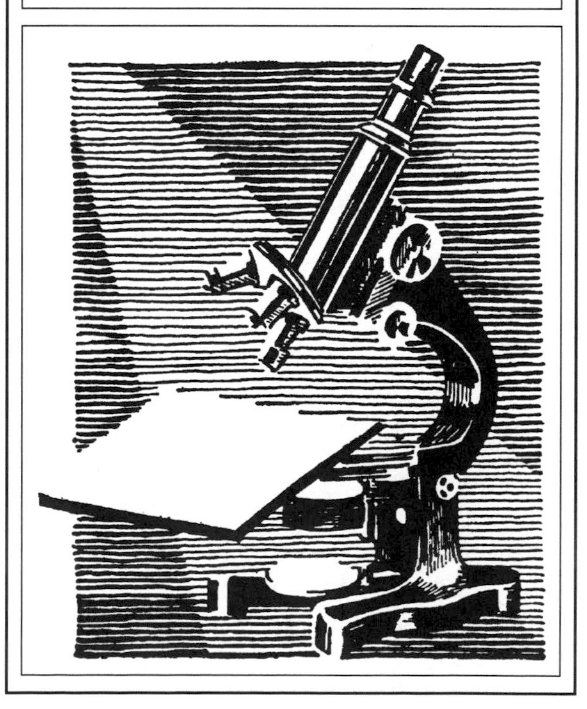

1930 Cellophane tape is patented. Richard Gurley Drew of St. Paul, Minnesota, receives patent No. 1,760,820 for "Adhesive Tape." Drew assigned his invention to the 3M Company (Minnesota Mining and Manufacturing), which introduced the tape for sale in September 1930.

1931 Man first enters the stratosphere. Auguste Piccard (a former pupil of Einstein's) and assistant Charles Kipfer take an 18-hour balloon ascension to 51,775 feet (nearly 10 miles high), in a pressurized cabin designed by Piccard. Among the equipment aboard is an electroscope, to detect/measure cosmic rays. When the hot sun brought the cabin near baking temperatures, the crew survived by licking water drops off the walls.

1957 In groundbreaking legislation against noise and vibration pollution, the Chicago City Council sets maximum sound pressure levels for manufacturing zones, for business districts, and for boundaries of residential areas.

1992 The Environmental Protection Agency reports statistics for pollutants released in 1990 in the United States. In that year, the nation's factories spewed 4.8 million pounds of toxic chemicals into the air, water, and soil. Officials further warn that the actual emissions may have been higher because many companies are not required to report their pollution every year, and new reporting techniques may reduce, on paper only, estimates of the actual toxic outpour.

1734 America's first legislation to protect fish is enacted by New York City. For "preserving fish in fresh water ponds," a fine of 20 shillings is set for anyone catching fish with any device (any net, machine, or "art") other than a rod, hook, and line.

1807 Louis Agassiz, one of history's greatest natural scientists and educators, is born the son of a minister in Mŏtier, Switzerland. At 19, while still a student in Munich, he assumed the task of classifying many Amazonian fish specimens when one of the collectors died. The ensuing publication established Agassiz as a first-class scientist. He became the world's preeminent fish expert in his day, but he was also known for studies of other animal groups and for his studies of glaciers (it was Agassiz who concluded that huge ice sheets once covered many now-temperate parts of Earth). In 1847 he became a professor of zoology at Harvard where he established its still-famous museum of comparative zoology. Agassiz was also a revered and revolutionary teacher; he pioneered replacing book-learning with experiential learning in the study of science, and thereby greatly influenced higher education in the United States. Ironically, he never accepted Darwin's theory of evolution; to his death he regarded each species as a divinely created "thought of God," rather than being the result of eons of organic interaction.

MAY 28

1848 Entrepreneuring explorers Alfred Russel Wallace (age 25) and Henry Walter Bates (age 23) reach the Amazon River in their first insect-gathering expedition. They planned, and succeeded, in covering their costs by selling tropical specimens to private collectors and to the British Museum in London. Western science knew little about tropical life-forms at the time. Both men were to make even deeper contributions to knowledge: Bates developed a now-accepted theory to explain mimicry, and Wallace created the theory of evolution independently of Charles Darwin (it was Wallace who coined the phrase"survival of the fittest").

1881 "An act to prevent the adulteration of food or drugs" is passed by New York State today. It goes into effect the following August. It is the nation's first state or federal law that attempts to ensure the purity of ingested substances on a blanket scale. Infractions are categorized as misdemeanors, with a $50 fine for the first offense and a maximum $100 fine for ensuring offenses.

1892 The Sierra Club is organized, in San Francisco.

1992 The World Wildlife Foundation announces that "Mrithi," the mountain gorilla who has starred in several movies including *Gorillas in the Mist*, has been killed in Rwanda. He is the first known gorilla fatality in recent political struggles between the Rwandan government and the Rwandese Patriotic Front.

1900 The name "escalator" is registered in the U.S. Patent Office.

1929 *On With The Show*, the first talking picture completely in color, is exhibited at the Winter Garden in New York City. Produced by Warner Brothers and filmed in Vitaphone Technicolor, the movie's cast included Joe E. Brown and Ethel Waters.

1930 Astronomer Frank Donald Drake,known for his studies of radio signals from outer space and his quest for signs of extraterrestrial life, is born in Chicago. With Carl Sagan, Drake designed the greeting/informational plaques on *Pioneer 10* and *Pioneer 11*, the first artifacts to leave our solar system in search of alien intelligence.

1934 The Dionne quintuplets are born on a farm in Ontario, Canada.

1937 San Francisco's Golden Gate Bridge is opened to vehicular traffic, when President Franklin D. Roosevelt pushes a button in Washington.

1959 For the first time in space exploration, animals are launched hundreds of miles above Earth and recovered alive. Able and Baker (two one-pound female monkeys, a rhesus and a spider monkey, respectively) are sent up 300 miles in the nose cone of a Jupiter rocket, and are recovered off Antigua by frogmen 90 minutes later. Able perished several days later during an operation to remove an electrode implanted under her skin.

1716 Louis-Jean-Marie Daubenton, a pioneer in both anatomy and paleontology, is born in Montbard, France (see January 1, 1800, for a biographical sketch).

1781 Naturalist Henri Braconnot is born in Commercy, France, the son of a lawyer who died seven years later. Braconnot is known for isolating glucose directly from plant material (it had previously been retrieved only from starch).

1794 Johann Heinrich von Mädler, astronomer (he produced maps of the moon and Mars that were unsurpassed in detail for decades) and science popularizer (his 1841 *Popular Astronomy* went through six editions during his lifetime), is born in Berlin.

1898 Alfred Nobel's heirs sign a "reconciliation agreement," indicating their approval of final plans by lawyers and accountants to execute his will. The will created the Nobel Prizes, but for the 18 months following his death, it created only hassles and bitterness.

1909 The nation's first domestic relations court is established in Buffalo, New York, by state legislation passed today. The court officially opened the following January, but Simon Augustine Nash, Buffalo's Judge of Police Court, had unofficially been hearing domestic disputes privately in his chambers for some time.

1919 Einstein's theory of relatively is dramatically verified during a solar eclipse. Astronomers from the Royal Astronomical Society of London, in separate expeditions to Brazil and Africa, observe a bending of light from several bright stars as this light passes our sun. Einstein predicted such an effect (based on his hypothesis that light has mass and would therefore be pulled by the sun's gravity), and today's verification made him world famous, the most revered scientist since Newton.

1935 The nation's America's first federal sanatorium for drug addicts opens its doors to patients. The United States Narcotics Farm in Lexington, Kentucky, consists of one building on 11 acres, and had been dedicated four days earlier by Surgeon General Hugh Smith Cumming. Dr. Lawrence Kolb was the facility's first director.

1942 The all-time greatest-selling phonograph record (*White Christmas*, written by Irving Berlin) is recorded by Bing Crosby. Sales had reached nearly 200 million by 1990.

1953 Mount Everest (which has recently been described as "the world's tallest garbage heap" because of all of the trash visitors have left on it) is scaled for the first time by Tenzing Norgay and Sir Edmund Hillary.

1972 A mother at the University of Pennsylvania hospital in Philadelphia gives birth to nine babies. Only two other cases of nonuplets are known in medical history. Although there are reports of ten children being born at a single birthing, none of these reports were verified by medical authorities.

1985 An interesting case study in sociology occurs in Brussels, Belgium. Thirty-five people are killed and hundreds are injured in riots during a soccer match between England and Italy.

1994 A 55-year-old Englishman is mauled by a herd of cows near Tetsworth, 40 miles outside London. John Hine was walking his two golden retrievers in a remote pasture when the Jerseys attacked. Knocked to the ground, his leg broken, and his chest bruised, Hine was unable to get up, but he could call for help on his cellular phone. When the paramedics arrived they were led to the man by his barking dogs. "There were calves in the field, so I think the cows were being protective," explained John Willis of the ambulance service. "We called in the police helicopter, and when it arrived all the cows came over to see what was going on."

1423 Mathematician-astronomer Georg von Peurbach is born in Peurbach, Austria. He was instrumental in getting Arabic numerals adopted by European scientists for regular use.

1539 Spanish explorer Hernando de Soto lands in Florida. The expedition was the first European penetration of much of the American Southeast, and it included the "discovery" of the Mississippi River in 1541.

1832 Mathematical genius Evariste Galois is shot in a duel, and dies the next day (see May 25, 1832, for some details on his tragic life and death).

1848 The ice cream freezer is patented. William G. Young of Baltimore is awarded patent No. 5,601 for an "improvement in ice cream freezers."

1896 In the nation's first automobile accident a bicycle rider (Evylyn Thomas) is clipped by a Duryea "Motor Wagon" driven by Henry Wells of Springfield, Massachusetts. Thomas's leg is broken and Wells spends the night in jail.

MAY 30

1908 Swedish astrophysicist Hannes Alfvén is born in Norrköping. He shared the 1970 Nobel Prize for Physics for his role in the founding of plasma physics (plasma being gaseous aggregates of subatomic particles/ions, created when atoms break down in the presence of great heat).

1912 Biochemist-pharmacologist Julius Axelrod, winner of the 1970 Nobel Prize for Physiology or Medicine,is born in New York City. Axelrod was the first to find, identify, and name an enzyme that breaks down neurotransmitters once an impulse is passed from one nerve to another. In effect, this breakdown resets the nerve–nerve connection, in preparation for the next impulse that comes along.

1934 Cosmonaut Aleksei Arkhipovich Leonov, the first man to climb out of a spacecraft in orbit, is born near Kemerovo, Russia.

1964 Leo Szilard, one of the fathers of the Atomic Age, dies at 66 in La Jolla, California. He was a physicist in Germany until the Nazis took power, after which he went to Vienna, then England, and finally the United States. At Columbia University he was part of the group that urged Franklin Roosevelt in a famous 1939 letter to build an atomic bomb. At the University of Chicago he collaborated with Enrico Fermi to build history's first nuclear reactor. He was also a key scientist in the Manhattan Project that built the first atomic weapons. After World War II Szilard devoted himself to curbing nuclear weapons and using nuclear power peacefully.

1971 Space probe *Mariner 9* departs Cape Kennedy on a journey to Mars.

1991 The Associated Press publishes findings from an EPA (Environmental Protection Agency) study of secondhand smoke. It estimates that 53,000 nonsmoking Americans die annually from smoke inhalation. The figures were not compiled by the EPA itself, but rather by scientists commissioned by the Agency. The Agency itself has postponed indefinitely its official release of the document, in agreement with tobacco companies that more research is needed before publication.

1994 Ezra Taft Benson, former U.S. Secretary of Agriculture, dies in his Salt Lake City home at 94. He was one of only two cabinet members to hold his post for the duration of the Eisenhower administration, even though his years there were rocky. Most controversial was his rollback of government price support for agricultural products. This measure was unsuccessful in raising prices, and was rescinded. Many blamed Benson for subsequent defeats of Republican candidates in farm states. On leaving government, Benson devoted himself to church work and was elected president of the Mormons in 1985.

1911 The first Indianapolis 500 auto race takes place. Eighty-five thousand spectators watch the event; one fatality occurs; 6 of the 44 starters do not complete the race. The winning time (by Ray Harroun) is 6 hours 41 minutes 8 seconds (for a speed of 74.7 mph). This was the first long-distance race on a U.S. track.

1819 William Worrall Mayo, the patriarch of the most famous medical family in US history, is born near Manchester, England. He emigrated to the United States in 1845, took medical degrees from several schools and opened a soon-blossoming surgical practice in Rochester, Minnesota. There a disastrous tornado became a turning point for Mayo and for world medicine. After caring for casualties of the tornado with the Sisters of St. Francis, Mayo decided to build and run a new hospital in collaboration with his two sons and with the Sisters. St. Mary's Hospital opened on October 1, 1889. In the early 1900s this facility developed into the world-famous Mayo Clinic, by which time the elder Mayo had retired.

1831 Pasteur supervises the injection of virulent anthrax into sheep at Pouilly-le-Fort, France. His assistants have already inoculated some of the sheep with an experimental vaccine. Government dignitaries, the world press and hundreds of farmers witnessed the well-publicized experiment, and within three days the success of Pasteur's vaccine against a previously incurable disease made him an international celebrity.

1870 Asphalt pavement is patented. Professor Edward Joseph De Smedt of the American Asphalt Pavement Company in New York City receives patents Nos. 103,581 and 103,582 to cover his invention. He signed over his rights to the invention to the New York Improved Anthracite Coal Company. The material was first applied to William Street in Newark, New Jersey, on July 29, 1870. It was then called "French asphalt pavement."

1889 A dam breaks outside Johnstown, Pennsylvania. The horrific Johnstown Flood kills more than 2000.

1910 Elizabeth Blackwell, considered the first female doctor in modern medicine, dies at 81 in Hastings, England (see February 3, 1821, for a biographical sketch).

1970 History's deadliest avalanche claims the lives of 18,000 in Yungay, Peru.

1977 The trans-Alaska oil pipeline is completed after three years' intensive labor, when the final weld is made near Pump Station 3.

1985 The longest flight by a chicken on record (630 feet 2 inches) is accomplished by a "barnyard bantam" in Parkesburg, Pennsylvania.

1994 The U.S. Supreme Court hands down two rulings of biological significance. In a Washington State case, the justices ruled 7-2 that states can require hydroelectric power plants to provide some minimum amount of water flow so as to protect fish. In a Kentucky case, the court ruled (again 7-2) that "inappropriate" groups can be barred from government-sponsored activities, such as fairs, festivals, and parades. The case involved the Great Pumpkin Festival in Frankfort, Kentucky, in which an antiabortion group was refused a booth in 1990; at the previous year's booth the group handedout plastic fetuses to the crowd.

1994 Inky and Blinky become the first pygmy sperm whales to be released into the wild after suffering near-death strandings. The two were rescued in separate beachings; Inky had convalesced in Baltimore, and then was flown to Marineland in St. Augustine, Florida, where Blinky was already in recovery. The two are released in the Atlantic, about 30 miles off Cape Canaveral.

1790 President George Washington signs the first federal copyright law "for the encouragement of learning by securing the copies of maps, charts and books to the authors and proprietors of such copies during the times therein mentioned." Copyright protection extended over 14 years, and could be renewed as long as the copyright holder was still alive. Only U.S. citizens were eligible to obtain copyrights; this restriction lasted until 1891.

1638 North America's first recorded earthquake hits Plymouth, Massachusetts, at 2 PM. Exactly 250 years later (1888) the first seismographs in the United States are exhibited at the opening of the Lick Observatory on Mount Hamilton, California.

1796 Engineer designer-theoretician Sadi Carnot is born into a family of prominent politicians in Paris. Carnot was the first to measure the efficiency of engines (i.e., the amount of work done in relation to the heat used), and is therefore considered the founder of thermodynamics ("heat movement").

1831 Man reaches the magnetic North Pole. "The land at this place is very low," reports Scottish explorer James Clark Ross.

1849 Twins Francis and Freelan Stanley are born in Kingfield, Maine. Both were inventors, and in 1897 they began manufacturing the Stanley Steamer, history's best-known steam-driven car. The Stanley Motor Company continued production through World War I. In 1906 one of their cars was clocked at 127 mph, faster than any vehicle had ever gone before. Francis died in 1918, Freelan in 1940.

1854 Henry David Thoreau's *Walden* is published.

1880 The nation's first pay phone goes into service, in New Haven, Connecticut. Located in the Connecticut Telephone Company office in the Yale Bank Building, the telephone could be used after giving a coin to an attendant.

1881 Pasteur is informed that all of the unvaccinated sheep are sick, in the field trials of his anthrax vaccine. It is a great triumph for all of medical science, but Pasteur spends a sleepless night when a late telegram reports that one of the vaccinated sheep also seems ill.

1907 Sir Frank Whittle, inventor of the first practical jet engine, is born in Coventry, England.

1907 Police in Berlin begin enforcing a ban on motion pictures, after doctors declared the fluttering motion harmful to children's eyes.

1911 The *Titanic* first touches water. It is launched from the shipyard into the River Laffan in Belfast, Ireland.

1925 Lou Gehrig, for whom one of the muscular dystrophy diseases is named, pinch-hits for Pee Wee Wanninger in the eighth inning and then replaces Wally Pipp at first base for the New York Yankees. It is the start of Gehrig's historic and heroic streak of 2130 consecutive games played. It is this streak that earned him the nickname "The Iron Horse," and which made his death at just 37 all the more dramatic.

1968 Helen Keller dies at 87 in Westport, Connecticut.

1980 CNN, television's first all-news service, debuts.

1992 The Tennessee Supreme Court rules in favor of a man trying to stop his ex-wife from using their frozen embryos. The seven embryos were produced by doctors artificially uniting his sperm and her eggs; they have been frozen in a Knoxville clinic for 3½ years. After divorcing, the wife first announced plans to have the embryos implanted in her, but then said she would donate them to infertile couples. Today's ruling means she can't do anything with them without her ex-husband's permission.

1994 Israeli archaeologists announce finding evidence that hashish was used by primitive women, apparently to ease the discomfort of childbirth. A mixture of fruit, grasses, and hashish was found in the abdomen of a young woman, dead for at least 1600 years, with a full-term fetus still inside her.

1869 Thomas A. Edison, 22, receives his first patent. It is for a mechanical voting machine that would speed up the proceedings of Congress. When a congressman informed Edison that the device was not wanted because there was no desire to hasten the proceedings, Edison vowed never again to invent anything that had an uncertain demand or need.

JUNE 2

1686 The minutes of London's Royal Society report, "It was ordered that Mr. Newton's book be printed, and that Mr. Halley undertake the business of looking after it and printing it at his own charge." The book is the *Principia*, one of science's greatest works, and it possibly might never have been published without the encouragement and finances of astronomer Edmund Halley.

1787 Swedish chemist Nils Sefström, discoverer of the element vanadium, is born in Hälsingland.

1850 Physiologist Edward Schafer is born in Hornsey, England. With George Oliver in 1894 he demonstrated the existence of epinephrine (adrenaline), which spurred great interest in the nature and function of hormones; adrenaline later became the first hormone to be isolated and purified. Schafer also invented the prone pressure method of artificial respiration; called the "Schafer method," it became standard lifesaving procedure. Schafer's mentor was William Sharpey of University College, London; in 1918 Schafer changed his name to Sharpey-Schafer.

1854 Max Rubner is born the son of a locksmith in Munich. Rubner was to unlock a mystery himself; his painstaking measurements proved that the energy yield of foodstuffs was the same whether the material was burned or eaten by an organism.

1857 The chain-stitch sewing machine is patented by James Gibbs of Mill Point, Virginia.

1874 The cornerstone of the American Museum of Natural History is laid in New York City.

1881 Pasteur is elevated to the status of Messiah, when dignitaries and the world press observe the fabulously successful results of field trials of his anthrax vaccine, outside Melun, France.

1889 A hydroelectric power plant provides alternating current over a long distance for the first time in U.S. history. The power originates at the Willamette Falls Electric Company power plant at Willamette Falls, Oregon, and is carried to users in Portland, Oregon, 13 miles away.

1896 The first radio patent is issued to Guglielmo Marconi by England.

1924 Congress extends citizenship to American Indians.

1930 In a first of its kind, a baby on a ship passes through the birth canal while the ship passes through the Panama Canal. Mr. and Mrs. M. Niezes of Panama have their child on board the Dutch steamship *Baralt* as it moves through Gatun locks.

1932 Lou Gehrig becomes the first to hit four consecutive home runs in a major league baseball game, in the 20-13 victory of the New York Yankees over the Philadelphia Athletics. Exactly nine years later (1941) Gehrig dies from amyotrophic lateral sclerosis (ALS). It is one of the muscular dystrophy diseases, and is now called "Lou Gehrig's disease."

1954 The first vertical takeoff and landing by an airplane is accomplished by the Convair XFY-1.

1966 *Surveyor 1* becomes the second craft to achieve a soft landing on the moon, when it touches down in the Ocean of Storms and begins engineering tests and photographing. Exactly four months before, the Soviet *Luna 9* achieved the first soft landing.

1993 Navajo officials in Arizona announce that medicine men will be enlisted in the battle against a mysterious flulike illness that has already killed 13 people. Eventually the responsible organism was found to be a virus, a "hantavirus," transmitted to humans through rodents.

1875 "Don't touch anything!" Alexander Graham Bell screams to assistant Thomas Watson, who had accidentally mistightened a screw on their prototype telegraph. The mistake allowed Bell to hear a continuous sound through the device, which was a critical step in his development of the telephone.

1657 William Harvey, who revolutionized science and human thought by demonstrating that blood circulates, dies at 79 in London. His work was the first great study of human physiology.

1659 Scottish mathematician-astronomer David Gregory is born in Aberdeen.

1769 During the first scientific expedition to the Pacific, James Cook and the HMS *Endeavor* reach Australia, and subsequently establish its size and location.

1833 The first clipper ship, the *Ann McKim*, is launched in Baltimore. Exactly eighteen years later (1851) the most famous clipper ship, the *Flying Cloud*, sets off on its maiden voyage from Sandy Hook, New Jersey.

1856 The nation's first screw machine (for making pointed screws) is patented by Cullen Whipple of Providence, Rhode Island. His is patent No. 15,502.

JUNE 3

1873 Physiologist Otto Loewi is born the son of a wine merchant in Frankfurt am Main, Germany. He shared the 1936 Nobel Prize for discovering neurotransmitters. The idea for the critical experiment came to him at 3 AM on two successive nights in 1921; the first night he made some notes and went back to sleep, but was unable to read the notes the next morning. That second night the vision returned and he immediately began the experiment. It was history by 5 AM.

1899 Georg von Békésy is born the son of a diplomat in Budapest. He was first to determine what happens in the inner ear when sound waves hit the eardrum. In 1961 Békésy became the first physicist to win a Nobel Prize for Physiology or Medicine.

1929 Microbiologist Werner Arber is born in Granichen, Switzerland. He made recombinant-DNA techniques possible by exploring restriction enzymes in bacteria, and shared the 1978 Nobel Prize.

1948 Earth's first telescope with a lens 200 inches in diameter is dedicated at the Mount Palomar Observatory on Palomar Mountain, San Diego, California. It is officially named the Hale telescope, after Dr. George Ellery Hale, who conceived, designed, and promoted the instrument. Hale died long before it was completed. Glass for the enormous lens was poured into a huge ceramic mold at the Corning Glass Works in Corning, New York, on December 2, 1934. It was cooled slowly, over a period of 11 months. The 20-ton disk was then shipped to the California Institute of Technology where it was ground and polished carefully, over a period of 11 years. On February 1, 1949, the work paid off, when observations were made of the constellation Coma Berenices, over 6 billion trillion miles away.

1961 Mrs. Saadet Cor of Cegham, Turkey, gives birth to a 24-pound 4-ounce boy, the heaviest human birth on record.

1965 The first U.S. spacewalk is taken by Edward White outside *Gemini 4*.

1979 The world's worst oil spill from an oil well blowout at sea begins beneath the drilling rig *Ixtoc I* in the Gulf of Campeche off Mexico. Some 505,600 tons of oil enter the water, creating a 400-mile slick.

1994 On the second anniversary of the historic "Earth Summit" in Rio de Janeiro (in which scientists and politicians from 160 countries met to try to slow man's degradation of Earth), a truck overturns outside the Siberian village of Udachny, spilling huge containers of radioactive waste into the Taldyn River.

1994 Wire services announce that water has been discovered farther from Earth than it's ever been detected before. Scientists feel that where there is water, there may be life. "If this stuff is out there, then there is a good chance that there are living beings elsewhere," observed Jack Walsh of Berkeley's radio astronomy laboratory. The water was discovered in Markarian 1, a galaxy 200 million light-years away. Appropriately, it is located in the constellation Pisces—the fish.

1726 James Hutton, credited with founding the science of geology, is born the son of a merchant in Edinburgh.

JUNE 4

1741 Danish navigator Vitus Bering departs Kamchatka to explore the Bering Sea, the Bering Strait, and Bering Island (where he eventually died of scurvy). On board is naturalist Georg Wilhelm Stellar, who becomes the first European scientist to see a number of life-forms on and around Alaska. Among these is Stellar's sea cow, hunted to extinction 30 years after being discovered.

1756 French chemist Jean Chaptal, author of history's first book on industrial chemistry (and originator of the word "nitrogen"), is born the son of a small landowner in Nogaret.

1844 Three fisherman go to the Icelandic island of Eldey and kill the last two great auks on Earth.

1877 Chemist Heinrich Wieland is born the son of a chemist in Pforzheim, Germany. He won the 1927 Nobel Prize for studies of steroid chemistry, and is noted for studying the conversion of food into energy.

1886 History's first successful gas-driven car (the Motor-wagon, built by Germany's Karl-Friedrich Benz) has completed its maiden test run in Mannheim, announces the local newspaper under "Miscellaneous."

1892 The ecological society, the Sierra Club, incorporates in San Francisco.

1896 Henry Ford, 32, takes an ax and rips out the door frame of the workshed behind his house on Bagley Street in Detroit. It is 2 in the morning. Having just completed his first car, he is about to take it on its first test run—once he gets it out of his garage.

1897 Good news and bad news for the forests and wildlife of the United States: Congress enacts legislation that allows the President to create forest reserves, but provides that "no public forest reservation shall be established except to improve and protect the forest, secure favorable conditions of water flow, and furnish a continuous supply of timber for the use and necessities of citizens of the United States." This law has been U.S. forestry policy ever since, under which most U.S. forests have disappeared.

1926 Robert Earl Hughes is born in Monticello, Missouri. He weighed 203 pounds at 6 years of age, and in February 1958 he weighed 1069 pounds. It remains the heaviest-ever precise weight for a human.

1931 The first rocket-powered glider flight in U.S. history is accomplished by pilot William G. Swan in Atlantic City. At an altitude of 100 feet Swan ignites the rockets, shooting the 200-pound craft 1000 feet forward. The next day Swan attached 12 rockets to the glider, and stayed in the air for 8 minutes.

1962 Explorer-zoologist William Beebe dies at 84 at the Simla Research Station near Arima, Trinidad (see July 29, 1877, for a biographical sketch).

1989 "Frankencycle," history's largest bicycle (with a front wheel diameter of ten feet), is ridden for the first time, in California.

1991 Nineteen-year-old Anissa Ayala receives a lifesaving bone marrow transplant at the City of Hope National Medical Center in Duarte, California. The donor is Anissa's 13-month-old sister Marissa, who was conceived specifically to provide the marrow.

1993 Portugal's Environment Minister Carlos Borrego gets himself into hot water while addressing a conference on water quality at Minho University in Braga. Speaking about an ecological disaster in the Alentejo region that killed 16 when aluminum got into the water supply, Borrego asked the audience (and a national audience several days later when his remarks were broadcast over TSF radio), "Do you know what they do with people who die in the Alentejo these days? They take them to the recycling plant to recover the aluminum." His resignation was accepted the day after the radio broadcast.

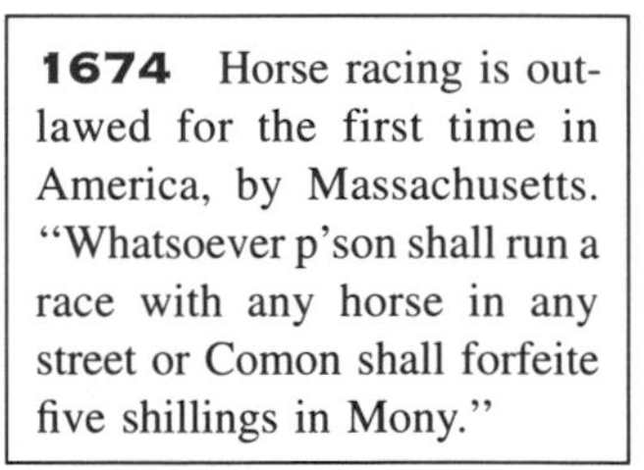
1674 Horse racing is outlawed for the first time in America, by Massachusetts. "Whatsoever p'son shall run a race with any horse in any street or Comon shall forfeite five shillings in Mony."

1760 Finnish chemist John Gadolin, discoverer of the element yttrium, is born the son of a physicist-astronomer in Åbo. Gadolin made his discovery on a 1794 field trip to a rock quarry in Ytterby, Sweden. This was the first of a family of 15 elements called the "lanthanides" or "rare earth elements" to be discovered.

1799 Legendary naturalist Alexander von Humboldt and Aimé Bonpland depart La Coruña, Spain, for a five-year scientific expedition in the South American jungles.

1846 The first U.S. telegraph line opens, between Philadelphia and Baltimore.

1862 Ophthalmologist Allver Gullstrand is born the son of a physician in Landskrona, Sweden. He won the 1911 Nobel Prize for Physiology or Medicine for studying how light is refracted in the eye.

JUNE 5

1877 New York State passes the first U.S. law to prevent the cheap and newly developed margarine from overwhelming the butter market. The legislation levies a tax on oleomargarine "for the protection of dairymen and to prevent deception in sales of butter." In 1886 a nationwide tax on margarine was passed.

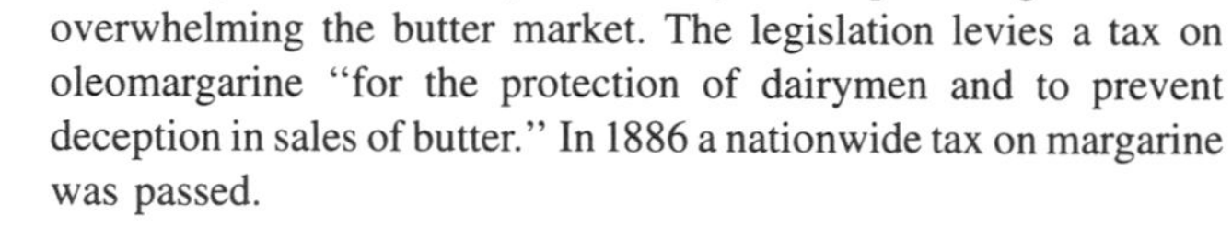

1887 Anthropologist Ruth Fulton Benedict is born in New York City.

1898 All of the heirs finally agree on the settlement of Alfred Nobel's will, which created the Nobel Prizes.

1900 Physicist Dennis Gabor, creator of holography (the photographic process that produces three-dimensional images), is born in Budapest. He won the 1971 Nobel Prize for his invention, the idea for which came to him while watching a tennis match at Rugby, England, in 1947.

1914 The oldest alligator on record arrives at the Adelaide Zoo, Australia. She was at least 66 when she died in the zoo after 64 years in captivity.

1938 The first machine capable of producing speech sounds is exhibited at the Franklin Institute in Philadelphia by Bell Telephone scientists. The device is named Pedro the Voder.

1946 The artificial sponge is first sold (for medical and surgical use). Under the trademark Oxycel, the material is marketed as an absorbent wound-cover by Parke Davis & Company of Detroit. The artificial sponge (made of processed cellulose) was invented in 1936 by Dr. William Orlin Kenyon.

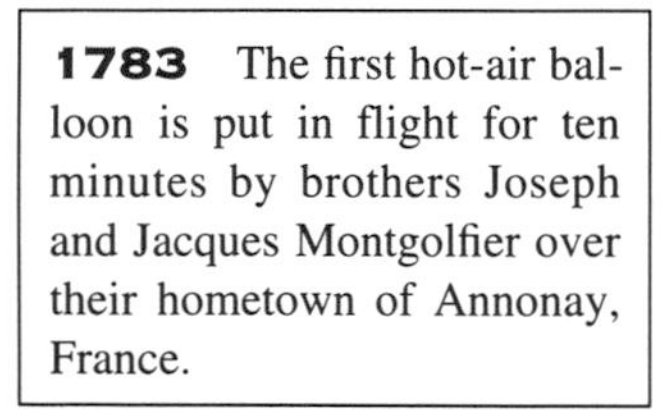

1783 The first hot-air balloon is put in flight for ten minutes by brothers Joseph and Jacques Montgolfier over their hometown of Annonay, France.

1980 All Europe is covered in a white cloud, created by the Mount St. Helens eruption 18 days before in the western United States.

1981 In a brief note in the newsletter of the CDC (the U.S. Centers for Disease Control in Atlanta) Dr. Michael Gottlieb describes a strange disorder that he has observed in five homosexual men. It is history's first publication on AIDS (although the disease was not yet named).

1991 Min-Chueh Chang, coinventor of the birth control pill, dies at 82 in Worcester, Massachusetts. Among his other accomplishments was pioneering *in vitro* fertilization, the process that produces test-tube babies. Today is also the tenth anniversary of the birth of the first test-tube twins (1981; Stephen and Amanda Mays were born by Cesarean section in Melbourne, Australia).

1992 "Highly coercive. Ghoulishly coercive." This was the response of one lawyer to today's proposal by lawyer Ellis Rubin to allow prison inmates to donate organs in return for a reduction in their sentences.

1992 Just one year and one day after receiving a lifesaving bone marrow transplant, 20-year-old Anissa Ayala is married in Redlands, California. The star of the ceremony is actually Anissa's 2-year-old sister, Marissa, who donated the marrow. Marissa was conceived specifically to provide the cancer-fighting substance.

1436 Regiomontanus is born the son of a miller in Königsberg, Germany. Although an ardent astrologer and strongly opposed to the notion that Earth moved, Regiomontanus achieved widespread fame in his own life (he was an adviser of Pope Sixtus IV), and made lasting contributions to astronomy. He was the first to scientifically study a comet (later called Halley's Comet), and he composed tables of planetary motion that many navigators, including Columbus, used.

1819 Physiologist Ernst von Brücke is born in Berlin. Best known as an early influence on Freud, Brücke was an eminent researcher (into vision, speech, and muscle) who helped bring physical/chemical methods and animal experimentation to medical research. He spent the last 42 years of his life as a professor of physiology at the University of Vienna.

1822 Dr. William Beaumont visits the wounded trapper Alexis St. Martin for the first time; it is the beginning of a historic investigation of digestion (see April 25, 1853, for a review of the work).

1832 Social philosopher Jeremy Bentham dies at 84 in London (see February 15, 1748, for a biographical sketch).

1833 Andrew Jackson becomes the first U.S. president to ride a railroad (on a pleasure trip to Baltimore), and exactly 148 years later (1981) history's worst railroad disaster occurs in Bihar, India, when a wreck kills at least 800 at the Bagmati River.

1882 The electric iron is patented. H.W. Seely of New York City receives patent No. 259,054 for an "electric flatiron."

1891 History's first polar expedition to include a woman sets sail from New York City. Josephine Diebitsch Peary is on board the *Kite* with the group led by her husband Robert Edwin Peary, bound for the North Pole. They never made it. Eventually a 1909 expedition by Robert Peary did reach the Pole, although an unresolved dispute remains as to whether this was man's first trip there. Exactly 23 years earlier on this date (1868) polar explorer Robert Falcon Scott was born in Devonport, England.

1925 Automobile magnate Walter Percy Chrysler founds the Chrysler Corporation. Automobile designer Louis Chevrolet dies on this date in 1941.

1933 The first drive-in movie theater opens on a 10-acre plot in Camden, New Jersey.

1961 Psychiatrist Carl Jung dies at 85 in his mountain retreat at Küsnacht, Switzerland.

1991 FDA Commissioner David Kessler slams companies for putting "low-fat" labeling on products that inherently contain large amounts of fat. Bologna and vegetable oil are two examples cited. The speech, made at a Washington, D.C., conference of the Center for Science in the Public Interest, comes after a recent crackdown on the misleading use of the word "fresh" in labeling. The FDA seized 12,000 gallons of Proctor and Gamble's "Citrus Hill Fresh Choice" orange juice after the company did not change the name.

1992 The formation of a new environmental organization, the International Green Cross, is announced in Rio de Janeiro by Mikhail Gorbachev. He reads a prepared statement at the Global Forum of Spiritual and Parliamentary Leaders, which is holding its own meetings at the same time that representatives from 160 nations are meeting at Rio's Earth Summit. Gorbachev will be the Green Cross's first chairman. Specific details are not provided about the group's headquarters, mission, or projects.

1850 Levi Strauss produces his first pair of jeans.

1692 An earthquake hits Port Royal, Jamaica, and within two minutes most of the town and its inhabitants slide into the ocean. In 1959 it became the first archaeological site explored by the *Sea Diver*, the research submarine designed by Edwin A. Link.

1811 One of the Simpsons is born in Bathgate, Scotland. It is Sir James Young Simpson, a pioneer in alleviating the pain and danger of childbirth.

1862 Physicist Philipp Lenard. A Nobel laureate in 1905, Lenard was a rabid anti-Semite who convinced Hitler that Einstein's work was worthless.

1870 An automatic signaling system, designed to prevent railroad trains from running into each other, is patented by Thomas S. Hall of Stamford, Connecticut. A railroad car made of steel and designed to carry mail first went into service on this date in 1905 on the New York, Salamanca and Ohio RR line.

JUNE 7

1873 Anatomist-anthropologist Franz Weidenreich is born in Edenkoben, Germany. The rise of the Nazis forced the Jewish Weidenreich to emigrate, first to Chicago, then to Peking, and finally to New York City. This odyssey was important to the history of anthropology; while in Peking, Weidenreich combined extensive anatomy experience with exceptional powers of observation to produce world-famous descriptions of a newly discovered human ancestor, *Sinanthropus pekinensis*, which has since become "Peking man." Weidenreich spent the rest of his career studying human evolution. He developed the theory that bipedalism, increased brain size, and reduced facial area were interconnected developments.

1994 Scientists report finding North America's oldest known site of animal butchering and ivory harvesting. It was discovered under the Aucilla River, near Tallahassee, Florida, by archaeologists David Webb and James Dunbar. Some 12,200 years ago prehistoric man hacked up mastodons and sliced out their tusks (possibly for tools) in the area. Cells, steroids, and stomach contents were also recovered from the behemoths' remains.

1892 The self-healing bicycle tire is patented. John F. Palmer of Chicago receives patent No. 476,680 for a tire in which the rubber tread is compressed in such a way that punctures tend to close themselves, rather than blowing open. Palmer's tire was manufactured in the same year by the B.F. Goodrich Company.

1896 Chemist Robert S. Mulliken is born the son of an MIT chemistry professor in Newburyport, Massachusetts. Mulliken won the 1966 Nobel Prize for his analysis of atomic structure (in which electrons exist in orbitals rather than shells) according to the principles of quantum mechanics.

1920 The Supreme Court upholds the 18th Amendment, Prohibition.

1953 In Chekiang, China, history's heaviest quintuplets (with a combined weight of 25 pounds) are born, and in Brookhaven, New York, scientists discover evidence that they have artificially produced one of the lightest entities known, the subatomic "hyperon."

1980 The nation's first solar-cell power plant is dedicated by Utah Governor Scott Matheson at the Natural Bridges National Monument. The 266,029 solar cells generate up to 100 kilowatts to provide power for six residences, water sanitation equipment, maintenance facilities, and the Monument's visitors' center. The nearest power line is 38 miles away. Costing $3 million, the facility was jointly produced by the National Park System, MIT's Lincoln Laboratory, and the Department of Energy.

1625 Giovanni Cassini, the last great astronomer to deny that Earth moves around the sun, is born in Perinaldo, Italy. His positive contributions were many, including the first nearly correct estimate of the distance from Earth to the sun and the discovery that Saturn's ring is a double ring (the two rings are separated by a dark band, which is still called Cassini's division).

1758 Optician John Dollond reports to the Royal Society that Newton was wrong, and achromatic (i.e., blur-free) lenses *can* be made. This development was extremely important for telescopes and microscopes.

1772 Gowan Knight, a true pathfinder, dies at 58 in London. He was a scientist and inventor whose work in magnetism led to innovations in the compass that are still used today. He developed a powerful technique for magnetizing metals (which he exhibited before the Royal Society in 1744), which he introduced to the crude and inaccurate needles then used in compasses. He introduced the now-common rhomboidal shape to needle design, and invented new means of suspending the needle. His improvements were readily accepted by England's Royal Navy.

1786 The nation's first ice cream advertisement appears in New York City. A "Mr. Hall" of 76 Chatham Street (now Park Row), is reportedly the first commercial maker of the dessert in the United States.

1803 Naturalist Robert Brown, 29, ends his plant-gathering during the two-year exploration of Australia by the HMS *Investigator*. Brown (most famous for discovering Brownian motion) collected 3900 specimens, and started publishing the results in 1810 as *Prodromus Florae Novae Hollandiae*. Although now considered a classic in systematic botany, the first volume sold poorly and Brown never finished the work.

1869 Frank Lloyd Wright, one of architecture's great figures, is born in Richland Center, Wisconsin.

1869 The suctioning vacuum cleaner is patented. Ives W. McGaffey of Chicago receives patent No. 91,145 for a "sweeping machine." It is a hand-powered device intended for light surface cleaning.

1916 Francis Crick, one of the two men who unraveled the mystery of DNA, is born in Northampton, England. He started out as a physicist (during World War II he developed magnetic mines), but from the late 1940s onward he devoted his engineering skill to determining the structures of large molecules found in living things. Crick's life, and the history of biology, changed forever when the young American James Watson ("regarded, in most circles, as too bright to be really sound" wrote Crick later) walked into Crick's Cambridge lab in 1951. The two worked out DNA's structure and the way DNA passes hereditary information from cell to cell and generation to generation. Despite adulation and accolades, including the 1962 Nobel Prize for Physiology or Medicine, Crick remained modest about his work. "Then there is the question, what would have happened if Watson and I had not put forward the DNA structure? This is 'iffy' history.... If Watson had been killed by a tennis ball I am reasonably sure I would not have solved the structure alone, but who would?"

1959 Mail is carried by missile for the first time. Three thousand pieces of mail are sent in a Regulus missile from the offshore submarine USS *Barbero* to a naval station in Jacksonville, Florida. The 100-mile distance is covered in 10 minutes.

1983 The first-ever test-tube triplets are born in Adelaide, Australia.

1990 The supertanker *Mega Borg* explodes in the Gulf of Mexico, creating an enormous oil slick. Scientists later scatter 100 pounds of laboratory-grown, oil-eating bacteria on the mess. It is history's first use of "bioremediation" in open waters.

1992 "One of the most important accomplishments" of the International Earth Summit in Rio de Janeiro is the creation of an agency to monitor environmental treaties and actions that follow the Summit. That agency, the UN Commission on Sustainable Development, is finally established tonight in Rio after two years of negotiations.

1637 René Descartes's *Discourse on Method* is published. The book presented analytic geometry to the world, and is now considered a benchmark in science, mathematics, and philosophy.

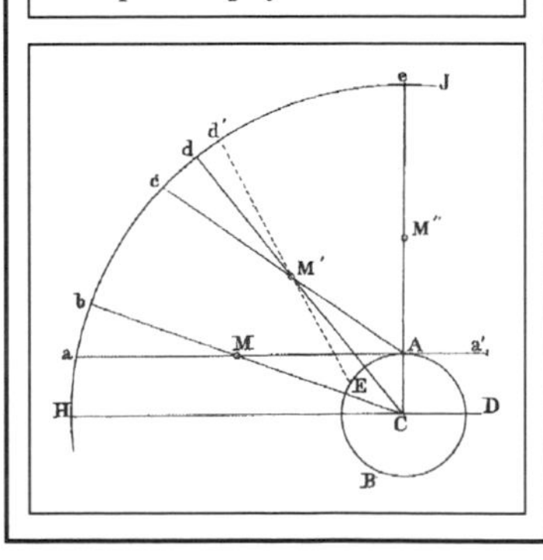

1781 George Stephenson, inventor of the first commercially-successful locomotive, is born in Wylam, England. His family was not well off, but his father's job as a steam-pump attendant introduced Stephenson at an early age to mechanical gadgetry.

1790 Nine days after President George Washington signed the original U.S. copyright law, the first book is entered for copyright protection. It is the first to be copyrighted under the new law. *The Philadelphia Spelling Book arranged upon a plan entirely new, adapted to the capacities of children and designed as an immediate improvement in spelling and reading the English language* is registered by its author John Barry in the clerk's office of the first district of Pennsylvania. The book was printed by Carey, Stewart and Co. of Philadelphia. It was also sold as *The American Spelling Book.*

1842 One of history's great oceanographic expeditions (and the first scientific expedition funded by the U.S. government) returns to New York harbor. Only two of the original six ships complete the 90,000-mile four-year "United States Exploring Expedition." The mass of scientific data was overshadowed by charges of cruelty against the voyage's commander, Charles Wilkes.

1850 Wilhelm Roux, a founder of experimental embryology, is born in Jena, Germany.

1869 Gregor Mendel, 46, presents his last paper on genetics, to the Brünn (Austria-Hungary) Natural Science Society.

1875 Physiologist Sir Henry H. Dale is born in London. He participated in two history-making studies. In 1910 he and colleagues showed that the allergic reaction to the fungus, ergot of rye, is caused by the presence of histamine in ergot. This work led to discoveries by others that histamine is found throughout the human body, and it plays a critical, controlling role in glands, blood vessels, and smooth muscles. Dale and Otto Loewi later found the key to another great problem in the control of bodily actions; the two showed that an impulse is passed from one nerve to another by the movement of chemicals (neurotransmitters) between them. Dale was the first to isolate such a chemical (acetylcholine, in 1914), and he and Loewi demonstrated in 1921 that it transmitted messages between nerves. They were Nobel Laureates in 1936.

1902 The nation's first restaurant with a mechanized, automatic system for dispensing food opens in Philadelphia. It is the Automat Restaurant, run by the Horn and Hardart Baking Company. It is the forerunner of later "automats," although this first facility used imported German machinery, whereas later automats used machinery that Horn and Hardart developed and patented themselves.

1905 Einstein publishes an article in *Annalen der Physik* applying Planck's quantum theory to light. Although no experimental work was involved, the analysis earned Einstein the Nobel Prize.

1908 The earliest known midair rescue occurs over Longton, Great Britain, when Miss Dolly Shepherd saves Miss Louie May from a balloon at 11,000 feet. Both return to Earth on Shepherd's parachute.

1911 Three men submit plans to the French Academy of Sciences for a device that records telephone messages, even if no one answers the call.

1967 The all-time human record for pill-taking begins. C.H.A. Kilner of Malawi begins consuming pills after his pancreas is surgically removed. By January 1, 1978, Kilner had swallowed 311,136 capsules.

1993 At the 160-nation Earth Summit in Rio de Janeiro rain forests and global warming top the agenda. The United States has increased its offer of financial aid to drastically slow rain forest destruction (a move that rain forest countries have opposed unless they have a larger say in how the money is spent), while delegates from the European Community pledge to sign an accelerated program to slow global warming (a move that the United States has opposed).

1994 A modern-day Boston Tea Party is held by tobacco growers protesting high taxes on their product. Numerous stalks and three bales of the crop are thrown into the Kentucky River from a bridge near Frankfort, as a crowd of 3000 looked on. "Like Americans of the 1770s, we tobacco farmers are being threatened with excessive taxes," observes Henry West from the pickup truck that held the bales.

1706 Optician John Dollond, the first to patent achromatic (i.e. blur-free) lenses, is born in London.

1832 Nikolaus August Otto, inventor of the four-stroke internal combustion engine, is born in Holzhausen, Germany. The pattern of piston movement in an engine is still called the Otto cycle. It was Otto's engine that made the airplane and automobile possible.

1842 Sailors disembark onto native soil (New York City) at the end of the first scientific expedition bankrolled by the U.S. government. It was the four-year Wilkes naval expedition, which explored and surveyed the South Sea and the Pacific Ocean.

1858 Robert Brown, botanist and discoverer of Brownian motion, dies at 86 in London. His keen powers of observation earned him a place in history. He recognized basic distinctions between gymnosperms (conifers) and angiosperms (flowering plants); he realized, and reported, that the nucleus is a fundamental component of plant cells; and he improved plant classification by establishing new groups and using seeds in classification. His observation skill also allowed him to detect and experiment with the irregular, minute movements of particles in solution. This action is called "Brownian motion" and is now known to be omnipresent in many molecules in the physical and living worlds.

1869 Machine-frozen food is transported over a considerable distance for the first time. A consignment of Texas beef (frozen by the compression device invented by John Gorrie) arrives by the steamship *Agnes* in New Orleans. The meat is served in hospitals and celebratory banquets in hotels and restaurants. Ironically, New Orleans is the very city in which Gorrie failed to get financial backing, shortly before he died broken and dispirited in 1855.

1902 The mailing envelope with a window is patented. Americus F. Callahan of Chicago receives patent No. 701,839 for his innovation.

1935 Alcoholics Anonymous is founded by William G. Wilson and Dr. Robert Smith in Akron, Ohio.

1943 The ballpoint pen is patented by H. Biro of Buenos Aires, Argentina.

1973 The largest-ever artificial object in space (the *US Radio Astronomy Explorer B* with an antenna that spans 1500 feet) is launched. The craft is also known as *Explorer 49*.

1991 England's House of Commons approves legislation to rid the country of pit bulls. Not a single member stood in opposition. The legislation also covers the Japanese tosa breed. (See May 21, 1991 for further details.)

1993 A man in Maidenhead, England, is cleared of animal cruelty charges for leaving his tropical fish "Home Alone." David Sharod is still swimming in legal bills (more than $3000) after battling the Royal Society for the Prevention of Cruelty to Animals, who brought the charges after Sharod left his fish and his home unattended for three days to run a pub for a friend.

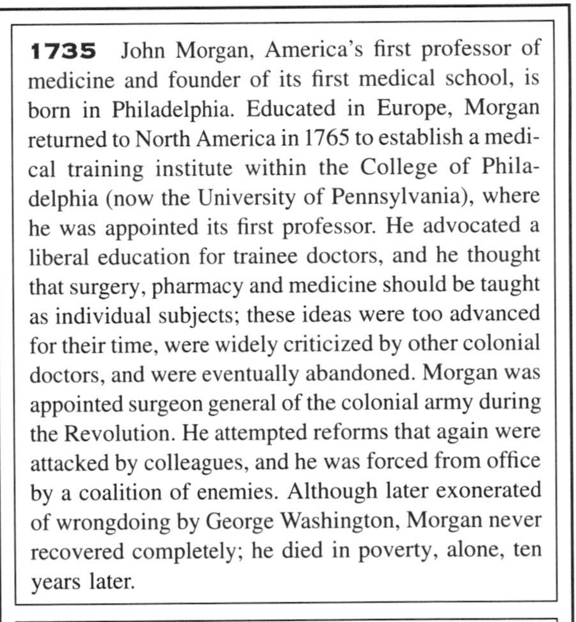

1735 John Morgan, America's first professor of medicine and founder of its first medical school, is born in Philadelphia. Educated in Europe, Morgan returned to North America in 1765 to establish a medical training institute within the College of Philadelphia (now the University of Pennsylvania), where he was appointed its first professor. He advocated a liberal education for trainee doctors, and he thought that surgery, pharmacy and medicine should be taught as individual subjects; these ideas were too advanced for their time, were widely criticized by other colonial doctors, and were eventually abandoned. Morgan was appointed surgeon general of the colonial army during the Revolution. He attempted reforms that again were attacked by colleagues, and he was forced from office by a coalition of enemies. Although later exonerated of wrongdoing by George Washington, Morgan never recovered completely; he died in poverty, alone, ten years later.

1292 English scholar Roger Bacon, one of the first to advocate experimentation and mathematics as tools of scientific advance, dies at about 72 in Oxford.

1770 Captain James Cook discovers Australia's Great Barrier Reef when his research ship, HMS *Endeavor*, runs into it.

1793 The first U.S. patent for a stove is granted to Robert Haeterick of Pennsylvania. This was not the first stove invented in the United States, however. That distinction goes to Benjamin Franklin, who not only refused to patent his invention, but also wrote a book describing how others could freely copy his design, which is known to this day as the "Franklin stove."

1827 Alfred Newton, responsible for England's first bird-protection laws, is born in Geneva, Switzerland. Newton was the first zoology professor at Cambridge University.

JUNE 11

1834 Charles Darwin first sees the Pacific Ocean during the voyage of the *Beagle*.

1847 Explorer-geographer Sir John Franklin dies of starvation at 61 near King William Island on an expedition in the British Arctic Islands (see April 16, 1786, for a biographical sketch).

1895 The first "real" automobile race begins in Paris. It was won two days later in Bordeaux by Emile Levassor driving a two-seater Panhard-Levassor vehicle. His average speed was 15 mph.

1857 Leland Ossian Howard is born in Rockford, Illinois. A pioneer in the "biological control" of insects (in which natural enemies, rather than pesticides, are used for controlling pests), Howard headed the U.S. Department of Agriculture for more than 30 years, described 47 new groups of parasitic wasps and ants and 20 new species of mosquitoes, and did important research in medical entomology (the study of insects), which led to his discovery that houseflies carry and transmit many diseases.

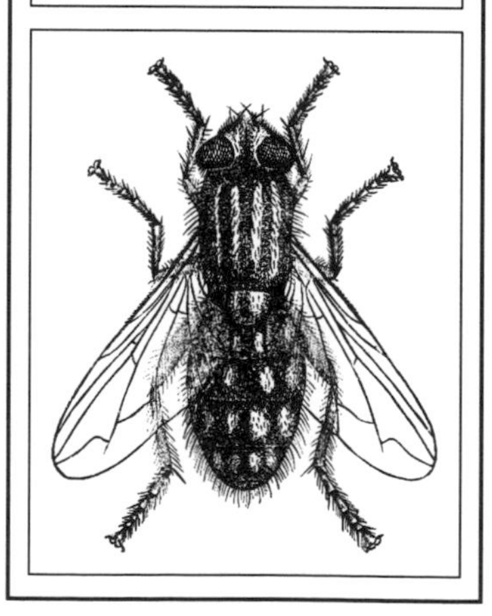

1910 Jacques-Yves Cousteau, oceanographer, conservationist, and inventor of scuba breathing apparatus, is born in St. André-de-Cubzac, France.

1912 An airplane takes off from a hotel roof for the first time in U.S. history. At 2:35 PM Silas Christoferson flies from a 170-foot wooden runway on top of the Multnomah Hotel in Portland, Oregon. He is not yet licensed, and he elected to attempt his feat in a rainstorm.

1948 Possibly the first animal in outer space, a nine-pound rhesus monkey named Albert is sent 37 miles into the skies above New Mexico on a V-2 rocket. The reentry parachute failed, but Albert had perished from breathing difficulties even before the launch.

1963 The first human lung transplant is conducted by Dr. James D. Hardy in Jackson, Mississippi.

1985 A historic right-to-die case ends when Karen Ann Quinlan, 31, dies in a nursing home in Morris Plains, New Jersey, ten years after being taken off artificial life support.

1991 "The government should have stuck to their original decision to destroy all pit bull terriers. They should never allow this sort of thing to happen again to anyone," proclaims 54-year-old Frank Tempest, who is missing his nose and one ear after a pit bull attack, in response to the previous day's passage by England's House of Commons of a bill that banned pit bulls from England.

1993 The U.S. Supreme Court rules that religious groups have a constitutional right to sacrifice animals.

1993 Steven Spielberg's *Jurassic Park* debuts. It sets motion picture income records and sparks a wave of popular interest in dinosaurs.

1993 More than 30 people in Baza, Spain, suffer severe eye injuries when they stare at the sun, expecting to see the Virgin Mary. They were following the directive of an 18-year-old seer and faith healer, Esteban Sanchez Casas. He later claimed that those injured were nonbelievers who had looked at the sun on the wrong day.

1819 Charles Kingsley, one of the first churchmen to accept Darwin's theory of evolution, is born in a vicarage in Devon, England (see January 23, 1875, for a biographical sketch).

1829 Abner Doubleday invents the game of baseball in Cooperstown, New York—maybe. The year, person, and place are still disputed by historians. In 1939 a museum devoted to the sport opens on this date.

1837 British inventors William Cooke and Charles Wheatstone patent the electromagnetic telegraph. Their device was put in public service in 1839, five years before the more-famous Morse telegraph.

1843 Astronomer Sir David Gill is born the son of a watchmaker in Aberdeen, Scotland.

1849 The forerunner of modern gas masks is patented (No. 6,529) by Lewis Phectic Haslett of Louisville, Kentucky. His "inhaler or lung protector" removed dust and other material from air through a filter.

1893 Sergius Winogradsky, 37, presents evidence to the French Academy of Sciences that bacteria are the critical agents in processing nitrogen into a form usable by living creatures.

1899 Biochemist Fritz Albert Lipmann, discoverer of coenzyme A (a critical molecule in the conversion of food into energy), is born in Königsberg, Germany. He shared the 1953 Nobel Prize in Physiology or Medicine with fellow metabolism scientist Hans Krebs.

1912 Communication and attitude psychologist Carl Iver Hovland is born in Chicago.

1913 The first animated cartoon made with modern techniques is released. *The Artist's Dream* (also known as *The Dachsund*) was produced by John Randolph Bray (who patented the process), and showed a dog eating sausages until it exploded.

1953 A 3-D advertisement first appears in a U.S. newspaper. It is an ad for carpets. It appears in the *Daily Freeman* of Waukesha, Wisconsin, for the decorating firm of Hale-Frame Associates. The next day, a department store in Los Angeles runs six full-page 3-D ads in the *Los Angeles Times*.

1971 The FDA bans the dumping of human excrement on railroad tracks. The ruling is another victory for the lobbying efforts of Ralph Nader.

1979 The *Gossamer Albatross* flies across the English Channel. The airplane is powered only by human energy, in the form of a pedaling mechanism operated by cyclist Bryan Allen.

1994 Nicole Brown Simpson and Ronald Goldman are murdered on the steps of Simpson's Los Angeles home. The grisly nature of the crime, the glamour of the victims, and the celebrity of the prime suspect (sports hero O.J. Simpson) caused unprecedented public focus on spouse abuse, racism, and DNA.

1994 The American Medical Association (AMA) selects an African American as its president for the first time in its 147-year history. The selectee is Dr. Lonnie R. Bristow, a 64-year-old California internist who was the AMA board chairman and its only black member.

1982 Ethologist (animal behaviorist) Karl von Frisch dies at 95 in Munich, Germany. Frisch opened the eyes of science to the sensory and communication capacities of "simple" animals. He demonstrated that fish can detect and distinguish sounds with greater acuity than humans can; he also showed that fish can distinguish colors and brightness. He established the ability of bees to tell the difference between various odors and tastes, and discovered their famous "waggle dance," in which individual bees communicate the direction and distance of food to hive-mates by a series of "dancing" movements. Frisch also showed that bees' perception of polarized light can be used as a compass, even when the sun is not visible. He shared the 1973 Nobel Prize for Physiology or Medicine with fellow ethologists Niko Tinbergen and Konrad Lorenz.

1773 Thomas Young, one of the first to study vision scientifically (he discovered how the eye focuses, described the cause of astigmatism, and developed a theory of color vision that bears his name), is born the prodigy son of a Quaker banker in Milverton, England.

1854 Sir Charles Parsons, inventor of the steam turbine, is born the son of an astronomer in London.

1868 Physicist Wallace Sabine, founder of the science of architectural acoustics, is born the son of a farmer in Richwood, Ohio.

1870 Bacteriologist Jules Bordet is born in Soignies, Belgium. He received the 1919 Nobel Prize for Physiology or Medicine for discovering immunity chemicals in blood; this founded the science of serology (which is the study of the serum portion of blood) and is the basis of diagnostic tests for many diseases.

JUNE 13

1877 Pasteur travels to the slaughterhouses of Chartres and takes blood samples of sheep, cows, and horses that have died of anthrax. It is the unglamorous start to his development of a vaccine.

1895 History's first "real" automobile race ends in Bordeaux, France (see June 11, 1895, for details).

1906 The first federal consumer protection law in U.S. history is enacted. 34 Stat. L. 260 is "an act forbidding the importation, exportation or carriage in interstate commerce of falsely or spuriously stamped articles of merchandise made of gold or silver or their alloys."

1911 Physicist Erwin Wilhelm Müller is born in Berlin. At 40 he emigrated to the United States and took a post at Penn State. Five years later he invented the field ion microscope, which produced the first-ever photographs of individual atoms.

1915 Physicist Henry Moseley departs England for the battlefields of Turkey. Two months later he died at Gallipoli. Although just 27, Moseley had already revolutionized chemistry and physics by developing X-ray analysis of elements and the concept of atomic number. He was certainly headed for a Nobel Prize.

1925 An object in motion is televised for the first time. It is a scale-model windmill with its blades turning. It is viewed by the U.S. Secretary of the Navy, the director of the Bureau of Standards, and the acting Secretary of Commerce. The windmill is in Bellevue, D.C., at radio station NOF, and the picture is telecast to the laboratory of Charles Francis Jenkins in Washington, D.C.

1944 The V-2 rocket is first launched by Germany against England.

1970 "During 27 days of sailing so far, oil lumps in varying sizes have been observed uninterruptedly every day ..." is the depressing telegraph message from Thor Heyerdahl in the Pacific Ocean on the *Kon-Tiki*, during the raft's voyage from South America to Asia. The trip proved that Polynesians may have come from South America, and it proved that modern man was already polluting the oceans.

1927 Charles Lindbergh receives a ticker-tape parade in New York City for his transatlantic flight.

1983 Eleven years after launch, *Pioneer 10* crosses the orbit of Neptune and becomes the first space vehicle to leave our solar system.

1993 "Deke" Slayton, one of the country's original seven astronauts, dies of cancer at age 69 in his home outside of Houston. Slayton first voyaged into space in the historic 1975 *Apollo–Soyuz* mission in which Soviet and U.S. craft rendezvoused in space. That voyage went smoothly ("I haven't done anything my 91-year-old aunt up in Wisconsin couldn't have done equally well," joked Slayton from the capsule) until reentry, when the cabin filled with harmful gas, causing crew member Vance Brand to pass out. Slayton remained calm, put on his mask, and put one on the unconscious Brand, who then recovered.

1994 A federal jury in Anchorage, Alaska, decides that the 1989 *Exxon Valdez* oil spill was caused by recklessness on the parts of the ship's captain, Joseph Hazelwood, and the parent company, Exxon Corp. The ruling opens the door to lawsuits by 10,000 fishermen, Alaskan natives, and property owners adversely affected by the 250,000 barrels of oil that entered the Alaskan waters.

1726 Thomas Pennant, one of the renowned zoologists and nature-writers of his time, is born in Downing, Wales. His writing was exceptional in its readability. His 1766 *British Zoology* spurred animal research, especially in ornithology. He traveled extensively throughout Europe, mostly on horseback, and carefully recorded his observations on the flora, fauna, local peoples, and antiquities.

1736 French physicist Charles-Augustin de Coulomb is born in Angoulême. Nine years as a military engineer in the West Indies irreparably damaged his health, after which he retired to research. In 1777 he invented a torsion balance, which he later adapted to electrical measurements. Coulomb's law describes the relationship between distance, charge, and electrical force, and established that electrical force follows rules similar to those of gravity.

1800 The Battle of Marengo begins. It was to become Napoleon's first victory as head of state, and highlighted the serious provisioning problems that led Napoleon to establish rich prizes for industrial inventions. One of the winners was Nicholas Appert, a cook who discovered that heating sterilizes and preserves food.

1834 Sandpaper is patented. Isaac Fisher, Jr., of Springfield, Vermont, receives four patents today to cover his abrasive.

1858 Darwin begins a section entitled "Pigeons" in the book that will become *Origin of Species*. Four days later the work is interrupted when a bombshell arrives from Alfred Russel Wallace (see June 18, 1858).

1862 John Nef (known for helping establish the modern system of chemical notation) is born the son of a textile mill foreman in Herisau, Switzerland.

1904 Margaret Bourke White, one of history's greatest photographers, is born in New York City.

1919 The first nonstop flight across the Atlantic begins at 4:13 PM in St. John's, Newfoundland. The Vickers Vimy Bomber lands 16 hours later in Derrygimla Bog, Ireland.

1951 A live human birth is televised for the first time in U.S. history at an Atlantic City meeting of the American Medical Association.

1967 *Mariner V* departs Cape Kennedy for a "flyby" of Venus, where it recorded information about the planet's atmosphere, radiation, and magnetism.

1972 The EPA issues an executive order banning the insecticide DDT after December 31.

1993 A woman in Vancouver, British Columbia, delivers the last two babies in a set of triplets—the first was born a month and a half earlier! Joanne March, a 29-year-old dental hygienist from Kelowna, is exhausted but happy at the "medical miracle."

1994 A fisherman in Essex, England, catches a large cod with a Roman coin dating back to 200 AD in its belly.

1994 Earth is struck with a 4.5-billion-year-old meteor. About the size of a grapefruit, the rock was traveling 60 miles per second when it crashed in a field outside St. Robert, Quebec. It was seen as a streak of fire, and heard as a sonic boom.

1868 Karl Landsteiner, discoverer of human blood groups, is born the son of a newspaper publisher in Vienna. As well as finding and naming the well-known ABO blood classification (this work made blood transfusions routine rather than risky), Landsteiner also discovered the "Rh factor" (this work explained the cause of some birthing and pregnancy malfunctions, when the child's Rh factor does not match the mother's). As well as its medical applications, blood typing has been important to the genetic and legal fields. Landsteiner was the sole recipient of the 1930 Nobel Prize for Physiology or Medicine.

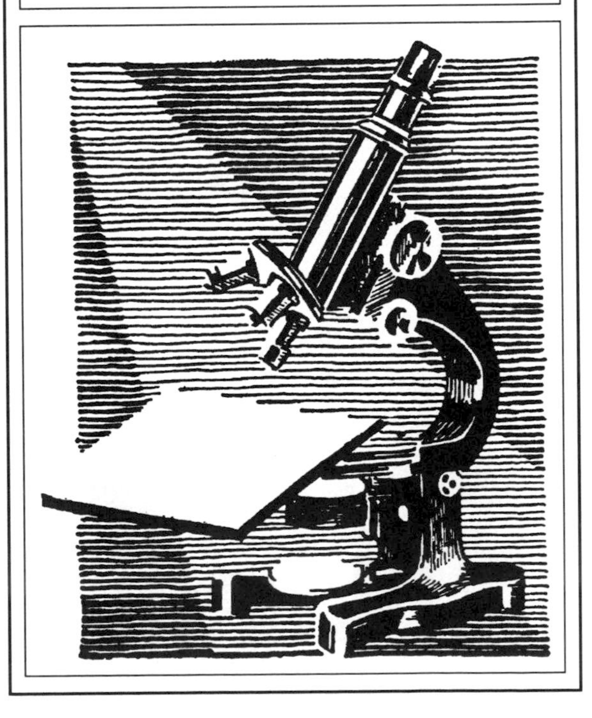

763 BC The first eclipse in recorded history is seen over Nineveh, Assyria.

1667 History's first blood transfusion involving a human occurs in Paris. It also involves a sheep. Physician Jean Denis, 37, transfers about 12 ounces of lamb's blood to an ailing young man, who seems helped by the procedure. Several others died from the same operation, and Denis was charged with murder. He was acquitted because the court believed he was trying to help people; he soon quit medicine forever.

1752 In Philadelphia Benjamin Franklin allegedly conducts his kite experiment demonstrating that lightning contains electricity. Historians disagree about the dates of this famous experiment, or whether it ever occurred. In October Franklin described the experiment in a letter to Peter Collison; in December this letter was read to the Royal Society of London, which effectively announced his discovery to science.

1785 History's first aviator becomes history's first air fatality. Pilâtre de Rozier (who first rode a balloon over Paris in 1783 with compatriot Marquis d'Arlandes) dies today in a fiery crash attempting to cross the English Channel.

JUNE 15

1835 The nation's first Patent Commissioner is appointed. He is Henry Leavitt Ellsworth, appointed by President Andrew Jackson. Ellsworth served until April 1845, when he left to become a real estate agent.

1844 Charles Goodyear patents vulcanization, a process that keeps rubber firm in warm weather and pliable in cold weather. It revolutionized the rubber industry. Goodyear began his experiments in 1934 in debtors' prison, and he later discovered the process by total accident. Goodyear died deep in debt, having been forced to spend huge sums of money he didn't have contesting infringements on his patent.

1902 Groundbreaking psychologist Erik Erikson is born in Frankfurt am Main, Germany. A major turning point in his life came in 1927 when he was invited by Anna Freud (daughter of Sigmund Freud and a famous psychoanalyst in her own right) to teach art, history, and geography at a small Vienna private school. Under Anna's influence Erikson underwent psychoanalysis and became an analyst. His specialty was the influence of culture and environment on personality development. In 1933 he joined the staff of the Harvard Medical School. Fieldwork with Indian tribes convinced him that different cultures have different ways of developing their members' personalities, but that each person goes through the same series of eight stages, or "crises," on the journey to maturity. *Childhood and Society* (1950) describes this theory, which is still widely accepted. Erikson is also known for analyzing historic figures from a psychological viewpoint, including Mahatma Gandhi, Martin Luther, and Adolf Hitler.

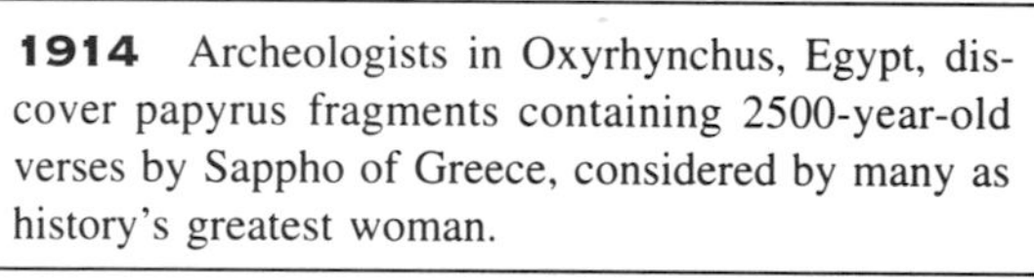

1914 Archeologists in Oxyrhynchus, Egypt, discover papyrus fragments containing 2500-year-old verses by Sappho of Greece, considered by many as history's greatest woman.

1971 Biochemist Wendell M. Stanley dies at 66 in Salamanca, Spain. He was a corecipient of the 1946 Nobel Prize for Chemistry for purifying and crystallizing a virus. It was the first time anyone had obtained such a pure sample of the tiny structures. Stanley worked on the tobacco mosaic virus. Once viruses were obtained in pure form, their structure could be determined. Subsequent workers showed viruses to be nothing but DNA wrapped in protein, unable to conduct life processes independently of the host cells that they infect. These revelations forced science, and philosophy, to reconsider the definition of life.

1994 A poll by the Gallup Organization finds that 5% of U.S. teens have attempted suicide, and another 12% have come close to trying it.

1994 The California gray whale is removed from the U.S. endangered species list. This is the first time a marine mammal has recovered strongly enough to be removed from the list.

1775 The granddaddy of the U.S. Army Corps of Engineers is born. On this day, the day before the Battle of Bunker Hill, the Continental Congress authorizes the "Army Engineering Department of the Continental Army." The personnel consists of one chief engineer (at $60 per month) and two assistant engineers (at $20 per month). The first chief engineer was Colonel Richard Gridley. Polish and French volunteers did the bulk of the physical work; the first major projects were building fortifications at West Point and Yorktown. The Department was dissolved with the successful conclusion of the Revolutionary War, but was reconstituted in 1802.

1801 Mathematician-physicist Julius Plücker is born the son of a merchant in Elberfeld, Germany. In the 1850s he discovered that magnetism influences the electrofluorescent glow in a vacuum tube; whatever the mysterious glow was, its production involves an electric charge. This was one of the first experiments in subatomic particle physics.

JUNE 16

1816 *Frankenstein* is born. During a night rainstorm at his villa in Geneva, Switzerland, Lord Byron reads several creepy tales, and then suggests that his guests create their own. One of the guests is 19-year-old Mary Wollstonecraft, who wrote the sci-fi classic.

1854 Carl Friedrich Gauss, "the Mozart of mathematics," is thrown from a carriage while visiting a railroad construction site between Cassel, Germany, and his hometown of Göttingen. Gauss suffers serious injuries, and dies the following February. It is the first, last, and only time in more than 20 years that he ventured from his beloved Göttingen.

1902 Geneticist Barbara McClintock is born in Hartford, Connecticut. She won the 1983 Nobel Prize for Physiology or Medicine for discovering that genes are not necessarily fixed on chromosomes, but may "cross over" (swap places with genes on other chromosomes during cell division). Such movement increases diversity in resulting cells and in a parent's offspring, and therefore may make individuals and populations more adaptable to changing environmental conditions. McClintock did her research on the genes in corn. Curt Stern, a corecipient of the 1983 Prize, discovered the same phenomenon in fruit flies. Crossing over seems to occur in the cells of all living things.

1903 Ford Motor Company incorporates.

1922 The first U.S. helicopter flight of importance occurs. The machine, driven by Henry Adler Berliner, rises to a height of seven feet on three occasions. The event takes place in College Park, Maryland, and is viewed by agents of the U.S. Bureau of Aeronautics.

1962 *The New Yorker* magazine begins publication of *Silent Spring* in three installments.

1963 Valentina Tereshkova becomes the first woman in space, on *Vostok 6*.

1975 A 14-leaf clover is discovered near Sioux Falls, South Dakota. It sets a record for most leaves for the legendary legume.

1991 A "vegetarian terrorist" hits a McDonald's Restaurant in Mesa, Arizona, and kidnaps the life-size plastic statue of Ronald McDonald (valued at $8000). Calling himself "Butch Cassidy," the desperado is protesting the lack of nonmeat sandwiches in the restaurant chain.

1994 The American Medical Association (AMA) announces new bookkeeping practices (lumping salaries of private practitioners with government and newly licensed doctors) intended to reduce the appearance of extravagant salaries. It is a cosmetic, on-paper change only. Dr. Nancy W. Dickey, the AMA's secretary-treasurer, openly admits this. "Yeah, it's an accounting change.... Now the physician looks less like he's gouging America."

2186 The longest-duration solar eclipse in the last 1469 years will be observed in the mid-Atlantic.

1897 German chemist Georg Wittig is born in Berlin. His 1979 Nobel Prize was won for studies of phosphorus-containing organic compounds.

1800 William Parsons is born the third Earl of Rosse in York, England. His earldom was in Cork, Ireland. He was known for humane and generous treatment of his tenant farmers during the 1846 potato famine, and for building the world's then-largest telescope (an unwieldy, 72-inch monster named "Leviathan"). Although the Irish climate prevented much use of the telescope, Parsons was able to discover the Crab Nebula, and he was the first to observe distant clouds that are now known to be distinct galaxies.

1832 Sir William Crookes is born the eldest of 16 children of a London tailor. Crookes discovered the element thallium in 1861, and in 1875 his skill in producing vacuums created the Crookes tube, the device with which Roentgen discovered X rays. On several occasions Crookes saw X-ray effects, but failed to realize their significance.

1836 The first U.S. school of homeopathic medicine (in which a disease is fought by treatments that induce the disease's symptoms) is chartered. The school is the North American Academy of the Homeopathic Healing Art, founded in 1835 in Allentown, Pennsylvania. It became known as the Allentown Academy. Instruction was in German; the first president and chief instructor was Constantine Hering, a German immigrant.

1891 The first solo transatlantic sailing race begins in Boston. One of the two entrants sunk seven times and finally abandoned the race; the other entrant (the *Sea Serpent*, piloted by Si Lawlor) won by default.

1894 The first U.S. polio epidemic officially begins. The first of 123 cases in the Rutland–Wallingford area of Vermont is reported.

1920 Nobel-Prize-winning biologist François Jacob is born in Nancy, France. He elucidated the mechanism that translates genetic information into protein formation.

1928 Amelia Earhart departs Newfoundland the first woman passenger on a transatlantic flight.

1928 The great explorer-geographer Roald Amundsen departs Bergen, Norway, for the Arctic Ocean on a mission to rescue survivors of the airship *Italia* that has crashed over the North Pole. Amundsen was never heard of again. It was actually the South Pole that made Amundsen famous; in 1911 he was the first person to reach the Pole, beating the English party of Robert Falcon Scott by less than a month.

1950 The first human kidney transplantion is performed in Chicago by Dr. Richard Harold Lawler. The healthy kidney was taken from a cadaver and put into 44-year-old Ruth Tucker, who lived for five more years.

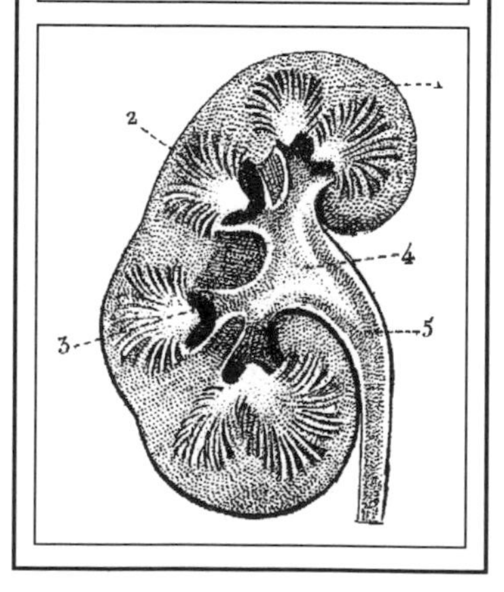

1993 The syringe-in-Pepsi hoax begins to unravel. Police in several states announce arrests of persons who fraudulently claimed to have found life-threatening objects in soda cans, including hypodermic needles, wood screws, and machine parts. Other felons voluntarily recanted their claims before they could be arrested. A vindicated Pepsi Company ran nationwide ads in the next few days to announce that there is no substance to charges of faulty, dangerous manufacturing. "Pepsi is pleased to announce … nothing," the ads' headline ran.

1994 "We want our customers to know that our popcorn is OK to eat," says Marc Pascucci, a vice president with the Loews-Sony chain of movie theaters, about the new "healthier choice" popcorn introduced this week. In April, the Center for Science in the Public Interest released a report claiming that a medium bucket of popcorn made with coconut oil had more than double the daily recommended amount of fat. Popcorn sales plummeted. The new, leaner snack is the result. "This is a customer driven business. If that's what they want, we'll provide them with that," said an executive with United Artists theaters, which had just added air-popped popcorn to their menu of snacks.

1994 In one of the most bizarre few hours in television and legal history, O.J. Simpson is arrested outside his Los Angeles home for murdering his ex-wife, Nicole Brown Simpson, and a friend, Ronald Goldman. The arrest follows a surrealistic cruise by the fugitive Simpson through California highways in full view of police, the media, hundreds of curbside spectators, and millions of television watchers. The case had enormous repercussions in both the social and biological sciences; it focused unprecedented public attention on DNA, racism, and domestic violence.

1818 Pietro Secchi, a pioneer in the application of both photography and spectroscopy to astronomy, is born the son of a cabinetmaker in Reggio, Italy. He surveyed the spectra of 4000 stars in the 1860s, showing for the first time that stars differ in more than position, brightness and color; by revealing their different spectra, he showed the stars to be of different chemical composition.

1845 Alphonse Laveran is born in Paris, the son of a military surgeon. A major force in the development of tropical medicine (he established the Laboratory of Tropical Disease within Paris's Pasteur Institute in 1907, and founded the Société de Pathologie Exotique in 1908), Laveran pioneered the study of human diseases caused by protozoa (one-celled animals). His discovery of the organism that causes malaria won him the 1907 Nobel Prize for Physiology or Medicine. It was the first time an animal, rather than a bacterium, was shown to cause disease.

1865 Edmund Ruffin, a founder of soil science, pens a note ("I cannot survive my country's liberty"), and then kills himself at age 71 in Amelia County, Virginia. He was despondent over the defeat of his beloved Confederacy in the Civil War (see January 5, 1794, for a biographical sketch).

1858 Darwin receives a bombshell in the mail. It is a letter and a paper by Alfred Russel Wallace that perfectly describes Darwin's own theory of evolution. Darwin had been contemplating and writing on the theory for over 25 years, but had yet to publish a single word.

1878 The U.S. Congress passes an "act to organize the Life Saving Service." The Service, introduced by Sumner Increase Kimball in 1871, was originally part of the Treasury Department, but then was merged with the Revenue Cutter Service in 1915 to become the U.S. Coast Guard.

1908 *Nature* publishes a brief letter, "Distant Electric Vision," by A.A. Campbell Swington. It contains the fundamentals of television transmission, 16 years before the first TV picture is seen.

1936 The nation's first bicycle traffic court is held, in Racine, Wisconsin. The first two judges are Sergeant Wilbur Hansen and Officer Alphonse Castabile of the police department. Sessions were held regularly on Saturday mornings. The following May the city's Common Council passed legislation requiring all bicycles to be registered with the police department.

1983 Sally Ride becomes the first U.S. woman in space, aboard the *Challenger*. She simultaneously becomes the youngest American in space (at age 32 years 23 days). The first-ever woman in space was Valentina Tereshkova in 1963, who was also the youngest-ever (at age 26).

1993 Pepsi-Cola Co. announces the end of a hoax that has plagued the company for weeks, and unveils an advertising campaign involving full-page ads in several nationwide newspapers. "Pepsi is pleased to announce ... nothing," will be the ads' headline. The action comes after police in several states announced arrests of persons who fraudulently claimed to have found life-threatening objects in soda cans; hypodermic needles, wood screws, and machine parts are among the reported items. Other people voluntarily recanted their claims before they could be arrested.

1993 The U.S. Supreme Court hands down two health-related rulings. In an Arizona Case involving the separation of church and state, the Court rules 5-4 that deaf students in private parochial schools can be provided with interpreters who are publicly funded. In a Nevada case, the Court rules 7-2 that an inmate's exposure to secondhand smoke in prison may violate his constitutional rights.

1799 Astronomer William Lassell is born in Bolton, England. A successful brewer, Lassell ground his own telescope lenses and pursued the science as a hobby. He discovered satellites around Neptune, Saturn, and Uranus.

1623 Mathematician-physicist Blaise Pascal is born in Clermont-Ferrand, France. Pascal was a child prodigy, but was denied access to mathematics books by his mathematician father, who wanted the boy to learn history first. However, legend has it that after he was told that geometry was the study of shapes and forms, he independently discovered (without books or training) the first 32 theorems in Euclid, in the correct order. His dumbfounded father then allowed him to study mathematics. His career saw important work in geometry, hydrodynamics, and pressure. He invented the syringe and the hydraulic press, and discovered Pascal's law (describing pressure changes of fluid in a container). Pascal was always sickly and weak in physical constitution and died at 39. His last decade of life was devoted to religious philosophy, not science. As a philosopher he also produced influential writings (most notably his *Pensées*). Autopsy revealed that Pascal had gone through life with a deformed skull.

1799 Alexander von Humboldt and Aimé Bonpland reach Tenerife (the largest Canary Island) on their scientific expedition to South America (see June 5, 1799, for background). The pair climb an active volcano during their six days there. Humboldt's notes on the change in vegetation with altitude helped him found the science of plant geography. The two also descended into the volcano and had their clothes burned with sulfurous vapors, apparently a first in scientific volcanology.

JUNE 19

1820 Sir Joseph Banks, explorer, naturalist, and financier of scientific endeavors, dies at 77 in Isleworth (London), England (see February 13, 1743, for a biographical sketch).

1902 Construction of the Panama Canal will be taken over by the United States. The Senate authorizes President Roosevelt to buy the rights and equipment from the French company that has abandoned the project. In 1904 Congress decided on a design for the Canal, which opened ten years after that.

1903 Lou Gehrig is born in New York City. Lou Gehrig's disease is named in his honor.

1906 Sir Ernst Boris Chain, developer of penicillin as an antibiotic (and the first to chemically assay it) is born the son of a chemist in Berlin. He shared the 1948 Nobel Prize for Physiology or Medicine.

1907 The "Nickelodeon" at 433-435 Smithfield Street in Pittsburgh opens today, becoming the world's first theater devoted solely to showing movies. Harry Davis renovated an empty store and installed 96 seats from other theaters. *Poor But Honest* and *The Baffled Burglar* are among the first showings. Profits accumulated immediately, over $1000 the first week.

1922 Nobel Prize-winning physicist Aage Bohr is born the son of Nobel Prize-winning physicist Niels Bohr in Copenhagen. The pair helped build the atomic bomb during World War II, and both are famous for mathematically refining our knowledge of atomic structure.

1934 The Federal Communications Commission is created by an act of Congress (48 Stat. L. 1064). The following June a seven-person board was appointed, with Eugene Octave Sykes as the first chair. He served until March. The FCC mission is to oversee national and international communications by satellite, radio, television, wire, or cable.

1977 The Trans-Alaska Pipeline begins carrying oil from the Arctic Ocean to Prince William Sound.

1981 The heaviest orange on record (5 pounds 8 ounces) is exhibited in Nelspruit, South Africa. It is the size of a human head, and it was stolen.

1993 The Spanish newspaper *El Pais* announces that more than 30 people recently suffered severe eye injuries when they stared directly at the sun, expecting to see the Virgin Mary. (See June 11, 1993.)

1783 Discoverer of morphine, German apothecary Friedrich Wilhelm Adam Sertürner was born on June 19, 1783. His initial reports of the drug's properties were challenged, so Sertürner conducted a series of experiments in which he and three friends took the substance themselves. These tests won acceptance for his work, but may have had sad consequences, as Sertürner apparently became insane in midlife. He died at age 57.

1819 The USS *Savannah* reaches Liverpool, becoming the first ship to cross an ocean by steam.

1859 Christian Freiherr, who introduced "gestalt" to psychology, is born in Rodaun, Austria.

1861 Sir Frederick Gowland Hopkins, codiscover of vitamins, is born into an unhappy childhood in Eastbourne, England. Following a succession of miscellaneous jobs, Hopkins used a small inheritance to obtain a medical degree, after which he began teaching and research (see May 16, 1947, for further biographical details).

1875 Reginald Crundall Punnett is born in Tonbridge, England. He wrote the first textbook on modern genetics (*Mendelism*, 1905), cofounded the *Journal of Genetics*, codiscovered a number of genetic processes (including linkage, sex determination, sex linkage, and autosomal linkage), and created the ubiquitous Punnett square (a simple diagram that predicts the chance of certain traits being passed from parents to their offspring).

1877 O.C. Marsh publishes the first description of a gigantic dinosaur (the largest known at the time), which he names *Titanosaurus*. Two months later Edward Drinker Cope publishes a paper on the same animal, which he names *Camarasaurus*. It is another battle in the lifelong war between the two fabled paleontologists. The beast is now called *Atlantosaurus*.

1895 Caroline Willard Baldwin becomes the first U.S. woman to receive a Doctor of Science degree. Cornell University of Ithaca, New York, presents the award to her for her physics dissertation, "A Photographic Study of Arc Spectra." She later married and changed her name to Mrs. Charles T. Morrison.

1909 (A SUNDAY) The country's first balloon honeymoon on record takes place. At Woods Hole, Massachusetts, Roger Noble Burnham marries Eleanor Howard Waring, and the couple then ascends in the balloon *Pittsfield* (piloted by William Van Sleet) at 12:40 PM. By 4:30 PM the honeymoon is over, when the ship lands in an orchard in Holbrook, Massachusetts.

1920 Microbiologist Dmitry I. Ivanovsky, who discovered viruses, dies at 55 in Rostov-on-Don, Russia. He was commissioned in 1890 to study mysterious diseases that were killing tobacco crops in the Crimea. He determined that some agent in sap could transfer the disease from plant to plant, and then (through detailed filtering and microscope work) he concluded that some invisible parasite, much smaller than any known bacterium, was the culprit. He assumed, and reported, that he was working with just a supersmall bacterium, but he had actually discovered a whole new form of life: the virus.

1943 The New Quebec (formerly Chubb) Crater is discovered in northern Ungava, Canada. With a 6.8-mile circumference and a depth of 1325 feet, it is one of Earth's largest craters.

1947 Everglades National Park is established in tropical Florida.

1963 The Soviet Union and United States sign an agreement establishing a "hot line" communication link.

1993 Dr. Herbert Goldstein releases the findings of his court-ordered psychological analysis of the parties involved in the famous baby-swap case in which Kimberly Mays was accidentally raised by a family to which she was biologically unrelated. When the swap was discovered years later, both families fought for Kimberly's custody. Goldstein's report agrees with Kimberly, that she should be "divorced" from her biological parents (the Twiggs) and stay with the man who raised her (Robert Mays); to do otherwise would cause "major psychological trauma" to the already-troubled teen. The court ultimately agreed—and Kimberly ultimately disagreed, running away from Mays to live with the Twiggs.

1994 Besicorp Group Inc. of Kingston, New York, announces that it will collaborate with Samsonite to produce solar-powered briefcases. Photovoltaic cells will be installed in the walls of the cases, and the power generated can then be used to operate computers, cellular phones, and other electronic machinery. Karen Keatar of Besicorp says the system should be adaptable to any type of luggage.

1622 America's first prohibition officers are authorized in a proclamation by Virginia Governor, Sir Francis Wyatt. "We do ordaine an officer for that purpose to be sworne in every plantacion, to give information of all such, as shalbe so disordered: the moiety of the forfeitures to be given the sd officer so informing, or for default in him to any other that shall informe, and the other to the publique Threasury."

1633 Galileo's sentence for heresy (saying Earth moves around the sun) is read at the Inquisition in Rome.

1805 Charles Thomas Jackson, one of science's more colorful, and more tragic figures, is born in Plymouth, Massachusetts. A curious dilettante, Jackson studied medicine, chemistry, and geology (he conducted several surveys of New England rock formations from 1837 to 1844). During experiments with ether he breathed himself unconscious. In 1844 he freely told William Thomas Green Morton about his experiments, and instructed Morton on administering the gas. Morton started using it as a surgical anesthetic in 1846, and became famous for the innovation. Jackson received no recognition, and a long, bitter feud resulted. At the same time Jackson pursued a similar quarrel with Samuel Morse over priority in the invention of the telegraph, and Jackson also claimed himself the inventor of the explosive, guncotton. Jackson went completely insane in 1873, and remained so until death in 1880.

1808 Gay-Lussac and Thenard announce the isolation of the element boron to the French Academy of Sciences, beating Humphry Davy's announcement of the same feat by nine days.

1834 Cyrus Hall McCormick patents his grain-reaping machine. It was history's first practical reaper and it changed the face of agriculture, but McCormick was unable to sell any until 1841.

1859 The nation's first rocket patent is granted to Andrew Lanergan of Boston for an "improvement in exhibition rockets."

1869 The first state board of health in the United States is established in Massachusetts. The legislature approves Chapter 420, Acts of 1869, which creates the "Massachusetts State Board of Health and Vital Statistics." Henry Ingersoll Bowditch was the Board's first chairperson, and George Derby its first secretary.

1913 A woman (Mrs. Georgina "Tiny" Broadwick, 18) makes a parachute jump from an airplane for the first time, over Griffith Park, Los Angeles. She lands in a barley field.

1852 Friedrich Froebel, developer of the kindergarten, dies at 70 in Marienthal, Germany. Ironically, his own childhood was a disaster (see April 21, 1782, for a biographical sketch).

1939 Lou Gehrig retires from baseball because of a mysterious disease that now bears his name.

1948 The LP (or long-playing record) is demonstrated by inventor Peter Goldmark of CBS Laboratories.

1948 The first computer with a stored program (the Manchester University Mark I) makes its first run through a complete program.

1982 John Hinckley, Jr., is found innocent by reason of insanity in the shooting of President Reagan.

1991 Researchers at Vanderbilt University have developed a male contraceptive pill that is ready for testing, according to wire reports today. The pill will be an oral version of synthetic hormones that produced complete, but reversible, sterility when injected into healthy men. Gabriel Bialy of the National Institutes of Health estimates that a safe and effective male "pill" should be widely available by the year 2000.

1611 Geographers Henry Hudson and son are set adrift by mutineers in Hudson Bay.

1799 Over four years in the making, the first prototypes of the meter and the kilogram are deposited in the French National Archives in Paris.

1864 Mathematician Hermann Minkowski is born in Aexotas, Russia. He was one of Einstein's teachers, and seized on Einstein's 1905 writings to develop the notion of reality as a four-dimensional space (now called "Minkowski space") of length, height, width, and time. This theory was then used by Einstein to build his general theory of relativity, but Minkowski had died by then.

1871 Psychologist William McDougall is born in Chadderton, England. In the early 1900s he was a major force in bringing Darwinian and biological perspectives to human psychology, especially to human social behavior. At that time, psychology was still firmly rooted in philosophy. McDougall's *Physiological Psychology* (1905) and *Introduction to Social Psychology* (1908; its 30th edition appeared in 1960) were especially influential.

1887 Biologist-philosopher Sir Julian Huxley is born in London, grandson of the legendary T.H. Huxley. Julian was knighted in 1958 for his own accomplishments, including reorganization of the Regent's Park zoo, the first directorship of UNESCO, and researching various biophenomena.

1914 Women are admitted to the U.S. College of Surgeons for the first time. Dr. Alice Gertrude Bryant, Dr. Florence West Duckering, and 1063 men are inducted into the College in a Philadelphia ceremony.

1914 A demonstration is given in London of sending photographs over 600 miles via telegraph wire.

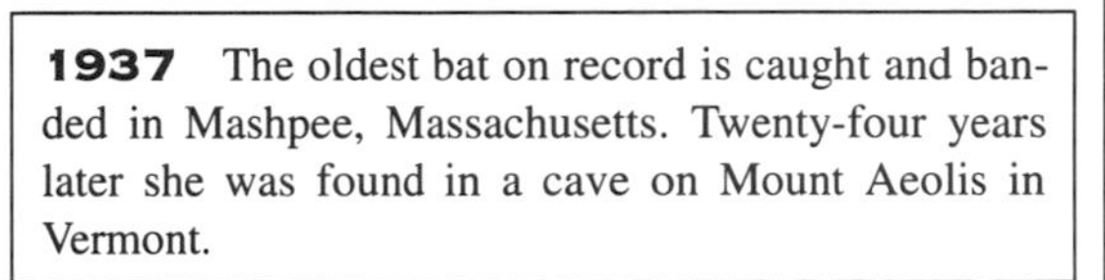

1937 The oldest bat on record is caught and banded in Mashpee, Massachusetts. Twenty-four years later she was found in a cave on Mount Aeolis in Vermont.

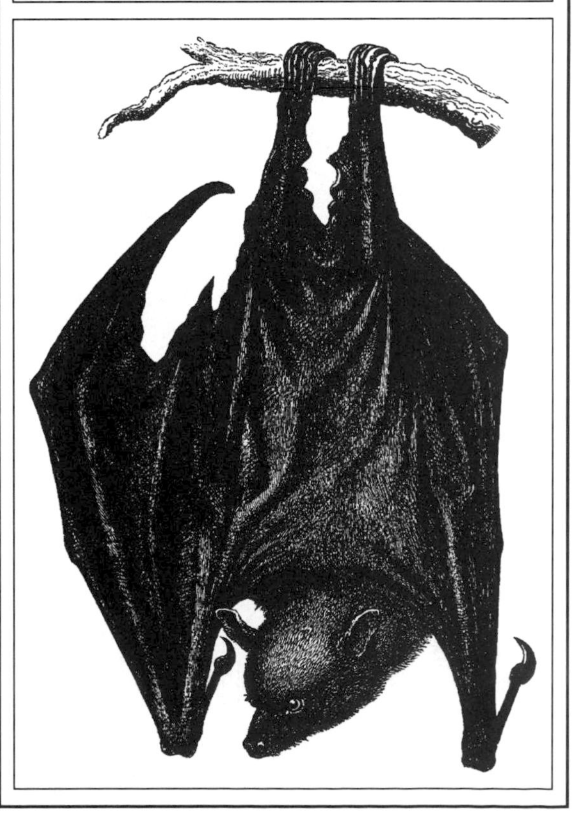

1956 The physics paper showing that "conservation of parity" is not valid for weak interactions within atoms reaches the offices of *Physical Review*. The document destroyed long-held beliefs and won the 1957 Nobel Prize for authors Tsung-Dao Lee and Chen Ning Yang.

1969 For the third time in the century, Cleveland's Cuyahoga River bursts into flames. Thousands of gallons of oil from an unknown source had been polluting the waterway. City residents joked that the Cuyahoga is the world's only body of water that should be declared a fire hazard.

1973 The first expedition in which medicine was practiced in outer space (by an earthling) is completed. *Skylab 2* splashes down in the Pacific Ocean, about 840 miles off San Diego after a 28-day flight. Crew members "Pete" Conrad and Paul Weitz were regularly examined by Navy commander Dr. Joseph Peter Kerwin, the first medic in orbit.

1978 Discovery of Charon is announced. It is the only known moon of Pluto, the coldest and smallest planet in our solar system.

1990 The northern spotted owl is declared endangered, and soon is on the cover of *Time*.

1993 Alfred J. Deskiewicz of Mercer Island, Washington, loses his attempt to make tobacco giant Philip Morris, Inc., pay for his stop-smoking treatment. Judge Linda K. Jacke throws his suit out of court because the relevant statute of limitations has expired, not because the claims lack merit.

1750 Geologist Dieudonné Dolomieu is born an aristocrat in Dolomieu, France. Extensive travels as a warrior in the Knights of Malta allowed him to gather an excellent mineralogy collection. He was also the foremost volcanologist of his day. It is said that he began his *magnum opus* during a two-year, solitary-confinement imprisonment by enemies within the Knights of Malta; he supposedly made a pen of wood and soot and wrote in the margins of a Bible. The common mineral dolomite is named for him.

1797 Frederic-Guillaume Joachim saws a tree trunk in half on the bed of Germany's Oder River, thus successfully testing the world's first true diving suit. Invented by Klingert of Breslau, the suit was an airtight, domed cylinder made of tin. It contained glass-covered eyeholes, breathing tubes running to the surface, and openings for the arms and legs.

1860 Sir Baldwin Spencer, the first scientist to study Australian anthropology, is born in Stretford, England.

JUNE 23

1868 The first practical typewriter is patented by Christopher L. Sholes, who coined the term.

1893 Captained by the legendary Fridtjof Nansen, the *Fram* departs Christiana (now Oslo), Norway, on a mission to study arctic currents and to be the first to the magnetic North Pole. The ship became frozen in ice for 35 months and never reached the North Pole, but it did get to the highest latitude man had ever reached.

1894 Sex researcher Alfred Charles Kinsey is born in Hoboken, New Jersey. Like Freud, Kinsey started out as a zoology researcher who switched to studying man. Kinsey moved from teaching at Harvard to Indiana University where he founded the Institute for Sex Research. He and colleagues interviewed 18,500 adult males and females to determine human sexual habits. It was history's most extensive, most scientific exploration of the topic, but it has been criticized because it relied on people's reports of their own behavior.

1912 Applied mathematician Alan Turing is born in London. He designed the first British computers and was a pioneer in the field of artificial intelligence, in which machines mimic human thinking.

1931 Wiley Post and Harold Gatty depart New York on the first round-the-world flight in a single engine plane. Seven years later on this date (1938) the Civil Aeronautics Authority is established by Congress. The date is also the anniversary of the first flight across North America in a 24-hour span; this was accomplished in 1924 by Army Lieutenant R.L. Maugham in a Curtiss PW8 pursuit plane.

1775 *Impenetrable Secret*, the first book to be made in America with American ink, paper, and type, is advertised for sale in the *Pennsylvania Mercury*. "Just published and printed with types, paper and ink manufactured in this Province" reads the ad by Story and Humphreys, the publishers.

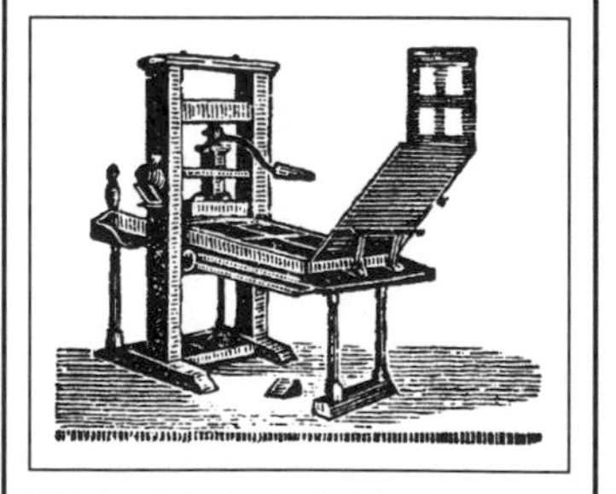

1926 The country's first national lipreading tournament takes place. It is run in Philadelphia during the seventh annual convention of the American Federation of Organizations for the Hard of Hearing.

1938 The nation's first public aquarium "for monsters of the deeps" (Marineland in St. Augustine, Florida) opens. Its two tanks are 11 and 18 feet deep.

1955 The first animated feature filmed in CinemaScope, Walt Disney's *Lady and the Tramp*, debuts.

1976 The nation's first public debate over banning recombinant DNA experiments takes place in the town hall of Cambridge, Massachusetts, halfway between Harvard and MIT.

1993 Astronauts on the *Endeavor* delay a jet firing so as to avoid a collision with space junk. It is the fourth time in two years that man-made trash floating in space forced avoidance actions by a shuttle flight. In today's incident, the craft was avoiding a one-ton booster rocket from a Soviet flight. There are now hundreds of thousands of pieces of junk floating in space that have been discarded by rockets since 1958. Even tiny objects are hazardous. "A pea at 17,500 mph is going to hurt," observed Don Kessler at the Johnson Space Center in Houston.

1994 The Associated Press reports a Philadelphia study on secondhand smoke: Nonsmokers who live with a pack-a-day smoker have a sharply increased chance of getting emphysema, asthma, and bronchitis. This study followed a recent EPA action that classified secondhand smoke as a carcinogen.

1497 File under "There goes the [ecological] neighborhood": Pristine North America is seen for the first time (in recorded history) by a European. The English expedition of John Cabot sees land, probably off present-day Canada.

1777 Oceanographer-explorer Sir John Ross is born in Balsarroch, Scotland. He is best known for several expeditions in search of the Northwest Passage, a navigable waterway between the Atlantic and the Pacific.

1784 Baltimore is the site of the nation's first manned balloon flight. Thirteen-year-old Edward Warren takes off in a 35 × 20-foot balloon made of multicolored silk patches; air is heated in a stove under the balloon. The owner of the craft, Peter Carnes, attempted a flight himself the following July, only to light the balloon on fire.

1795 Psychophysiologist Ernst Heinrich Weber is born the son of a theology professor in Wittenberg, Germany. He became interested in science under the influence of the important physicist Ernst Chladni, a neighbor and family friend. Weber's law describes the relationship between the perceived and the actual intensity of physical stimuli.

1839 The first photography exhibition takes place in France, showing the work of inventor Hippolyte Bayard, who had developed a method of making good prints from photographic negatives.

1839 Meat king Gustavus Swift is born in West Sandwich, Massachusetts (see March 29, 1903, for a biographical sketch).

1859 The Red Cross is born. During a business trip Henri Dunant, 31, witnesses the horrific Battle of Solferino, in which many of the 40,000 casualties died of poor medical treatment. Dunant begins a crusade for the formation of international relief organizations, which culminated in a multinational Geneva Convention in 1864. Dunant was awarded the first Nobel Prize for Peace, in 1901, and cofounded the YMCA. Unfortunately, his business suffered, and Dunant spent many years in obscurity and poverty.

1881 Sir William Huggins takes the first photographs of a comet's light spectrum.

1883 Victor Francis Hess, discoverer of cosmic rays from outer space, is born the son of a forest warden in Schloss Waldstein, Austria. In 1936 he shared the Nobel Prize for Physics.

1900 Erich Tschermak von Seysenegg reports discovering the fundamental laws of genetics to the German Botanical Society. Unbeknown to him, the monk Gregor Mendel had made the same discoveries decades before.

1915 Cosmologist Sir Fred Hoyle is born in Bingley, England. He is famous for theories on the chemistry surrounding the birth and death of heavenly bodies.

1940 A political convention is televised for the first time. The honor goes to the Republican convention in Philadelphia. It nominates that year's losing candidate, Wendell Wilkie. It is the GOP's 22nd convention. It is broadcast by W2XBS, the NBC affiliate in New York.

1963 The home videotape recorder is first demonstrated, at the BBC News Studio in London.

1991 A bloodless diabetes test is under development, announce researchers with Futrex, Inc. of Gaithersburg, Maryland. The new procedure is like holding your hand over a flashlight; the user simply inserts a finger into the device, and blood glucose level is shown by the skin's absorption of light.

1527 Paracelsus publicly burned the books of Galen and Avicenna, the "sacred cows" of ancient and medieval medicine. One of the most colorful, and influential, characters in medical history was Theophrastus Bombastus von Hohenheim (1493–1541). He is known as "Paracelsus," a name he boastfully gave himself because he considered himself "better than Celsus" (Celsus was the greatest of Roman doctors). His outrageous self-promotion earned him many enemies, caused him to move constantly throughout Europe, Africa, and Asia, and delayed the acceptance of his important innovations.

1783 Lavoisier announces the composition of water to the French Academy of Sciences.

1864 Physical chemist Hermann Nernst is born the son of a judge in Briesen, Prussia, just 20 miles from Copernicus's birthplace. He developed the famous Nernst equation at age 25; it relates electric potential to chemical properties in electric cells. He won the 1920 Nobel Prize for Chemistry for discovering the third law of thermodynamics (that changes in entropy approach zero at absolute zero temperature).

1898 Botanist-naturalist Ferdinand Cohn dies at 70 in the town of his birth, Breslau, Prussia.

1903 Marie Curie goes before an examination committee for her Ph.D., and is awarded the Nobel Prize later in the year.

1906 Architect Stanford White is shot to death atop his most famous creation, New York's Madison Square Garden, by his lover's husband.

JUNE 25

1907 Nuclear physicist Johannes Hans Daniel Jensen is born a gardener's son in Hamburg. His 1963 Nobel Prize came for developing the notion that atomic protons and neutrons are arranged in shells.

1964 The first commercial picturephone (in which callers can both see and hear each other) opens to the public. The charge is $16 or $21, depending on the location, for the first three minutes. Outlets are in New York, Washington, and Chicago. The device was a product of Bell Laboratories.

1964 The fastest-ever airplane flies for the first time. It is the X-15A-2, powered by liquid oxygen and ammonia; in 1967 the craft achieved a speed of 4520 mph.

1984 Social philosopher-scholar Michel Foucault dies at just 57 in Paris. He is best remembered for his studies of the written and unwritten rules that control human societies. His first major interest was the history of mental illness and its therapy, during which he developed the idea of societal "principles of exclusion" (such as distinctions between the sane and insane, the haves and have-nots). His *Madness and Civilization* (1965) dealt with classification of insanity in the 1600s. Foucault then turned his attention to the history of modern prisons (*Discipline and Punishment: The Birth of the Prison* appeared in 1977). Finally he examined sexuality (his *History of Sexuality*, 1978, studied Western attitudes toward sex from the ancient Greeks to the present).

1990 The Supreme Court announces its first decision in a "right to die" case. Family members can be barred from ending the lives of persistently comatose relatives who have not made their wishes known. This specific case involved a battle between the state of Missouri and the relatives of Nancy Cruzan.

1994 The meaning of life for fish is the topic in the second annual "Great Think-Off" in New York Mills, Minnesota. Peter Hilts, high school teacher and part-time fisherman, surmises, "the life cycle of a fish is meaningful, the survival drive of a fish is meaningful and in a fish you can find a rhythm that is common to all of life." His sentiments impress the audience, who vote him the Think-Off winner, nosing out a Zen Buddhist who argued that life is neither meaningful nor meaningless.

1994 Beaches around Cape Town, South Africa, are coated with a black, crusty oil slick. Thousands of penguins on a nearby island are also covered in oil. The events are not unusual these days, but in this case the ship carrying the oil sank 11 years before the slick appeared. An airplane discovered that the oil was emerging from a spot in Saldanha Bay where the Spanish tanker *Castello de Bellver* sank in 1983.

1951 CBS in New York City broadcasts history's first TV show that is in color and paid for by ads. It starts at 4:35 PM, it is broadcast to Boston, Philadelphia, and Washington, D.C., and has 16 sponsors. Among the program's stars are Ed Sullivan, Robert Alda, Garry Moore, and Arthur Godfrey.

JUNE 26

1274 Persian scholar Nasir ad-Din at-Tusi dies at 73 in Baghdad (see February 18, 1201, for a biographical sketch).

1730 Astronomer Charles Messier, referred to as "my little comet ferret" by Louis XV, is born the 10th of 12 children in Badonvillier, France. He was the first Frenchman to spot the famous 1758 return of Halley's Comet (predicted by Halley 50 years before). This inspired him to become one of history's greatest comet-hunters. In 1781 he published a catalogue of fuzzy objects that might be comets (Messier's poor telescope prevented him from making certain determinations at the time); the work provided names still in use today, and was a standard reference for all subsequent cometologists.

1824 Mathematician-physicist-inventor Lord Kelvin is born as William Thomson, the prodigy son of an eminent mathematician in Belfast. His many achievements include the Kelvin temperature scale, with his calculation of absolute zero at −273°C. He is buried in Westminster Abbey, a few feet from Newton.

1826 Psycho-anthropologist, Adolf Bastian, who created the concept of the "collective unconscious," is born in Bremen, Germany (see February 2, 1905, for a biographical sketch).

1886 Fluorine, the most active of the elements, is first isolated, by Ferdinand Moissan in Paris.

1900 Walter Reed spends his first full day as director of the medical compound in Quemado, Cuba, where the cause of yellow fever was discovered within months.

1914 Physicist-astronomer Lyman Spitzer, Jr. is born in Toledo, Ohio. Interest in the formation of new stars led him to invent equipment for achieving nuclear fusion.

1923 The first in-flight refueling using a pipe between two aircraft takes place over San Diego.

1943 Karl Landsteiner, who discovered human blood groups, dies at 75 in New York City (see June 14, 1868 for a biographical sketch).

1945 The United Nations charter is signed by 50 countries in San Francisco.

1951 The first color television program (commercial) to be seen daily begins broadcasts today. It is Ivan T. Sandersons's "The World Is Yours." It presents Earth's natural wonders on CBS channel 2 from 4:30 to 5:30 PM. The very first commercial television program in color made its debut the previous day; it was a variety program on the same network, and starred Arthur Godfrey, Ed Sullivan, and Robert Alda.

1992 The last major issue preventing total reunification of East and West Germany was a bioethical one, and the German Parliament settles it today: Women are granted the more liberal abortion rights that prevailed in East Germany, as opposed to the stricter regulation of West Germany. Although a pregnant woman must be counseled at least three days before the abortion, she now has complete freedom of choice through her third month of pregnancy. The law passes 357-284 (with 16 abstentions) after 14 hours of speeches by more than 100 politicians.

1994 An exhibit devoted to the history of perfume opens in the Open Air Biblical Museum in Nijmegen, the Netherlands. Researchers searched the Old and New Testaments in efforts to re-create the fragrances of the ancients. Perfumes were mainly oil-based, with basil, cinnamon, and myrrh among the most common ingredients. In all, 15 different fragrances were concocted for the exhibit. What was Delilah wearing when she asked Samson to cut his hair? What fragrance did Cleopatra wear for Marc Antony? These were some of the questions facing researchers.

1894 Annie Londonberry begins the first-ever bicycle trip around the world by a woman. She starts pedaling today from the State House in Boston, and ended her trip on September 12, 1895, in Chicago. She collected a $10,000 bet and $5000 in lecture fees.

1767 French astronomer Alexis Bouvard is born into poverty in Contamines. His mathematical skill won him a position as Pierre Simon Laplace's assistant at the Paris Observatory. It was Bouvard who did most of the computations in Laplace's immortal *Mécanique céleste*. He also discovered and mapped eight new comets.

1806 Augustus De Morgan, the mathematician who laid the foundation for George Boole's invention of symbolic logic, is born the son of an English colonel in Maduri, India.

1869 Embryologist Hans Spemann is born the son of a book publisher, in Stuttgart, Germany. He tackled one of the great biological mysteries: How does an embryo develop from a single, invisible cell into a highly organized organism with trillions of cells? Spemann spent his life researching newt embryos. He discovered embryonic induction, that is, in the early stages of life, embryo cells are highly plastic and their final form is determined by their location in the embryo. For example, if cells that would normally become skin are transplanted to an area of nerve development, the cells become nerve. Spemann was the sole recipient of the 1935 Nobel Prize for Physiology or Medicine.

1877 English physicist Charles Barkla is born in Widnes. His major contributions (recognized with the 1917 Nobel Prize) stemmed from discovering that gases scatter the X rays, and that the degree of scattering depends on the density of the gas. This provided great insights into both the rays and the gases.

1880 Helen Keller is born in Tuscumbia, Alabama. Nineteen months later she was stricken with a disease that left her deaf and blind, which eventually led to a historic career devoted to helping the handicapped (see June 1, 1968, for a note on her death).

1939 The final isolation of the chemical that will become the first successful antihemorrhagic drug begins in the Wisconsin lab of Karl Paul Link. The chemist doing the work is H.A. Campbell; and today it is his 30th birthday, the chemical is called H.A. (for "hemorrhagic agent"). It will later be called dicumarol, and a derivative was made into the rat poison warfarin.

1955 A seat belt law is first enacted in the United States. Illinois passes the law, which requires new cars to have holes in the frame where seat belts can be attached.

1969 A riot ensues when police raid the Stonewall Inn, a gay bar in New York's Greenwich Village. The incident is considered the birth of the homosexual rights movement.

1985 Legendary Route 66 (connecting Chicago and California) is decertified.

1994 The "Great Possum Debate" of Tumwater, Washington, is over—for the time being. District Judge C.L. Stilz drops animal cruelty charges against 39-year-old Steven Garity, who went out of his way to run over a family of possums. The previous March a policeman driving behind Garity witnessed Garity suddenly change lanes and squash a mother possum and the young that were clinging to her. Garity's lawyer later claimed his client was actually trying to avoid hitting another possum, but Garity's statement at the scene told a different story. "They're pests. All's they do is eat dog food and cat food, and be a pest. They just hiss at you."

1994 The Associated Press reports the findings of a study from the Centers for Disease Control on tobacco-use deaths. Among the results: the nationwide rate of such fatalities is definitely falling; the drop in the death rate is related to better health care.

1652 America's first traffic law of any kind is enacted today in New Amsterdam (New York City). "The Director General and Council of New Netherland in order to prevent accidents do hereby ordain that no Wagons, Carts or Sleighs shall be run, rode or driven at a gallop within this city of New Amsterdam, that the drivers and conductors of all Wagons, Carts and Sleighs within this city (the Broad Highway alone excepted) shall walk by the Wagons, Carts or Sleighs and so take and lead the horses, on the penalty of two pounds Flemish for the first time, and for the second time double, and for the third time to be arbitrarily corrected therefor and in addition to be responsible for all damages which may arise therefrom."

1824 Pierre Paul Broca, the first to scientifically match a specific ability with a specific brain region, is born in Sainte-Foy-la-Grande, France. The region controls speech, and is now called Broca's area.

1834 Congress allocates funds ($5000) for the nation's first federal geological survey. The area between the Missouri and Red Rivers will be charted, under the direction of George William Featherstonhaugh.

1860 "The Lord hath delivered him into my hands," whispers biologist Thomas Henry Huxley to a companion as they listen to Samuel Wilberforce, the bishop of Oxford, attack the theory of evolution that Charles Darwin has recently published. It is part of a famous debate on the theory that took place before a crowd of 700 in Oxford, England. As now, the argument was heated; when the official speakers were finished, many in the audience were yelling. Among those denouncing the theory was Robert Fitzroy, who captained the 1830s *Beagle* voyage during which Darwin developed the theory.

JUNE 28

1873 Transplant surgery was advanced tremendously by the innovations of Alexis Carrel, born the son of a textile manufacturer on this day in Lyon, France. He became a skilled surgeon, but suddenly quit the field in 1904, when he went to Canada to become a cattle rancher. In 1906 he returned to medicine to become a researcher at New York's Rockefeller Institute, and stayed there until his retirement in 1939. He is remembered for his technique of delicately suturing blood vessels together, and for his methods of keeping tissues and organs alive outside of the body. Carrel received the 1912 Nobel Prize for Physiology or Medicine.

1906 The beautiful and brilliant Marie Goeppert Mayer is born into a long line of professors in Kattowitz, Germany (now Poland). She shared the 1963 Nobel Prize for Physics for suggesting that protons and neutrons are arranged in shells in the atomic nucleus.

1912 Astronomer Carl Weizsäcker, famous for theories of universe formation, is born in Kiel, Germany.

1914 Austrian Archduke Francis Ferdinand and wife are assassinated in Sarajevo. This sparks World War I, which in turn produced great advances in aviation and chemical warfare, among other fields.

1914 German physicist Werner Kolhörster sends an instrument-laden balloon to 30,000 feet and confirms the preposterous assertion by Victor Hess that Earth's atmosphere is alive with cosmic rays.

1938 The nation's first aerial tramway ski lift is dedicated on Cannon Mountain in Franconia, New Hampshire. Giant cables carry two 27-person cars 40 feet above the trees.

1951 *Amos N' Andy* debuts as the first series in television history with an all-black cast.

1980 The first successful solar aircraft, Paul McCready's *Solar Penguin*, makes its initial test flight. The flight begins at 8:25 AM from Sheafter Airport in southern California. The craft, later named *Solar Challenger*, weighs 55 pounds, while the 32-year-old teacher who serves as the pilot weighs 97.

1992 The world's first baboon-to-human liver transplant is conducted at Pittsburgh's Presbyterian University Hospital. The recipient is a 35-year-old man, dying from liver failure resulting from hepatitis B. The donor is a 15-year-old male baboon from a Texas laboratory that was sacrificed in Pittsburgh to provide the organ. A human liver would not have worked because the hepatitis would have destroyed it, whereas baboon livers seem immune to the disease.

1993 "Fantastic!" yells Kirk Bloodsworth, 32, as he walks out of the Maryland House of Correction, a free man after nine years' imprisonment for a horrible crime against a young girl. DNA testing showed he was innocent. He had been convicted twice, on the basis of eyewitness testimony.

1565 Physician-naturalist Conrad Gesner, 49, about to die during a plague outbreak, completes *De Rerum fossilium* ("On fossil Objects"). It is a landmark in paleontology. It was the first book of its kind to systematically use illustrations, the first to extensively mention museum collections, and the first to urge and use cooperation between scientists.

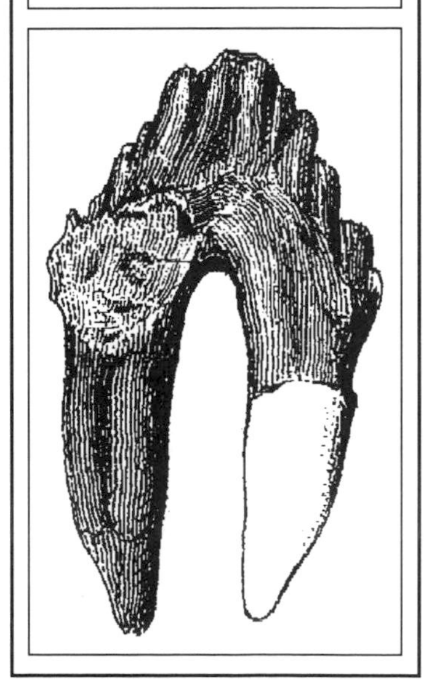

1833 Norwegian chemist Peter Waage is born the son of a ship's captain in Flekkefjord. An ardent supporter of temperance, Waage is remembered for his collaboration with brother-in-law Cato Guldberg in first describing the law of mass action (1864), which states that the speed of a chemical reaction is proportional to the concentrations of the reactants.

1837 Horace Mann is appointed head of the Massachusetts State Board of Education; it was established by a law passed two months before, becoming the nation's first state board of education. Mann's position was first called "secretary," later changed to "commissioner." His annual salary was $1000.

1860 English physician Thomas Addison dies at 67 in Bristol, England. He was the original describer of Addison's disease (later to afflict President John F. Kennedy), in which atrophy of the adrenal gland produces a hormone imbalance. This was the first time a physician was able to correlate a set of disease symptoms with an irregularity in an endocrine gland. Addison was also the first to describe pernicious anemia, also known as Addison's anemia.

1861 William James Mayo, one of the founders of the world-famous Mayo Clinic, is born in Le Sueur, Minnesota. He was the first son of William Worrall Mayo, the surgeon who founded St. Mary's Hospital in Rochester, Minnesota, after a tornado devastated the town in 1889. The elder Mayo and his two sons (William James and Charles Horace) performed all of the hospital's surgery through the early 1900s. Virtually all patient needs were handled under one roof. The facility became the Mayo Clinic and was the forerunner of modern group medical facilities.

1868 George Hale is born in Chicago, son of the elevator manufacturer who made the city's great skyscrapers possible. Hale himself achieved fame for lofty endeavors, as the creator of several of history's largest telescopes. Hale also invented the spectroheliograph (which enabled photographing single spectral lines around the sun), and he discovered magnetic fields within sunspots (1908). This was the first time magnetism had ever been associated with any heavenly body.

1895 Precisely 35 years and 1 day after his famous debate with Bishop Wilberforce over evolution, biologist Thomas Henry Huxley dies at 70 in Eastbourne, England.

1906 Mesa Verde National Park (Colorado) and Platt National Park (Oklahoma) are established. Olympic National Park (Washington) is established on this date in 1938.

1927 The first astronomy observation from an airplane occurs when a twin-engine British Imperial Airways plane flies above the London fog to photograph a total eclipse of the sun.

1948 South Africa bans interracial marriages.

1948 The U.S. Congress establishes airmail parcel post service. It begins the following September 1. The first pound cost $0.80 to ship, and $0.65 for each additional pound. The first air transport of letters began in 1911, and a regular air service was established between Washington and New York in 1918.

1952 The research ship *Galathea* returns to Copenhagen after bottom-sampling Earth's deepest ocean trenches for the first time.

1954 The Atomic Energy Commission revokes the access of J. Robert Oppenheimer to classified information.

1993 A shipment of 312 five-foot-long boa constrictors arrives from Colombia at Miami International Airport. A customs officer notices a strange bulge in one, and has it X-rayed. A foreign object is discovered and removed: it is cocaine wrapped in a condom. In all, 80 pounds of the drug was seized from the snakes, only 63 of which survived the ordeal.

1994 A horrible traffic accident begins a most unusual patient identification saga. On this day a woman in her mid-thirties steps off a Chicago bus into the path of a tractor-trailer. For three weeks, this "Jane Doe" lies in a hospital slipping in and out of a coma, with no one knowing her identity—until an astute and creative nurse, Cheryl Wolfinger gets the idea to sing to the patient. She sings *The Nifty Fifties*, a nursery rhyme that teaches the names of the fifty states. The heavily sedated patient perks up at the mention of Arizona. Wolfinger then starts naming all of the cities in Arizona that she can think of, and hits a winner with Scottsdale. The next day doctors have a name (Peggy Scott), a phone number, and a call to relatives.

1616 Galileo, 52, leaves Rome after a six-month visit, during which he met with Cardinal Bellarmine of the Inquisition and Pope Paul V. Galileo was warned not to "hold or defend" Copernicus's heliocentric theory, but it was not until 1633 that he was tried and convicted for the heresy.

1808 Humphry Davy announces his isolation of the element boron in England, nine days after Thenard and Gay-Lussac announced the same feat in France. The substance now has many uses, including water softener, Pyrex glass, rocket fuel, and roach poision.

1817 Sir Joseph Dalton Hooker, celebrated botanist and one of Darwin's earliest champions, is born in Halesworth, England, the son of a director of the Royal Botanic Gardens in Kew, outside London.

1858 Darwin's theory of evolution is first submitted in writing to a scientific society for publication. Geologist Charles Lyell and botanist Joseph Hooker submit to London's Linnaean Society three documents: an 1839 description of the theory that Darwin had written and set aside for 19 years, an 1857 letter by Darwin to Harvard's Asa Gray describing the theory, and a third rendition of the theory, this time written by naturalist Alfred Russel Wallace, who independently formulated the theory during an attack of malaria in the jungles of Malaysia. The documents were read to a Society meeting the following day.

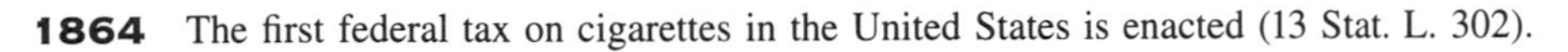

1864 The first federal tax on cigarettes in the United States is enacted (13 Stat. L. 302).

1869 The electric stove is first patented in the United States. William S. Hadaway, Jr., of New York City receives patent No. 563,032 for a one-ring coil that provided uniform surface heat.

1908 History's greatest known meteor explosion occurs 33,000 feet above Siberia. The force of a 10- to 15-megaton bomb is released, and 1500 square miles of Earth is devastated.

1926 Biochemist-geneticist Paul Berg is born in New York City. He shared the 1980 Nobel Prize for Chemistry for developing recombinant DNA techniques, in which genes from one organism are spliced into the genes of another, to create new species with new abilities.

1940 The U.S. Fish and Wildlife Service is established.

1971 *Soyuz 11* returns to Earth. The crew of three has perished.

1987 The world's largest planetarium—the dome diameter is 88 feet 7 inches—is completed in Miyazaki, Japan.

1992 Several days after the world's first baboon-to-human liver transplant, an angry protest is held outside Pittsburgh's Presbyterian University Hospital where the operation took place. "Did the baboon consent?" and "Animals are not ours to experiment on" read some of the placards. The crowd is dramatically silenced by an emotional Robert Winter, 34 years old and suffering from liver disease. "I didn't ask for this, but I've got it and I'm fighting it just like that guy laying up there.... Do you know what it's like to have liver disease?" he asks.

1994 An American bald eagle named "Hope" is released back into the wild after months of rehabilitation from a broken wing. She just clears a group of pine trees, and then soars over the marshlands in Maryland's Blackwater National Wildlife Refuge. The release is part of a ceremony in which the U.S. Fish and Wildlife Service announces its decisions to upgrade the eagle's status from "endangered" to "threatened." Today there are an estimated 8000 mating adults in the lower 48 states; 20 years ago there were just 800.

1795 Chemist Joseph Caventou is born in Saint-Omer, France. In the footsteps of his pharmacist father, Caventou formed a partnership with Pierre-Joseph Pelletier and isolated a number of important substances, including strychnine, chlorophyll, quinine, and caffeine.

1770 Lexell's Comet comes within 745,000 miles of Earth; it is the nearest known pass of a comet, except for 1910 when Earth is thought to have passed through the tail of Halley's Comet.

1791 The first internal revenue tax in the United States goes into effect. Established by a law passed the previous March, the tax was levied on carriages and distilled spirits.

1796 In the English country town of Berkeley, Dr. Edward Jenner inoculates eight-year-old James Phipps in both arms with deadly smallpox material. Jenner's own father had died of smallpox. But Phipps survives, because Jenner had already inoculated him with cowpox in May. This is history's best-known vaccination, and today's breathtaking experiment with Phipps proved its value.

1858 The Wallace–Darwin theory of evolution is first published, as an oral presentation to the Linnaean Society in London by geologist Charles Lyell and botanist Joseph Hooker. Neither Darwin nor Wallace is present.

JULY 1

1818 Ignaz Philipp Semmelweis is born the son of a prosperous merchant in Budapest. Semmelweis was a law student when he attend an anatomy lecture with a friend; it inspired him, and changed his life. Semmelweis was intrigued by puerperal fever ("childbed fever"); it was virtually absent in mothers who gave birth at home, but it killed as many as 30% of mothers who delivered in hospitals. Semmelweis outraged colleagues by concluding that doctors themselves transmitted the disease. He had those working under him wash their hands in strong chemicals between patient examinations. Deaths from the fever dropped dramatically, which only strengthened the jealousy and hostility of other doctors. Following some political activities in 1848, Semmelweis was fired from a Vienna clinic, but he secured employment in a Budapest hospital, where, again, his methods practically erased puerperal deaths. After he left Vienna the rates of childbed fever once again soared. The years of conflict took their toll, and Semmelweis died in an insane asylum—of puerperal fever.

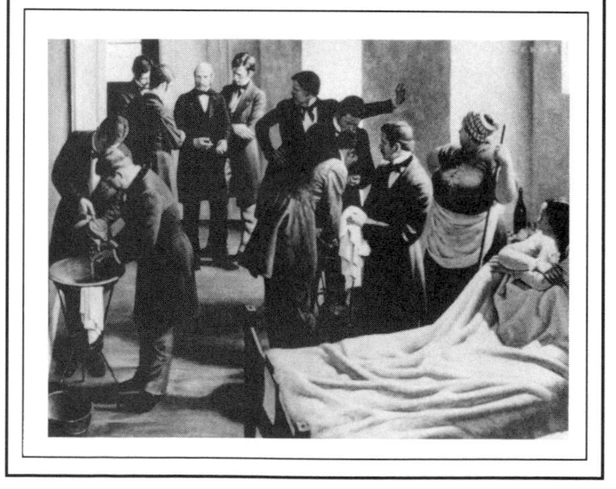

1862 Polygamy is banned in the United States.

1966 Medicare—the U.S. health insurance program—goes into effect.

1968 Sixty nations sign the Nuclear Nonproliferation Treaty.

1983 Visionary-engineer-architect R. Buckminster Fuller dies at 87 in Los Angeles. He came from a family of New England nonconformists, most notable of which was Margaret Fuller, a great aunt who founded *The Dial*, the unofficial organ of the Transcendental Movement. After two expulsions from Harvard and a stint in the Navy, Fuller went into business with his father-in-law manufacturing unique modular homes. But after the company failed and his four-year-old daughter died, Fuller found himself one night in Chicago on the verge of suicide. In a flash of inspiration (he called it "a blind date with principle") he committed his life to the nonprofit designing of things that maximized the social use of limited natural resources. He is especially remembered for designing the "Dymaxion" (called "the first streamlined car," it was omnidirectional, all-terrain, and bumpered all-around) and designing the geodesic dome (still the only large building that can be theoretically set on the ground as a single structure and that has no limiting dimensions). He also designed floating cities, underwater cities, expandable paper domes, a system of map-making, and die-stamped prefabricated bathrooms.

1994 A new process for reviewing Medicare claims is started in Florida; it doesn't take long for some interesting scams to surface. "Bass Orthopedic" consisted of two rented mailboxes and a phone number; it got $2.1 million from Medicare. "Med E O Diagnostic," using a rented West Dade mailbox, had collected $332,939 for the care of already-dead patients.

1566 Nostradamus dies at 62 in Salon, France. He first achieved prominence as an innovative physician during outbreaks of the plague in 1546 and 1547, the year of his first prophecy. Nostradamus is now famous for his prophecies relating to modern-day politics and world events.

1729 The movement of electric current is discovered. England's Stephen Gray finds that corks at the end of a glass tube become charged when he creates a static electric charge in the tube. Whatever the "electric fluid" is, it transfers itself from the glass to the cork. He later attached an 800-foot cord to the cork and found that a feather would stick to the far end of the cord once the glass tube was charged.

1842 The first iron ship built in the United States is registered. The *John Randolph*, 122 tons, was built in Savannah, Georgia, by John Caut for Gazaway Bugg Lamar. The metal plates were made in Birkenhead, England, and riveted together after shipment to Georgia. The vessel operated commercially out of Savannah.

1862 Sir William Bragg (winner with his son of the 1912 Nobel Prize for Physics for developing X-ray crystallography) is born in Wigton, England, son of a master mariner who became a farmer.

1893 Physicist Sir Franz Simon is born into a well-to-do family in Berlin. When Hitler seized power, Simon fled to Oxford, England, where he did his most important work: developing ultracold temperatures. His team reached 20 millionths of a degree above absolute zero just before his 1956 death.

1906 Physicist Hans Albrecht Bethe is born the son of a professor in Strassburg, Germany. Like a number of great scientists and thinkers, Bethe fled Germany when the Nazis came to power. He settled at Cornell University in New York. In 1939, from his knowledge of hydrogen atoms, he calculated the energy production by the sun. He was the chief theoretician in the Manhattan Project, which built the first nuclear weapons. In 1967 he received the Nobel Prize for Physics for his work on energy production by stars. When he was asked what dangers the honor of the Prize might hold, he replied, "Well, I think that the main danger is that from now on I may feel I can only do important work."

1931 Stephen Moulton Babcock, the father of scientific dairying, dies at 87 in Madison, Wisconsin. He was an agricultural chemist who studied in both the United States and Germany, finally settling for good at the University of Wisconsin. He developed the Babcock test (introduced in 1890), which was a simple method of gauging the amount of butterfat in milk. The test helped detect, regulate, and prevent milk adulteration, spurred development of dairy methods, and improved the production of cheese and butter.

1937 Famed aviator Amelia Earhart disappears in the Pacific.

1971 Oregon is the first state to enact antilitter legislation. Soft drinks and beer are targeted by the law: Pull-tab cans and nonreturnable bottles are banned from the state.

1976 The Supreme Court rules that the death penalty is neither cruel nor unusual.

1987 The first hot-air balloon flight across the Atlantic begins in Sugarloaf, Maine. It ends 31 hours 41 minutes later when Richard Branson and Per Lindstrand reach the coast of Ireland.

1993 Conservationists hail the approval of $2 million by the House Appropriations Committee for the initial purchase of land in central Florida for the nation's first-ever plant refuge. The area is home to some 20 endangered plant species. This is the first time that plants will be the primary concern of preservation funding; previously, plants were saved only as a by-product of legislation to save some popular, cuddly, or otherwise appealing animal. Environmentalists have called this traditional approach "the Bambi syndrome."

1900 The first directed flight by man occurs. A balloon/dirigible invented by Ferdinand Zeppelin ascends over Lake Constance, Germany, carrying a gondola, an engine, propellers, and a steering mechanism.

1839 The nation's first state "normal school" (i.e., for the education of teachers) opens in Lexington, Massachusetts. Only three pupils take advantage of the offer of free tuition.

1879 Scientist-philosopher Alfred Korzybski is born in Warsaw, Poland (see March 1, 1950, for a biographical sketch).

1898 Canadian Joshua Slocum becomes the first to single-handedly sail around the world. He reaches Fairhaven, Massachusetts, aboard his 36-foot-long sloop *Spray*. He had departed Boston 3 years 2 months 2 days before. His *Sailing Alone Around the World* (1909) is now a classic.

1903 The world is encircled by telegraph cable for the first time. On this date the last link in the chain, a line across the Pacific Ocean from Hawaii to the Philippines, is spliced and complete. Tomorrow, July 4, President Teddy Roosevelt sends a message westward from his home in Oyster Bay, New York. Eleven minutes later the message reaches Clarence Hungerford Mackay, president of the Commercial Pacific Cable Company. Mackay then sends a reply eastward, and it reaches Roosevelt in nine minutes.

1909 Gifford Pinchot, chief of the U.S. Forest Service, signs an agreement with W. D. Board, president of the University of Wisconsin's Board of Regents, to establish a center for forestry education and research.

1913 A common tern (*Sterna hirundo*) is banded by ornithologists at Eastern Egg Rock, Maine. In August 1917, the same bird is found (dead) at the mouth of the Niger River, West Africa, making it the first bird ever to have been recorded as flying across the Atlantic Ocean.

1918 The Migratory Bird Treaty Act between the United States and England (on behalf of Canada) is ratified.

1935 Harrison "Jack" Schmitt, the first geologist to walk on the moon, is born in Santa Ria, New Mexico. In the 1972 *Apollo 17* mission, Schmitt and Eugene A. Cernan spent three days on the lunar surface.

1938 The steam locomotive *Mallard* reaches a speed of 125 mph in Stoke Bank, Britain. It remains the greatest verified speed attained by a steam locomotive.

1956 The first ship outfitted for hurricane research, the 125-feet converted cutter *Crawford*, is put in service by the Woods Hole Oceanographic Institution. It carries 8 scientists and 14 seamen.

1920 Dr. William Crawford Gorgas, who made the Panama Canal possible by killing a bug, dies at 65 in London. Gorgas was sent to Panama in 1904 to oversee mosquito control; his methods were so effective that both malaria and yellow fever, transmitted by mosquitoes, were eradicated.

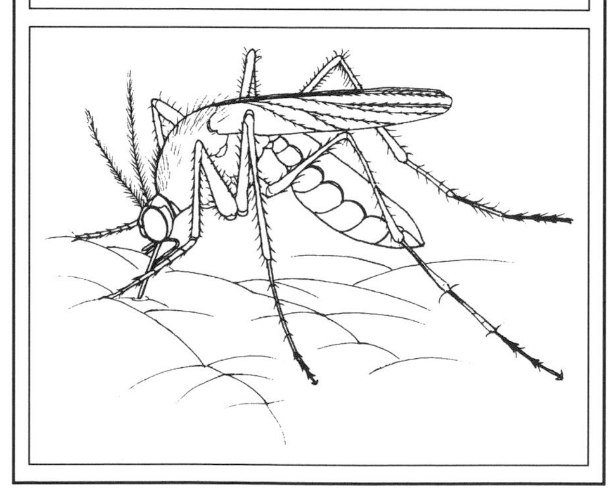

1987 Richard Branson and Per Lindstrand become the first to cross the Atlantic Ocean in a hot-air baloon—almost. The pair had to leap into the ocean off the coast of Ireland before their craft reached land.

1991 A four-day asteroid convention ends in San Juan Capistrano, California. The scientists discussed ways to detect and divert Earth-bound killer asteroids. The chance of any one such body hitting Earth is very small, but the loss of life would be so great if one did hit, that statisticians now estimate that death-by-asteroid has a greater probability than death by airplane crash.

1992 The Fourth of July weekend is off to a great start. A nuclear plant in Omaha, Nebraska, shuts down when thousands of gallons of coolant flow onto the floor of a containment building after an electrical failure. The next day a transformer of a nuclear plant in Delta, Pennsylvania, explodes.

1995 The Golden Gate Bridge in San Francisco is the scene of its 1000th known suicide.

JULY 4

1054 An exploding supernova is scientifically recorded for the first time, by Chinese and Korean astronomers. Rock paintings in North America suggest that Indians in Arizona and New Mexico also saw it. Fragments of the blast are visible today as the Crab Nebula.

1753 Aeronaut-inventor Jean-Pierre-François Blanchard is born into poverty in Les Andelys, France. He became famous for daring balloon flights at the dawn of aviation. He and John Jeffries were the first to fly across the English Channel (1785; this flight was also the first transport of air mail), and he invented the parachute (also in the 1785). The first parachute jump in history was made by a dog that Blanchard tossed out of a balloon.

1826 Thomas Jefferson, scientist, inventor, agriculturist (and third U.S. president), dies in Virginia. John Adams (the second president) dies in Massachusetts. James Monroe (the fifth president) dies exactly five years later in New York.

1828 The cornerstone is laid for the Tremont House in Boston, the first U.S. hotel to install bathrooms and toilets.

1845 Henry David Thoreau moves into a little cabin at Walden Pond near Concord, Massachusetts. His account of his two years there, *Walden*, is now a classic in the literature on ecologically sound living.

1868 Astronomer Henrietta Swan Leavitt is born the daughter of a minister in Lancaster, Massachusetts. In 1912, after five years studying a group of stars called the Magellanic clouds, Leavitt discovered a relationship between brightness and the period of light fluctuation in a particular species of star, now called Cepheid stars. This relationship was used to calculate distance, and for the first time it became clear just how vast the universe is.

1885 Nine-year-old Joseph Meister is walking to school in the remote Alsace village of Meissengott when he is attacked and bitten 14 times by a rabid dog. The boy is taken to the laboratory of Louis Pasteur, who has been developing a vaccine for the killer disease. Meister arrives just in time to be the first human to be inoculated, and cured, with a rabies vaccine. As an adult, Meister became the gatekeeper of the Pasteur Institute in Paris. When the Nazis overran the city in 1940, soldiers demanded access to Pasteur's crypt. Meister committed suicide rather than comply.

1906 Chemist-meteorologist Vincent Schaefer is born in Schenectady, New York. In 1946 he began the science of experimental meteorology by seeding clouds with dry ice and producing snow.

1906 "A real-life Indiana Jones," Henry Ward, is fatally struck by an automobile in Buffalo, New York. It is that city's first automobile death. Ward had traveled the world and was famous as a supplier and taxidermist of animal specimens for museums and exhibitions. He founded the huge science supply company Ward's Natural Science Establishment, Inc. Ironically, after a life of globetrotting, Ward perished just 80 miles from his birthplace.

1925 The most visible, most conspicuous advertising sign in history is switched on. It is a 68-foot-high ad for the Citroën automobile, attached to the Eiffel Tower in Paris. It can be seen 24 miles away.

1928 Jean Lussier becomes the first human to go over Niagara Falls in a rubber ball. Lussier built the device himself at a cost of $1485; it weighed 750 pounds, and had oxygen tanks and padding.

1934 Marie Curie dies of leukemia caused by radiation poisoning.

1992 A flaming crow starts a 200-acre fire at the Department of Energy's Nevada Test Site, 105 miles northwest of Las Vegas. It takes a day to control the blaze. "A crow landed on a transformer and it caused an arc of some type," explained DOE spokesman Darwin Morgan.

1997 The *Pathfinder* spacecraft lands on Mars, in an ancient riverbed, and begins sending back pictures. A roving robot is soon released, and it performs chemical analysis of Martian rocks.

1862 Mathematician Charles L. Dodgson begins inventing a tale for a friend, Alice Liddell, during a boat trip and picnic on the Thames River at Oxford. Dodgson used the pen name Lewis Carroll when he published the tale as *Alice in Wonderland*.

1794 Sylvester Graham, inventor of the graham cracker, is born in West Suffield, Connecticut. Young Graham went through a series of jobs before becoming a Presbyterian minister in 1826. But rather than preaching, he put his energies into developing healthy diets and life-styles. He was known for advocating use of a coarse, unsifted wheat in baking, which he made into the graham cracker in 1829. Graham founded a famous commune at Brook Farm, near Boston, where the "Grahamites" slept on hard mattresses, took cold showers, and ate a vegetarian diet. Graham was once attacked by a mob of butchers and bakers, a sign of how popular he was.

1805 Robert Fitzroy is born a descendant of Charles II near Bury Saint Edmunds, England. He commanded the surveying/scientific circumnavigation of Earth by the *Beagle* (1831–1836) on which Darwin sailed. Fitzroy is famous for fighting Darwin and the theory of evolution, but in fact he won prizes as a geographer and was appointed head of Britain's Meteorologic Office, from which he issued the first weather reports to be published in a daily newspaper.

JULY 5

1820 Scottish engineer William Rankine is born in Edinburgh. His first teacher was his father, an army officer; Rankine must have learned well because by age 14 he had read Newton's *Principia*—in the original Latin. His major contribution was preparing thermodynamic theory in a form that working engineers could use. His *Manual of the Steam Engine* (1859) introduced much terminology and notation still used in engineering. He also created the Rankine temperature scale, and popularized the term "energy."

1888 Neuroscientist and Nobel laureate Herbert Gasser is born the son of an immigrant physician in Platteville, Wisconsin (see biographical note on May 11, 1963).

1891 John Howard Northrop, the first to crystallize pepsin and several other enzymes, is born the son of a Columbia University zoology instructor in Yonkers, New York. His work helped establish the structure and role of enzymes in the workings of the body. Northrop shared the 1946 Nobel Prize for Chemistry with James B. Sumner and Wendell M. Stanley.

1927 DNA explorer Albrecht Kossel dies at 73 in Heidelberg, Germany. He switched his career interest from botany to medicine at the urging of his father, and then changed again to biochemistry under the influence of chemist Ernst Hoppe-Seyler. In 1867 one of Hoppe-Seyler's previous students, Johann Miescher, had extracted a chemical, "nuclein," from the nucleus of pus cells. It was Kossel who did detailed analysis of the structure of nuclein, thereby providing the foundation for the modern understanding of DNA. Kossel was the sole recipient of the 1910 Nobel Prize in Physiology or Medicine.

1944 The first U.S. airplane to be powered by rockets takes its maiden flight. Called the *Rocket Ram*, the craft is piloted by Harry Crosby, who lies flat on his back in the cockpit to be able to withstand the thrust. It was built by Northrup Aircraft, Inc., of Hawthorne, California, and used monoethylaniline as fuel. The plane was originally tested as a glider the previous October.

1946 A new standard in the economic use of materials is set: The bikini bathing suit debuts at a Paris fashion show. The designer was Louis Reard.

1948 National health care goes into effect in Britain.

1951 Bell Telephone Laboratories of Murray Hill, New Jersey, announces the invention of the junction transistor by Dr. William Shockley. The device revolutionized electronics, and therefore all of technology and culture. As a tiny amplifier and rectifier, the transistor replaced the bulky and slow radio tube, and made possible the miniaturization of all electronic control equipment, in TVs, radios, hearing aids, satellites, bombs, and many others. Computers that filled a room in 1950 now fit in a pencil.

1978 The first Polish astronaut, Major Miroslaw Hermaszewsk, lands in Kazakhstan in a *Soyuz* spacecraft after completing his maiden voyage.

1982 The nation's second shuttle, the *Challenger*, is taken to its launch site (the Kennedy Space Center in Florida) atop a Boeing 747 jet.

1994 The Associated Press reports the passing of William C. Taylor, 95, and his wife Jessie, 94. They had been married since 1929, and both died of heart trouble within hours of each other on the same floor of a Boca Raton, Florida, hospital.

1686 Antoine de Jussieu is born in Lyon, France. He came from a family of noted scientists, including three uncles who were famous botanists. Jussieu wrote many papers on botany, zoology, and human anatomy.

1781 Sir Stamford Raffles, a founder of the London Zoo (and its first president), is born at sea off Jamaica, the son of an "improvident" merchant captain.

1785 Sir William Jackson Hooker, first director of the Royal Botanic Gardens at Kew, is born the son of an accountant in Norwich, England.

1817 Embryologist-histologist Rudolph Kölliker is born the son of a banker in Zurich. He became a doctor, but his genius was in preparing once-living material for the microscope. He was the first to isolate smooth muscle, one of the first to interpret tissues as groups of cells, and he determined that sperm, eggs, and elongated nerve fibers were all cellular in origin. He also theorized that evolution proceeds in discontinuous jumps, a notion that is just coming into vogue today.

JULY 6

1858 The nation's first shoe-making machine is patented. Lyman Reed Blake of Abington, Massachusetts, receives patents No. 29,561 and 29,775 for his stitching machine, which revolutionized shoe manufacturing. The device, called the "McKay sewing machine," provided an efficient, fast way of attaching the outsole to the insole and upper. It was first used in 1861 in the factory of William Porter & Sons in Lynn, Massachusetts.

1869 John Wesley Powell departs with a party of nine on the first scientific exploration of the Colorado River. Three of the men later abandon the trip in the face of massive rapids, and are killed by Indians.

1885 The first rabies vaccination is given by Louis Pasteur (see further description on July 4).

1897 U.S. patent No. 586,025 is awarded for a combination food grater, slicer, and mouse and fly trap.

1897 Otto Wilhelmi of Düsseldorf, Germany, is granted a patent on a rocket that releases advertising leaflets when it explodes.

1903 Swedish biochemist Axel Hugo Theorell is born in Linköping. He followed in his father's footsteps and became a physician, until an attack of polio forced him to abandon his practice and concentrate on laboratory research into chemicals produced by the body. Theorell was the first to isolate the muscle protein myoglobin in crystal form, and he won the 1955 Nobel Prize for Physiology or Medicine for research into the structure of enzymes.

1905 The first international exchange of fingerprints involving the United States occurs today, when the prints of the habitual offender John Walker (a.k.a. Captain John Pearson) arrive in St. Louis from Scotland Yard in London.

1928 The first all-talking motion picture, *Lights of New York* produced by Warner Brothers, debuts at the Strand Theater in New York City.

1932 Kenneth Grahame, author of the animal classic *The Wind in the Willows*, dies at 73 in Pangbourne, England.

1994 A couple who are both HIV positive have been permitted to adopt a child, according to a report today by the Associated Press. The HIV status of Charles and Mary (last name withheld by authorities to protect the child) was known by workers in Florida's Department of Health and Rehabilitative Services when they recommended the adoption, but the information was not passed on to Pinellas Circuit Judge Horace Andrews, who had the final say in allowing the adoption.

1944 History's worst fire at a circus erupts in the main tent of the Barnum & Bailey Circus in Hartford, Connecticut; 168 perish. The anniversary of Barnum's birth was celebrated the day before.

1568 William Turner the father of English botany, dies in London at about 60 (his exact birth date is unknown).

1802 A comic book is first published in the United States. "The Wasp," edited by Robert "Rusty-Turncoat" Rusticoat, is a political satire magazine. Volume 1, No. 1, dated today, contains four pages.

1843 Physiologist-cytologist Camillo Golgi is born the son of a physician in Corteno, Italy. Originally planning a career in psychiatry, Golgi switched to researching the cell after reading the works of pathologist Rudolf Virchow. Golgi's greatest contribution to science came in 1873 when he developed an entirely new way of staining material for microscope viewing; previous stains used organic dyes, whereas Golgi used silver salts for staining. Never-before-seen details of all kinds of tissue became visible. Golgi, of course, was the first of many to benefit from this innovation. He discovered cell parts that still bear his name: the Golgi complex and Golgi bodies (which he first detected in the brains of barn owls in 1898). He further confirmed that nerve cells do not actually touch, but are separated by small spaces (synapses). He also discovered that several different species of parasite are responsible for different varieties of malaria. Golgi was awarded the 1906 Nobel Prize for Physiology or Medicine.

1861 Nettie Maria Stevens, who discovered that the X and Y chromosomes determine the sex of a human, is born in Cavendish, Vermont. Stevens originally worked as a zoologist, investigating the sperm and eggs of planaria and other simple animals. She went on to study sex determination in beetles, flies, plants, and mosquitoes. In 1906 she and Edmund Beecher Wilson independently determined that two X chromosomes make a female human, and an X and a Y a male. This finding was of great significance for genetics because it firmly established a relationship between heredity and cell contents.

1898 The United States annexes Hawaii. Sixty years later President Eisenhower signs the Alaska Statehood Act.

1907 The first transplantation of human nerve tissue occurs. Dr. Walter Jacoby successfully transfers 1.7 inches of nerve into the right hand of a 35-year-old manual laborer in Munich, Germany.

1930 Construction begins on Boulder Dam, later called Hoover Dam.

1935 A system for identifying people by the pattern of blood vessels in their eyes is presented to the annual meeting of the International Association of Chiefs of Police in Atlantic City, New Jersey. Dr. Isadore Goldstein, an ophthalmic surgeon at Mount Sinai Hospital in New York City, developed the system with Dr. Carleton Simon, former Deputy Police Commissioner of New York. Identification is based on the positioning of the four major veins in the eye's retina.

1770 A kangaroo is first described by a Western scientist. Joseph Banks reports observation of the animal in his journal, during the exploration of Australia by the Captain Cook-commanded *Endeavor*.

1981 *Solar Challenger* becomes the first solar aircraft to cross the English Channel. Pilot Stephen Ptacek took off near Paris and landed just outside Canterbury, England, 5 hours 23 minutes later.

1991 Ash from the eruption of Mt. Pinatubo in the Philippines completes its girdle around Earth.

1994 A Norwegian study in the *New England Journal of Medicine* indicates that environment has a much greater role in the production of birth defects than previously suspected. Women who produce a child with a birth defect have a very high probability of producing another child with that same defect—unless the mother moves to a new town! The likelihood is cut in half by the change of environment. The finding is dramatic, unexpected, and as yet unexplained. "It suggests there are things out there that we just have not been clever or lucky enough to find so far," said Allen J. Wilcox, one of the authors.

1994 Flirting will *not* be banned in the public schools of Syracuse, New York, according to director of pupil services Thomas Colabufo. Until today, flirting was classified as sexual harassment by the schools' new policy statement. Following negative publicity, a public hearing, and a flood of media calls, the behavior was removed from the list of taboos.

1810 Physiologist Gabriel Valentin is born in Breslau (now Wroclaw, Poland). Among his discoveries were the presence of cilia in the vertebrate oviduct and the digestive activity of pancreatic juices. He became the first Jewish professor in a German-speaking university (University of Bern, 1836).

1838 Ferdinand Zeppelin, inventor of the flying machines that bear his name, is born in Constance, Germany. As expected of a nobleman, he became a military officer. In the 1860s he traveled to the United States to observe the Civil War; during this trip he made his first balloon flight.

1856 The machine gun is patented. Charles E. Barnes of Lowell, Massachusetts, receives patent No. 15,315 for "an improved automatic cannon." Like the more popular and faster Gatling gun (which was patented six years later), the device was crank-operated, with the speed of firing controlled by the speed of cranking.

1857 Psychologist Alfred Binet, creator of the IQ test, is born in Nice, France.

1861 Sir John Arthur Thomson, Scottish naturalist and science popularizer, is born in Salton, East Lothian (see biographical note on February 12, 1933).

1889 History's last bare-knuckle boxing match is a 75-round marathon between John L. Sullivan and Jake Kilrain in Richburg, Mississippi. Gloves and the Marquess of Queensberry rules were soon adopted.

1892 The nation's first (and still preeminent) psychology society, the American Psychological Association, is organized at Clark University in Worcester, Massachusetts. "The object of this society shall be to advance psychology as a science." Professor Granville Stanley Hall was the first president and Dr. Joseph Jastrow the first secretary and treasurer. The first scientific meeting of the APA was held on December 27, 1892, at the University of Pennsylvania in Philadelphia.

1894 Soviet physicist Pyotr Kapitsa is born in Kronshtadt. In the 1920s he worked in Rutherford's lab in England, where he pioneered the production of large magnetic fields (he became the first foreigner elected to the Royal Society in two centuries). He was awarded the 1978 Nobel Prize for work on low temperatures.

1895 Soviet physicist Igor Tamm is born the son of an engineer in Vladivostok. His Nobel Prize came in 1958 for explaining Cherenkov radiation (a "wake" of light created when high-energy particles exceed the speed of light in a transparent medium like water).

1939 "If men and women are to understand each other, to enter into each other's nature and mutual sympathy, and to become capable of genuine comradeship, the foundation must be laid in youth." This was one of the insights of physician-essayist-sex researcher Havelock Ellis, who dies on this day in Washbrook, England, at age 80. The son of a sea captain, Ellis spent several years as a teacher in Australia before returning to England at age 20 to study medicine. Two years later he met and collaborated with iconoclast George Bernard Shaw in a literary society, the "Fellowship of the New Life." Ellis's major academic work was the seven-volume *Studies in the Psychology of Sex* (1897–1928), which did much to raise public awareness and dialogue about sex and sexual problems. It also raised outrage. A judge called the work "a pretense, adopted for the purpose of selling a filthy publication." The work was banned for sale to all but medical professionals until 1935. As well as a sex educator, Ellis was known as a champion of women's rights.

1994 Professional golfer John Daly claims in a newspaper interview that the use of cocaine is common on the PGA tour. He calls for random drug testing, as exists in other pro sports. His colleagues are not unanimous in their agreement; "It's a load of rubbish," responded one pro to the charges. Daly is a confessed alcoholic who made his statements during the Anheuser-Busch Classic in Virginia.

1994 Astronomers have now found the heaviest known metals in outer space, according to a report in today's *Science*. Lead, krypton, and arsenic are among the elements detected with the Hubble Space Telescope as it viewed a giant gas cloud 400 light-years away. The pattern of light emission from the cloud indicates its contents. Up until now, zinc was the heaviest metal to have been detected in space. Such metals are thought to be created when stars explode.

JULY 9

1792 The nation's first college professor of agriculture was Samuel Latham Mitchill, appointed on this day as a Professor of Natural History, Chemistry, Agriculture and related sciences at New York City's Columbia University.

1819 Sewing machine inventor Elias Howe is born the son of a farmer in Spencer, Massachusetts.

1831 Wilhelm His, histologist and founder of histogenesis (the science of the embryonic origins of tissues), is born in Basel, Switzerland. In 1865 he invented the microtome, a tool now used world-wide to slice material into extremely thin pieces so it can be viewed through a microscope. In 1886 he used the tool in his discovery that each nerve fiber comes from a single nerve cell; this was a major plank in the "neuron theory" (that the nerve cell is the basic unit of the nervous system). His three-volume *Human Embryonic Anatomy* (1880–1885) was history's first comprehensive study of the human before birth. His's son, Wilhelm His, discovered the bundle of His in the heart.

1845 Astronomer Sir George Howard Darwin, one of the first to propose that the moon was once part of Earth, is born in Down, England, in the house where his father, Charles Darwin, worked and died.

1856 Electricity wizard Nikola Tesla, who made alternating current practical, is born in Smiljan, Croatia, the son of a clergyman and a mother who invented gadgets for the kitchen.

1856 Avogadro's number is up. The famous physicist dies at 79 in Turin, Italy.

1872 The doughnut machine is patented. John F. Blondel of Thomaston, Maine, receives patent No. 128,783 for a spring-loaded device that removes dough from the doughnut center.

1877 The Bell Telephone Company is founded, on "the most lucrative patent in history."

1911 Physicist John Wheeler is born in Jacksonville, Florida. His theoretical calculations helped create the atomic bomb and the concept of "black holes" (a term he coined).

1926 Physicist Ben Roy Mottelson is born in Chicago. He shared the Nobel Prize in 1975 for determining the shape of the atomic nucleus.

1943 Clifford Beers dies at 67 in the Butler Hospital in Providence, Rhode Island. He was born in New Haven, Connecticut, the home of Yale University. Beers began suffering mental illness as a student there, and underwent a full-blown breakdown after graduation. He spent three years in abusive asylums. When finally released, he wrote *A Mind that Found Itself* (1908), and began a lifelong campaign for more understanding and better treatment for the insane. Also in 1908 he formed the Connecticut Society for Mental Hygiene, the nation's first such organization. He was a cofounder of the International Committee for Mental Hygiene in 1920. His work on *A Mind that Found Itself* was so intense and passionate that his family had him temporarily reinstitutionalized when he started writing it.

1966 "Namu," the first killer whale kept alive in captivity, dies in Seattle one year after capture.

1993 The American Psychiatric Association releases its latest judgment on premenstrual syndrome, stating that the ailment is real, it is serious, and it is worthy of more research, but it is not a mental illness. In the latest edition of the *Diagnostic and Statistical Manual of Mental Disorders* the Association calls the disorder PMDD (premenstrual dysphoric disorder), pointing out that it is "quite rare, found in 3 to 5 percent of menstruating women." In a related judgment, the Association removed another psychological problem, "self-defeating personality disorder," from the *Manual*.

1993 The first in a long line of guilty pleas is registered in the syringes-in-Pepsi-cans hoax that recently swept across the nation. Christopher J. Burnette, 25, of Williamsport, Pennsylvania, says he put a used insulin syringe in a can of Diet Pepsi because he was depressed. Eventually, over 40 liars in 20 states were arrested for falsely claiming to have found a variety of dangerous objects in Pepsi products, including sewing needles, wood screws, hypodermic needles, and machine parts. Motives ascribed to the reports were practical jokes, attention-seeking, and greed. Tampering with a consumer product is a federal offense, punishable by up to five years in jail and $250,000 in fines.

1797 Edward Jenner submits his first paper on vaccination to the Royal Society, which rejects it.

1832 Alvan Clark, the first to find evidence that white dwarfs are stars, is born in Fall River, Massachusetts. Like his father he was a lens grinder and telescope maker, and he became world famous for his skill. In 1897 he supervised construction of the 40-inch Yerkes telescope, which remains the world's largest refractor telescope. He passed away in the same year.

1902 Chemist Kurt Alder is born the son of a teacher in Königshütta (now Chorzow, Poland). In 1950 he shared the Nobel Prize for developing the Diels-Alder reaction, which allows scientists to create dienes (chemical compounds that contain two double bonds between molecules).

1913 The highest-ever temperature in the Western Hemisphere (134°F) is recorded in Death Valley.

JULY 10

1920 Physicist Owen Chamberlain is born in San Francisco. In 1959 he shared the Nobel Prize with collaborator Emilio Segrè for detecting the antiproton.

1921 Eunice Kennedy Shriver, the sister of John F. Kennedy, is born in Brookline, Massachusetts. In 1961 she and husband Sargent (then the first director of the Peace Corps) opened their Maryland backyard to several mentally retarded children. It was the start of the Special Olympics. "A couple of parents told me they couldn't get their kids into camp," she recalled. "We had some land, and we decided to use it—teaching the kids to swim and so on. And I began to look into how people with mental retardation were treated, even by our own government. For example, in 1961, no jobs were open to the retarded in the Civil Service. So I made some remarks about that, and President Kennedy called me. 'Listen,' he said. 'If you're going to blast me, give me a little notice.'"

1925 The Scopes "Monkey Trial" begins in Dayton, Tennessee. A jury is selected.

1925 The state news agency Tass is established in Russia.

1931 Medical quack William Horatio Bates dies in New York City. He developed and marketed the Bates Method of Eye Relaxation for curing many eye defects—simply by relaxing one's gaze. Nearsightedness, farsightedness, glaucoma, cataracts, and syphilitic iritis were among the troubles he claimed to cure. Thousands in the early 1900s believed him, including the famous Aldous Huxley, author of *Brave New World* and (ironically) *The Doors of Perception*, *The Art of Seeing*, and *Eyeless in Gaza*.

1962 The first privately owned satellite, *Telstar 1*, is launched into orbit by NASA from Cape Canaveral at 4:35 AM. The device was built by Bell Telephone Laboratories for AT&T; it was 3 feet in diameter, weighed 170 pounds, and launched with a three-stage Thor-Delta rocket that weighed 112,000 pounds. It revolutionized communications on Earth.

1985 Coca-Cola Co. announces it will resume selling the old-formula Coke.

1992 Scientists report discovering a gene for learning. Two studies by three different research groups in today's *Science* concern "knockout mice," which lack a particular gene; otherwise normal in most behaviors, these mice are "much impaired" in their ability to return home in a water-filled maze. The missing gene means that a specific chemical is not produced, a chemical that is abundant in certain parts of the brains of mice and men. This is the first time a gene has been linked to learning in any creature.

1929 Paper money in its present size is first issued in the United States. Some 823 million pieces of the larger, old style money (valued at $4997 million) was in circulation at the time. Paper money of any kind was first used in the country's colonies on February 13, 1690, by Massachusetts, to pay soldiers who had fought against Quebec. Wooden money was used one time in the country's history, in the 1930s in the town of Tenino, Washington, when the Citizens Bank of Tenino failed in December 1931, and the town had a shortage of cash with which to conduct business. The Chamber of Commerce arranged for the issuance of wooden money in various denominations, made from three-ply Sitka spruce. Cedar was later used.

1677 Leibniz's version of calculus is published.

1732 French astronomer Joseph Lalande is born the son of a postal official in Bourg-en-Bresse. He was headed to a career in law until he happened to lodge next to an observatory. In 1801 he published a catalogue of 47,000 stars; one of these (called "Lalande 21185") turned out to be the third nearest our sun and one of the first known to have a planet.

1770 The German Academy of Sciences is founded in Berlin.

1862 Health care pioneer Charles Frederick Menninger is born in Tell City, Indiana.

1906 The Northwest Passage is first navigated, by Norway's Roald Amundsen. The Passage is the longsought water route between the Atlantic and Pacific Oceans across North America.

JULY 11

1916 Soviet physicist Aleksandr Prokhorov is born in Australia. He and partner Nikolay Basov developed maser theory, and shared the 1964 Nobel Prize with Charles Townes who did the same work independently in the United States. The maser spawned the laser.

1927 Physicist Theodore Maiman, designer and builder of the first laser, is born the son of an electrical engineer in Los Angeles.

1934 The first seven members of the Federal Communications Commission are appointed. The first chairman is Eugene Octave Sykes. The body was created by 48 Stat. L. 1064, passed on June 19, 1934; its mission is to oversee international and interstate communications by satellite, radio, television, wire or cable.

1948 Anatomist-anthropologist Franz Weidenreich dies of a heart attack at 75 in New York City (see June 7, 1873, for a biographical sketch).

1962 The first transatlantic TV transmission via satellite. The United States receives a seven-minute speech by the French minister of communications, followed by Yves Montand singing "La Chansonette." The broadcast originated from Pleumeur-Bodou in Brittany; it was received at 7:35 PM EDT, and was relayed through the satellite *Telstar 1*, launched the previous day. That craft was the first privately owned satellite in orbit around Earth.

1979 *Skylab I* returns to Earth as a fiery mass of debris raining down on the Indian Ocean and Australia.

1991 The U.S. Senate passes a bill that reinstates the federal death penalty for a variety of crimes: assassinating the President, hijacking an airplane, murdering a chicken inspector. Murdering a horse inspector is equally covered. Simply *finding* a horse inspector would be a trick, as there are less than two dozen nationwide. The USDA was caught off-guard by the bill's protection. "I can assure you that attacks on federal meat inspectors are few and far between," said a spokesman. Ohio Senator Howard Metzenbaum was also not very impressed with the new regulation, observing that the murdering of poultry inspectors is "hardly the cause of the crime epidemic we face today."

2126 The comet Swift-Tuttle will make its closest-ever pass by the sun, according to calculations made and announced in 1992 by Harvard astrophysicist Brian Marsden. He added, however, that a slight error in his calculations or in the orbits of the comet or of Earth "could cause the comet to hit the Earth." The resulting cataclysm would be larger than the comet crash that wiped out the dinosaurs, according to the theory of many. Marsden further added that because the orbits of Earth and the comet cross each other "sooner or later, it will hit us."

1991 Rejection-proof transplants are finally here, according to a report by scientists at Massachusetts General Hospital in today's *Science*. "It's a nifty trick," said Dr. Paul Lacy, a transplant specialist, about the Massachusetts technique of pretreating cells with antibodies prior to transplanting them. Dr. Camillo Ricordi, director of cellular transplantation at the University of Pittsburgh, assessed the procedure, "I think this a great step. I think the final victory will be through a combination of pretreatments."

1817 Author-naturalist Henry David Thoreau is born in Concord, Massachusetts, the third child of a "feckless" businessman and his "bustling, talkative" wife. After graduating from Harvard, Thoreau got a teaching job at his old grammar school, but quit after just two weeks. He then founded his own school with his brother John; the school survived three years, until John became ill. In 1839 a canoe trip down the Concord and Merrimack rivers with John convinced Thoreau that he really wanted to be a "poet of nature." Thoreau wrote many essays and several books, which were largely ignored and sold poorly during his lifetime. His *Walden; or, Life in the Woods* has since become a classic ecological analysis of man living in harmony with nature in an industrial society.

1843 Explorer-naturalist David Douglas dies at just 36 in the Sandwich islands (now Hawaii). More plant species are named for him than other botanist; the best known of these is the Douglas fir.

1851 Photography inventor Louis Daguerre dies at 61 in France (see biographical note on, November 18, 1789).

1854 Photography inventor George Eastman is born into poverty in Waterville, New York. After patenting photographic film and the Kodak camera he was no longer in poverty.

1859 Patent No. 24,734 is issued to William Goodale of Clinton, Massachusetts, for inventing a machine to make paper bags. It is the nation's first patented paper bag machine.

1882 The first elevator with an electric light is installed in the Blue Mountain House, Blue Mountain Lake, New York.

1895 Visionary-engineer-architect R. Buckminster Fuller is born in Milton, Massachusetts (see biographical note on July 1, 1983).

1908 William Jennings Bryan becomes the first presidential candidate to be shown on motion pictures. The screening takes place in New York City, at Hammerstein's Roof at 42nd and Broadway.

1922 *France II*, the largest sailing vessel ever built, wrecks off New Caledonia in the Pacific. Launched in 1911 from Bordeaux, France, the ship was 418 feet long, weighed 5806 gross tons, and had five masts.

1972 The nation's first clinic to specialize in acupuncture opens. The Acupuncture Center of New York begins seeing patients today, then closes shortly thereafter "due to legal complications," then reopens in December as the Acupuncture Center of Washington, in Washington, D.C.

1977 Man sinks lower than he's ever gone before—11,749 feet straight down, to be exact. The mark is reached at the Western Deep Levels Mine at Carletonville, South Africa. The rocks at this depth are 131°F. So far, no human has gone deeper.

1994 There are not many sea turtles left on Earth. On this date, in the dead of night, one female turtle carefully digs a hole on Fort Lauderdale Beach and lays some 100 eggs in it. On discovering the nest the next day, college students record its location and place a protective wire cage around it. Three times a day the students check that the nest is safe, until August 22 when the cage should have been opened to let the hatchlings out. The cage was not opened, and the young turtles perished.

1994 Two primate smugglers are released from a Miami prison after just ten days because they helped authorities arrest Victor Bernal, 58 and a former zoo director in Mexico, for violating the U.S. Endangered Species Act.

1813 Physiologist Claude Bernard is born the son of poor vineyard workers in the French village of Saint-Julien. Bernard discovered much about how bodies work, and in the late 1850s he orginated the unifying concept of homeostasis (that organ systems act together to maintain a constant environment within the body. He is pictured here outside the old farmhouse in Saint-Julien where he wrote the 1865 classic, *An Introduction to the Study of Experimental Medicine*, during a period of convalescence.

1527 Mathematician John Dee is born in London. His lectures and writing (including the first English translation of Euclid) spurred revival of interest in math in England. In 1551 he became the astrologer to Queen Mary Tudor, but was imprisoned shortly thereafter for being a magician. He was released in 1555.

1629 Anatomist-reproduction scientist Caspar Bartholin dies at age 44 in Sorø, Denmark (see February 12, 1585, for a biographical sketch).

1822 Astronomer Heinrich Arrest is born the son of an accountant in Berlin. He discovered several comets and the 76th asteroid, but he is most famous for suggesting to his doctoral supervisor, Johann Galle, which star chart to use in searching for a planet that existed beyond Uranus, but which had not yet been detected. Neptune was discovered that night.

1826 Stanislao Cannizzaro is born the youngest of ten children of a judge in Palermo, Sicily. A fiery politician, he served in the Italian senate and was sentenced to death for aiding the Sicilian revolt in 1848. He was also a key delegate to a very important science congress, the First International Chemical Congress in 1860 in Karlsruhe, Germany, where he explained the distinction between atomic and molecular weights; this was critical in standardizing chemical formula-writing. He also discovered the Cannizzaro reaction (a conversion of an aldehyde to alcohol and acid).

1836 The United States adopts the present numbering system for patents. Prior to this date, 9957 patents were issued without numbers. Patent No. 1 is issued today to John Ruggles of Thomaston, Maine, for "traction wheels for locomotive steam-engine for rail and other roads." Ruggles happens to be the chairman of the Senate's Committee on Patents.

1929 History's first talking picture in Esperanto (an international language based on words common to many European languages) is made today in the Paramount Studios in New York City. The four-minute epic stars Germaine Chomette and Henry W. Hetzel. It was exported to 16 countries and presented to the 22nd annual Convention of the Esperanto Society of North America.

1793 Charlotte Corday stabs the French revolutionary Jean-Paul Marat through the heart as he sits in his bath. It is the end of one of science's saddest stories, in which a jealous, tyrannical Marat used his political power to have "the Newton of chemistry," Antoine Lavoisier, beheaded. Marat was a sloppy and self-deluded scientist, and his application to the French Academy of Sciences was soundly refused in 1780. Lavoisier was one of his main critics. Marat became a journalist and politician during the French Revolution, but he always harbored a grudge against the Academy and especially against the superstar Lavoisier, whom Marat repeatedly attacked in print. Just before entering Marat's bathroom to murder him, Corday is asked what business she had with Marat. "I should like to interview Monsieur Marat, so as to put him into a condition that will serve France well," she said.

1945—A FRIDAY THE 13TH The first atomic bomb arrives partly assembled at its test site in the New Mexico desert. By Sunday it is complete and hoisted atop the tower from which it will be dropped.

1978 Lee Iacocca is fired as president of Ford Motor Company.

1983 *Pioneer 10* becomes the first spacecraft to leave our solar system, as it passes the orbit of Neptune.

1991 Not only is beef a potential health hazard to those who eat it, but its production also harms the environment, according to a report released today by the Worldwatch Institute of Washington, D.C. Among the ecological problems caused directly or indirectly by cattle farming are a huge production of methane gas, rain forest destruction, water contamination, destruction of wildflowers, and depletion of water supplies.

1992 Six companies are found negligent by a Baltimore jury in the nation's largest-ever consolidation of asbestos claims. The 8555 plaintiffs combined their suits against a total of fourteen companies for making or selling the once-common building material. Most of the plaintiffs are former shipbuilders, pipe fitters, steel workers, and other craftsmen who worked with or near asbestos insulation and fireproofing. It is now known that the material causes fatal respiratory problems.

1610 Ferdinand II of Tuscany is born into the Medici family. In 1654 he invented a sealed thermometer (which was an improvement on the open design of Galileo), but he was most important as a patron of Galileo, Steno, and other scientists.

1770 A kangaroo is first killed by a European (or at least a written record of such an event is first made). It is just one week after kangaroos were discovered by the westerners.

1791 The home, library, and laboratory of chemist Joseph Priestley (discoverer of oxygen) are destroyed in Birmingham, England, by an anti-French mob on the second anniversary of Bastille Day. Priestley was an outspoken humanitarian who defended the principles underlying the French Revolution.

1800 Jean-Baptiste Dumas, French organic chemist and early sponsor of Pasteur, is born the son of a town clerk in Alais. As well as achieving many technical "firsts," Dumas led a revolt against Berzelius's well-accepted theory of chemical structure based mainly on electrical polarity. Dumas's "type theory" prevailed.

1801 Physiologist Johannes Peter Müller is born in Koblenz, Germany. One of his earliest scientific triumphs was an enormous study of human and animal vision, during which he discovered that each sense organ responds to external stimuli in only one manner, regardless of the stimulus (for example, poke an eye and you still see a flash of light). This finding was important to philosophy as well as science, because it showed that animals can know their world only through the limitation of their sense organs. Müller also did important work on nervous systems, reproductive organs (he discovered the Müllerian duct), blood, and cancer (his 1838 microscope study of cancerous cells was the foundation for the modern discipline of pathological histology). Some historians credit Müller and François Magendie as the founders of the science of physiology.

1867 Alfred Nobel illustrates the stability of his new invention, dynamite, before English businessmen by standing in a quarry while sticks of dynamite are thrown down on the rocks around him.

1914 Robert Goddard patents the liquid fuel rocket.

1921 Sir Geoffrey Wilkinson is born in Todmorden, England. He shared the 1973 Nobel Prize for Chemistry for studies of hydrogen-to-metal bonding and for determining the structure of ferrocene.

1929 Sir Baldwin Spencer dies at 69 while on an anthropological expedition in the wilds of Tierra del Fuego at the southern tip of South America. Baldwin was the first professional scientist to study the habits, rituals, and beliefs of the aborigines of Australia. He advanced the theory that their social organization was similar to that of prehistoric man. He wrote a number of books and was knighted in 1916.

1951 A sports event is televised in color for the first time. It is a horse race. CBS broadcasts the $15,000 Molly Pitcher Handicap at the Monmouth Park Jockey Club in Oceanport, New Jersey.

1965 Mars is photographed close up for the first time. In fact, it is the first time any planet has been photographed from beyond Earth's atmosphere. The 544-pound *Mariner IV* spends eight hours photographing the Martian regions of Arcadia, Cebrenia, and Amazonis. The camera is 10,500 miles from Mars and 134,000,000 miles from Earth when the photos are taken and sent back to Earth.

1992 Forty-nine pilot whales strand themselves in shallows beneath the lighthouse at Seal Rocks, 125 miles north of Sydney, Australia. Several die quickly and several get free quickly, but the majority are tended for days by a small army of fishermen, tourists, soldiers, environmentalists, and surfers.

1994 A Japanese study reports that the Japanese have the world's longest life expectancy. This is the eighth straight year Japan has ranked first, according to the Health and Welfare Ministry in Tokyo. Japanese females born in 1993 can expect to survive for 82.51 years (0.29 year longer than 1992 babies) and males can expect a life span of 76.25 years (up 0.16 year from 1992). Women from France, Switzerland, Iceland, and Sweden will be the next most long-lived, while men from Iceland, Sweden, Hong Kong, and Israel will be the best survivors.

1662 The "Royal Society of London for the Promotion of Natural Knowledge" is granted a charter from Charles II. The Royal Society is one of history's oldest, and still most prestigious, scientific bodies.

1809 Friedrich Henle, one of history's outstanding anatomists, is born the son of a Jewish merchant in Fürth, Bavaria (see May 13, 1885, for a biographical sketch).

1864 Alfred Nobel receives a Swedish patent for nitroglycerin. He founds his first company on it.

1871 Max Bodenstein, one of the first to postulate the existence of chemical chain reactions, is born in Magdeburg, Germany, the son of a brewer.

1904 The first Buddhist Temple in the United States is established. Surprise, surprise, it pops up in California, at a meeting room in Los Angeles. The chief priest in Rinbon Izumeda; the congregation is largely of the Shin Sect.

JULY 15

1904 Physicist Pavel Cherenkov, discoverer of the nuclear phenomenon that bears his name ("Cherenkov radiation"; 1934), is born to a peasant family in Voronezh, Russia. He was to share the 1958 Nobel Prize with Ilya Frank and Igor Tamm, who explained what he had found.

1914 Naturalist-author Gavin Maxwell is born in Elrig, County Wigtown, Scotland.

1916 William Boeing founds Pacific Aero Products, which eventually becomes the Boeing Company.

1921 Bruce Merrifield, who invented a simple, ingenious method of stringing amino acids together (which is what nature does to build proteins), is born in Fort Worth, Texas. His technique proved extremely important, and earned him the 1984 Nobel Prize for Chemistry.

1940 The world's tallest human on record, Robert Pershing Wadlow, dies at age 22 in Manistee, Michigan, at 1:30 AM. At death he was 8 feet 11.1 inches tall and still growing. His coffin was 10 feet 9 inches long. At age 9 he could carry his father, the mayor of his hometown of Alton, Illinois, up the stairs. Wadlow underwent a huge growth spurt after a double hernia operation at age 2. He studied to become a lawyer, but had to drop out because walking between classes was too difficult for him. He did some circus work and traveled on behalf of the company that made his shoes (size 37AA). Despite enormous physical problems and the social status of a freak, Wadlow remained cheerful and good-natured.

1943 Astronomer Jocelyn Bell Burnell, discoverer of pulsars (stars whose output of radiation follows a regular, pulselike beat), is born the daughter of an architect in Belfast, Northern Ireland.

1922 A platypus is exhibited for the first time outside its native Australia, at the New York Zoological Park at Bronx Park, New York. It was sent from Australia via San Francisco. It is a true mammal (having fur and mammary glands) but it is like a duck in that it has a bill and webbed feet and lays eggs. Its unusual structure led some to believe it was a hoax after it was discovered in 1797.

1954 On its 38th anniversary, the Boeing Company first tests the Boeing 707, the first jet airliner.

1975 Russian and U.S. astronauts blast off in different spacecraft (*Soyuz* and *Apollo*), later to rendezvous in orbit.

1992 A collision of elephants topples a barricade and injures several spectators at the Tarzan Zerbini International Circus in Lafayette, Indiana. The pachyderms were being led around in a circle when one stopped suddenly; he was hit by the next in line and knocked into the barricade, which then went flying into the crowd. Most of those injured were hurt by other people scrambling to flee the scene.

1994 Japan ends its trade in endangered sea turtles. Up until today's announcement, Japan (which imported 35 tons of hawksbill sea turtles in 1993) had claimed that trinkets and eyeglass frames made of sea turtle shells involved "traditional crafts," and therefore were exempt from international treaties.

JULY 16

1741 Alaska is first seen by the Danish explorer Vitus Bering.

1746 Giuseppe Piazzi, the monk who explored the heavens, is born in Ponte de Valtellina, Italy. In 1814 he produced an enormous map/catalogue of 7646 stars, and he determined that most of them are moving in relation to the sun. On January 1, 1801, he discovered the first (Ceres) of an entirely new class of extraterrestrial object—the asteroid (also known as planetoids or minor planets). The 1000th planetoid discovered was named after him.

1867 Reinforced cement is patented by the French gardener Joseph Monier.

1872 Roald Amundsen, the first man to the South Pole, the first through the Northwest Passage, and the first to locate the magnetic North Pole, is born in Gravning, Norway.

1877 Béla Schick, destined to save many lives by developing the Schick test for diphtheria, is delivered prematurely by his mother's uncle, Sigismund Telegdy, in Boglar, Hungary. Telegdy was a dedicated country doctor who became a major influence in Schick's selection of a career in medicine.

1888 Fritz Zernike, winner of the 1953 Nobel Prize for Physics for developing the phase-contrast microscope, is born the son of a headmaster in Amsterdam.

1916 Russian zoologist-microbiologist Elié Metchnikoff dies at 71 in Paris (see May 15, 1845, for a biographical sketch).

1926 Underwater color photographs are taken for the first time, off the Tortugas of the Florida Keys. Dr. William Harding Longley of Goucher College and Charles Martin of the National Geographic Society take the photos for *National Geographic* magazine. The camera is in a specially made brass case with a plate-glass window.

1927 Anthropology legend L.S.B. Leakey publishes his first scientific paper ("Stoneage Man in Kenya Colony") in *Nature*.

1935 The nation's first automatic parking meters are installed, in Oklahoma City. Carlton Cole Magee devised the "Park-O-Meter"; the Dual Parking Meter Company of Oklahoma City sold and installed them. Parking spaces are 20 feet wide, marked off by white paint, and have a nickel-operated machine in each space.

1945 The first atomic bomb is exploded shortly before dawn in New Mexico.

1969 The first manned mission to the moon is launched from Cape Kennedy.

1993 At an Orlando, Florida, news conference, the U.S. Humane Society calls for a release of whales and dolphins from the "horrors of captivity" in marine parks and aquariums. Humane Society President Paul Irwin cites a five-year study of five species of marine mammals in captivity: Accidents, disease, and low birth rates were among the findings. Early deaths, stress, ulcers, and abnormal behavior patterns are often seen in captive animals of many species. Also today, the animal rights group People for the Ethical Treatment of Animals begins weekend-long demonstrations in nine cities to publicize the issue and to promote the movie *Free Willy*.

1994 The first mountain-sized fragment from the comet Shoemaker-Levy 9 slams into Jupiter, creating a fiery explosion 1200 miles wide. The mountain was traveling at 130,000 mph when it hit. It is a "once-in-a-millennium spectacular."

1874 Physician Joseph Goldberger is born in Girált, Austria-Hungary. When he was six his family emigrated to the United States. Goldberger entered the Public Health Service and was sent to the tropics to study yellow fever and typhus. He caught both diseases. His great triumph (see September 12, 1915) was conquering the disease pellagra by determining its cause to be a dietary deficiency, in a famous experiment with Mississippi jail inmates.

709 BC A total eclipse is first observed and convincingly recorded, in Chu-fu, China.

1827 Sir Frederick Augustus Abel, coinventor of the explosive cordite, is born in the Woolwich section of London. After studying chemistry as one of the original 26 students of A.W. von Hoffman at the Royal College of Chemistry, Abel devoted himself to military chemistry. In 1889 with Sir James Dewar he created cordite, which was an important innovation in warfare for two reasons: because it was gelatinous, it could be squeezed and formed into cords (hence the name) and other shapes; and it was smokeless, and therefore allowed generals a clear view of battle scenes. Cordite made gunpowder obsolete; the Spanish–American War was the last major conflict to use it.

1850 A star other than the sun is photographed for the first time. A professional photographer named Whipple attaches a daguerreotype plate to the eyepiece of a 15-inch telescope at the Harvard Observatory in Cambridge, Massachusetts, to photograph the star Vega. Overseeing the operation is William Cranch Bond, the Observatory's first director.

JULY 17

1861 Paper money is first authorized to be issued by the federal government. Congressional acts passed on this date and on August 5, 1861, authorize $50 million to be issued in three denominations: $5 (Hamilton), $10 (Lincoln), and $20 (Liberty). The money is called "demand notes" because they are redeemable on demand at various government subtreasuries.

1887 Dorothea Lynde Dix, "godmother of the insane," dies at 85 in her room at the New Jersey State Asylum. She founded the institution and many like it throughout the nation. She called this one her "first-born child" (see April 4, 1802, for a biographical note).

1894 Cosmologist-priest Georges Lemaître, formulator of the big bang theory, is born in Charleroi, Belgium.

1938 One of history's great aviation blunders occurs when Douglas "Wrong Way" Corrigan departs New York for California. He flies to Ireland by mistake.

1955 Disneyland opens in Anaheim, California.

1955 Nuclear power provides the electricity for the lighting of an entire town (Arco, Idaho) for the first time.

1959 Archeologist L.S.B. Leakey wakes up with a fever, so he stays in camp while his second wife, Mary, goes fossil-hunting in a part of Olduvai Gorge named after Leakey's first wife, Frida. Accompanied by two dalmatians, Mary makes one of anthropology's greatest finds: *Zinjanthropus*. The skull is nearly complete, and many tools are later found in the area. Dated at 1.75 million years old, this hominid caused a rewriting of man's prehistory and evolution.

1975 The Russian spacecraft *Soyuz* docks with the U. S. *Apollo* in outer space for the first time.

1989 The Stealth bomber is first tested.

1992 Rhode Island's first drive-thru condom shop opens in a renovated Fotomat hut in Cranston. Above the motto, "Caring enough to protect yourself," is the shop's menu of various colors, styles, and flavors. Unfortunately, arrival of the shop's inventory was delayed for weeks, and today's grand opening coincides with the holy Feast of the Madonna Della Civita, celebrated by the many devout Catholics in the town. "Burn it!" yelled one passerby at the shop. Others had vandalized the hut before it opened.

1992 The record for gasoline nonconsumption is set by a group of students from Lycée St. Joseph La Joliverie, Nantes, France, who built a car that achieves, on this date, 7591 miles per gallon.

1994 The German media reports that during a raid for counterfeit money in Baden-Württemberg, authorities discovered a quantity of weapons-grade plutonium in the possession of criminals. It is the first known case of fissionable material falling into unauthorized hands. Even though the quantity (one-fifth of an ounce) is not enough to make a bomb, it is enough to poison a water supply. Later investigations revealed that the material came from Russia, and high-ranking Soviet officials were implicated in its smuggling.

JULY 18

1964 The Great Fire of Rome begins. It lasts several days. Contrary to legend, Nero did not actually play a fiddle during the blaze. He recited poetry.

1627 The first known spring of oil in America is described in a letter by the Franciscan missionary Joseph de la Roche d'Allion. The spring was found near Cuba, New York, and was marked on a French map in the same year by missionaries François Dollier de Casson and René de Brehant de Galinée.

1635 Robert Hooke, coiner of the term "cell" in biology, discoverer of Hooke's law in physics, and an investigator of many natural phenomena, is born the son of a clergyman in Freshwater, Isle of Wight.

1877 Thomas Edison shouts "Halloo!" into his prototype phonograph; it is history's first recording.

1908 Cleveland becomes the first large U.S. city to pass fireworks legislation.

1920 History's biggest weed is discovered near the Juanita River, Pennsylvania. It is a mat of box huckleberry covering approximately 100 acres. It is estimated to be 13,000 years old.

1932 Canada and the United States sign a treaty to develop the St. Lawrence Seaway.

1964 *Sealab I* is lowered to the ocean floor for the first time, in 192 feet of water off Bermuda.

1968 Physiologist Corneille Heymans dies at 76 in Knokke, Belgium. In his late twenties he started working with his father at the University of Ghent in researching how blood pressure and blood content affect breathing and heart activity. He succeeded his father there as professor of pharmacology in 1930. Working on anesthetized dogs, he discovered tiny sense organs (called pressoreceptors) in neck arteries that are sensitive to blood pressure, and that control breathing and heart rate. Near these receptors, in the aorta, he discovered other sense organs that respond to the oxygen content in the blood, and that also influence breathing by sending messages to the brain's medulla. He won the 1938 Nobel Prize for Physiology or Medicine.

1986 Woods Hole Oceanographic Institution releases the first pictures of the sunken *Titanic*.

1990 Karl Menninger, "dean of U.S. psychiatry," dies at 96 of abdominal cancer in Topeka, Kansas. After graduating with honors from Harvard Medical School and two years psychiatric work with Ernest Southard in Boston, Menninger joined his father's medical practice in 1920 in Topeka, and four years later was joined by brother William. The trio developed a group practice modeled on that of the Mayo family in Minnesota (which the elder Menninger, Charles Frederick, visited in 1908). The brothers' strong interest and training in psychiatry, and Topeka's lack of a hospital for the insane, convinced them to add mental illness to their specialties. They revolutionized psychiatric care by uniting psychoanalytic theory with a therapeutic social environment. Their methods attracted worldwide attention and students, and in 1931 the Menninger Sanitarium became the first facility to be approved for the training of psychiatric nurses. The Menninger Foundation (1941), the Menninger School of Psychiatry (1945), and the Center for Applied Behavioral Sciences (1974) were formed to further education, cure, and research in medicine and behavior. Among other things, Karl is remembered for his compassion, his belief in rehabilitation, his plain approach and plain language in the many books he wrote, and his forceful personality (which included a hot temper).

1994 After making headlines by exposing the huge fat content found in movie popcorn, the Center for Science in the Public Interest today reports similar findings for Mexican food.

1994 A former Mexican zoo director and an accomplice are sentenced in Miami to 70 days in prison for trying to illegally purchase primates and smuggle them to a zoo in Mexico. The pair were originally arrested in the Miami airport in a sting in which one Fish and Wildlife officer posed as a gorilla. The felons were apprehended just after they offered nearly $100,000 to buy the agent-gorilla and a real baby orangutan. "I am an environmentalist," claimed the defendant Victor Bernal, 58, at his trial. "You don't have to convince me that you're an environmentalist," countered the judge. "The question is, since you have been found guilty, what do I do with you now?"

1779 Jean-Paul Marat, 36, writes to the French Academy of Sciences, claiming to have research that overthrows Newtonian physics. The complete rejection of his work drives him from science to journalism/politics, where he instigates one of the greatest tragedies in science history: the guillotining of Lavoisier (see further note on July 13, 1793).

1846 Edward Charles Pickering, creator of the first photographic atlas of the entire sky (and inventor of the meridian photometer which allowed light from many stars to be compared and recorded simultaneously), is born in Boston, the great grandson of one of Washington's cabinet members.

1865 Charles Horace Mayo, called a "surgical wonder" by his colleagues, is born in Rochester, Minnesota. He was the second son of William Worrall Mayo, a surgeon who founded St. Mary's Hospital in Rochester, Minnesota, after a tornado devastated the town in 1889. The elder Mayo and his two sons (William J. and Charles Horace) performed all of the hospital's surgery through the early 1900s. Virtually all patient needs were handled under one roof. The facility became the famous Mayo Clinic.

1870 The Franco-Prussian war erupts. The great Louis Pasteur, 47, searching the battlefields for his wounded son, is horrified at hospital conditions, and begins a campaign to improve medical cleanliness. *Sans* a medical degree, Pasteur is elected to the French Academy of Medicine.

1882 Francis Maitland Balfour, a founder of modern embryology, dies at just age 30 trying to climb an unconquered portion of Mont Blanc, Switzerland. He was the younger brother of English Prime Minister Arthur James Balfour. His zoology research focused on the prebirth development of vertebrates. In positions at Cambridge and the marine lab in Naples, Italy, Balfour made many discoveries about the origins of sexual organs, the kidneys, and the nervous system; his findings also helped explain the prehistoric evolution of vertebrates. It was Balfour who suggested in 1880 that vertebrates should be a subclass of the phylum Chordata (which comprises all species with a flexible rod, called a notochord, running down their backs at any stage in their lives). He was also the first to suggest that adrenaline affects nervous activity. His historic *Treatise on Comparative Embryology* was published shortly before his mountain climbing accident. Edinburgh and Oxford had just offered him positions, but he had refused both, deciding instead to stay at Cambridge in a special professorship that had been created just for him. These decisions announced, he left for Switzerland.

1921 Rosalyn Yalow, inventor of the radioimmunoassay, is born in New York City. Her innovative procedure (which garnered the 1977 Nobel Prize for Physiology or Medicine) allows detection of antibodies and other biologically active compounds in the body in quantities that are so small that no other procedure can identify them.

1957 The nation's first rocket with a nuclear warhead is given its maiden test firing over Yucca Flat, Nevada. The *Genie* is an air-to-air missile with its own built-in guidance system, and was built by the Douglas Aircraft Company of Santa Monica, California. It is fired today from an F89 Scorpion jet.

1969 Astronauts Neil Armstrong, Edwin Aldrin, and Michael Collins go into orbit around the moon, prior to man's first moonwalk.

1985 Christa McAuliffe is chosen from 11,000 applicants to be the first schoolteacher in space, on board the ill-fated *Challenger* the following January.

1990 Officials of the Everglades National Park in Florida announce that mercury poisoning is rampant among the already-endangered Florida panther.

1994 The Associated Press reports results from a four-year study of sharks in the Atlantic Ocean and Gulf of Mexico. A dangerously high level of mercury has been found in shark meat that is usually used for human consumption. When that meat is eaten by humans, the mercury is transferred and accumulates in their systems. Nerve damage, blindness, reproductive disorders, and death are all consequences of mercury poisoning.

1969 The first transatlantic solo rowboat trip ends today at 1:48 PM in the surf off Hollywood, Florida. John Fairfax started his journey from Las Palmas, Canary Islands, on January 20. His craft, the 22-foot *Britannia*, was initially swept off course to the Cape Verde Islands of West Africa.

JULY 20

1534 Cambridge University Press receives its Royal Letters Patent to print and sell books. It is now the world's oldest publishing company.

1801 The country's first cheese factory, a cooperative of dairy farmers in Cheshire, Massachusetts, begins pressing a cheese at the farm of Elisha Brown, Jr. A month later the cheese weighed over a half-ton. It was loaded on a six-horse wagon and carried to the White House, where it was presented to President Thomas Jefferson on January 1, 1802.

1804 Sir Richard Owen, coiner of the word "dinosaur" (in 1842), is born the son of a merchant in Lancaster, England. He obtained a medical degree at the University of Edinburgh, but soon thereafter took an assistant's job in London's Museum of the Royal College of Surgeons, where he fell in love with comparative anatomy. He dissected every animal he could get. In 1852 he discovered the parathyroid gland in the throat of a rhinoceros; this was not discovered in man for many years. He undertook a massive study in the forms of animal teeth in the 1840s, which brought him to studying fossilized and extinct forms. Among his "firsts" were the first description of the newly extinct moa bird of New Zealand, the first description of *Archaeopteryx* (the first known fossil bird, which Owen reconstructed incorrectly), and the preparation of the first public display of reconstructed dinosaurs (done for the 1854 exhibition at the Crystal Palace in London). Owen was violently opposed to Darwin's theory of evolution; he wrote many anonymous articles attacking the theory, and provided material to its opponents.

1816 Sir William Bowman, who discovered that urine is a by-product of blood filtration, is born in Nantwich, England. He also made important discoveries about the eye and muscle, but his most important work was on the kidney, part of which is named for him (Bowman's capsule).

1836 Physician-inventor Sir Thomas Clifford Allbutt is born in Dewsbury, England. His most famous contribution to science was the most-used medical instrument in history: the clinical thermometer. Introduced in 1867, it replaced a foot-long instrument that took 20 minutes to yield a patient's temperature. Allbutt's device enabled routine and fast tracking of fevers and other diseases. He was knighted in 1907.

1844 Jonathan Walker arrives in Pensacola, Florida, under arrest for stealing slaves. He is found guilty and is placed in a public pillory and pelted with rotten eggs. He is sentenced to a fine of $600 and one year imprisonment for each slave stolen. "SS" (for "slave stealer") is branded into the palm of his right hand. It is the first time that branding is used to punish a federal crime.

1897 Chemist Tadeus Reichstein is born the son of an engineer in Wloclawek, Poland. He shared the 1950 Nobel Prize for Physiology or Medicine for determining the structure and effects of steroids.

1919 Sir Edmund Hillary, conqueror of Earth's highest mountain (Mt. Everest, in 1953), is born in Auckland, New Zealand.

1969 A man (Neil Armstrong from the *Apollo 11* mission) first lands on the moon.

1975 Archaeologist-treasure hunter Mel Fisher loses his son and daughter-in-law in the search for the *Atocha* (a Spanish galleon that sunk in 1622 in a hurricane off Key West). Ten years later, to the day, his party finds the wreck, packed with history and $400 million in gems, gold, and silver.

1976 A robot (*Viking I*) first lands on Mars.

1992 The level of mercury in the once-pristine Florida Everglades is steadily climbing, according to a year-long study by University of Florida engineering professor Joseph Delfino. The levels have been rising over the last five years. "In some places, it is as high as a threefold increase, and in others, it's only a 20-percent increase. Right now, it is not possible to determine the sources." Health officials have advised that people limit their consumption of fish in various areas in and around the state, because mercury and other heavy metals tend to accumulate in animals at the tops of food webs.

1620 Astronomer-priest Jean Picard, the first to accurately measure a degree of longitude (thereby producing the most accurate estimate of Earth's size since Eratosthenes 19 centuries before), is born in La Flèche, France. He was the first to use the telescope not only for observation but also for precise measurements of small angles; this allowed his famous estimate of Earth's size (published in 1671), which was used by Newton to explain gravity.

1694 Chemist Georg Brandt is born in Riddarhyttan, Sweden, the son of an apothecary-turned-metallurgist. Brandt was history's first major chemist to renounce alchemy, and in his later years he crusaded against charlatans who used simple chemistry to dupe others. Brandt discovered and named cobalt, the first metal unknown to ancient man.

1810 Chemist-physicist-instrument maker Henri Regnault is born in Aix-la-Chapelle, France. His father was an army officer who died two years later in Napoleon's invasion of Russia; his mother died shortly after that. He lost his own son in the Franco-Prussian War. Regnault discovered carbon tetrachloride, recalculated the specific heats of many substances with his own apparatus, reworked Boyle's law (showing it only worked for perfect gases under perfect conditions), and determined the exact temperature of absolute zero.

1816 Paul Julius Reuter, founder of Reuter's News Agency, is born in Hesse, Germany.

1864 The earliest known description of radio communications is written by Dr. Mahlon Loomis of Virginia. It is his 38th birthday. Exactly eight years later, less one day, he receives a patent on the system.

1880 America's first air compressor explosion in the United States occurs. It is during the construction of the Hudson River tunnel between New York and New Jersey. On this date, just 360 feet from the Hoboken shaft, the blast blows a hole through the tunnel's roof. Twenty workers died in the resulting flood. The mishap delayed the completion and opening of the tunnel for 28 years.

1919 Pilot John Boettner becomes the first American to join aviation's "Caterpillar Club" (members have parachuted safely from an aircraft about to crash). At 1200 feet Boettner bails out of the Goodyear balloon, *Wing Foot*, which subsequently crashes into a building at the corner of La Salle Street and Jackson Boulevard in Chicago. Three die and twenty-eight others are injured.

1959 Zoologist E. Newton Harvey dies at 71 in Woods Hole, Massachusetts. From 1911 until retirement three years before his death, he taught at Princeton. He is known for discovering what makes fireflies glow. In the early 1900s, using chemical extracts from glowing shrimp, he determined the chemical pathway that is responsible for "bioluminescence" in many species, including fireflies, squid, bacteria, fungus, jellyfish, and worms.

1969 Neil Armstrong and Edwin Aldrin depart the moon's surface having been the first humans to visit there.

1971 The highest-ever count of airborne fungus spores is made near Cardiff, Wales. Approximately 6 million spores per cubic foot is recorded.

1990 After 18 months, a name change, and $30 million of repairs, the *Exxon Valdez* is refilled with oil for the first time since its historic collision with an Alaskan reef causing it to dump millions and millions of gallons of oil into the once-pristine Prince William Sound.

1992 A Kansas state judge rules that life begins at conception. Sedgwick county District Judge Paul W. Clark agrees with abortion protester Elizabeth Ann Tilson, and acquits her of criminal trespass charges arising from a protest at a Wichita clinic.

1992 A spokesman for the Environmental Protection Agency says that recent oil gushes from the mountains around Los Angeles were caused by earthquakes ten days before. The oil has been seeping into rivers and threatening wildlife.

1587 A second English colony is established in the New World, on Roanoke Island off North Carolina. The settlement later vanishes mysteriously.

1784 Friedrich Wilhelm Bessel, the first to accurately measure the distance to a star other than the sun, is born the son of a civil servant in Minden, Prussia. He determined the location of 50,000 stars.

1802 Marie-François-Xavier Bichat, one of the fathers of histology, dies at age 30 after fainting and falling down a flight of stairs in his laboratory (see biographical note on November 14, 1771).

1822 Gregor Mendel, discoverer of the laws of genetics, is born into poverty in the small town of Heinzendorf, Austria (now Czechoslovakia). He helped his father tend fruit trees as a child, which along with a natural interest in mathematics, formed the foundation for his great insights. On his 25th birthday, he was ordained a subdeacon at the monastery in Brünn (now Brno, Czechoslovakia), where he single-handedly conducted his genetics experiments in a small pea patch.

JULY 22

1887 Gustav Hertz, nephew of physicist Heinrich Hertz, is born in Hamburg. Gustav received the 1925 Nobel Prize for Physics for establishing the quantized nature of the atom (i.e., that energy within the atom exists in discrete packets, or "quanta").

1893 Karl Menninger, "dean of U.S. Psychiatry," is born in Topeka, Kansas (see July 18, 1990 for a biographical sketch).

1933 The first solo flight around the world is completed in New York City by Wiley Post, in a Lockheed Vega monoplane named the *Winnie Mae*. The flight took 7 days 18 hours 49 minutes.

1964 The U.S. Navy's *Sealab I* submerges 192 feet to the bed of the Atlantic Ocean, 30 miles off Hamilton, Bermuda. It is the second descent in several days, but this time with people aboard. The aquanauts (Robert E. Thompson, Lester E. Anderson, Robert A. Barth, and Sanders W. Manning) surface ten days later, thereby setting a longevity record for living under the sea. In the vessel they breathed a mixture of helium (80%), oxygen (4%), and nitrogen (16%), and had to spend an entire day in a decompression chamber before returning to the surface.

1969 Aldrin, Armstrong, and Collins leave their orbit of the moon and head back to Earth, after being the first humans to walk on the moon.

1979 Samuel Jackson ("Sam") Snead becomes the first golfer to shoot below his age on a PGA tour event. On this date, at the "Quad Cities Open" in Coal City, Illinois, 67-year-old Snead shoots a 66. It is the last day of the four-day tournament. Twenty years before, Snead had become the first pro golfer to shoot below 60 for 18 holes.

1992 Unhealthy levels of mercury have been found in people who regularly eat fish from the Florida Everglades, announces a grass-roots research organization, the Florida Mercury Alliance. Hair samples from eleven adults and three children, analyzed in Boston, showed trace amounts of the toxic metal in six adults and unacceptable amounts in five. In the case of mercury in children "there are no known safe levels," said neonatal nurse Rosalyn Scherf. Mercury is troublesome because it causes nerve damage and death, because it accumulates and is not removed from the body, and because it is especially drawn to fatty tissue, which comprises much of the brain.

1994 The nation's 499th and newest wildlife preserve has been established just 50 miles from its very first. Both are in central Florida. The Associated Press reports today that the newly opened Lake Wales Ridge National Wildlife Refuge is only one quarter acre now, but it is expected to grow to 19,000 acres; it consists of a rare scrub habitat with species found nowhere else on Earth. A U.S. Fish and Wildlife director calls the habitat "one of the rarest and most severely threatened vegetation communities in North America." In 1903 President Teddy Roosevelt established the 3-acre Pelican Island National Wildlife Refuge on the nearby Indian River in hopes of protecting the breeding areas of several bird species.

1775 Physicist Étienne Malus, discoverer and namer of "polarized light," is born the son of a Paris bureaucrat.

1829 The typewriter is first patented in the United States. William Austin Burt of Mount Vernon, Michigan, is granted a patent for his "typographer." The first letter written on the device was sent the previous May to Martin Van Buren, the secretary of state, by John P. Sheldon, editor of the *Michigan Gazette*.

1847 Hermann von Helmholtz presents "On the Conservation of Force" to the Physical Society of Berlin. The paper is a landmark in both physics and physiology.

1871 Ovide Decroly, creator of the "Decroly method" of educating both normal and handicapped children, is born in Ronse, Belgium.

1880 Hydroelectric power is provided commercially for the first time in the United States. The Grand Rapids (Michigan) Electric Light and Power Company begins providing electricity to the Wolverine Chair Company.

JULY 23

1902 Ant scientist T.C. Schneirla is born in Bay City, Michigan. Through a series of tropical expeditions in the 1930s and 1940s, Schneirla studied the fabled marches of army ants. He found that ant colonies operate on a two-phase cycle: roaming for 16 days, stationary for 20. When moving into a new territory, raids also have a two-phase aspect, with peak raiding activity occurring in the morning and again in the afternoon. Sudden weather changes can lead to sudden activity bursts. Raids are not controlled by the availability of prey, but by the level of excitement in the colony. In 1947 Schneirla became a curator of the American Museum of Natural History in New York City, and shortly thereafter he imported from Mexico the first army ants to enter North America.

1906 Swedish chemist Vladimir Prelog is born in Sarajevo, Bosnia-Herzegovina. A corecipient of the 1975 Nobel Prize, he is famous for determining the structure of antibiotics by X-ray diffraction and for establishing rules to specify whether an asymmetric molecule is "right-handed" or "left-handed."

1904 The ice cream cone is invented, according to some accounts, by Charles E. Menches at the Louisiana Purchase Exposition in St. Louis.

1937 A pituitary hormone has been crystallized for the first time, announces today's issue of Science. The feat was accomplished by Yale's Abraham White, Hubert Catchpole and Cyril Long.

1968 Physiologist Sir Henry H. Dale dies in Cambridge, England, at 93. He participated in two history-making studies. In 1910 he and colleagues showed that the allergic reaction to the fungus, ergot of rye, is caused by the presence of histamine in ergot. This work led to discoveries by others that histamine is found throughout the human body, and that it controls glands, blood vessels and smooth muscle. Dale and Otto Loewi later found the key to another great issue in bodily control: communication between nerves. The two showed that an impulse is passed from nerve to nerve by the movement of chemicals (neurotransmitters) between them. Dale was the first to isolate such a chemical (acetylcholine, in 1914), and he and Loewi demonstrated in 1921 that it transmitted messages between nerves. They were Nobel laureates in 1936.

1993 Florida wants to be known for its golf courses—but not when they're part of insane asylums. Today, Health and Rehabilitative Services Secretary Jim Towey orders the sale or lease of a 9-hole, 3100-yard course within the 620-acre Florida State Hospital at Chattahoochee. "We're not the Department of Health, Rehabilitation and Golf," says Towey.

1994 The nation's longest space flight in more than 20 years ends happily—except for 90 newts, 300 fish, and a handful of flies, which perished during the 15-day mission of *Columbia*. The two surviving newts are not headed to a happy fate; they will be dissected in a few days to observe the effects of extended weightlessness. Goldfish, jellyfish, urchins, and toads were also part of the flight. Chiaki Mukai not only became the first Japanese female in space, she also set the longevity record for any female in orbit; she described her experience as "a wonderful memory."

1794 Zoologist Christian Pander, who discovered the three layers of the vertebrate embryo, is born the son of a wealthy banker in Riga, Latvia. His description of these layers in 1817 founded the science of embryology.

1843 Astronomer-photography inventor Sir William Abney is born the son of a clergyman in Derby, England. He created several special photographic emulsions that enabled new views of heavenly events, including (1887) the first view of infrared emissions by the sun. His innovations in infrared technology also allowed him (1882) the first correlation between light absorption and the structure of organic molecules. A century later his methods were used to determine the contents of cosmic gases and dusts.

1844 A steam engine pump is first patented in the United States. Henry Rossiter Worthington of New York City receives patent No. 3,677 for a device that supplies water to steam engine boilers.

1866 After a very slow start as a student, O.C. Marsh becomes the world's second (and the nation's first) professor of paleontology, when Yale University creates a chair and a professorship. Neither salary nor teaching duties are involved.

JULY 24

1895 The first dream that Freud ever analyzed in detail is dreamt on this night, by Freud himself.

1912 Niels Bohr, 26, temporarily leaves England to get married in Copenhagen. In the hurried days before departure, he jotted down some preliminary thoughts (later to win him the Nobel Prize) in a memo to Ernest Rutherford about the structure of the atom. During his honeymoon, Bohr wrote a paper on the absorption of alpha particles.

1913 Biophysicist Britton Chance is born in Wilkes-Barre, Pennsylvania. He was the first to prove experimental proof of the existence of the enzyme–substrate complex (by which enzymes control life by controlling chemical reactions within cells).

1945 The Potsdam Conference of superpowers (i.e., Russia, England, and the United States) adjourns. Truman mentions to Stalin that the United States has been developing an "unusual weapon" (meaning the atomic bomb, which Stalin's spies had already informed him about).

1946 An atomic bomb is exploded underwater for the first time. The device is dropped by a plane 7000 feet above the Pacific Ocean, three miles from Bikini Island. It sinks 90 feet and explodes, sinking ten target ships, including the battleship *Arkansas*.

1952 French surrealist poet André Breton tours the Pech-Merle cave in Cabrerets, France, which boasts dramatic and numerous prehistoric paintings of extinct mammoths—as well as a bar, jazz music, and electric lighting, all of which were installed by the town's mayor, Monsieur Bessac, whose wife is selling tickets at the cave entrance. The juxtaposition of the old and new is too much for Breton, who runs his thumb over one of the paintings. The color readily comes off on his skin. "Forgery, Monsieur," he says with a chuckle to Bessac, who becomes enraged, slugs the poet, and runs him out of the cave. Bessac later sued Breton, and won, but was awarded much less than asked. Bessac, an ex-blacksmith, was not really disturbed because the incident inflated tourist attendance.

1957 Famed biologist J.B.S. Haldane emigrates to India, in protest of England's invasion of Egypt.

1969 *Apollo 11* splashes down in the Pacific after man's first walk on the moon.

1975 An *Apollo* spaceship splashes down in the Pacific with astronauts Thomas Stafford, "Deke" Slayton, and Vance Brand, after the first-ever docking with a Soviet *Soyuz* ship.

1994 The Justice Department reports that one in every six violent crimes in the United States occurs at the workplace.

1951 African "killer bees" escape from the laboratory of Dr. Warrick Kerr in São Paulo, Brazil. They begin interbreeding with native bees "within minutes." The hyperaggressive animals have since spread throughout the Western Hemisphere.

1575 Astronomer-priest Christoph Scheiner, who determined how the eye focuses (by changing lens shape), is born in Wald, Germany. His most famous astronomy work involved no new discovery at all; in 1611 he observed sunspots, and reported them as small bodies around the sun (so as not to violate church dogma by conflicting with Aristotle's views). He claimed to have seen the spots before Galileo. The incident began a feud between Galileo and the church, which ended with Galileo's well-known Inquisition ordeal.

1832 The first railroad accident in the United States occurs. In Quincy, Massachusetts, four visitors are invited to ride to the top of a large inclined plane in an empty car after watching a demonstration by "Granite Railways" of how effective railroad cars are in moving tremendous loads of stone. The cable snaps with the people on board, sending the car and the people over a 40-foot cliff. One passenger dies, and the others are hurt.

1847 Physician Paul Langerhans, discoverer of the islets of Langerhans in the pancreas, is born the son of a physician in Berlin.

JULY 25

1871 A carousel is first patented in the United States. Patent No. 117,336 is awarded to William Schneider of Davenport, Iowa, for a two-story amusement that was neither practical nor successful.

1884 Davidson Black is born in Toronto, the son of a lawyer. Black was an anthropological Sherlock Holmes. In a 1927 fossil-gathering expedition in Chou-k'ou-tien, China, he discovered a single humanoid molar tooth, and from that alone he deduced the existence of a small-brained human ancestor. He called this missing link *Sinanthropus pekinensis* (this translates to "China man of Peking" and is now known as "Peking man"). In 1929 other teeth were found, and in 1930 skulls, body bones, tools, and campfire sites were discovered, confirming Black's incredible deduction.

1909 Frenchman Louis Blériot makes the first plane flight across the English Channel. "England is no longer an island," warned one writer, as war clouds gathered throughout Europe.

1920 Rosalind Elsie Franklin is born into a banking family in London. Although she received relatively few accolades during her brief life, it was her X-ray diffraction studies of DNA that gave Watson and Crick important clues about its structure, the description of which won them Nobel Prizes. Franklin died of cancer at just age 37, shortly after a convalescence in Crick's home.

1639 Thomas Tompion is baptized in Northill, England. The foremost clockmaker of his time, Tompion made some of the first watches with mainsprings, and he produced the first compact, flat watches.

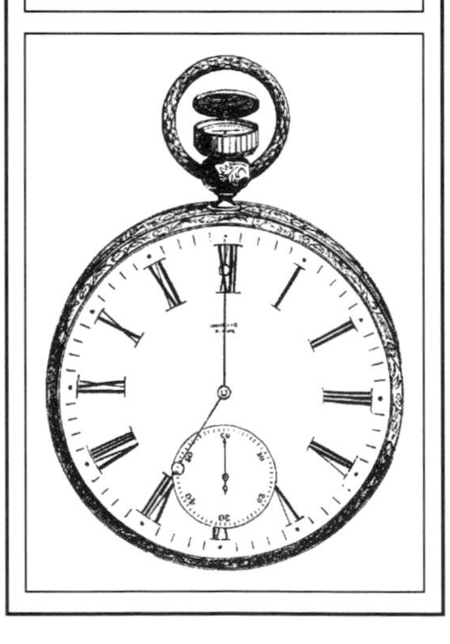

1956 The Italian liner *Andrea Doria* sinks off Nantucket after a famous collision with the *Stockholm.*

1963 Russia, Britain, and the United States initial a treaty to end nuclear tests in space, the atmosphere, and underwater.

1978 Louise Joy Brown, the first test-tube baby, is born in Oldham, England.

1983 The first test-tube ape is born in San Antonio, Texas. She is a black baboon named E.T.

1984 Svetlana Savitskaya becomes the first woman to walk in space, conducting three hours of experiments outside the orbiting *Salyut 7.*

1991 Astronomers report in today's issue of *Nature* that they have detected the first known planet outside our solar system. Although this and previous claims of such a planet are not universally accepted, many scientists are impressed with this latest find. "They've got something here that is significant, no question about it," said a Princeton University professor about the discovery by scientists in Manchester, England. The planet is about twelve times heavier and two to three times larger than Earth. It is in the direction of the constellation Sagittarius, near the center of the Milky Way.

1994 Professional football player Bryan Cox of the Miami Dolphins sues the National Football League for making him play in a 1993 game against the Buffalo Bills. According to the suit, filed in New York City, the racially hostile environment of Buffalo made him a lesser player and caused him to drink heavily.

1815 Neurologist-embryologist Robert Remak is born in Posen, Germany (now Poznan, Poland). He was the son of a Jewish merchant in an anti-Semitic society, which may have indirectly pushed him into important biological discoveries. After obtaining a medical degree and researching the structure of nerves under the supervision of the famed Johannes Müller, Remak was barred from teaching by anti-Jewish laws. He therefore devoted himself to further research. He discovered nerve cells in the heart (now called "Remak's ganglia") and nonmedullated nerve fibers, and was one of the first to treat nervous disorders with electrotherapy. It was Remak who gave the names "ectoderm," "mesoderm," and "endoderm" to the three parts of the vertebrate embryo.

1863 Paul Walden, discoverer of the Walden inversion (in which some organic molecules are "turned inside out" through a series of chemical reactions), is born the son of a Latvian farmer.

1875 Carl Gustav Jung, a pioneer in psychotherapy and the exploration of man's subconscious, is born the son of a pastor in Kesswil, Switzerland.

1892 The United States's first triple-screw cruiser is launched. The 7475-ton *Columbia* (which was actually the fourth U.S. cruiser named *Columbia*, but the first to have such a complicated drive mechanism) was built by William Cramp and Sons of Philadelphia. It is 413 feet long and carries a crew of 475. G.W. Sumner is its first captain.

1894 Futurist Aldous Huxley is born in Godalming, England.

1895 Famous scientists Pierre Curie and Marie Sklodowska marry in Sceaux, France. They spend their honeymoon bicycling in the countryside.

1895 The tallest woman on record, Jane ("Ginny") Bunford, is born in Bartley Green, England. When she died on April 1, 1922, she was 7 feet 7 inches tall (although she had developed curvature of the spine, without which she would have stood 7 feet 11 inches). Her growth became abnormal at age 11 after a head injury. Her skeleton is now preserved in the museum of the Medical School at Birmingham University, England.

1903 The first car trip across the United States (by a nonprofessional driver in his own car) is over. Dr. Horatio Nelson and mechanic Sewell K. Crocker reach New York City, after departing San Francisco the previous May 23. Of the 63 days required for the journey, 44 were spent traveling and 19 were spent waiting for parts and supplies. The escapade won Nelson a $50 bet.

1930 The first nonstop car trip from New York to Los Angeles, in reverse gear, begins.

1945 The Allies deliver their last warning/ultimatum to Japan, and the atomic bomb reaches Tinian Island in the Pacific (where it will be loaded onto the *Enola Gay*, the plane that dropped it on Hiroshima).

1958 *Explorer IV* is launched, and confirms the presence of much high-energy radiation in outer space. On the same day, the U.S. Navy sends a Stratolab balloon up 80,000 feet. Along with two human pilots, the craft carries 10,000 insects (including fleas, fruit flies, and honey bees), to examine the effects of cosmic rays on mutation rates. The results were inconclusive.

1978 The northernmost piece of land on Earth is discovered. Uffe Petersen of the Danish Geodetic Institute comes on the 100-foot-long "islet of Odaaq Ø," 438.9 miles from the North Pole.

1990 George Bush signs the Americans With Disabilities Act; it goes into effect two years later.

1993 NASA releases a 45-page study reporting a growing feeling among shuttle workers that simple and honest mistakes will be punished harshly. Therefore, such mistakes are less likely to be reported. The study was ordered in April after a missing pair of pliers was found stuck to a *Discovery* rocket booster after it returned from space.

1994 The Associated Press reports that residents and tourists in Daytona Beach are upset because driving on the beach has been restricted to daylight hours between 8 AM and 6 PM, to protect sea turtle hatchlings.

1759 French mathematician-cosmologist Pierre-Louis Moreau de Maupertuis dies at 60 in Basel, Switzerland. Disastrous feuds with Voltaire and Leibniz are thought to have hastened his death. Maupertuis was the spoiled child of wealthy parents. From 1715 to 1723 he was a musketeer in the French army, resigning to become a mathematics instructor at the French Academy of Sciences. He was a great admirer of Newton, who had recently died, and in 1732 he introduced Newton's theory of gravity into France. In 1736 he led an expedition to Lapland to estimate Earth's curvature, which verified Newton's guess that Earth was an oblate sphere (a globe flattened at the poles). In 1744 he enunciated the "principle of least action" (energy and moving bodies tend to follow the most economical paths in nature). He was elected to major scientific societies in England, France, and Germany, but his abrasive personality earned him powerful enemies. The loud, public dispute with Voltaire (who mercilessly called him the "earth flattener") caused him to leave Berlin for Basel in 1753.

1775 The first surgeon general of the Continental Army, Benjamin Church, takes office. His salary is $120 a month. He served only until mid-October, and on November 1 was jailed for treason. He was later cleared of all wrong doing by George Washington, but his life was ruined by that time.

1801 Astronomer George Airy is born in Ainwick, England. He is known as much for several failures as for many accomplishments (for which he was knighted in 1872) and for his petty, domineering personality.

1848 Physicists Roland Eötvös (who determined the mass of Earth and the force of gravity) and Friedrich Dorn (who discovered radon) are born in Europe.

1866 The first successful transatlantic cable is completed. The New World and the Old World are now connected by telegraphic communication. It was the third attempt, all three of which were financed by paper merchant Cyrus Field. In 1858 a cable was successfully laid between Ireland and Newfoundland, but it was rendered inoperable when an operator sent too much voltage through it. In 1865 another cable of 2200 miles was laid, but this snapped.

1870 Bertram Borden Boltwood is born the son of a lawyer in Amherst, Massachusetts. He established that radioactive elements turn into other radioactive elements when they decay; from this he created the theory of radiometric dating (on which carbon-14 dating is based).

1881 Chemist Hans Fischer is born the son of a dye chemist in Höchst-am-Main (now part of Frankfurt), Germany. Coloration was a major portion of Fischer's research. He determined what makes blood red and plants green—and ultimately discovered it was the same molecule in both cases. Through the 1920s, working at the University of Munich, he worked-out the chemical structure of heme, a major portion of hemoglobin (the coloring agent in blood). In the 1930s he tackled the chemical structure of chlorophyll, the green coloring agent in plants. Fischer must have been overwhelmed when he saw that both heme and chlorophyll are built from the same pyrrole molecule (consisting of four carbons and one nitrogen atom). He won the 1930 Nobel Prize in Chemistry, but took his own life in despair in 1945 after World War II air raids destroyed his Munich laboratory.

1909 Orville Wright tests the U.S. Army's first airplane (a Wright biplane, named *Miss Columbia*), flying himself and a passenger for 1 hour 12 minutes in Dayton, Ohio. The first official test flight occurred on July 30, and in August the government purchased the plane for $30,000.

1918 The *Socony 200*, the country's first concrete barge and her first reinforced-steel concrete barge for shipping oil, is launched into Flushing Bay, New York. The craft was built by the Fougner Shipbuilding Company for New York's Standard Oil Company. It is 98 feet by 31 feet by 9½ feet.

1940 Bugs Bunny makes his film debut in the Warner Bros' cartoon *A Wild Hare*. Elmer Fudd costars.

1967 President Lyndon Johnson appoints the Kerner Commission to determine the cause of urban rioting. Exactly one year later, militant H. Rap Brown tells a Washington news conference, "Violence is necessary. It is as American as cherry pie."

1993 Scientists at a genetics meeting in Bar Harbor, Maine, report "the most important medical finding in the latter part of the 20th century": fortifying foods with folic acid will dramatically reduce birth defects. "It's like the Salk vaccine—it's that magnitude of importance," said Dr. Godfrey Oakley, director of the birth-defects division at the U.S. Centers for Disease Control in Atlanta.

1565 Physician-naturalist Conrad Gesner completes the massive *De Rerum fossilium*. The event has been called "the birth of paleontology."

1774 In Vienna, Dr. Franz Mesmer begins a series of treatments, using magnets, on the heart ailments of socialite Franzl Oesterline. The therapy is successful, helping launch Mesmer's fame (see further discussion on March 5, 1815).

1840 Edward Drinker Cope, discoverer of about 1000 extinct species of vertebrates, is born into a wealthy family in Philadelphia. He exhausted much of his personal fortune, and much of his health, in a battle to outshine rival O.C. Marsh. Their feud is legendary.

1867 U.S.-Argentinian astronomer Charles Dillon Perrine is born in Steubenville, Ohio. He discovered the first of Jupiter's outer satellites, beyond the four large satellites discovered by Galileo three centuries before.

1869 Czech physiologist Jan Evangelista Purkinje dies at 81 in Prague. He was the first to call the living material in cells "protoplasm," the first to use a microtome (a device for automatically and easily making very thin slices of tissue for microscope viewing), and the first to scientifically describe the effects on humans of camphor, turpentine, belladonna, and opium. He created the world's first independent department of physiology, in 1839 at the University of Breslau, Prussia. He discovered the Purkinje effect (as light intensity decreases, red is perceived to fade faster than blue, which explains why dimly lit areas look more blue than red), Purkinje cells (large, multibranched nerves in the brain), and Purkinje fibers (the tissue that conducts the pacemaker impulse over the heart). He also discovered sweat glands and suggested that fingerprints could be used for identification.

1898 A big day in the history of the Mayo Clinic. Charles William Mayo, one of its shapers, is born. Both he and uncle William James Mayo (another founder) died on this date (in 1968 and 1939 respectively). All of these events occurred in Rochester, Minnesota.

1898 A big day in the history of the maser (the forerunner of the laser). In Rymanów, Austria (now in Poland), Isidor Isaac Rabi is born. His research on molecular beams led to the development of the maser by Charles Hard Townes, who was born on this date in 1915. Both men received Nobel Prizes.

1922 Jacques Piccard, who journeyed deeper into the ocean than any before or since, is born in Brussels. His record descent—nearly seven miles down—occurred in 1960 in the Marianas Trench off Guam, in a vessel built by his father Auguste Piccard.

1925 Physician Baruch S. Blumberg is born in New York City. During studies of why some ethnic groups are immune to diseases that ravage others, he discovered a fragment of the hepatitis B virus in the blood of Australian aborigines. This fragment provides a natural immunity; its discovery led to blood-screening methods, a vaccine, and the Nobel Prize.

1933 The singing telegram is introduced. The Western Union Telegraph Company of New York City adds the innovation to its services, despite opposition by at least one of its major executives. It is immediately popular.

1992 Actress Cybill Shepherd and a brain cancer patient testify before Congress, pleading that the French abortion pill RU-486 be allowed into the United States. Research has revealed many potential benefits of the pill, including helping to cure brain tumors, endometriosis, depression, Cushing's disorder, and cancer of the breast, ovary, and prostate. The cancer patient, J. David Grow, testifies to having "about a 75 percent chance of prolonging my life" if allowed RU-486. "I have no other option for treatment."

1993 The *Journal of the American Medical Association* reports that workers in smoke-filled restaurants are more likely to get cancer than those in smoke-free environments. Waiters and bartenders breathe six times the secondhand smoke that a secretary does, and have one and a half times the likelihood of getting lung cancer. Heretofore, scientific attention has been on the hazards to customers in such places, but today's report shows that employees are also imperiled.

1865 The nation's first dental Code of Ethics is proposed to the American Dental Association. Dr. John Allen presents the proposal to the ADA's fifth annual convention, in Chicago.

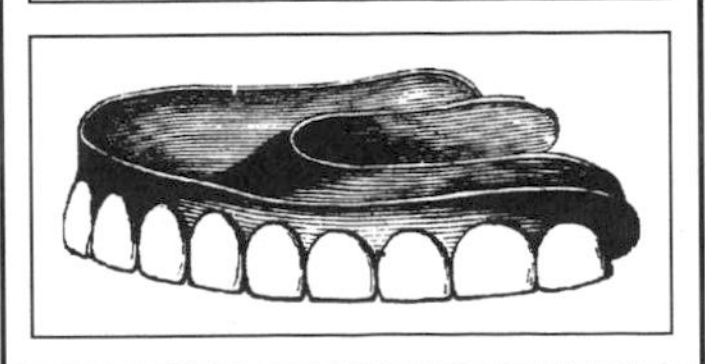

1609 Galileo demonstrates the first telescope able to provide astronomical observations.

1773 The first schoolhouse west of the Allegheny Mountains is completed, in Schoenbrunn, Ohio. Started on December 22, 1772, it was built by Moravian missionaries. The Reverend David Zeisberger was its first teacher.

1877 Explorer-zoologist William Beebe is born in Brooklyn, New York. His earliest inspirations were a love of birds and the novels of Jules Verne. He became the director of several departments at the New York Zoological Gardens, and conducted numerous tropical research expeditions during his long career. His most famous journey was straight down, in the ocean off Bermuda in a specially built, icy, leaky, and dangerous submarine. He and Otis Barton descended to 3028 feet, a world record at that time (1934). Beebe once compared wild animals with art masterpieces; the latter could always be re-created if lost, "but when the last individual of a race of living things breathes no more, another heaven and another earth must pass before such a one can be again."

JULY 29

1890 Vincent van Gogh, a psychological enigma and the epitome of "tortured genius," dies at 37 in Auvers, France, two days after shooting himself.

1910 Biochemist Heinz Fraenkel-Conrat is born the son of an eminent gynecologist in Breslau, Germany. In 1950 he succeeded in taking apart a virus (separating the protein from the RNA), and showed the RNA to be the active, infectious agent. This forced philosophers to reconsider the definition of "life," and it focused science's attention on nucleic acids, such as RNA, as the keys to heredity.

1914 Coast-to-coast telephone service begins in the United States; a call is made between New York and San Francisco.

1933 The cornerstone of the first federal sanatorium for drug addicts is laid in Lexington, Kentucky. The United States Narcotics Farm received its first patient two years later.

1957 The International Atomic Energy Agency is established.

1958 The nation's first space agency, NASA, is established. President Dwight Eisenhower signs the National Aeronautics and Space Act. It states that the organization will be headed by a civilian, with an annual salary of $22,500. Thomas Keith Glennan was sworn in as the first NASA administrator on August 19, 1958.

1962 Sir R.A. Fisher, creator of the analysis of variance and other statistical tools, dies at 72 in Adelaide, Australia. His *Statistical Methods for Research Workers* (1925) was in print for over 50 years.

1982 The Soviet *Salyut 6* returns to Earth after five years in orbit.

1992 Hardee Memorial Hospital in rural Wauchula, Florida, closes its doors and files for bankruptcy. Fourteen years earlier it was the site of a still-mysterious, still-famous baby swap, in which Kimberly Mays was somehow taken from Ernest and Regina Twigg and given to Robert Mays; Mays's real daughter was raised by the Twiggs, and Mays raised Kimberly. Discovery of the swap produced a protracted and twisted series of court cases. Part of the reason for today's closing is the hospital's liability in the incident.

1994 The first creature to emerge from the ocean to solid land may have done so in Pennsylvania; today's *Science* reports the discovery of North America's oldest known fossil—at least 365 million years old—and one of Earth's oldest known amphibians. It was found in Clinton County, Pennsylvania, by a team led by Ted Daeschler of Philadelphia's Academy of Natural Sciences. "Every limbed animal sprang from this animal or one similar to it," he said. Analysis of the heavy shoulder bone convinced scientists that the animal walked on land. "This animal could do push ups. It could lift itself up on its feet," said coauthor Neil Shubin. "That can't be explained if it only lived in water."

1868 Medical pioneer John Elliotson dies at 76 in London. He was one of the first British doctors to advocate use of the stethoscope, to advocate lectures as part of clinical training, and to advocate the therapeutic use of hypnosis. In 1849 he founded his own "mesmeric" hospital, having been forced to resign from London's University College Hospital for his defense of hypnosis.

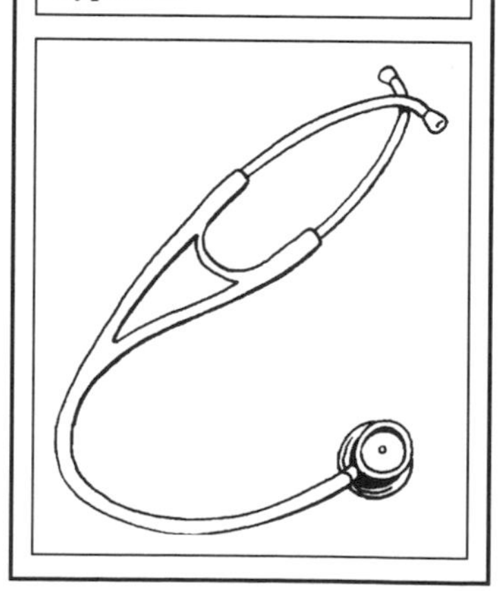

1641 Physician Reinier de Graaf, who discovered where eggs are formed in females, is born in Schoonhoven, the Netherlands. The egg-producing structures in the female ovary are now known as "Graafian follicles." Graaf was the first to use the word "ovary," and the first to report changes that the ovary undergoes during the ovulatory cycle.

1863 Henry Ford is born on a farm near Dearborn, Michigan. The automobile industry and Earth's ecology will never be the same.

1889 Vladimir Zworykin, "the father of television," is born the son of a riverboat merchant in Mourom, Russia. His iconoscope (developed by 1938) was the first practical TV camera. Zworykin also coinvented (in 1939) the first practical electron microscope, which provided magnifications many times greater than any optical microscope available.

1898 The United States sees its first automobile advertisement (in a national magazine). "Dispense with a Horse" is the heading of the one-column ad in *Scientific American* for the products of the Winton Motor Car Company of Cleveland.

1909 The first airplane purchased by the U.S. government is given its first official test flight. Built by the Wright brothers, *Miss Columbia* was 28 feet long, had a wingspan of 36 feet 4 inches, and weighed 740 pounds. Three days after the test flight the plane was purchased for $25,000, plus a bonus of $5000 because the plane exceeded the specified speed of 40 mph.

1930 Davidson Black announces to the Geological Society of China that a second specimen of Peking man (one of the first "missing links" discovered) had been found.

1941 Edward Benjamin, the world's greatest documented sword swallower, is born in Binghamton, New York. Under his stage name "Count Desmond," he swallowed eight swords at once (each 23 to 26 inches long) on ABC television in March 1978.

1943 The modern, explosion-powered ejector seat is first tested. It was developed in Sweden, and first tested in a Saab automobile.

1945 The cruiser USS *Indianapolis* is torpedoed and sunk in the Pacific, just four days after delivering the atomic bomb to Tinian Island (its last stop before Hiroshima).

1946 A rocket attains an altitude of 100 miles for the first time. At White Sands Proving Grounds in New Mexico a captured, modified German V-2 rocket is sent 104 miles into space.

1947 After 93 days at sea, the crew of the *Kon-Tiki* first sees Polynesia.

1965 Medicare is signed into law by President Lyndon Johnson.

1971 Alfred Worden becomes the "most isolated human" in history. During the *Apollo 15* lunar mission Worden sits alone, 2233.2 miles from the nearest person, while David Scott and James Irwin explore the moon for two days.

1988 The record for the greatest distance that any object has been propelled by human power is set today outside Austin, Nevada, when Harry Drake shoots an arrow 6141 feet 2 inches.

1994 After seven years behind bars at Tampa's Lowery Park Zoo, a 31-year-old female Asian elephant named Tillie swats a handler to the ground with her trunk, then delivers a fatal kick to the handler's head. Tillie is calm and obedient after the incident. In the following summer at least three different circus elephants in the United States run amok and injure humans "for no apparent reason."

1718 Physicist John Canton is born in Stroud, England. His father was a weaver who pulled John out of school at an early age to join the family business. Largely self-educated, he became an eminent scientist. Among his achievements: in 1762 he proved that water *is* slightly compressible; in 1749 he developed a new method of preparing artificial magnets (which won him the Copley Medal and election to the Royal Society); and in 1752 he became the first in England to verify Franklin's hypothesis that lightning and electricity are identical.

1790 The first patent issued by the U.S. government is granted to Samuel Hopkins of Vermont for a process of making potash and pearl ashes. The document is signed by Edmund Randolph, Thomas Jefferson, and George Washington. It is one of three patents issued that year.

1800 Chemist Friedrich Wöhler is born in Eschershein, Germany, the son of a veterinarian. Wöhler was a gynecologist until one of his professors, Leopold Gmelin, persuaded him to pursue chemistry. His work was to change science—and philosophy. In 1828, he created urea in the lab while working on a totally unrelated project. Never before had man artificially reproduced a chemical made by the body of a living thing. This work destroyed the distinction between organic and inorganic chemicals, and it vanquished the doctrine of vitalism, which held that only the bodies of living things had the ability to produce organic molecules.

1803 Naval engineer-inventor John Ericsson is born the son of a mine inspector in Långbanshyttan, Sweden. His greatest triumphs were creating the screw propeller (which allowed steam to power warships for the first time) and building the armored turret warship (he built the Civil War's *Monitor*; Napoleon turned down plans for the vessel, but *Monitor*-type ships were highly effective in maintaining the blockade of the South during the War). Ericsson changed naval warfare forever.

1859 Pathologist Theobald Smith is born the son of a tailor in Albany, New York. Working with Texas cattle fever, he was the first to prove (1892) that disease can be spread by bloodsucking insects.

1860 Anton Eiselsberg, a founder of modern gastrointestinal surgery, and the first in Europe to remove a spinal cord tumor, is born in Schloss Steinhaus, Austria.

1861 The 12-month record for rainfall is set in Cherrapunji, India. 1041.8 inches has fallen in the last 365 days.

1912 The U.S. government passes its first motion-picture censorship law. Carrying a penalty of $1000 and/or one year of hard labor, the law prohibits "the importation and the interstate transportation of films or other pictorial representations of prize fights."

1918 Automobile maker Francis Edgar Stanley dies at 69 in Ipswich, Massachusetts. In 1897 he and twin brother Freelan began manufacturing the Stanley Steamer, history's best-known steam-driven car. The Stanley Motor Company continued production through World War I. In 1906 one of their cars was clocked at 127 mph, faster than any vehicle had gone before. Freelan outlived Francis by 22 years.

1953 The FCC announces that television network sales have surpassed those of the four national radio organizations for the first time in history.

1964 The first closeup photographs of the moon by a U.S. craft are taken and transmitted back to Earth by *Ranger VII*. The 4316 pictures are taken in 17 minutes, before the craft crashes near the moon's Sea of Clouds.

1971 Man uses a vehicle on another planet for the first time. Astronauts D.R. Scott and J.B. Irwin from *Apollo XV* ride the four-wheeled LRV (Lunar Roving Vehicle) around Hadley Hills, a 1200-foot-deep canyon on the moon. Because the vehicle is electric, no pollution from burning gas is left.

1990 Gene therapy is approved for use on humans for the first time. A federal panel approves the revolutionary technology (in which curative genes are inserted into a patient's cells) for use against a lethal skin cancer and an immune disorder in children. "What we're doing today is adding gene therapy to vaccines, antibiotics and radiation in the medical arsenal," says NIH official Gerard McGarrity. "Medicine has been waiting thousands of years for this."

1744 Jean-Baptiste Lamarck who coined the terms "vertebrate" and "invertebrate" (and popularized the word "biology"), is born the youngest of 11 children of an impoverished baron in Bazentin-le-Petit, France. Lamarck was nearly 50 before he decided what he wanted to be in life, which was a naturalist and author. Until then he had tried careers in the church, in the military, and in medicine. While a soldier on the French Riviera, he realized an interest in plants, which led to an interest in insects and in all of nature. His major work was a classification system for the invertebrates. His massive, seven-volume *Natural History of Invertebrates* (1815–1822) is recognized as the birth of modern invertebrate zoology. During this work Lamarck came up with a theory to explain how life-forms evolved into the creatures on Earth today. His idea, now called Lamarckism, was that characteristics changed from generation to generation depending on how much they were used. This theory was discarded when Darwin's theory was announced, but Lamarck was the first major scientist to propose a plausible and complete explanation for changes in species over time.

1774 Joseph Priestley isolates oxygen in the lab in his home in Calne, England. Exactly 100 years later (1874), the city where he once lived and worked, Birmingham, England, unveils a monument to Priestley and his discovery of oxygen. Little is made of the fact that a mob in Birmingham destroyed his home, lab, and library in 1791.

1779 Lorenz Oken, a naturalist and early advocate of international scientific congresses, is born the son of a poor farmer in Bohlsbach, Germany. He was also an ultranationalist and a staunch defender of the notion that the skull developed from a backbone; this theory was wrong, but it helped establish the concept of evolutionary change of species.

1790 The first census of the United States is completed. Authorized by Congress the previous March, the census cost $44,377, and employed 17 marshals and 650 assistants. The count revealed 3,939,326 citizens in the 16 states and the Ohio territory. Virginia was the most populous state (747,610), Rhode Island the least (68,825). New York City (33,131), Philadelphia (28,522), and Boston (18,320) were the three most populous cities.

1793 The new Republic of France accepts the recommendation of scientists that the metric system be based on measurements of Earth's size.

1818 Maria Mitchell, the first professional female astronomer in the United States, is born in Nantucket, Massachusetts. She was also the first female member of the American Academy of Arts and Sciences.

1885 Chemist George Charles de Hevesy is born into a wealthy family in Budapest. After getting his Ph.D. in Freiberg, Germany, he went to work at England's University of Manchester with Ernest Rutherford, studying the nature of radioactivity. He used this training to develop a unique and revolutionary method of studying chemical processes within the bodies of living things. He inserted radioactive forms of various chemicals into the bodies of living organisms; because they were radioactive the movement and processing of these molecules were easily tracked. For his development of radioactive "tracers," Hevesy was awarded the 1943 Nobel Prize for Chemistry. He also codiscovered (with Dirk Coster) the element hafnium.

1893 The first U.S. patent relating to breakfast foods is issued. Henry D. Perky and William H. Ford of Watertown, New York, receive patent No. 502,278 for a machine that makes fine wheat filaments for "Shredded Wheat Biscuits."

1992 A Denver jury awards Susie Quintana $6.5 million in a suit against United Blood Services of Albuquerque, New Mexico, who negligently provided AIDS-infected blood to Quintana in a transfusion. The plaintiff also won $105,000 for emotional distress and a further $1.5 million for her husband. The awards came posthumously. Quintana died during her lawyer's closing arguments, just as the jury was about to deliberate (although jurors were not told of her death until after their decision was reached).

1993 China's *Guangming Daily* newspaper reports that remains of Genghis Khan's palace have been discovered by archaeologists in the Ningsxia Autonomous Region in northern China. Khan died more than 750 years ago, in 1227; the palace remains have been dated back to 1271. Scientists think it is the Khan's palace because the floor is covered in yellow-glazed tiles, and only the emperor's family was allowed to use yellow.

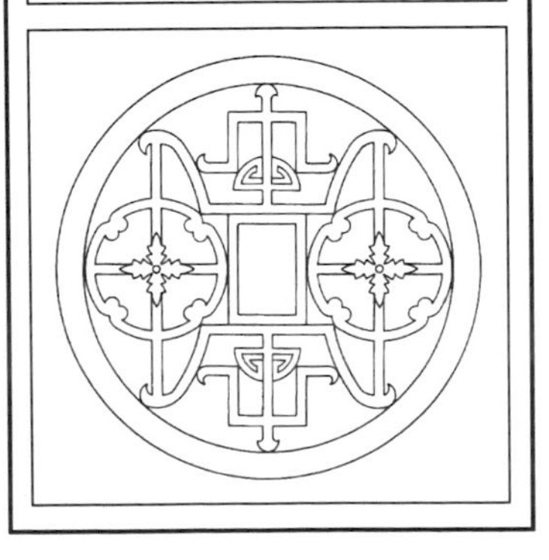

1744 The aurora borealis is seen at Cuzco, Peru. Never before or since have "northern lights" been seen so close to the equator.

1788 German chemist Leopold Gmelin is born into a dynasty of eminent scientists in Göttingen. He coined the terms "ester" and "ketone" and discovered potassium ferrocyanide. His best-known accomplishment was the massive and historic *Handbook of Chemistry* (first edition in 1817), the first systematic treatment of chemistry since Lavoisier made it an exact science in the 1700s.

1791 For the first time in U.S. history, a patent is issued jointly to a father and son. Samuel Briggs Sr. and Jr. of Philadelphia receive the award for a machine that makes nails.

1820 Physicist John Tyndall, who showed why the sky is blue, is born in Leighlin Bridge, Ireland. His research centered on the movement of heat and light through gases. He discovered the Tyndall effect (a light beam cannot be seen from the side when it moves through a clear liquid or gas, but when dust and other particles are in the medium, the light beam can be seen). The scattering of sunlight by atmospheric dust (which is especially dramatic for wavelengths in the blue end of the spectrum) accounts for the overall blue appearance of the sky.

AUGUST 2

1835 Inventor Elisha Gray, Alexander Graham Bell's greatest rival, is born in Barnesville, Ohio. Both men filed statements with the Patent Office on the same day (February 14, 1876) claiming to have invented the telephone.

1841 The word "dinosaur" is coined. Anatomist Richard Owen, 38, uses the term for the first time in a 2½-hour presentation, entitled "Report on British Fossil Reptiles," to the annual meeting of the British Association at Plymouth, England. Owen proposes that a separate classification be established for the extinct beasts.

1922 Alexander Graham Bell dies at 75 in Beinn Bhreagh, Nova Scotia. He was born into a family of eminent speech teachers (his father, Alexander Melville Bell, wrote *Standard Elocution*, which passed through some 200 editions in English). The younger Bell first achieved notoriety by teaching the deaf to speak by adapting methods that his father had developed for the nondeaf. In his midtwenties he opened his own Boston school to train teachers of the deaf. It was actually his passion to help the hearing-impaired that led him to invent the telephone.

1942 Albert Einstein writes to President Roosevelt, informing him of the feasibility of an atomic bomb. This marks the beginning of the Manhattan Project, which produced the first nuclear weapons.

1946 Atomic energy is first produced and delivered for peacetime use. On this date, the Clinton Laboratories of Chicago delivers a tiny amount of radioactive carbon-14 to the Barnard Free Skin and Cancer Hospital of St. Louis. One millicurie of the white powder (an amount too small to be seen with the naked eye) is shipped in aluminum and steel containers.

1990 A terrible auto accident leads to death—and a new legal definition of life. In Oklahoma today, an intoxicated Treva La Nan Hughes drives into the car of Reesa Poole, who is pregnant and scheduled to deliver in four days. The fetus perishes in the accident. Four years later the state Court of Criminal Appeals rules that Hughes can be charged with homicide in the death of the fetus. Prior to this case, a person had to be born before homicide could be charged. In effect, the event and the trial produced a new definition of "human being."

1994 Marge Schott, owner of baseball's Cincinnati Reds, appeals to the Cincinnati City Council to allow smoking throughout Riverfront Stadium during ball games. Smoking is currently permitted only on concourses and in private boxes. Schott blames smoking restrictions for an attendance slump (although other stadiums with greater smoking restrictions have experienced attendance increases). The Associated Press reports Schott's appeal on the same day that it reports a study that found deaths and heart attacks from secondhand smoke to be much more prevalent than previously thought.

1776 German chemist Friedrich Strohmeyer is born the son of a medical professor in Göttingen. He developed one of history's first laboratory courses in chemistry, but his greatest claim to fame was the discovery of the element cadmium in 1817. One of Strohmeyer's most famous pupils was Leopold Gmelin (see next entry).

AUGUST 3

1492 In the middle of hurricane season, Columbus sets sail from Palos, Spain, with the *Niña*, *Pinta*, and *Santa Maria*. Most of the 120 men in the crews are from prison. Columbus is seeking a route to China, but instead "discovers" the New World.

1750 America's first book on teaching methods is finished. Originally in German, it was not published for another 20 years. It was written by Christopher Dock and entitled *A Simple and Thoroughly Prepared School-Management clearly setting forth not only in what manner children may best be taught in the branches usually given at school, but also how they may be well instructed in the knowledge of godliness.*

1805 Mathematician-astronomer Sir William Rowan Hamilton is born the son of a lawyer in Dublin, Ireland.

1811 Elish Graves Otis, inventor of the safety elevator (which does not crash even if its cable is cut completely), is born the son of a prosperous farmer in Halifax, Vermont.

1851 Physicist George FitzGerald, the first to suggest a way to produce radio waves, is born in Dublin, Ireland. In 1895 he was the first to realize a theory (now called the Lorentz–FitzGerald contraction) stating that any moving object shortens along its line of motion. Einstein grafted this theory into his own special theory of relativity.

1857 Louis Pasteur delivers a paper to the Lille Society in Lille, France, announcing that he has discovered what makes wine ferment. Pasteur asserts that biochemical action of tiny organisms causes fermentation.

1909 Psychobiologist Neal E(lgar) Miller is born in Milwaukee, Wisconsin.

1909 Robert Stroud, age 19, pleads guilty to killing a man in a fight over a prostitute. He enters the federal penal system and never leaves. He was not a good inmate, and later murdered a fellow prisoner and a guard. Mostly in solitary confinement he began studying and researching bird diseases, eventually becoming a recognized authority. *Stroud's Digest of the Diseases of Birds* appeared in 1942, the same year he was transferred to Alcatraz Prison, where he got the nickname "Birdman of Alcatraz."

1914 An oceangoing ship enters the Panama Canal for the first time. It crosses successfully, but problems with the locks were discovered. The official opening of the Canal came weeks later.

1915 Biophysicist Donald R. Griffin, who discovered how bats "see" at night, is born in Southampton, New York. In 1938, while still a student at Harvard, Griffin got his hands on a recently invented device that translates "ultrasonic" sounds (i.e., beyond human hearing) into audible sounds; with this he discovered that bats constantly emit high-frequency sounds that they use as radar to hunt prey and avoid obstacles. During World War II, Griffin helped develop communication equipment, insulated clothing, and night-viewing apparatus. In the later part of his career he explored intellectual experience in nonhuman animals; *The Question of Animal Awareness* appeared in 1976.

1958 For the first time in naval history, a sailor reenlists while under the North Pole. James Robert Sordelet of Fort Wayne Indiana (an electrician's mate first class on the submarine *Nautilus*) reenlists on this day, as the ship becomes the first to cross the North Pole underwater.

1970 The record for most failures on a driver's test ends. Mrs. Miriam Hargrave (born on this day in 1908) of Wakefield, England, finally passes on her 40th attempt. Over the eight years she has been trying for her learner's permit, she has spent $720 on 212 driving lessons. She can no longer afford to buy a car.

1994 A rocket blasts off from Cape Canaveral containing a time capsule with audiovisual messages from 40,000 earthlings. The capsule is designed to orbit Earth indefinitely, until discovered by future humans or visitors. "Dear Whoever," begins one of the messages, "I am Andy Resch. I will express the conflict between man and nature with a drawing. The drawings below are before and after man came. I hope you understand what I am saying." The 11-year-old's first drawing shows an oak tree with a bird on one of its limbs; the second drawing shows a stump and the bird lying belly-up on the ground.

1992 The Associated Press reports that a dolphin (held captive in Florida's Clearwater Marine Science Center) has learned to paint. "Sunset Sam," a 14-year-old bottlenose dolphin who is blind in one eye and has liver disease, learned quickly how to hold a brush with his teeth and how to touch it to the canvas.

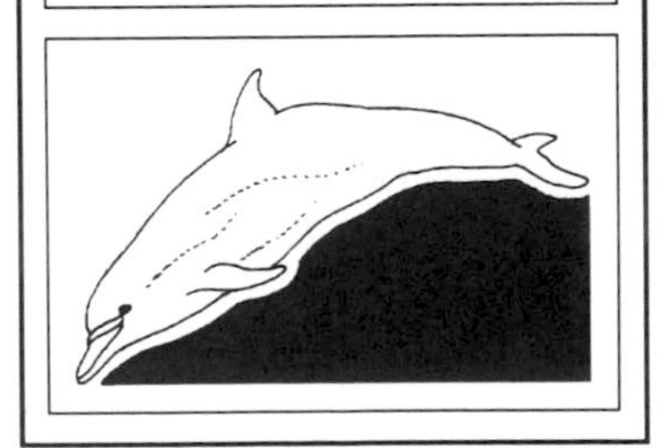

1815 German physician Carl Reinhold Wunderlich, the first to realize that fever is merely a symptom and not a disease in itself, is born in Sulz.

1834 Mathematician-logician John Venn, creator of the famous "Venn diagrams," is born in Hull, England.

1841 Ornithologist-naturalist-novelist, W(illiam) H(enry) Hudson is born near Buenos Aires, Argentina, into a family of sheep farmers who had emigrated from New England. His childhood (fondly described in *Far Away and Long Ago*, 1918) was spent observing nature on the Argentine pampas. Illness at 15 permanently made him withdrawn, studious, and introspective, which led to his reading Darwin's *Origin of Species*, a lifelong influence on his thinking. In his late 20s Hudson moved to England, where, after years of poverty, continued ill-health, and undocumented wandering, he married a much-older woman (in 1876, at age 35) and settled in a London house that she inherited. There he wrote a number of novels, each full of the power and glory of nature, but largely devoid of strong character development. He also wrote nonfictions about English and South American birds and on the natural history of the English countryside; these brought him recognition and fueled a "back-to-nature" movement in Western society that peaked in the 1920s and 1930s. His enduring *Green Mansions* (1904) is the story of Rima, a half-man, half-bird forest dweller. Rima is enshrined in a 1925 sculpture by Jacob Epstein that was erected as a tribute to Hudson in the bird sanctuary of London's Hyde Park.

1899 Embattled Secretary of Agriculture Ezra Taft Benson is born in Whitney, Idaho. He was one of only two cabinet members to hold his post during the entire Eisenhower administration, even though his years there were rocky. Most controversial was his rollback of government price support for agricultural products. This measure was unsuccessful in raising prices, and was rescinded. Many blamed Benson for subsequent defeats of Republican candidates in farm states. On leaving government, Benson devoted himself to church work and was elected president of the Mormons in 1985.

1912 Raoul Wallenberg (credited with saving more lives than any other person in history) is born in Stockholm. His feat came during World War II, when he used diplomatic ties to save nearly 100,000 Hungarian Jews from the Nazis.

1937 An okapi is first imported to the United States. The unfortunate 21-month-old animal was captured in the Belgian Congo and put aboard the Red Star liner *Pennland* the previous July. It is delivered to the Bronx Zoo, New York City. Discovered in 1900, the animal looks like a mahogany-colored cross between a zebra and a giraffe, with horns and a huge tongue. It is extremely shy, living in small family groups in dense forests and emerging to feed only at night—unless it is in a zoo.

1977 The U.S. Department of Energy is created. President Carter signs the Department of Energy Organization Act. The new department has cabinet status, with James R. Schlesinger serving as the first secretary of energy. He oversees 20,000 employees among its subdepartments, including the Federal Power Commission, the Federal Energy Administration, and the Energy Research and Development Administration.

1987 The Federal Communications Commission votes to rescind the Fairness Doctrine, which required radio and TV stations to provide balanced coverage of controversial topics.

1993 Dr. Jack Kevorkian assists 30-year-old Thomas Hyde in a "merciful suicide" on Belle Isle in Michigan's Detroit River. It is the second time the physician attended a "medicide" since Michigan declared the practice illegal. Hyde inhaled carbon monoxide after years of battling Lou Gehrig's disease. Kevorkian took the opportunity to blast the medical profession. "They're politicians first, businessmen second, and they ought to be ashamed of themselves to have human beings like Thomas Hyde suffer immensely, unable to move any muscle, cannot speak, cannot swallow, have pain in addition to all that, and they turn their heads because 'We've got to discuss this a little more.' "

1994 Wire services report that 78-year-old land developer John V. Howard will donate 137 acres in the Florida Keys to a federal deer refuge, rather than pursuing plans to build a hotel complex there. "I don't need it," said Howard. "And I have enough money to last me forever. I thought it would be nice to be altruistic."

AUGUST 5

1540 Scholar Joseph Justus Scaliger is born the son of a physician-philosopher in Leiden, Holland. Among his researches was the creation of an encyclopedia of the calendars and time calculations of previous civilizations. This work placed chronology on a firm scientific basis for the first time.

1802 Mathematical genius Niels Henrik Abel is born on the Norwegian island of Finnøy, the son of an alcoholic minister. Born into poverty, Abel stayed that way until death at 26. By then he had achieved international recognition; Abel's theorem, Abelian functions, and Abelian integrals are now standard tools in mathematics. Perhaps his greatest feat was determining in 1824 the solution to equations of the fifth degree, thereby solving a famous, age-old problem in algebra. Abel proved it was *impossible* to solve such equations algebraically. He sent his proof to the great mathematician Friedrich Gauss, who thought it was just another crackpot shot-in-the-dark attempt, and tossed the paper aside. By 1829 Abel's talent had been realized, and the University of Berlin offered him a professorship; he died of tuberculosis two days before notification reached him at home.

1858 The first telegraph message across the Atlantic is sent from Ireland to Newfoundland at 2:15 PM. It announces the arrival of one of the cable-laying ships.

1859 Biochemist Thomas Osborne is born the son of a banker in New Haven, Connecticut. He was one of the first to discover that proteins come in many forms and that their makeup and nutritional value depend on the amino acids in them. Osborne discovered vitamin A (but Elmer McCollum published his independent discovery three weeks before, and therefore got the credit), and he discovered that cod-liver oil is especially rich in this vitamin (thereby condemning many children for many years to this disgusting nutrition supplement).

1914 An electric traffic light is installed for the first time on a street in the United States, at the corner of Euclid and East 105th Street in Cleveland. Erected by the American Traffic Signal Company, the device incorporates red and green lights and buzzers.

1930 Neil (Alden) Armstrong, the first man to touch the moon, is born in Wapakoneta, Ohio.

1940 Physician-explorer Frederick Cook dies at 75 in New Rochelle, New York. He was a surgeon on Robert Peary's first Arctic expedition (1891–1892). In 1908 he claimed to be the first to reach the North Pole, creating a public fight with Peary who got there in 1909. Cook's own Eskimo companions said that he had stopped hundreds of miles short of the Pole and that he had faked verifying photographs. Cook was imprisoned in 1923 for mail fraud; he was paroled six years later, and received a presidential pardon in the year of his death.

1962 An "occultation" (a blocking of the observation of a heavenly body when another object in space comes between the observed object and the observer) of the radio-emitting star 3C-273 is observed from Australia. The data generated lead Maarten Schmidt to deduce and announce the existence of quasars in 1963. Quasars, or "quasi-stellar objects," are objects beyond our own galaxy that have a starlike appearance; the source and nature of their energy emissions are poorly understood.

1963 The United States, USSR, and Britain sign a treaty in Moscow banning nuclear tests in the atmosphere, space and underwater.

1969 Space probe Mariner 7 flies by Mars, sending back data and photographs.

1994 "Speedy" Atkins finally comes to rest. He's been dead for 66 years. His body was embalmed in 1928 by funeral director A.Z. Hamock in Paducah, Kentucky, who was experimenting with preservative fluids, trying to duplicate the success of ancient Egyptians. Although he never revealed his formula, Hamock was proud of his results, and regularly displayed Atkin's body.

1994 Opera may not be everyone's cup of tea, but in this case it proves fatal. A group of singers from Denmark's Royal Theater rehearse Wagner's *Tannhauser* today in the Copenhagen Zoo, 300 yards from a rare okapi (a giraffe relative, native to Africa), which goes into shock and subsequently dies. "She started hyperventilating, went into shock, and collapsed," said a zoo spokesman.

1667 Johan Bernoulli, a founder of a dynasty of Swiss mathematicians, is born the son of a pharmacist in Basel. He applied the newly invented mathematical calculus to a variety of problems. Perhaps his greatest creation was his son, Daniel Bernoulli, a giant in mathematics and other sciences.

1766 William Hyde Wollaston, the first to isolate and name the elements palladium and rhodium, is born one of 17 children of a clergyman in East Dereham, England. His achievements were many, in many branches of science.

1774 Conscientious objectors arrive in America for the first time. Led by Ms. Ann Lee, a party of nine Shakers reach New York City from Liverpool, England, on the *Mariah*, refusing to aid England in the war against the colonies. By 1776 they settled in Watervliet, New York, but were soon jailed for treason because they refused to aid the colonies against England. They received no trial. They were all released in 1780.

1775 The great chemist Claude-Louis Berthollet announces to the French Academy of Sciences that he has abandoned his belief in phlogiston, a theory of matter that dominated science for about a century.

AUGUST 6

1881 Nobel laureate Sir Alexander Fleming is born in Lochfield, Scotland, the seventh of eight children of a farmer (see March 11, 1955, for a biographical sketch).

1890 In the 1770s Benjamin Franklin performed the first experiments on electrocution as a means of ending life; he killed chickens, a lamb, and a ten-pound turkey. On this date, a person is electrocuted for the first time as punishment in the United States. William Kemmler, a.k.a. John Hart, is executed in Auburn, New York, for the first-degree murder of Matilda Ziegler on March 29, 1889. The chair was designed by Dr. Alphonse David Rockwell, and the autopsy was performed under the direction of Dr. Carlos Frederick MacDonald three hours after the execution.

1941 The longest coma on record begins. Six-year-old Elaine Esposito of Tarpon Springs, Florida, has her appendix removed, and never regains consciousness. She died, still comatose, 37 years 111 days later.

1945 Hiroshima.

1954 David Grandison Fairchild, globetrotting explorer-botanist, dies at 85 in Coconut Grove, Florida. While a USDA administrator, he supervised the introduction of many useful plants, including mangos and avocados, into North America. He wrote a number of books, including his 1938 autobiography, *The World Was My Garden*. The Fairchild Tropical Garden in Coral Gables houses his collection of trees and other plants.

1991 A performing killer whale dies suddenly and prematurely in Orlando's Sea World after becoming ill overnight. The animal was three-fourths of the way through a 17-month pregnancy. It was the fifth whale death at a Sea World amusement park in the last two years. "We have no evidence that the animal died of anything improper on Sea World's part," said Humane Society investigator Ken Johnson. "But our concern is that a large number of animals have died there, and that causes us to further address the question of captivity." Sea World parks are part of the Anhaeuser-Busch empire.

1994 "Music soothes the savage beast." On this date, music kills a not-so-savage beast. In the Copenhagen Zoo a six-year-old okapi (a relative of the giraffe, native to Africa) dies from "severe stress" after a group of opera singers rehearsed Wagner's *Tannhäuser* 300 yards from the okapi's enclosure. "She started hyperventilating, went into shock, and collapsed," said a zoo spokesman. "We did all we could." Wagner could not be reached for comment.

1809 Poet Alfred, Lord Tennyson is born in Somersby, England. His is the earliest birth of anyone whose voice was recorded.

1495 Holy Roman Emperor Maximilian issues an Edict declaring syphilis to be a God-sent punishment for blasphemy.

1844 French mineralogist Auguste Michel-Lévy, pioneer in microscopic petrology (the study of rocks), is born in Paris.

1859 Services are first held in an all-deaf church in the United States. St. Ann's Church for Deaf-Mutes, New York City, is presided over by Episcopal priest Thomas Gallaudet. He is the son of Thomas Hopkins Gallaudet, founder of the country's first school for the deaf.

1859 The fifth edition of Darwin's *On the Origin of Species* is published.

1888 The revolving door is patented. Theophilus Van Kannel of Philadelphia receives patent No. 387,571 for a "storm door structure."

1903 Anthropologist L(ouis) S(eymour) B(azett) Leakey is born the son of English missionaries in Kabete, Kenya. He is the first white baby ever seen in Kikuyuland. His fossil discoveries in East Africa proved both that man is far older than previously thought and that man's evolution began in Africa, not Asia. He also hired Jane Goodall and Dian Fossey to perform groundbreaking research on the behavior of man's closest relatives, the great apes. Leakey is the only white man with a statue in his honor in Nairobi.

1912 Cosmic rays from outer space are discovered by 29-year-old physicist Victor Hess. Just after dawn he ascends outside Prague in a hydrogen balloon (highly flammable) with an electroscope, an instrument that detects radioactivity. When he lands six hours later near Berlin, he has discovered that gamma rays are plentiful several miles up in the atmosphere. Ridicule and criticism were his rewards.

1947 The balsa wood raft *Kon-Tiki* reaches the Tuamotu Islands (crashing into a reef there) in the South Pacific. The craft had left Peru 101 days before. The 5000-mile voyage supports the hypothesis of anthropologist Thor Heyerdahl that ancient man spread from South America to Polynesia in rafts.

1951 A *Viking* rocket establishes a new altitude record for a man-made object, 136 miles high. This is the seventh launch of a *Viking*, and surpasses by 20 miles the best height of man's first major rocket, the German V-2.

1959 *Explorer 6* is launched. It collects data on the newly discovered Van Allen radiation belts around Earth, and it transmits the first televised cloud-cover picture of Earth.

1963 Scientists end a week-long series of measurements in space, in which they discover the lowest-ever temperature recorded in the upper atmosphere (−225.4°F at an altitude of 50–60 miles over Sweden).

1973 Virginia park ranger Roy C. Sullivan is struck by lightning for the fourth time. His hair is set on fire and his legs are seared. Eventually Sullivan was hit by lightning seven times (the only human on record to be hit so many times) until he took his own life, apparently after being rejected in love.

1979 Highly enriched uranium leaks from a top-secret nuclear fuel plant near Erwin, Tennessee. One thousand people are dosed with radioactivity well in excess of a normal year's intake.

1991 The Florida Everglades National Park no longer contains any female Florida panthers, announces Joe Podgor of the watchdog group Friends of the Everglades.

1992 The EPA announces that lawnmowers are a significant source of air pollution.

1992 A treaty to ban chemical weapons forever is finally complete, after 24 years of negotiations. The breakthrough comes at the 39-nation Conference on Disarmament in Geneva.

1982 The largest single coral colony on record (over 52½ feet long) is discovered in Sakiyama Bay off Iriomote Island, Okinawa, by Dr. Shohei Shirai.

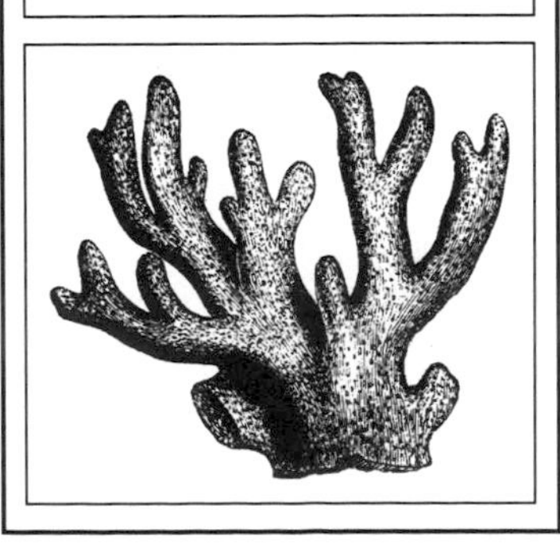

1709 A hot-air balloon flies for the first time on record, indoors at the Casa da India in Terreiro do Paço, Portugal. The craft was invented by Father Bartolomeu de Gusmão (born in Santos, Brazil, in 1685).

1793 One hundred twenty-seven years after its founding, the French Academy of Sciences is closed by the revolutionary Assembly amid financial difficulties and charges of elitism.

1797 Founded by Drs. Edward Miller and Elihu Hubbard Smith, *The Medical Repository* (a "depository of facts and reasonings relative to Natural History, Agriculture and Medicine") is first published. It is the nation's first medical and first scientific periodical. Dr. Samuel Latham Mitchill is the first editor. It appeared until 1824, and was printed by T. & J. Swords, printers to the Faculty of Physicians of Columbia College in New York City.

1799 Nathaniel B. Palmer, discoverer of Antartica, is born in Stonington, Connecticut.

1802 Germain Henri Hess, the founder of thermochemistry (analysis of heat involved in chemical reactions), is born the son of an artist in Geneva, Switzerland.

1854 Patented today are the metal cartridge bullet (by Daniel Baird Wesson of the Smith and Wesson Company) and the Millington Graduator (a tool for graduating carpenter's squares).

1876 Edison patents the mimeograph machine.

1897 Organic chemist Viktor Meyer dies at 48 in Heidelberg, Germany. He became a master of his craft early in life, obtaining his Ph.D. *summa cum laude* at just age 19. Ironically, one of his most memorable discoveries followed the failure of a demonstration he was giving to a class of undergraduates in 1882; he was showing the students a color test for benzene, but on this day he was using benzene derived from benzoic acid, a source he'd never used before. The test failed; ensuing experiments led him to discover thiophene, an organic molecule almost identical to benzene. Among other accomplishments, Meyer coined the term "stereochemistry" (the study of molecules that are identical but that have different spatial configurations). Meyer ended his own life in a "chemist's suicide," drinking prussic acid during a fit of depression over ill-health and constant pain from neuralgia.

1897 Kiyoshi Shiga discovers and names the bacterium that causes dysentery.

1899 The nation's first home refrigerator patent is issued. Albert T. Marshall of Brockton, Massachusetts, receives patent No. 630,617 for "an automatic expansion-valve for refrigerating apparatus."

1901 Ernest Orlando Lawrence, inventor of the "atom smasher" cyclotron, is born in Canton, South Dakota, the grandson of Norwegian immigrants. In 1929 he conceived of an entirely new process for propelling neutrons at great speeds for the purpose of bombarding and dissecting atomic nuclei. His first cyclotron appeared in 1930 (actually built by one of his students, M. Stanley Livingston); a handful more had been built by decade's end. He received the 1939 Nobel Prize for physics.

1902 Theoretical physicist P(aul) A(drien) M(aurice) Dirac is born in Bristol, England, the son of a Swiss immigrant schoolteacher. A Nobel laureate in 1933, Dirac is famous for his work in quantum mechanics, for deducing/predicting the existence of antimatter, and for theorizing that the electron spins and that its behavior can be described by wave equations.

1991 Although expensive, it should be possible to transform Mars into a place suitable for plants and animals, including man. This is the conclusion of NASA scientists in a report in *Nature*. The first step would be erecting factories on Mars to extract minerals from the surface; CFCs (chlorofluorocarbons) would be produced and pumped into the atmosphere to create a greenhouse effect. This should cause Mars's temperature to climb above freezing in 100 to 10,000 years.

1991 The second most lethal cancer in the United States, colon cancer, is a big step closer to eradication. Independent research teams at Johns Hopkins and the University of Utah announce in *Cell* and *Science* magazines that the abnormal gene responsible for inherited colon cancer has been isolated. The next step is finding a drug to combat its effects.

1776 Lawyer-turned-physicist Amedeo Avogadro is born into a family of lawyers in Turin, Italy. He is famous for Avogadro's law (equal volumes of different gases contain the same number of molecules) and Avogadro's number (6.0221367×10^{23}). Avogadro died before his law and his fame were established.

1803 Robert Fulton, 37, exhibits his first prototype steamboat on the River Seine in Paris.

1819 Dental surgeon William Thomas Green Morton, the first to successfully demonstrate gaseous anesthesia to the medical profession, is born in Charlton, Massachusetts.

1854 Henry David Thoreau's *Walden* is published.

1859 The escalator is patented. Nathan Ames of Saugus, Massachusetts, receives patent No. 25,076 for revolving stairs on an endless belt. The Otis Elevator Company was the first to actually produce such a device for public use. "Escalator" was trademarked May 29, 1900, by Otis.

1893 The country's first bowling magazine, *Gut Holz*, debuts. It was printed in German in New York City until May 19, 1894, when it was renamed *Bowlers' Journal*.

1897 Crystallographer Ralph Walter Graystone Wyckoff, the first to develop a technique allowing the electron microscope to provide 3-D images, is born in Geneva, New York. He also developed ultracentrifuges strong enough, for the first time, to isolate viruses. With this tool he helped develop the first *in vitro* vaccine against a viral disease (sleeping sickness in horses).

1916 Lassen Volcanic National Park is established in California.

1919 Ernst Haeckel, who coined the term "ecology," dies at 85 in Jena, Germany. The son of a government official, Haeckel began in medicine, but one year after he devoted his career to zoology and comparative anatomy. He championed the Darwinian concept of evolution, which he introduced to Germany. Haeckel once said, "Nothing is constant but change! All existence is a perpetual flux of 'being and becoming'!"

1936 Scottish paleontologist Robert Broom arrives in Sterkfontein, South Africa, in search of fossils to verify Raymond Dart's 1924 discovery of *Australopithecus*, thought to be a direct ancestor of *Homo sapiens*. Broom immediately goes to the regularly dynamited limestone quarry, where the manager, G.W. Barlow, "rather thought" he had seen some bones and skulls that Broom described. Just nine days later, Broom had most of a nearly perfect skull of an *Australopithecus*.

1944 The first Smokey Bear forest conservation poster appears.

1945 Nagasaki.

1961 The first 100-megaton bomb is under construction, announces Nikita Khrushchev.

1991 The Louisville Gas and Electric Company halts construction of a switching station on the banks of the Ohio River at New Albany, Indiana, because an Ice Age midden has been found on the site.

1991 A new, painless, and simple test to diagnose heart attacks has been discovered, report wire services today. It is a breath test being developed at Loyola University and the University of Illinois. Patients simply breath into a device that measures pentane, a gas that is elevated in persons suffering a heart attack. This gas is not elevated in people with other internal problems that have similar symptoms. "This is a very clever use of an old technology," said one researcher. "It's elegantly simple."

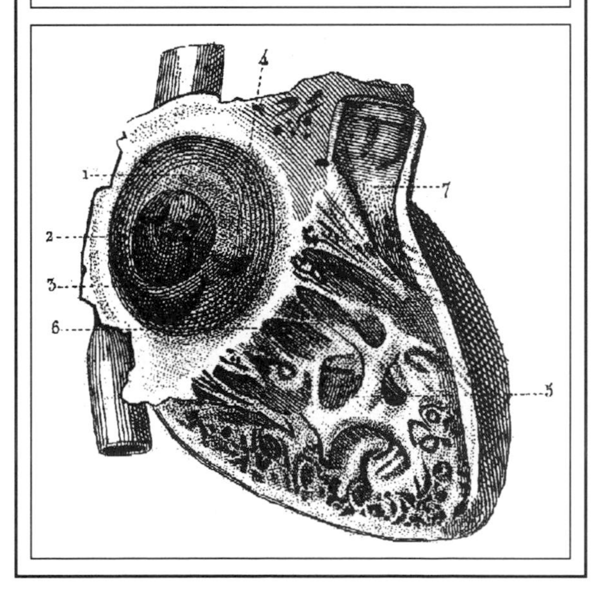

1709 Explorer Johann Gmelin, discoverer of permafrost (the permanently frozen ground that covers 20–25% of Earth's land surface), is born the son of an apothecary in Tübingen, Germany.

1740 John Frere, a founder of prehistoric archaeology, is born the son of a landowner in Roydon Hall near Diss, England. His discovery in 1790 of Stone Age flints was largely ignored because, at the time, most experts believed that Earth was created in 4004 BC; Frere suggested the implements and the people who had used them were much older than that.

1846 The Smithsonian Institution is chartered by Congress. A $500,000 bequest from English scientist James Smithson established the institute in Washington, D.C.

1869 A motion picture projector is patented by O.B. Brown of Malden, Massachusetts. It is the nation's first patent on such a device.

AUGUST 10

1878 The first serious and licensed home study course in the United States is organized, by the Literary and Scientific Circle of the Chautauqua Institution in Chautauqua, New York. Its correspondence School of Theology was chartered in 1881 by New York State.

1889 A meteor is captured photographically for the first time in U.S. history. At the Harvard College Observatory in Cambridge, Massachusetts, a Gundlach camera is attached to an 11-inch telescope; the shutter is left open for 13 hours 15 minutes. The meteor is pictured as a straight, dense line.

1902 Biochemist Arne Wilhelm Tiselius, who developed electrophoresis to the point where it is now standard procedure in analyzing the chemical makeup of complex substances, is born the son and grandson of mathematicians in Stockholm. He won the 1948 Nobel Prize for Chemistry.

1909 Laminated glass is patented by French artist-chemist Edouard Bénédictus.

1915 Physicist-genius Henry Moseley is killed at age 27 at the battle of Suvla Bay, Gallipoli, Turkey. He had already made fundamental discoveries about the atom's structure, and would surely have won the Nobel Prize.

1921 Franklin D. Roosevelt is stricken with polio while vacationing at his summer home on the Canadian island of Campobello. FDR was to become the disease's most famous victim, dying with it on April 12, 1945. Ironically, this triumph by the disease led to man's triumph over the disease; Roosevelt's death was a huge boost to the March of Dimes (which fights the disease), and the announcement that the Salk polio vaccine was effective was made on the tenth anniversary of FDR's death.

1990 The *Magellan* space probe enters Venus's orbit.

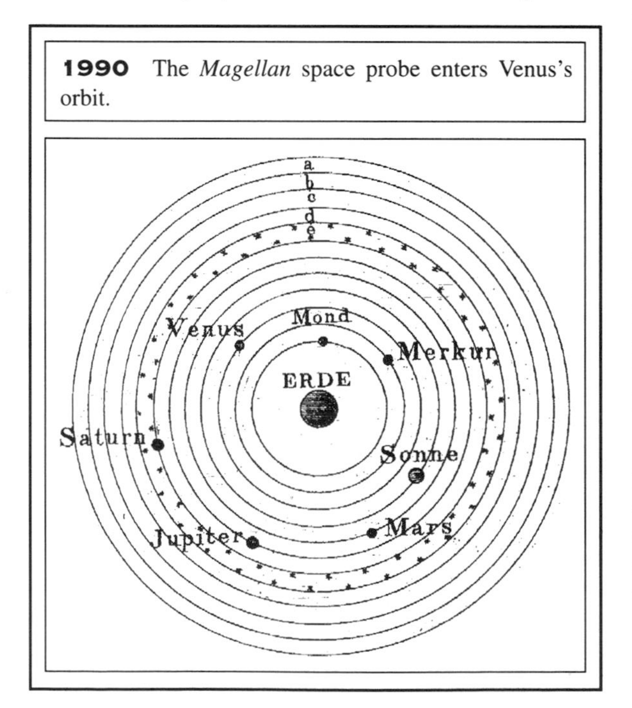

1991 The tallest-ever structure (the 2120-foot-tall Warszawa Radio tower outside Warsaw, Poland) collapses during renovations.

1992 Scientists with the National Oceanic and Atmospheric Administration release a study showing that people are more often hit with lightning at the end of a thunderstorm than at the beginning; lightning in the middle of a storm is least likely to strike humans. Ironically, when lightning bolts are least frequent (1–30 flashes per hour) they are most likely to hit people.

1993 "The Holy Grail of medicine," a blood substitute, is patented. Enzon, Inc., of Piscataway, New Jersey, chemically combined an altered form of hemoglobin (PEG-hemoglobin) with plastic-based fibers to create a substance that can carry oxygen and still be tolerated by the body.

1836 Chemist Cato Maximilian Guldberg, formulator of the law of mass action, is born the son of a minister in Christiania (now Oslo), Norway. This law, announced in 1864 in collaboration with Peter Waage, is a cornerstone of modern chemistry. It describes the effects of temperature, concentration, and mass on chemical reactions.

1858 Physician-pathologist Christiaan Eijkman is born the son of a schoolmaster in Nijkerk, Holland. His demonstration in the 1890s that beriberi is caused by poor diet led to the discovery of vitamins. Eijkman shared the 1929 Nobel Prize for Physiology or Medicine.

1861 Cardiologist James Bryan Herrick, the first to observe and describe sickle-cell anemia, is born in Oak Park, Illinois.

1874 The sprinkler head (for automatic fire sprinklers) is patented. Henry S. Parmelee of New Haven, Connecticut, receives patent No. 154,076 for a perforated head combined with a water valve held closed by a spring that melts.

1877 Asaph Hall, 47, abandons his search for satellites of Mars. "Try it just one more night," his wife suggests when Hall announces his decision at home. Sure enough, on the next night he discovers one of Mars's two moons, and several days later he discovers the other one.

1906 A process for producing sound-on-film "talkie" motion pictures is first patented, by Frenchman Eugene Augustin Lauste.

1907 History's first long-distance car rally (from Peking to Paris) is won in Paris by Prince Borghese, driving an Itala automobile. The event started on June 10.

1909 The radio distress signal SOS is first sent by a U.S. ship. The single-screw steamship *Arapahoe* loses her engines 21 miles off Cape Hatteras at 3:45 PM, and sends both SOS and the previous standard CQD. The cry for help is received and acknowledged by R.J. Vosburg, radio operator of station HA on Cape Hatteras. Foreign registered vessels had been using SOS for some time before it was adopted by the United States.

1934 Otis Barton and William Beebe break the surface off Bermuda after setting a new depth record (well over a half mile) for a descent into the ocean. It is actually the first time anyone has gone beyond the ocean's surface layer. The pair discovered a host of new fish species.

1963 "Interest in hair today has grown to the proportions of a fetish. Think of the many loving ways in which advertisements refer to scalp hair—satiny, glowing, shimmering, breathing, living. Living indeed! It is as dead as rope." This is the view of dermatologist Dr. William Montagna of Brown University, discussing the decrease of hairiness in human evolution, in a New York *Herald Tribune* interview.

1972 Microbiologist Max Theiler dies at 73 in New Haven, Connecticut (see January 30, 1899, for a biographical sketch).

1992 Two more agricultural workers fall victim to killer manure fumes. A young man and his uncle are overcome by gases from a manure pit on a farm near Hastings, Minnesota. It is the second such incident in a week in that state. Three days earlier a man and his son suffered the same fate on a Canby farm.

1994 Diesel fuel finally stops leaking from the *Columbus Iselin*, a 170-foot ecology research ship that ran aground on a fragile coral reef inside of Florida's Looe Key National Marine Sanctuary. In all, 200 gallons of toxic liquid was spilled. At the time of the accident the ship was investigating ways to manage oil spills in reefs.

1673 Richard Mead, a pioneer in preventive medicine, is born in London. He was the physician to Isaac Newton and King George I, and his personal library contained over 10,000 volumes.

1585 A letter is written in English in America for the first time in recorded history. Ralph Lane, commander of Raleigh's first colony, writes four letters from Porte Ferdynando in present-day North Carolina. The first letters in any language written in America were probably written by crew members on Columbus's 1492 voyage, including several by Columbus himself.

1759 Botanist Thomas Andrew Knight is born in Ludlow, England. He did important research into the formation of bark and the flow of sap in plants, but is best known for pioneering studies of geotropisms (i.e., the effects of gravity on growth) in roots and stems.

1851 Isaac Singer patents the sewing machine.

1883 The quagga, a type of zebra, goes extinct. The last dies in the Amsterdam zoo.

1887 Physicist Erwin Schrödinger, founder of quantum wave mechanics, is born in Vienna. His father was a prosperous industrialist, and Schrödinger was educated at home in his early years. He saw action during World War I, and decided to abandon physics for philosophy when the war was over; but the town where he planned to work was lost in the armistice, so Schrödinger remained a physicist. It was not until his 30s that he produced the theories and papers for which he is famous and a Nobel laureate.

1897 Otto Struve, the last of four generations of famous astronomers, is born in Kharkov, Russia. He discovered interstellar matter (the thin gas between stars) and the plentiful existence of various elements (especially hydrogen) in outer space.

1920 Terence James MacSwiney, lord mayor of Cork, Ireland, begins a hunger strike to protest his imprisonment by the British. He died 2½ months later. Through letters to his wife and careful notes on his condition by British doctors, MacSwiney's physical demise provided science with a unique, detailed case study in nutrition and starvation.

1923 A portable motion picture camera is advertised for the first time in the United States. It is the Victor Cine Camera, manufactured by the Victor Animatograph Company, Inc., of Davenport, Iowa.

1865 The beginning of antiseptic surgery. Using disinfectant chemicals and procedures of his own design, Joseph Lister treats the compound fracture of a boy, James Greenlee, in the Glasgow Royal Infirmary. It is the first of several cases that formed the basis of Lister's reports on his methods, which dramatically reduced the rates of infection and mortality in hospitals, and changed health care forever.

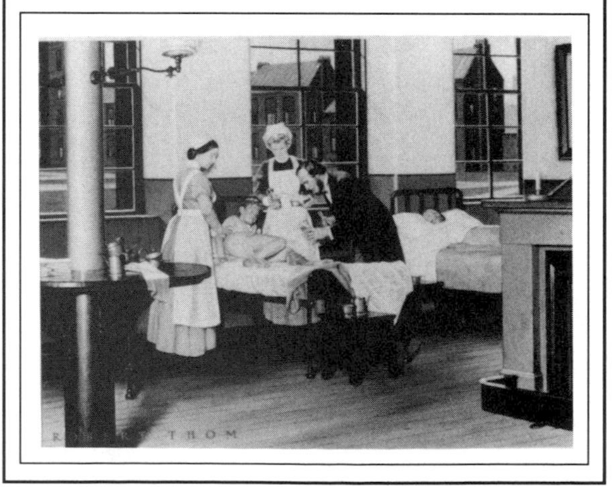

1953 Russia tests its first H-bomb.

1955 Biochemist James B. Sumner dies at 67 in Buffalo, New York (see November 19, 1887 for a biographical sketch).

1960 The first telecommunications satellite, *Echo I*, is launched by NASA. It was a 100-foot-diameter aluminum balloon that passively reflected radio waves back to Earth—until it was burst by meteorites.

1991 The editors of *Gourmet* magazine report a slight error in a recent recipe for "Aunt Vertie's sugar cookies." Instead of wintergreen extract, the printed recipe called for wintergreen oil, which is sold at drugstores for the care of sore muscles; if eaten it can cause nausea, convulsions, and death.

1991 Addiction researchers at the annual meeting of the American Psychological Association in San Francisco report that caffeine withdrawal is "a real phenomenon," and one reason that many people continue to drink coffee is to prevent the withdrawal symptoms (including headaches, lethargy, and depression). Some 90% of North American adults regularly consume caffeine.

1625 Erasmus Barholin, discoverer of the double refraction of light (i.e., the bending of light in two different directions simultaneously), is born into a dynasty of physicians-anatomists in Roskilde, Denmark.

1814 Physicist Anders Ångström, discoverer of hydrogen in the sun (in 1862), is born the son of a lumbermill chaplain in Lögdö, Sweden. He pioneered spectroscopy of the heavens, and was extremely precise in measuring the size of light waves. In 1905 his name was given to a unit of length in the metric system, the angstrom, which is 1/10,000,000,000 of a meter.

1844 Johann Friedrich Miescher, discoverer of nucleic acids (DNA and RNA are nucleic acids), is born the son of a physician in Basel, Switzerland. Miescher first found the substance in pus.

1865 Ignaz Philipp Semmelweis (who discovered the cause and prevention of childbed fever and who introduced antiseptic procedures to surgery) dies at 47 in Vienna (see July 1, 1818, for a biographical sketch).

AUGUST 13

1869 Led by John Wesley Powell, the first scientific party to explore the Colorado River enters the Grand Canyon (which Powell named).

1889 The pay phone is patented. William Gray of Hartford, Connecticut, receives patent No. 408,709 for a "coin-controlled apparatus for telephones." He had filed his application exactly one year earlier. The device was first installed in 1889 in the Hartford Bank. In 1891 Gray teamed with (Francis) Pratt and (Amos) Whitney, to manufacture phones and install them in stores on a rental basis; the store got 10% of the take, Pratt, Whitney, and Gray got 25%, and the phone company got the rest.

1918 Double Nobel laureate Frederick Sanger is born the son of a physician in Rendcombe, England.

1942 The Manhattan Project is established. Its mission was to build the first nuclear weapons.

1946 Carl E. Weller of Easton, Pennsylvania, patents the soldering gun. Patent No. 2,405,866 is issued for an "electrical heating apparatus"; the rights to the invention are assigned to the Weller Manufacturing Company. Weller's patent was reissued in February 1953.

1961 "I knew that there was something in the nature of homesickness called nostalgia, but I found that there is also a homesickness for the earth. I don't know what it should be called but it does exist. There is nothing more splendid...than Mother Earth on which one can stand, work and breathe the wind of the steppes." These are the remembrances of Major Gherman Titov of his recent 434,960-mile flight around Earth, in a New York *Herald Tribune* interview published on this date.

1993 Alzheimer's researchers report in *Science* that they have found a gene linked to one form of the incurable disease (that is thought to afflict 2.5 million Americans). The gene, called "apolipoprotein-E," has been studied for years because it is important in cholesterol processing. Duke scientists have now discovered that if a person has a particular form of this gene, there is a 90% chance that he will eventually develop Alzheimer's. The discovery caused "a great deal of excitement" because "it could become a very important diagnostic tool," according to Dr. Zaven Kachaturian at the National Institutes of Health.

1992 What killed the dinosaurs? The best evidence yet for the meteor theory is reported in *Science* magazine. This theory holds that the Earth was hit with an enormous meteor or asteroid; the resulting dust storm blocked the sun, which then led to dinosaur extinction. Walter Alvarez, who codeveloped the theory with his Nobel laureate father, Luis, reports new studies of an enormous crater in the Yucatan Peninsula, Mexico, which indicate that it was created by a meteor collision 65 million years ago, just when the dinosaurs disappeared. The crater is 110 miles wide, and is Earth's largest. The new dating of the crater origin is called a "smoking gun" by theory proponents because it indicates that the meteor collision and the dinosaur extinction occurred at the same time.

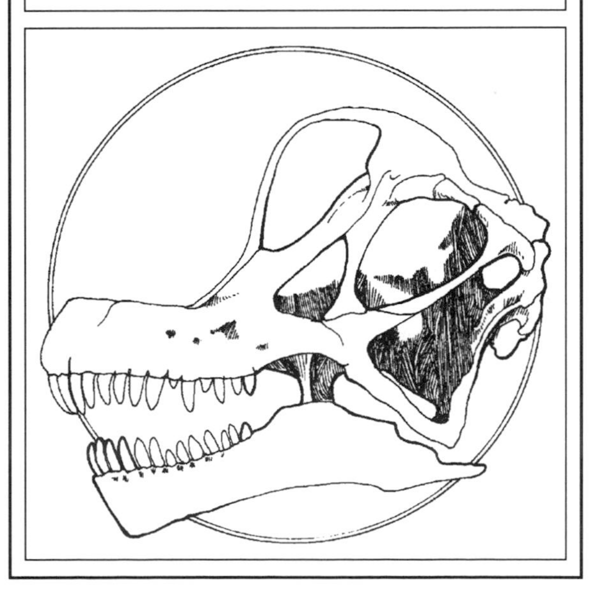

1457 The earliest exactly dated printed work is completed. It is the Psalter printed in Mainz, Germany, by Johann Fust (c. 1400–1466) and Peter Schöffer (1425–1502), who had been Gutenberg's chief assistant.

1777 Physicist Hans Christian Oersted, famous for one experiment, is born the son of an apothecary in Rudkøbing, Denmark. In 1819 he was conducting a classroom demonstration when he happened to bring a compass near an electrified wire, and discovered the compass needle oriented at right angles to the flow of electric current. It was the first demonstration of the interconnection of electricity and magnetism.

1785 Benjamin Franklin, 79, attempts the first measurements of temperature within the Gulf Stream. Franklin is on his last transatlantic journey, and is hoping to dedicate the remainder of his life to scientific researches.

1840 Psychiatrist Richard Krafft-Ebing is born in Mannheim, Germany. He coined the terms "paranoia," "sadism," and "masochism," and is famous as a pioneer in the scientific study of sexual aberration. His 1886 *Psychopathia Sexualis* opened the door to the theories of Freud.

AUGUST 14

1860 Pasteur, 37, and assistant Émile Duclaux disprove the theory of spontaneous generation (which stated that life can arise spontaneously from lifeless matter). The two researchers open ten flasks (filled with sterilized nutrient solution) in the streets of Paris, then reseal the flasks. They do the same thing with other flasks deep within the cellars of the Paris Observatory. Microbes start growing immediately in the street air samples, while the solutions in the Observatory flasks stay pure and lifeless. This proved that air itself is not sufficient to grow microbes. It is another great triumph for simplicity and for Pasteur.

1886 Physicist Arthur Jeffrey Dempster, discoverer of uranium-235, is born in Toronto. The fuel for the first nuclear weapons, uranium-235 is the most famous isotope in history.

1909 Earth's most massive natural bridge is discovered. It is the Rainbow Bridge in Utah; it spans 278 feet and is 22 feet wide (Earth's longest natural bridge spans 291 feet, but is only 6 feet wide in one section; it is also in Utah).

1910 An airplane carries three passengers for the first time in history. Charles Foster Willard pilots a Curtiss biplane at Mineola, New York, with R.F. Patterson, A. Albin, and Harry Willard aboard.

1936 The fastest animal on Earth over long distances (1000 yards or more) is the pronghorn antelope of the western United States. On this date, at Spanish Lake, Oregon, the record is set when a buck is clocked at 61 mph.

1936 The last public hanging in U.S. history takes place in Owensboro, Kentucky, in front of a crowd of 10,000–15,000.

1945 Japan surrenders, thereby ending World War II. It is nuclear weapons' shining hour.

1959 Earth is first seen by television. The paddlewheel satellite *Explorer 6* sends back a television telecast showing all of the planet's 20,000-square-mile area.

1979 The longest-lasting rainbow on record (over three hours) is seen in northern Wales.

1994 Massage eases trauma, according to a report by a researcher at the University of Miami School of Medicine. Tiffany Field, of the University's Touch Research Institute, announces her findings in San Francisco to a meeting of the American Psychological Association. She studied 60 Florida children who had been traumatized by Hurricane Andrew in 1992. Those who were massaged twice a week were more relaxed, less anxious, and less depressed than victims who watched relaxing videos with a therapist.

2126 In 1992 Brian Marsden, a Harvard astronomer, predicted that the comet Swift-Tuttle may hit Earth on this date. Marsden actually predicted that the comet would miss Earth, but a collision could occur if his calculations are off by just a little.

1796 Botanist John Torrey, the first to classify many of the plants in North America, is born in New York City. Torrey's Peak in Colorado is named for him.

1814 The first U.S. building to be designed specifically as a museum and art gallery opens at 225 North Holliday Street in Baltimore. Designed by Robert Cary Long, Sr., "Peale's Baltimore Museum and Gallery of the Fine Arts" is operated by Rembrandt Peale, son of Charles Willson Peale, well-known painter of U.S. patriots and founder of the Peale Museum in Philadelphia. Rembrandt's brothers (Raphaelle, Rubens, and Titian) were also painters. In 1830 the building was sold to become Baltimore's first City Hall.

1865 Japanese physicist Hantaro Nagaoka, one of the first to theorize that the atom has a positively-charged center surrounded by negatively charged electron satellites, is born in Nagasaki.

1875 The greatest concentration of animals ever recorded is discovered in Nebraska. It is a swarm of Rocky Mountain locusts covering 198,600 square miles. It flies over the state for the next 10 days. An estimated 12.5 trillion locusts, with a combined weight of 27.5 million tons, are in the formation.

1885 The bacterium *E. coli* first appears in print, in the German medical journal *Fortschritte der Medizin*, in an article entitled "The Intestinal Bacteria of the Newborn and Infants." The author, pediatrician Theodor Escherich, discovered the organism in the messy diaper of a Munich baby. From this humble beginning, *Escherichia coli* has become a superstar of modern science, now reigning as the number one subject in recombinant DNA research. Furthermore, because it appears in human feces, tests for *E. coli* are basic to the analysis of foods, beverages, and drinking water.

1892 French physicist Louis de Broglie is born into an aristocratic family (his great-great-grandfather died on the guillotine) in Dieppe. De Broglie won the 1929 Nobel Prize for discovering the wave nature of electrons.

1896 Biochemist Gerty Theresa Radnitz Cori is born in Prague, exactly four months to the day before husband-to-be Carl Ferdinand Cori is born in the same city. The pair shared the 1947 Nobel Prize for Physiology or Medicine.

1913 Albert Schweitzer is forced to perform major surgery before the medical ward is built in his now-famous mission-hospital in Lambaréné, Gabon. A man named Ainda begs the 38-year-old doctor (who had given up a promising career as an organist to devote himself to the service of others in Africa) to operate on a strangulated hernia. The operation takes place in the bedroom of a colleague's children; Schweitzer's wife administers the anesthetic. It is Schweitzer's first hernia operation. The patient recovers completely.

1914 The Panama Canal opens to traffic.

1935 Famed aviator Wiley Post and beloved humorist Will Rogers die in a plane crash near Point Barrow, Alaska.

1989 The fastest plane climb in history is achieved by Heinz Frick of British Aerospace at the Rolls-Royce test flight center in Filton, Bristol, England. He takes a Harrier GR5 jet from a standing start to a height of 39,370 feet in 2 minutes 6.63 seconds.

1994 The international terrorist-for-hire, "Carlos the Jackal," is behind bars after a 20-year manhunt. The 44-year-old Venezuelan (whose real name is Ilich Ramirez Sanchez) is arrested in Khartoum, Sudan, and immediately flown to Paris. The Jackal is linked to a number of spectacular acts through the 1970s and 1980s, including the 1975 kidnap of 11 OPEC oil ministers in Vienna.

1994 A link between suicide and teen smoking has been found, reports psychologist Kenneth Carter to a Los Angeles meeting of the American Psychological Association. Adolescents who smoke are 18 times more likely to have attempted suicide than nonsmokers. This does not mean that smoking leads to suicide, observes Carter (of the federal Centers for Disease Control and Prevention), but that smoking may simply be a result of the underlying depression that also leads to suicide. Whatever the cause of the link, Carter says of a student who seems down and also smokes a pack a day, "I'm going to be a lot more worried about that student."

1848 The dentist's chair is patented. Waldo Hanchett of Syracuse, New York, receives patent No. 5,711 for a piece of furniture that combines the needed features of a headrest and a back and seat that can be adjusted in height and position.

1821 Mathematician Arthur Cayley is born in Surrey, England. In addition to developing matrix algebra (important to Einstein's and Heisenberg's later theories in physics), Cayley was a lawyer, a renowned linguist, and a campaigner for women's access to college education.

1829 The original Siamese twins, Chang and Eng, arrive in the United States (in Boston aboard the ship *Sachem*) to be exhibited throughout the Western world.

1832 Psychologist Wilhelm Wundt is born the son of a minister in Neckarau, Germany. A pioneer in measuring the relationship between physical stimuli and perceived sensation, Wundt created the first ever university course in experimental psychology (1862), established the first laboratory devoted to this field (1879), and founded its first journal (1881).

1845 Physicist Gabriel Jonas Lippmann, inventor of color photography, is born in Hollerich, Luxembourg. Recipient of the 1908 Nobel Prize, he concluded his acceptance speech thusly, "Perhaps progress will continue. Life is short and progress is slow."

AUGUST 16

1858 Queen Victoria and President Buchanan exchange messages on the just-completed transatlantic cable. Weak signals caused the service to be suspended within a month, and the line was blown out shortly thereafter when efforts were made to send more electric current over the wire.

1863 Frederic Stanley Kipping, author of the first textbook restricted to organic chemistry, is born the son of a banker in Manchester, England. He devoted 40 years of his life researching silicon.

1893 Neurologist Jean-Martin Charcot dies at 86 of pulmonary edema, at Lake Settons, France. His father was a successful carriage-maker who was able to offer Jean-Martin the choice of a career in medicine or in art. Charcot chose the former. He revolutionized the treatment of neuroemotional disorders by his careful attention to symptoms and by his use of hypnotism. Freud's interest in the unconscious began while a student of Charcot's. "To take from neurology all the discoveries made by Charcot would be to render it unrecognizable," wrote Joseph Babinski, a renowned neurologist himself.

1896 Gold is found in the Yukon, at Bonanza Creek, Alaska.

1916 The United States enters into its first international treaty to protect migrating birds. In Washington, U.S. and British diplomats sign the Migratory Bird Treaty, designed to protect transients between Canada and the United States. President Woodrow Wilson signed the treaty on September 1, and it was publicly announced on December 8.

1904 Biochemist Wendell M. Stanley is born in Ridgeville, Indiana. He was a corecipient of the 1946 Nobel Prize for Chemistry for being the first to purify and crystallize a virus (see June 15, 1971, for a biographical sketch).

1920 A major-league baseball player is killed in a game for the first time. In New York City's Polo Grounds, Ray Chapman, shortstop of the Cleveland Indians, is beaned on the left side of the head by New York Yankees pitcher C. W. ("Willie") Mays.

1931 Twins with the lightest-ever birth weight (2 pounds 3 ounces total) are born in Peterborough, England, to Mrs. Florence Stimson.

1994 The Associated Press reports that history's first female condom is ready for sale and will soon be on drugstore shelves. Called "Reality," it comes at a good time (with female AIDS and teen pregnancy rates rising), but it may be a hard sell because it is costly ($2.75 each), it comes with extensive directions (two typewritten pages containing eleven diagrams), and even the manufacturer admits that the product looks "funny."

AUGUST 17

1648 Famous scientist-mathematician René Descartes, 52, flees Paris during political upheavals, and seeks refuge in Sweden under the patronage of young Queen Christiana (whose desire to be tutored in predawn hours accidentally killed Descartes in 1650).

1673 Anatomist Reignier de Graaf dies of the plague in Delft, the Netherlands, at age 32. He had a short but important career (see biographical note on July 30, 1641).

1771 At his laboratory in Leeds, England, Joseph Priestley "put a sprig of mint into air in which a wax candle had burned out." He "found that, on the 27th of the same month, another candle could be burned in this same air." Thus, Priestley had discovered that plants replenish the supply of oxygen in the atmosphere.

1835 The wrench is first patented in the United States, by Solyman Merrick of Springfield, Massachusetts. The first practical and successful U.S. wrench was invented by Daniel C. Stillson of Somerville, Massachusetts, who whittled his first prototype from a block of wood.

1859 Officially dispatched and stamped mail is first carried by balloon in the United States. The flight is a dismal failure. Headed for New York City, John Wise departs from Lafayette, Indiana, in the *Jupiter* with 123 letters and 23 circulars in a pouch. After just 23 miles he is forced to land in Crawfordsville, Indiana. In another balloon flight, Wise tried to fly mail from St. Louis to New York City, but threw it overboard during a storm.

1893 Chemist Walter Noddack, codiscoverer of the element rhenium (the last nonradioactive element to be found), is born in Berlin.

1932 An era in neurosurgery ends. In Boston Harvey Cushing performs his last operation.

1933 Russia's earliest rocket, the semi-liquid-fueled GIRD-IX, is given its first test flight.

1936 Anthropologist Robert Broom is handed the second-discovered skull of man's ancestor *Australopithecus* by a quarry manager in Sterkfontein, South Africa.

1958 The first U.S. moonshot is launched. It is a *Pioneer* rocket on a Thor missile, and it explodes shortly after takeoff. It was not given a number. *Pioneer 1* was launched in October, but it failed to escape Earth's gravity.

1978 The first successful transatlantic balloon flight ends when Maxie Anderson, Ben Abruzzo, and Larry Newman land *Double Eagle* outside Paris.

1993 A Roman Catholic priest decides to recant his advocacy of murdering abortion doctors. The Rev. David Trosch of Magnolia Springs, Alabama, unsuccessfully tried to run a newspaper ad with the words "Justifiable homicide?" under a picture of an abortionist about to be shot. Trosch's boss, Archbishop Oscar H. Lipscomb, ordered him to recant or resign. Trosch still plans to protest and is very pleased with the free publicity.

1994 The *American Journal of Public Health* publishes the first solid estimates of what caring for an Alzheimer's patient costs: $213,732 over the last four years of his life. On a national level, the public cost is $82.7 billion annually, in lost productivity and round-the-clock nursing. These estimates do not include $14,140 per patient in additional medical expenses.

1807 Robert Fulton demonstrates his steamboat *Clermont*, guiding it up the Hudson River from Greenwich Village to Albany. The trip took 32 hours, while contemporary sailing boats took up to four days to travel the same distance. A farmer who witnessed the event likened it to "the devil going up the river in a sawmill."

1685 Brook Taylor, a developer of calculus (founder of the branch of mathematics called the calculus of finite differences), is born in Edmonton, England.

1734 William Bull, of Charleston, South Carolina, was the first American-born physician; on this date Bull receives his M.D. degree from Leiden University, the Netherlands, for a thesis entitled "Colica pictonum."

1838 One of the first oceanographic expeditions to circle Earth (and the first scientific expedition outfitted the U.S. government) departs Hampton Roads, Virginia, under the leadership of Lieutenant Charles Wilkes.

1871 An engineless airplane is flown successfully for the first time. It is a model constructed by French engineer Alphonse Penaud, based on the plans of English engineer Sir George Cayley.

1908 Plant pathologist Sir Frederick Charles Bawden is born in North Tawton, England. His research specialty was viruses in plants. In 1937 he and colleagues discovered that the tobacco mosaic virus contains RNA; this was the first indication that nucleic acids occur in subcellular life forms, and was a major plank in the emerging notion that nucleic acids may be the most basic feature of things called "living."

1922 W(illiam) H(enry) Hudson (ornithologist, naturalist, and novelist) dies at 81 in London (see August 4, 1841, for a biographical sketch).

1931 A plant patent is first issued in the United States, to Henry F. Bosenberg of New Brunswick, New Jersey. It is for a climbing rose named New Dawn.

1974 The most absorbent material on record by the U.S. Department of Agriculture. It is called "H-span" or "Super Slurper," and is composed mostly of starch and plastic. When treated with iron, it retains 1300 times its own weight in water.

1990 Eight days after receiving a lifetime achievement award from the American Psychological Association, B(urrhus) F(rederic) Skinner dies at 86 of leukemia complications in Cambridge, Massachusetts. He will forever be remembered for the Skinner box (a highly simplified environemnt containing means of delivering a reward or punishment and means of detecting voluntary behavior, such as a lever for an animal to press), which Skinner used to develop original theories of learning. His experimental subjects were mainly rats and pigeons, but in *Walden Two* (1948) and *Beyond Freedom and Dignity* (1971) he applied his theories to the behavior of humans, arguing that man is simply another animal whose free will is largely controlled by the consequences of his previous actions and experiences. Critics argued vigorously against this view, but Skinner's scientific findings remain bulwarks in the fields of learning and behavior. "The real problem is not whether machines think, but whether men do."

1992 A Norwegian fishing net snags a 7½-foot, 730-pound leatherback turtle near the Arctic Circle. Catching endangered sea turtles in fishing nets is nothing new, but in this case the turtle had strayed from its usual waters—8000 miles away in the Gulf of Mexico. "I couldn't believe my eyes," said skipper Arild Olsen.

1992 The City Council of Sacramento, California, votes to amend its prohibition of livestock in residential zones to allow Deborah Denbaugh (a 20-year-quadriplegic) to keep her pet duck Andora. At the emotional Council hearing, the girl's mother pleaded that the duck given her daughter such a morale boost that she was able to speak and move for the first time in four years.

1872 The country's first mail-order catalogue is issued. It is a list of prices on a single sheet, 8 × 12 inches, with no illustrations. It is published by (Aaron) Montgomery Ward, who began his business with $2400 in a 12 × 24-foot room in Chicago. The next year the catalogue had four pages and nearly 400 items. By 1904, the Montgomery Ward catalogue weighed four pounds.

AUGUST 19

1580 Mathematician-bureaucrat Pierre Vernier, inventor of the vernier caliper (still the instrument of choice in many cases where great accuracy in linear measurement is needed), is born the son of a lawyer-bureaucrat in Ornans, France.

1745 Swedish mineralogist Johan Gottlieb Gahn, discoverer of manganese (1774), is born in Voxna.

1830 Chemist Lothar Meyer is born in Varel, Germany, the son of a doctor. He followed in his father's footsteps for a while, but then switched fields to his real love: chemistry. Independently of Mendeleyev, Meyer created a periodic table of the chemical elements (in which elements are classified according to their atomic weights) in 1868. But Mendeleyev is famous and Meyer obscure, because Meyer waited until 1870 to publish (Mendeleyev published in 1869), and because Meyer did not take the bold step of predicting that elements would be discovered (which Mendeleyev did with flair, daring, and accuracy).

1839 The birth of photography. The daguerreotype process of fixing a visual image on a chemically treated plate is announced to the French Academy of Sciences.

1848 Discovery of gold in California is announced in the *New York Herald*.

1856 Condensed milk is patented, by Gail Borden of Brooklyn, New York; an "improvement in concentration of milk" is awarded patent No. 15,553. The value of the invention was questioned by the Patent Office, and for a while this doubt seemed justified. Two condensaries in Connecticut failed in 1856 and 1857. Borden opened a third plant, the largest yet, in Wassaic, New York, in June 1861. It prospered and became the Borden Company.

1871 Airplane inventor Orville Wright is born in Dayton, Ohio. He and older brother Wilbur were sons of a minister, and they lived chaste lives, never smoking, drinking, or marrying.

1891 Astronomer Milton La Salle Humason is born in Dodge Center, Minnesota. His first professional employment was in the Mount Wilson Observatory—as a janitor. His talent was soon recognized by Edwin Hubble, and Humason went on to significant achievements. Most notable was his measurement of the rates of recession of some 800 galaxies (which enabled estimation of their distances). He also nearly discovered Pluto in 1919, but a freak flaw in a photographic plate caused him to miss it. Clyde Tombaugh made the discovery 11 years later. Tombaugh is in all of the textbooks, Humason in very few.

1957 Man takes a balloon higher than 100,000 feet for the first time. Major David Simons departs on this day from Crosby, Minnesota, and lands 32 hours later after reaching a height of 101,516 feet.

1958 The *Triton*, eventually to become the first submarine to circle the globe underwater, is launched in Groton, Connecticut. Equipped with two nuclear reactors (also a first), the *Triton* is 447 feet long.

1982 Svetlana Savitskaya of Russia becomes the second woman to orbit Earth.

1991 Suppressing emotions may suppress longevity, according to three research reports to a San Francisco meeting of the American Psychological Association. A 41-year Johns Hopkins study found that those who hold in tension at age 20 are twice as likely as more expressive peers to die by age 55.

1993 Scientists with the National Oceanic and Atmospheric Administration announce results from the first-ever observation of an undersea volcano eruption. Underwater cameras have been viewing the event 270 miles off the Oregon coast, about 1½ miles deep.

1897 In a remote, primitive laboratory in Secunderabad, India, Ronald Ross finally finds the malarial parasite in the gut of a mosquito. He began looking for it in 1892. His discovery proved that mosquitoes transmit malaria to man, and it won Ross the 1902 Nobel Prize for Physiology or Medicine.

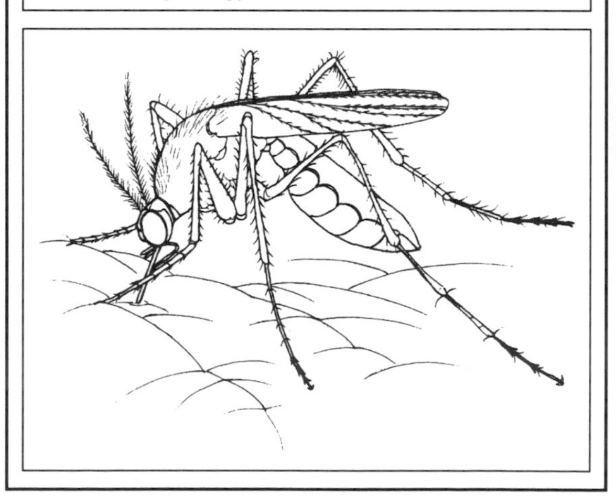

1601 Pierre de Fermat is baptized in Beaumont-de-Lomagne, France (his exact date of birth is uncertain). His father was a leather merchant. Although he was a lawyer by profession and only did mathematics in his spare time, Fermat is acknowledged as one of the great mathematicians of the seventeenth century, and has been called the "founder of modern prime number theory," "cofounder of the theory of probability," and "inventor of the differential calculus."

1779 Jöns Jacob Berzelius, a founder of modern chemistry, is born the son of a clergyman-schoolmaster (who died four years later) near Linköping, Sweden.

1831 Austrian geologist, Eduard Suess (who first hypothesized the existence of Gondwana, a single landmass that gave rise to Africa, Antarctica, South America, and Australia), is born in London, where his father ran a wool business.

AUGUST 20

1900 A motorbike with a two-stroke engine is patented in Paris by the Frenchman Cormery.

1910 The first pilot to shoot a gun from a U.S. airplane is Lt. Jacob Fickel, 29th Infantry, who today fires a rifle at the Sheepshead Bay Racetrack, New York City, from a single-seat Curtiss airplane.

1911 The first message sent around the world by commercial telegraph begins at 7:00 PM from the 17th floor of the New York Times Building in New York City. The message ("This message sent around the world") is received at the same station 16 minutes later, after traveling 28,613 miles. Relay stations are at the Azores, Gibraltar, Bombay, the Philippines, Midway, Guam, Honolulu, and San Francisco.

1915 Paul Ehrlich, medical giant in the fields of immunology and chemotherapy, dies of a stroke at 61 in Bad Homburg vor der Höhe, Germany. Born in Strehlen, Germany, into a Jewish family that was successful in business and industry, Ehrlich became a medical student and researcher, and soon began a long string of disease-fighting innovations and discoveries. His crowning achievement was developing the "magic bullet," a shot that cured syphilis. In fact, Ehrlich's development of artificial chemicals to combat disease is acknowledged as the birth of chemotherapy. He was awarded the 1908 Nobel Prize for Physiology or Medicine.

1993 "Oh give me a home, where the buffalo roam...." These lyrics probably weren't written about Lancaster, Pennsylvania, but on this date a male and female bison escape from a stockyard and temporarily roam the suburbs, crossing traffic, wandering through yards, and romping over a golf course. One unsuspecting woman recounted her close encounter, "I was just walking the dog. I went up to a stranger's door and said, 'Excuse me, can I come in because there's a [buffalo] there.'"

1953 Russia announces that she has successfully tested an H-bomb.

1956 The first and only issue of the *Walden Two Bulletin* is published. It was envisioned as a periodical devoted to the establishment of communities following guidelines in the book *Walden Two* by psychologist B.F. Skinner.

1968 Ant scientist T.C. Schneirla dies at 66 in New York City (see July 23, 1902, for a biographical sketch).

1977 *Voyager II* is launched. The unmanned space craft carries a 12-inch copper phonograph record containing music, sounds of nature, and greetings in dozens of languages.

1994 For the second year in a row, untrained, unsponsored Tarahumara Indians from northern Mexico astound everyone in Colorado's 100-mile "Leadville Trail 100" footrace. The Indians, wearing homemade rubber-tire sandals, take a number of the top ten spots. They were brought to the race by the nonprofit Wilderness Research Expeditions of Tucson, Arizona, to publicize the ecological plight of the Indians, whose diminishing habitat is under assault by logging, roads, water pollution, and drug wars.

AUGUST 21

1754 Scottish inventor William Murdock, a pioneer in steam power and the use of coal gas for illumination, is born in Old Cumnock.

1789 Baron Augustin-Louis Cauchy, one of the greatest of modern mathematicians, is born the son of a government official in Paris. His family fled Paris to escape the Reign of Terror (1793–1794), and went to the small town of Arcueil, where Cauchy first met two great scientists, the mathematician Laplace and the chemist Berthollet.

1813 Jean Sevais Stas is born the son of a shoemaker in Louvain, Belgium. Possibly the greatest chemical analyst of the nineteenth century, Stas produced measurements of atomic weight that were more accurate than any before, and he developed analytic procedures that were standard for the next century.

1816 Chemist Charles Gerhardt, one of the first to use equations systematically to represent chemical reactions, is born in Strasbourg, France. With Auguste Laurent, Gerhard presented a new classification of organic compounds; although ultimately rejected, the system forced a rethinking of organic molecule structure.

1826 Anatomist Karl Gegenbaur is born in Würzburg, Bavaria. Early in his career he came under the influence of mentors Köllicker and Virchow, and in turn Gegenbaur exerted great influence on one of his students, Ernst Haeckel (an early defender of Darwinism, and the man who coined the term "ecology"). Gegenbaur's *Elements of Comparative Anatomy* (1878) was the standard text in its day on evolutionary development of animals. This book advanced the idea that homology (i.e., structural similarity between corresponding organs in different species) holds the key to evolutionary history. Gegenbaur proved that eggs and sperm are single cells.

1878 The highest waterspout on record is sighted off Ryde, Isle of Wight, England. The height of the "Spithead waterspout" is measured with a sextant to be "about a mile" (or 5280 feet).

1879 Aviator Claude Grahame-White, founder of the first British flying school (in 1909 at Pau, France), is born in Bursledon, England.

1888 The first commercially successful adding machine is patented (No. 388,118) by William Seward Burroughs of St. Louis. He had originally applied for the patent in January 1886, the same month he incorporated his business as the American Arithmometer Corporation. In 1905 the organization was bought by the Burroughs Adding Machine Company, with capital stock worth $5 million.

1897 "I killed my last mosquito and rushed at his stomach!" So writes Ronald Ross about the climax of a five-year search for malarial parasites in the body of mosquitoes. From his primitive laboratory on a military outpost in Secunderabad, India, Ross is writing to his mentor in London, Patrick Manson. Ross did find his parasites, thus proving that mosquitoes transmit malaria to humans. Ross received the 1902 Nobel Prize for Physiology or Medicine.

1928 A patent for a submersible oil-drilling barge is filed by Louis Giliasso. It was built in 1933 and named for its inventor.

1992 Archaeologists have discovered an ancient Roman house while digging under the Leaning Tower of Pisa, according to wire service reports today. The structure was built between 200 and 300 AD, and was found complete with dinner plates and wine jugs.

1993 Scientists lose contact with the $980 million *Mars Observer* as it is about to reach the Red Planet. The fate of the craft remains unknown.

1994 Les Paul, inventor of the electric guitar, records a concert on the Phonogram, a sound-recording device invented by Thomas Edison over 100 years ago.

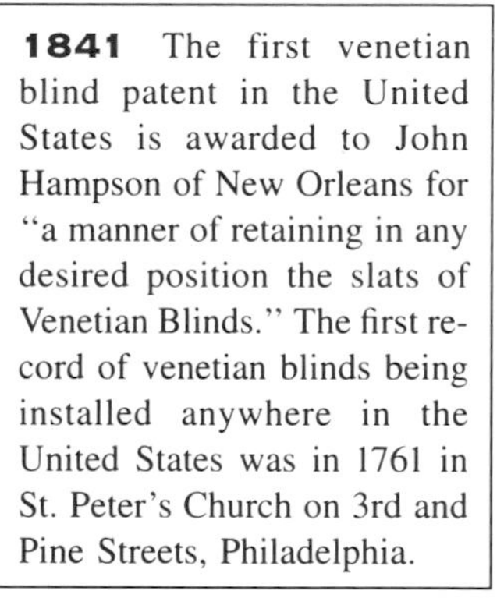

1841 The first venetian blind patent in the United States is awarded to John Hampson of New Orleans for "a manner of retaining in any desired position the slats of Venetian Blinds." The first record of venetian blinds being installed anywhere in the United States was in 1761 in St. Peter's Church on 3rd and Pine Streets, Philadelphia.

1647 French physicist Denis Papin, inventor of the pressure cooker, is born in Coudraies, near Blois. He constructed the device in 1679, and used it to prepare meals for England's King Charles II and for the Royal Society (to which he was admitted in 1680). Papin observed that steam raised the lid of the vessel, which suggested to him that steam could do work. From this he designed the first steam engine; although impractical, others built on his plans and revolutionized human industry and transportation.

1787 The first practical steamboat is demonstrated on the Delaware River to delegates of the Constitutional Convention. The inventor, John Fitch, had already been refused financial backing by the Continental Congress, but was able to build his contraption with private funds from Philadelphia backers.

AUGUST 22

1828 Physician Franz Joseph Gall dies at 70 in Paris. Gall was an important brain scientist who was the first to understand that the brain's gray matter is composed of nerve cell bodies, while the white matter is composed of fibers that carry nervous impulses. He also became convinced that different functions are controlled by different regions of the brain. This conviction led him and others to develop the specious practice of "phrenology" (in which the contours on a person's skull were felt so as to gauge personality and mental powers); but it also presaged Paul Broca's 1861 discovery that the power of speech is controlled by one specific brain center. Since then many other functional centers have been discovered within the human brain.

1834 Samuel Pierpont Langley, the man who almost invented the airplane, is born the son of a merchant in Roxbury, Massachusetts. Between 1897 and 1903, funded by the government, Langley built several devices that flew (using steam engines), but that could not carry a person. The last failure prompted a scathing editorial in the *Times* in New York, which predicted that man would not fly for 1000 years. Nine days later the Wright brothers conducted the first successful manned flight.

1771 America's first advertisement of a dwarf exhibition appears in the Massachusetts *Spy*. For one shilling a spectator could view a man who is 55 years old but just 22 inches high. The show takes place in the home of Widow Bignall, next to the King's Head Tavern in Boston.

1864 The first Geneva Convention is signed by 12 nations in Geneva, Switzerland. The Red Cross is founded.

1911 *Mona Lisa* (the masterwork of artist-inventor Leonardo da Vinci) has been stolen, announce authorities of Paris's Louvre Museum. The painting was recovered two years later in Italy.

1939 The aerosol can is patented. Julian Seth Kahn of New York City invented the first disposable can for dispensing liquids under pressure; today he receives patent No. 2,170,531 for an "apparatus for mixing a liquid with gas." The key to his invention is a cheap valve. His device was originally used for paints, poisons, and whipped cream (not in the same can).

1957 The first session of the Potomac (Environmental) Enforcement Conference is held. It concludes that the Potomac River in the nation's capital presents sight and odor problems, and is unfit for human or animal use.

1963 "Frustrate a Frenchman, he will drink himself to death; an Irishman, he will die of angry hypertension; a Dane, he will shoot himself; an American, he will get drunk, shoot you, then establish a million dollar aid program for your relatives. Then he will die of an ulcer." These are the observations of psychologist Dr. Stanley Rudin (Dalhousie University) on the correlation between nationality and psychological problems in a recent speech to the International Congress of Psychology, reported in today's *New York Times*.

1994 One hundred sea turtles are accidentally killed by ecology students trying to protect them (see July 12, 1994, for details).

1994 Prosecutors in the O.J. Simpson double-murder case play their seemingly most-damning card: DNA evidence. A trail of blood leading from the murder scene has the same genetic makeup as that of the defendant, announce prosecutors in court papers.

AUGUST 23

1769 The great anatomist Baron Georges Cuvier is born the son of a soldier in Montbéliard, France. His extensive studies/writings comparing the same structures in different species are said to have founded the science of comparative anatomy. Because he was the first to extend classification to fossil forms, he is also credited with founding paleontology.

1829 Moritz Cantor is born in Mannheim, Germany. His *Lectures on the History of Mathematics* (1880–1908) is possibly the finest work ever produced on mathematical development from prehistory up to 1799.

1859 The first U.S. hotel with an elevator opens. The six-story Fifth Avenue Hotel in New York City boasts the convenience, which uses an Archimedes screw mechanism to move the car up and down. The first U.S. elevator of any type was invented in 1850 by Henry Waterman, also in New York City; it was installed to carry barrels in the Cherry Street mill of Hecker and Brother.

1888 Philip Henry Gosse, inventor of the institutional aquarium, dies at 70 in Devon, England. He was born the son of an impoverished miniaturist in Worcester, England. This artistic background served Gosse well in his career as a self-taught naturalist; his illustrated writings elevated popular appreciation of nature, especially marine biology. Gosse established the world's first public aquarium in Regent's Park, London.

1904 Harry D. Weed patents snow chains for car tires. The Canastota, New York, inventor receives patent No. 768,495 for a "grip-tread for pneumatic tires." The product was first manufactured by the Weed Chain Tire Grip Company, which was eventually bought by the American Chain and Cable Company.

1931 Microbiologist Hamilton Othanel Smith is born in New York City. In 1970, while studying how one particular bacterium absorbs DNA from a virus, Smith and colleagues discovered an enzyme that breaks DNA at one specific point. This was the first "type II restriction enzyme," substances that now allow geneticists to make and break strings of DNA of their own design. Smith won the 1978 Nobel Prize.

1933 Boxing is first televised. Archie Sexton and Laurie Raiteri fight an exhibition match at Broadcast House in London.

1940 Speed monster Gary Gabelich is born in Long Beach, California. In 1970 he drove an automobile (rocket-powered) over 650 mph; this still stands as the fastest land speed measured with calibrated equipment.

1961 *Ranger I* is launched. It is the first in a program of nine launches intended to lay the foundation for a manned assault on the moon. *Ranger I* was supposed to orbit the moon, but the top rocket stage failed to ignite, and the satellite stayed in an Earth orbit.

1966 Earth is photographed from the moon for the first time. *Lunar Orbiter* 1 sends back 207 photos. This ship was the first U.S. vessel to orbit the moon, and it also photographed all nine primary landing sites for the Apollo missions that landed on the moon. It had taken off from Earth on August 10, and was eventually disposed of by crashing it into the moon's far side on October 29, to clear the way for subsequent ships.

1992 Diseased brain cells can be repaired with the use of the herpesvirus, announce scientists with the University of Pennsylvania and the Wistar Institute at a press conference. "The herpes virus [which normally causes cold sores] is uniquely able to transfer genetic material to a brain cell without killing it," says researcher Nigel Fraser. Once inside the brain cells, the new genes help the cells produce an enzyme, the absence of which results in a brain disorder called Sly disease, which leads to mental retardation. At this stage the work has only been on mice; "major advances" are needed before it will be tried on humans.

1993 "Naturemade Aphrodite," the oldest goat on record, dies. She lived 18 years 1 month. Owned by Katherine Whitwell of Moulton, England, she was bred for 10 years, and delivered 26 kids (including five sets of triplets and one set of quads).

1804 One of the earliest adventures in space science occurs. Jean-Baptiste Biot and Joseph Gay-Lussac load a balloon gondola with scientific equipment and small animals, and make an ascension that establishes that Earth's magnetic field is undiminished at great heights.

79 Long-dormant Mt. Vesuvius erupts, burying the Roman cities of Herculaneum and Pompeii in ash and lava, and killing approximately 20,000. When uncovered centuries later, the scene was an archaeological treasure.

1830 The American Institute of Instruction, the country's first national education association, adopts its constitution in Boston. Fifteen states are represented at the convention. The association incorporated on March 4, 1831; Francis Wayland, Jr. (president of Brown University) was AII's first president.

1831 "My dear Darwin, I shall hope to see you shortly fully expecting that you will eagerly catch at the offer which is likely to be made you of a trip to Tierra del Fuego & home by the East Indies." Thus begins one of the momentous letters in the history of science. Cambridge Professor J.S. Henslow informs Charles Darwin (age 22 and already having quit careers in medicine and the ministry) that he will be offered a job as an unpaid naturalist on the five-year scientific voyage of the HMS *Beagle*. Darwin accepted, although both his father and the ship's captain opposed his participation.

AUGUST 24

1869 The waffle iron is patented by Cornelius Swarthout of Troy, New York. It is patent No. 94,093.

1894 Rudolf Oskar Robert Williams Geiger, a founder of microclimatology (the study of climate conditions within a few meters of ground surface, examining the complex interactions between vegetation, heat, radiation, water, soil, air, and animals), is born in Erlangen, Germany.

1898 Cell biologist Albert Claude is born in Longlier, Belgium. He brought two powerful instruments to the study of the cell: the centrifuge (with which he separated the different organelles of the cell) and the electron microscope (which showed him the fine structure of the organelles). His accomplishments were rewarded with the 1974 Nobel Prize for Physiology or Medicine.

1930 Naturalist-nature illustrator Anna Botsford Comstock dies at 75 in Ithaca, New York. Her career in biology began at age 24 (1878) with her marriage to the eminent entomologist John Henry Comstock. She illustrated his lectures and books on insects. She went on to establish herself as a first-rate naturalist. In 1895 she developed a pioneering nature-study course in Westchester County, while on the New York State Committee for the Promotion of Agriculture. She then was hired to teach nature study at Cornell. As a member of several national societies, she helped form policies on nature education. Her *Handbook of Nature Study* (1911) saw 24 editions and was translated into eight languages.

1932 Amelia Earhart completes the first nonstop flight across the United States by a woman, and exactly 39 years later (1971) Brereton Sturtevant is sworn in as the first female examiner-in-chief of the U.S. Patent Office.

1968 France explodes its first thermonuclear H-bomb, in the South Pacific. She becomes the fifth nation to demonstrate the device.

1935 History's most durable radio program about sports debuts. The program is *The Tenpin Tattler*, first broadcast on WCFL, Chicago, now broadcast on WGN. It still has the same host, Sam Weinstein, making him radio's most durable host for any type of program.

1981 The filmless camera (replacing chemical solutions with electromagnetic imagery) is presented to the press. The device is the Mavica, made by Sony.

1989 Launched 12 years earlier, *Voyager 2* comes within several thousand miles of Neptune.

1994 The first group of refugees to return to Rwanda after its bloody civil war are met and attacked by a mob of Hutu extremists. One refugee is nearly killed and the rest scatter through the bush. The action causes the United Nations to temporarily halt further repatriations. This first group of bold emigrants were all personnel from the world-famous gorilla reserve, the Karisoke Research Center in northeastern Rwanda. While outside Rwanda, they lived on a gorilla preserve in Zaire. When attacked, they were trying to get back to Karisoke and its endangered animals.

1994 A 21-year-old woman with AIDS flags down a car driven by a 90-year-old man in West Palm Beach, Florida, and asks for money. Shortly thereafter the panhandler makes her way into the man's car, and bites him several times in the ensuing struggle. The man later developed AIDS, becoming the first documented case of AIDS transmission by biting.

79 Scholar-natural historian Pliny the Elder, 56, dies of asphyxiation while studying the eruption of Mt. Vesuvius.

1793 Philippe Pinel is appointed chief physician of Bicêtre, a Paris asylum for the insane. One of his first acts is to unchain the inmates. These and other reforms in his "moral therapy" changed mental health care forever.

1831 The first U.S. patent on a process for manufacturing bedsprings is issued to Josiah French of Ware, Massachusetts. The first box springs in the United States were imported from France in 1857 by James Boyle, a New York City bedding manufacturer.

1841 Emil Theodor Kocher, Swiss surgeon and 1908 Nobel laureate, is born in Bern. He invented/improved many surgical procedures, including the removal of the thyroid to treat goiter.

AUGUST 25

1900 Sir Hans Adolf Krebs, discoverer of the Krebs cycle (the citric acid cycle, in which energy in foods is harvested by the body), is born the son of a Jewish physician in Hildesheim, Germany. Hitler's rise forced him to England, where he won the 1953 Nobel Prize for Physiology or Medicine.

1908 Henri Becquerel, cowinner of the third Nobel Prize in Physics, dies at 55 in Le Croisic, France. His grandfather, father, and son were all eminent physicists; in 1891 Henri assumed the post of physics professor at Paris's Museum of Natural History, which both of his progenitors held. He also took over their research into fluorescence, which led in 1896 to an accidental discovery of explosive proportions: radioactivity. In late February Becquerel was experimenting with a uranium salt, trying to determine if sunlight would induce it to give off the mysterious X rays that Roentgen had discovered several months before. When the Paris skies became overcast, Henri put his experiment in a drawer, with a photographic plate wrapped in black paper sitting under some of the uranium crystals. After several days the sun still had not reappeared, but Becquerel became impatient and decided to develop the photographic plate anyway. To his amazement, the uranium had left a clear image, even though no sun had struck it. He had discovered that the energy was in the rocks themselves. Marie Curie (who shared the 1903 Prize with Becquerel and with her husband Pierre) later named this energy "radioactivity."

1916 The National Park Service is established within the Department of the Interior. Until now, the U.S. Army had managed the national preserves.

1940 The country's first parachute wedding is held. Arno Rudolphi and Ann Hayward take the plunge above the World's Fair amusement area in New York City. Attending are the Reverend Homer Tomlinson (of the Church of God, Jamaica, New York), the best man, the maid of honor, and four musicians, all of whom are suspended in parachutes 50 feet above ground.

1956 "We are recorders and reporters of the facts—not judges of the behavior we describe." Sex researcher Alfred Charles Kinsey dies at 62 in Bloomington, Indiana. Like Freud, Kinsey started out as a zoology researcher who switched to studying the behavior of man. Kinsey moved from teaching at Harvard to Indiana University where he founded the Institute for Sex Research. He and colleagues interviewed 18,500 adult males and females to determine human sexual habits. It was history's most extensive, most scientific exploration of the topic, but it has been criticized because it relied on people's reports of their own behavior.

1981 Four years and five days after launch, *Voyager 2* makes its closest approach to Saturn (100,250 miles from its center), and heads to Uranus.

1993 A widely used blood test to detect colon cancer recurrence is declared "worthless" in the *Journal of the American Medical Association*. The "carcinoembryonic antigen" (CEA) test misses many recurrences, it indicates the presence of malignancy where none exists, and when it is correct it is often too late to do any good. These are the findings of an eight-year, multistate study of 1216 cancer patients by the Mayo Clinic.

1994 "An ounce of prevention is worth a pound of cure" seems to be true in obesity, as in many other aspects of health care, according to a report delivered on this date by physiologist Barbara Hensen (University of Maryland) to the Seventh International Congress on Obesity in Toronto. Persons who avoid weight gain in middle age slash the risk of heart disease, and they may prevent diabetes completely.

1728 Mathematician Johann Heinrich Lambert, the first to prove that pi is irrational (i.e., cannot be expressed as the quotient of two integers), is born the son of a poor tailor in Mülhausen, Alsace. Among his other achievements were creating the first methods to measure light intensity accurately; the unit of intensity is called the lambert.

1740 Joseph-Michel Montgolfier, coinventor of the hot-air balloon, is born one of sixteen children of a successful paper manufacturer in Annonay, France.

1843 The first typewriter that actually typed is patented (No. 3,228) by Charles Thurber of Norwich, Connecticut. Known as "Thurber's Patent Printer," the device was invented as an aid for the blind. Inking was done by roller, however, which made the machine slow and impractical. In 1867 Christopher Latham Sholes invented the first successful typewriter. It was called the Type-Writer, a term coined by Sholes. The device was originally manufactured by E. Remington & Sons of Ilion, New York.

AUGUST 26

1873 A public school kindergarten is authorized for the first time in the United States, by the St. Louis Board of Education. In September 42 youngsters attend the opening of the Des Peres School. Susan Blow was the nation's first kindergarten teacher.

1873 Inventor Lee De Forest is born the son of a minister in Council Bluffs, Iowa. One of Lee's 300 patents was for the triode (the forefather of the transistor, the triode is the basis of the radio tube, which made radio and many other electronic devices practical).

1883 At 1 PM Krakatoa (east of Java) begins erupting in a series of increasingly violent explosions. The blasts are heard 2200 miles away in Australia. Ash is shot 50 miles into the atmosphere, and pressure waves are recorded around the entire planet. Fine dust from Krakatoa circled Earth several times, causing spectacular red sunsets for the next year.

1743 Antoine-Laurent Lavoisier, the founder of modern chemistry, sometimes called "the Newton of chemistry," is born the son of a prosperous lawyer in Paris. At his laboratory in the Royal Arsenal in Paris in the early 1790s, Lavoisier measures gas exchange during human breathing. It was one of many experiments that earned Lavoisier the title "the father of modern chemistry." In the foreground of this painting is his constant collaborator, his wife Marie Paulze; she recorded his data, translated English science writings, and illustrated his experiments and publications. Lavoisier and Marie's father were guillotined within minutes of each other during the French Revolution.

1906 Microbiologist Albert Bruce Sabin, creator of the Sabin polio vaccine, is born in Bialystok, Russia (now in Poland).

1909 "It's a man, it's a man!" shouts an excited worker in the fossil-hunting party of Otto Hauser in the Combe-Capelle grotto, France. The skeleton turns out to be an almost perfectly preserved example of Cro-Magnon man, 34,000 years old, from the dawn of the emergence of *Homo sapiens sapiens*.

1910 William James, a groundbreaker in both psychology and philosophy, dies at 68 in his summer home in Chocorua, New Hampshire. William was plagued by poor health throughout life, and he underwent a suicidal breakdown in his mid-20s. He had trouble deciding on a career. In 1872 he was hired as a physiology teacher by Harvard, but in 1876 took the radical step of switching fields to psychology, then in its infancy. James's monumental *The Principles of Psychology* (1890) did much to bring biology and physiology to the social sciences. James then turned his interest to the psychology of religion. His classic The *Varieties of Religious Experience* (1902) remains in perpetual reprinting. James ended his intellectual journey by founding the influential school of Pragmatism in the early 1900s.

1993 One of the great archaeological mysteries of the twentieth century—what happened to the golden treasure of ancient Troy after Soviet troops overran Berlin in 1945?—is solved today in Moscow. Authorities announce that the objects are safe and will soon be displayed.

1650 Englishmen venture west of the Allegheny Mountains for the first time. An expedition departs Fort Henry at the falls of the Appomattox River in Virginia. The eight-man party returns eight days later. Making the trip are Captain Abraham Wood, merchant Edward Bland, Elias Pennant, Sackford Brewster, two servants, and two Indian guides (Oyeocker, a Nottaway chief, and Pyancha, an Appamattuck war captain).

1667 A cyclone is recorded for the first time in American history, at Jamestown, Virginia. The storm caused "such violence that it overturned many houses burying in the ruins much goods and many people, beating to the grounds such as were any ways employed in the fields, blowing many cattle that were near the sea or rivers, into them, whereby unknown numbers have perished. To the great affliction of all people...the sea swelled twelve foot above the normal eight, drowning the whole country before it, with many of the inhabitants, their cattle and feed."

1850 Physicist Augusto Righi who unequivocally established that radio waves and light show the same properties, and therefore firmly established the existence of the electromagnetic spectrum, is born in Bologna, Italy.

1859 The first-ever oil rush starts when Edwin L. "Colonel" Drake strikes "black gold" in Titusville, Pennsylvania, at a depth of 69½ feet. It is the country's first successful oil well and the birth of its oil industry. Dozens of derricks are erected in the next few months.

1874 Chemist Carl Bosch, who achieved greatness in both the fertilizer and explosives industries, is born the son of an engineer in Cologne, Germany. He codeveloped the Haber–Bosch process of making ammonia—critical to both industries—from atmospheric nitrogen. For his achievement he was awarded a Nobel Prize in 1931.

1883 The moon appears green for the first time on record. The volcanic eruption of Krakatoa is blamed.

1900 "If there is anything to the mosquito theory, I should get a good dose of yellow fever!" writes 46-year-old surgeon James Carroll from Quemados, Cuba, to his boss in Washington, Walter Reed. Earlier in the day Carroll, a father of five, had allowed Dr. Jesse Lazear to apply an infected mosquito to his arm. Carroll indeed became very sick, proving Reed's theory that "yellow jack" is carried to man by the insect. This discovery produced an effective strategy to eradicate the disease. Although once close to death, Carroll did recover; Lazear was not so lucky, and he perished after being accidentally infected by one of the mosquitoes during the research.

1913 Biochemist Martin David Kamen, the first to isolate carbon-14 (a substance extremely important in dating ancient objects), is born in Ontario, Canada.

1939 The first flight by an airplane powered by a turbojet engine occurs at Marienehe, Germany.

1958 Ernest Orlando Lawrence, inventor of the cyclotron, dies at 57 in Palo Alto, California (see August 8, 1901, for a biographical sketch).

1962 *Mariner 2*, the first space probe to reach Venus, is launched.

1984 President Ronald Reagan announces that the first "citizen astronaut" will be schoolteacher Christa McAuliffe (an eventual casualty in history's worst shuttle disaster 1½ years later).

1992 Five-month-old infants can add and subtract, according to a series of experiments reported in today's *Nature* by psychologist Karen Wynn (of the University of Arizona at Tucson). A typical experiment went as follows: A baby observed one Mickey Mouse doll, the view of which was then blocked with a screen while an experimenter clearly added another doll to the first. When the screen was removed and the baby's view restored, the infant was surprised if there were three dolls instead of two, indicating that he knew that one plus one should equal two, not three. "The appearance of this paper is a notable event in the history of developmental psychology," declared Oxford psychologist Peter Bryant.

1994 Four women complete a cross-country trip in a van powered by used french-fry oil. "The Lard Car," driven by the self-proclaimed "Greasy Riders," left New York on August 10, and averaged 24 mph, the 1983 blue Chevy van reached San Francisco late on this Saturday night. "This is great, exciting," proclaims driver Sara Lewison at trip's end. Her copilots were Nicole Causino, Julie Konop, and Gina Todus. Their stunt was part of a "Fat of the Land" recycling project.

1609 Henry Hudson "discovers" Delaware Bay.

1749 The versatile genius Johann Wolfgang von Goethe is born in Frankfurt am Main, Germany. He was the son of a lawyer, and took a law degree himself in 1771, but never practiced. A prolific author and thinker right to the end, Goethe published the second part of his great drama, *Faust*, in the last year of his life. He is best known as a poet and dramatist, but he also wrote on biology and botany, he advanced a theory on the nature of color, and the abundant mineral goethite is named for him. He coined the term "morphology" to cover the scientific study of form.

1830 The first locomotive made for carrying passengers in the United States is given its first test in Baltimore. Designed and built by Peter Cooper, the 6-ton *Tom Thumb* carries 26 people 13 miles on tracks of the Baltimore and Ohio Railroad. The outgoing trip takes 1 hour 15 minutes, the return trip with 30 passengers takes just 61 minutes, including a 4-minute stop for water. In November a second passenger locomotive, the *Best Friend*, derailed during its maiden voyage.

AUGUST 28

1850 The first cable under the English Channel is ruptured, allegedly cut by fishermen who thought they had discovered algae filled with gold. The following year, a heavily reinforced cable was laid between Calais and Dover by the English ship *Blazer* under the direction of English engineer Jacob Brett (and financed by Frenchman Louis-Napoleon Bonaparte).

1878 Physician George H(oyt) Whipple is born in Ashland, New Hampshire. While studying how the body produces new blood cells, Whipple discovered in 1920 that dietary liver helped cure anemia in dogs; this led George Minot and William Murphy to a cure for pernicious anemia in humans, and the three shared the 1934 Nobel Prize for Physiology or Medicine.

1880 Charles Thomas Jackson, one of science's more colorful and more tragic figures, dies at 75 in Somerville, Massachusetts (see June 21, 1805, for a biographical sketch).

1964 The satellite *Nimbus 1* is launched from Point Arguello, California. It later provided the first high-resolution nighttime cloud-cover photographs of weather formations. It sent back photos of Hurricane Cleo its first day in orbit, and later viewed Hurricane Dora and Typhoon Ruby in complete darkness.

1798 The first successful vineyard in the United States is established by John James Dufour 25 miles from Lexington, Kentucky. The 630-acre plot is called "The First Vineyard" by Dufour. Colonists had attempted to establish vineyards as early as 1619.

1965 Nine aquanauts take up residence in the underwater laboratory/residence *Sealab II* off the California coast. They spend a month on the ocean floor. Their leader is M. Scott Carpenter, one of the original Mercury astronauts. Carpenter was the first human to experience the two most forbidding habitats known to man: outer space and the ocean. "I understand there is some euphoria under water ... there is some in space flight too," he later said. "The ocean is a much more hostile environment than space."

1981 John Hinckley, Jr., pleads innocent by reason of insanity in the shooting of President Ronald Reagan.

1981 The Centers for Disease Control announces a task force to investigate high rates of Kaposi's sarcoma and *Pneumocystis carinii* pneumonia in homosexual men. AIDS was later determined the culprit.

1992 A major record in horse racing falls at the Pocono Downs track near Wilkes-Barre, Pennsylvania: The nine-year-old gelding Treboh Joe losses his 167th race in a row, making him history's losingest horse. At retirement in 1994, Treboh Joe had just one win in 247 starts; his lifetime earnings totaled $16,438.

1994 The *Galileo* space probe passes the asteroid Ida, and discovers what seems to be an orbiting moon. It is the first time an asteroid moon has been sighted.

AUGUST 29

1809 Oliver Wendell Holmes is born the son of a minister in Cambridge, Massachusetts. Holmes is best known as an essayist-poet and father of a famous Supreme Court judge of the same name, but he was firstly a physician who coined the term "anesthesia" and who discovered the contagious nature of childbed fever.

1821 Archaeologist Gabriel Mortillet, the first to organize man's prehistoric cultural development into a sequence of epochs, is born in Meylan, France.

1828 The first U.S. patent for a brake of any kind is issued to Robert Turner of Ward (now Auburn), Massachusetts, for a "self-regulating wagon brake." Ten years later the first U.S. patent for a railroad brake was issued to Ephraim Morris (Bloomfield, New Jersey) for "eccentric brakes for cars."

1831 Charles Darwin, 22 years old and just graduated from Cambridge, returns home from a geology field trip and finds several letters informing him that he will soon be invited to serve as a naturalist on the round-the-world scientific voyage of HMS *Beagle*. It is a dream come true, but Darwin initially turned down the invitation.

1865 Neurologist-embryologist Robert Remak dies at 50 in Kissingen, Germany (see July 26, 1815, for a biographical sketch).

1876 Inventor Charles F(ranklin) Kettering is born the son of a farmer in Loudonville, Ohio. His name is enshrined in the Sloan-Kettering Institute for Cancer Research and other research foundations His fortune largely came from a series of inventions (most notably the electric starter for cars) that developed the automotive and other industries.

1896 Chop suey is invented. The chef for Chinese Ambassador Li Hung-chang creates the dish in New York City to appeal to both U.S. and Chinese tastes; it was unknown in Asia at the time. The chef, the ambassador, twenty-two servants, five valets, two other cooks, and a barber had just arrived in New York City the previous day and had been greeted by President Grover Cleveland.

1962 A major milestone in the environmental movement and in the life of Rachel Carson occurs when President John F. Kennedy publicly acknowledges the importance of Carson's *Silent Spring* at a press conference. Question: "Mr. President, there appears to be a growing concern among scientists as to the possibility of dangerous long-range side effects from the use of DDT and other pesticides. Have you considered asking the Department of Agriculture or the Public Health Service to take a closer look at this?" Kennedy: "Yes, and I know that they already are. I think particularly, of course, since Miss Carson's book, but they are examining the matter." The president soon ordered a federal study of the issue, which eventually led to the banning of DDT in the United States.

1965 Before *Gemini 5* splashes down today, Gordon Cooper speaks from the satellite 100 miles above Earth to M. Scott Carpenter, who is in *Sealab II* 205 feet beneath the surface of the Pacific Ocean. It is history's first conversation between an astronaut and an aquanaut.

1990 The *Federal Register* publishes a rule written by the National Marine Fisheries Service that bans feeding of dolphins by humans in the wild, and bans dolphin-feeding cruises. Such cruises are popular and profitable in the Gulf of Mexico. The decision is definitely not popular, but one that put the interests of the dolphins above the interests of charter-boat operators. Man is by far the greatest natural predator of marine mammals; bringing these animals closer to humans, as artificial feeding inevitably does, can result in any number of hazardous outcomes for the dolphins.

1991 The "first evidence for an association between stress and a biologically verifiable infectious disease" is published in the *New England Journal of Medicine*. Researchers in Salisbury, England, exposed humans to five different types of cold virus, and found that people in high-stress conditions were more likely to catch each type of cold than people with lesser degrees of stress in their lives.

1904 German surgeon Werner Forssmann is born in Schopfheim. He was the first to successfully catheterize the human heart (a procedure in which a tube is sent into the heart through a vein, for purposes of analyzing the structure and function of the organ; Forssmann developed the technique on his own body). He won the 1956 Nobel Prize for the achievement.

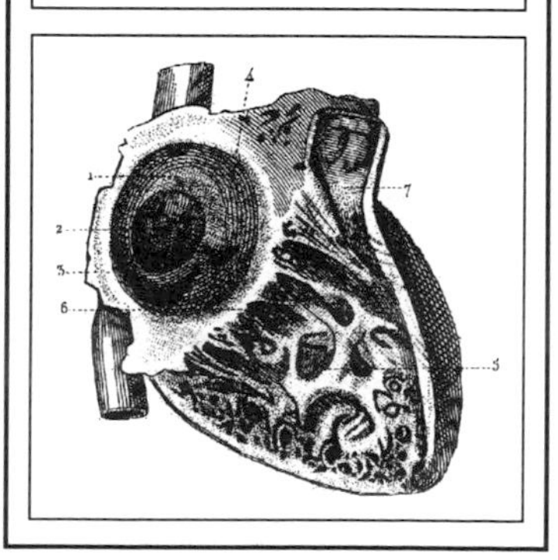

1779 Russian explorer Fabian Gottlieb von Bellingshausen, the first to circumnavigate Antarctica, is born in Arensburg on the island of Ösel.

1797 Mary Wollstonecraft Shelley, author of *Frankenstein*, is born in London.

1842 The nation's first federal tariff on the importation of opium is levied. A duty of 75 cents must be paid on each pound that is brought into the country. Prior to this, no tax was charged. A tariff to prevent the importation of obscene literature is also levied on this date. On April 1, 1909, it became illegal to bring opium into the United States "other than smoking opium for medicinal purposes."

1852 Jacobus Henricus van't Hoff, winner of the first Nobel Prize for Chemistry (in 1901), is born the son of a physician in Rotterdam, the Netherlands.

AUGUST 30

1871 Physicist Ernest Rutherford is born the second of twelve children of a wheelwright-farmer in Brightwater, New Zealand. Rutherford laid the groundwork for modern nuclear physics by exploring radioactivity, discovering the alpha particle, and advancing the modern theory of atomic structure. He won the 1908 Nobel Prize for Chemistry.

1884 The(odor) Svedberg (Nobel Prize winner in 1926 for Chemistry) is born in Fleräng, Sweden, the son of a civil engineer. Svedberg invented the ultracentrifuge (which generates forces that are hundreds of thousands of times greater than the force of gravity), thereby making possible extremely precise studies of proteins and other chemicals. He also helped develop the cyclotron and the process of electrophoresis, two other important advances in chemical analysis.

1890 A federal meat inspection law is enacted for the first time in the United States. 26 Stat. L. 414 states that bacon, salted pork, swine, cattle, and sheep for export are now subject to inspection.

1907 Engineer John William Mauchly, codesigner of ENIAC (the first practical electronic digital computer) and UNIVAC (the first data processor to use magnetic tape), is born in Cincinnati, Ohio.

1925 Eminent Canadian paleontologist William Edmund Cutler dies of blackwater fever during a research expedition in Tendaguru, Tanzania. It was Cutler's first trip to Africa. It was also the first professional fossil-finding expedition for Louis Leakey, 22, whose numerous discoveries later in his fabled career were to rewrite the history of man's origins. Leakey was born in Kenya and spoke several African languages fluently. On this first expedition, he managed all of the native personnel, he was to write several articles, and he learned much about fieldwork from Cutler.

1954 President Dwight Eisenhower signs a liberalized Atomic Energy Act, moving some control over atomic energy from the military to the civilian sector.

1963 The "Hot Line" communications link between Moscow and Washington becomes functional.

1983 Guion S. Bluford, Jr. becomes the first black American in space, flying aboard the shuttle *Challenger*. (The first-ever black in space was the Cuban Arnaldo Tamayo Méndez in September 1980).

1984 The shuttle *Discovery* makes its maiden voyage, after several postponements in the previous two months.

1987 The shuttle *Challenger* makes its first full-scale test firing (near Brigham City, Utah), after being redesigned in the wake of the January 1986 disaster in which seven astronauts perished.

1994 In a major political embarrassment, as well as a deep personal tragedy, the son of U.S. Surgeon General Joycelyn Elders is sentenced in Little Rock, Arkansas, to ten years in prison for selling cocaine.

1993 Simon LeVay of the Salk Institute in San Diego publishes in *Science* a comparison of brain anatomy of homosexual and heterosexual men; the homosexual hypothalamus is smaller than the heterosexual one. This finding supports the contention that sexual orientation is an inherent quality, decided at birth. Such a finding, if verified, could expand legal protection and societal acceptance of homosexuality.

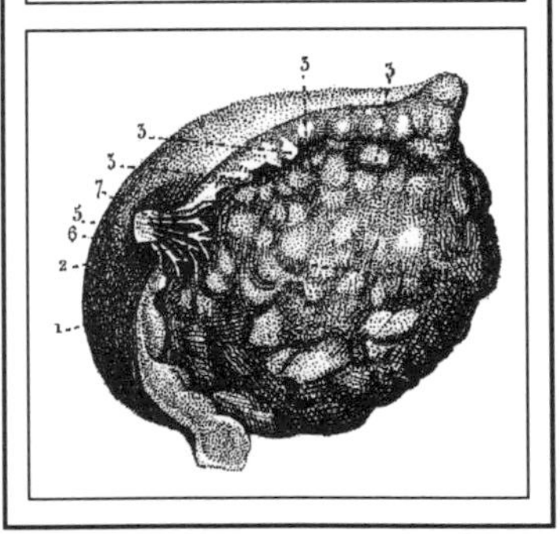

1663 Physicist Guillaume Amontons, famous for inventing scientific instruments (especially thermometers, hygrometers and barometers), is born the son of a lawyer in Paris. Like Edison, he was hard of hearing most of his life; both men used this to their advantage to shutout interruptions while creating.

1786 Chemist Michel-Eugène Chevreul is born in Angers, France (see April 9, 1889, for a biography).

1842 Micah Rugg patents the country's first nut and bolt machine of importance. His device (patent No. 2,766) trims the nut and bolt heads. Rugg and Martin Barnes had established a factory in Marion, Connecticut, in 1840, which produced 500 bolts a day. Prior to this each nut and bolt was individually hammered out by a blacksmith.

1877 The battleship *Temeraire* is completed at Chatham, Great Britain. Its sails the largest in the history of sailing. The fore and main yards are 115 feet long, the foresail weighs 2.23 tons (5100 feet of canvas was used), and the total sail area is 25,000 square feet.

1886 The first earthquake of importance in U.S. history rocks Charleston, South Carolina. Over 100 lives are lost, and 90% of the city's 6596 brick buildings are damaged.

1887 Thomas Edison patents his "Kinetoscope," a device for producing moving pictures.

1887 Chemist Friedrich Adolf Paneth is born the son of eminent physiologist Joseph Paneth in Vienna. Friedrich specialized in radioactive compounds, and with them he worked out one of the first methods to accurately date rocks (achieved in the 1920s), and developed the first radioactive tracer techniques (in 1912–1913 with George Charles de Hevesy).

1900 Private William Dean allows himself to be fed on by Mosquito Number 12 in the research facility of Walter Reed in Quemados, Cuba. Dean was the first volunteer to expose himself to a proven carrier of yellow fever.

1909 Chemotherapy is born—and syphilis is conquered. In his Frankfurt lab, Paul Ehrlich directs assistant Sahachiro Hata to inject an infected rabbit with "preparation 606" (it being the 606th chemical that Ehrlich's team concocted and tested). The compound works like a "magic bullet"; the sores on the rabbit clear up immediately. Already a Nobel laureate, Ehrlich's work marked the beginning of modern chemotherapy (a term Ehrlich coined).

1913 Radio astronomer Sir Bernard Lovell is born in Oldland Common, England.

1916 Radio astronomer Robert Hanbury Brown is born in Aruvankadu, India.

1955 A sun-powered car is demonstrated for the first time in the United States. Built by William G. Cobb of General Motors, the 15-inch "sunmobile" is powered by 12 selenium photoelectric cells that convert sunlight to electricity, which then powers a tiny electric motor connected to a drive shaft. The exhibition is held at the General Motors Powerama in Chicago.

1993 Inventor Nelson E. Camus demonstrates a urine battery. At an exhibition in Los Angeles, Camus and coinventor Edgardo Aguayo generate only enough power to light a small bulb, but they claim that urine, when mixed with lithium and soil, could run every appliance in an average home.

1993 Researchers at the National Institutes of Health announce that they have developed a skin patch test for Alzheimer's disease. The test is especially valuable because it can distinguish between the incurable Alzheimer's and other types of mental impairment that are treatable.

1821 Hermann von Helmholtz is born the son of a schoolteacher in Potsdam, Prussia. Helmholtz was responsible for fundamental discoveries and theories in many areas: physiology, optics, sensory perception, electrodynamics, and meteorology. He is best known for the law of conservation of energy, which he formulated in 1847.

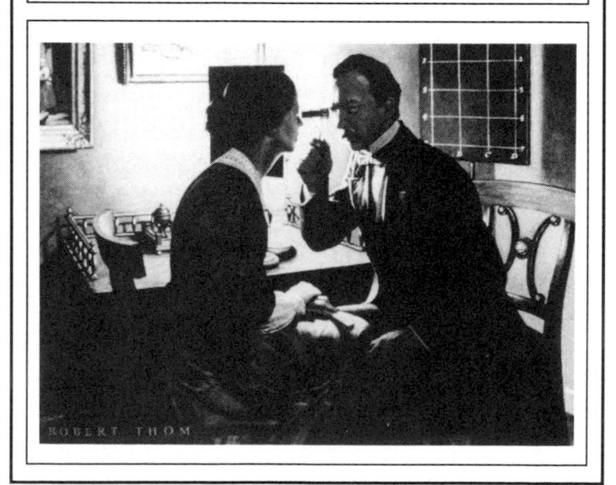

1836 The first women known to have crossed North America, Narcissa Prentiss Whitman and Eliza Hart Spalding, complete their journey in Walla Walla, Washington. They are accompanied by their husbands, Presbyterian missionaries Dr. Marcus Whitman and Rev. Henry Harmon Spalding. The party had crossed the Continental Divide on the Fourth of July, 1836.

1854 Naturalist-nature illustrator Anna Botsford Comstock is born in Otto, New York (see August 24, 1930, for a biographical sketch).

1869 Daily weather bulletins are first posted in the United States by Cleveland Abbe from an observatory in Cincinnati, using telegraphic reports from all over the country. A national weather service was begun shortly thereafter, and Abbe was its chief meteorologist until 1916.

1877 Francis William Aston is born the son of a merchant in Harborne, England. He discovered nonradioactive isotopes and invented the mass spectrograph, which is used for dispersing radiation so as to photograph or map the spectrum, and for which he was awarded the 1922 Nobel Prize for Chemistry.

SEPTEMBER 1

1897 The country's first municipal subway, in Boston's Tremont Street Subway, opens. Begun in 1895, the first line ran from Public Gardens to Park Street. The subway was built by the City of Boston for $4,369,000, and was leased annually for 4½% of this cost to Boston Elevated Railway.

1906 Chemist Karl August Folkers, leader of the research team that first isolated vitamin B_{12}, is born in Decatur, Illinois. The remarkable B_{12} is the only known treatment for pernicious anemia; it contains cyanide; it was the first naturally occurring compound known to contain cobalt; and the human body requires it in far smaller amounts than ordinary vitamins.

1914 The last passenger pigeon on Earth, a 29-year-old female named Martha, dies in the Cincinnati Zoo. The species is now extinct. Once "probably the most abundant bird in the whole world," the passenger pigeon had not been seen in the wild since 1899. Repeated attempts at captive breeding programs had failed completely. Martha remains on display at the National Museum in Washington, D.C.

1923 The greatest physical damage ever caused by an earthquake results from the one that strikes Japan's Kanto Plain today, destroying 575,000 dwellings in Tokyo and Yokohama and killing 150,000 people.

1931 The country's first public anthropology laboratory opens in Santa Fe, New Mexico. The Laboratory of Anthropology is directed by Jesse Logan Nusbaum.

1994 Avis car rental company announces that it will be the first to equip rental cars with hand controls in compliance with the Disabilities Act of 1990.

1994 The U.S. Fish and Wildlife Service announces that the northern spotted owl will continue to be listed as an endangered species. The battle between wildlife advocates and loggers continues. Gilbert Murray, president of the California Forestry Association, said, "The U.S. Fish and Wildlife Service and Secretary Babbitt have proved that good science and public policy are not high on their agenda." Richard Hoppe of the Wilderness Society responded, "I think commodity interests in this country are grabbing onto any and every straw that will promote their agenda of bringing down the federal environmental laws of this country."

1818 Samples from the ocean floor at a depth of over 1000 fathoms are taken for the first time. English explorer Sir John Ross, leading an expedition of the HMS *Isabella*, in search of a Northwest Passage from the Pacific to the Atlantic Ocean, uses a tool of his own invention to unearth unknown worm species and mud samples from the floor of Baffin Bay.

SEPTEMBER 2

1609 Henry Hudson enters New York Bay—at present-day New York City at the mouth of the Hudson River—during his *Half Moon* expedition in search of a Northwest Passage between the Atlantic and Pacific Oceans.

1851 Frenchman Pierre Carpentier applies for a patent on a machine to produce metal sheets that are ribbed, for strength, and galvanized, for rust resistance.

1853 Friedrich Wilhelm Ostwald, a founder of physical chemistry, is born the son of a master cooper, or cask and barrel maker, in Riga, Latvia. Among numerous achievements, Ostwald won the 1909 Nobel Prize for Chemistry for developing the theory of catalysis that still holds today, which states that catalysts do their work by lowering the "energy of activation" needed to drive chemical reactions. Ostwald first presented this theory in 1894 simply as casual comments on the abstract of another scientist's paper. Ostwald strongly believed that chemistry should concern itself only with measurable and observable phenomena, and therefore was one of the last to accept atomic theory.

1877 Chemist Frederick Soddy, one of the first to suggest the existence of isotopes, and coiner of the term "isotope," is born the son of a merchant in Eastbourne, England. In collaboration with Ernest Rutherford at McGill University from 1900 to 1902, he worked out an explanation of radioactive decay in which a precise sequence of new elements are formed when subatomic particles are emitted. Soddy won the 1921 Nobel Prize for Chemistry.

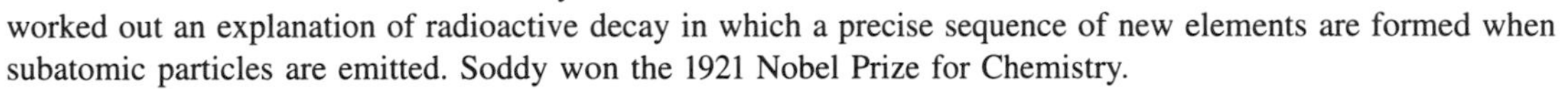

1917 Conservationist-animal welfare activist Cleveland Amory is born in Nahant, Massachusetts.

1922 Henry Ford posts signs in his Detroit auto factory that any worker will be fired if he has "the odor of beer, wine or liquor on his breath" or if he has any alcoholic beverage in his possession or in his house. "The Eighteenth amendment is a part of the fundamental laws of this country. It was meant to be enforced," proclaim the signs.

1937 A game preserve act is first passed by the U.S. Congress. It is "an act to provide that the United States shall aid the states in wildlife restoration projects," and its approval is soon followed by appropriation of $1 million to establish game preserves. The first project of the U.S. Fish and Wildlife Service was to stabilize water levels on 2000 acres around the Great Salt Lake in Utah.

1952 The world's first open-heart operation is performed in Minneapolis by the team of Drs. Floyd John Lewis, C. Walton Lillehei, and Richard Varco. They open a one-inch hole in the chest of a five-year-old girl. To slow the heart, the patient is cooled in an ice-water bath, also a medical first.

1971 President Richard Nixon appoints a five-person Advisory Panel on Timber and The Environment after a 5-month delay. Environmentalists interpreted this delay as an indication of his low priority for ecology, as does the majority of the panel. Despite a pledge that the Advisory Panel will have no member with "ties or commitments that might prejudice objective judgment," Nixon appoints Ralph Hodges, Jr., a professional lobbyist for the timber industry.

1985 It is announced that a U.S.-French team has located the wreckage of the *Titanic* 560 miles off Newfoundland. The luxury liner never completed its first transatlantic trip, and has been underwater for 73 years.

1991 Termites that gnaw through metal and concrete have reached California from Formosa, announces the Associated Press today. Furthermore, the destructive creatures are not affected by traditional soil treatments. The Asian termites resemble species that already exist in the United States, but they have harder jaws and bigger appetites. "They're not ingesting the asphalt or the steel or the iron for its nutritional value," said Harvey Logan, director of Pest Control Operators of California, Inc. "They're just going through it to get wood."

1993 Wire services carry the story of a 9-year-old boy in Perth, Australia, who discovered and then re-buried a 2000-year-old fossilized egg from the extinct elephant bird. He refuses to tell government authorities where it is buried unless he is permitted to sell it for $102,000. The government maintains that the egg is state owned because it was found on state-owned land. They offer the boy a reward of $17,000.

1732 Denis Diderot, compiler of history's first great encyclopedia, receives his master of arts degree from the University of Paris, at age 18.

1710 Swiss naturalist-reproduction scientist Abraham Trembley is born in Geneva. As a private tutor to wealthy families, Trembley had time to conduct numerous experiments, primarily on his favorite subject, the hydra, a tiny relative of the jellyfish. He demonstrated that a hydra cut in half will regenerate new bodies, and two hydra can be grafted together to form a single animal; this work foreshadowed modern medical practices of tissue grafting and regeneration. Trembley also discovered asexual budding in hydra—in which a new individual grows from the side of an old one—which was the first time this process was seen in any animal. He also was the first to see reproduction by simple cell division in algae and one-celled animals, or in effect, that these organisms multiply when they divide.

1814 "I really love my subject." These were the words that James Joseph Sylvester, born today in London, used to summarize his life in mathematics. Ironically, Sylvester had once quit mathematics, at age 27, after three months as a professor at the University of Virginia. He then worked as an insurance actuary, a lawyer, and a private mathematics tutor (Florence Nightingale was one of his pupils). In 1846 he met Arthur Cayley, also a disappointed mathematician-turned-lawyer. The relationship rejuvenated their interest in math, and they made mathematics history by jointly developing a number of areas in algebra.

SEPTEMBER 3

1821 A hurricane rips through New England. The pattern of fallen trees allows meteorologist William C. Redfield to become the first to correctly deduce that hurricanes are storms that whirl counterclockwise.

1869 Microchemist Fritz Pregl is born in Laibach, Austria. He won the 1923 Nobel Prize for developing techniques that allow structural analysis of chemical substances when only extremely small samples are available.

1898 The country's first municipal subway, Boston's Tremont Street Subway, opens its third station (see September 1, 1897).

1899 Sir Frank Macfarlane Burnet, virologist and 1960 Nobel laureate for discovering the nature of organ-transplant rejection, is born in Traralgon, Victoria, Australia.

1905 Carl David Anderson, discoverer of the meson and positron, is born in New York City. He was a corecipient of the 1936 Nobel Prize for Physics.

1992 The Associated Press reports a study in demography concluding that divorce is more likely for couples who cohabit before marriage. Rather than validating the notion of the 1960s and 1970s that living together improves compatibility, this latest study concludes that "cohabiting experiences significantly increase young people's acceptance of divorce." The research was conducted by sociologists at the Universities of Chicago and Michigan.

1907 Anthropologist-natural philosopher Loren Corey Eiseley is born in Lincoln, Nebraska.

1964 President Lyndon Johnson signs the Wilderness Act, establishing a National Wilderness Preservation System.

1976 The unmanned *Viking 2* lands on Mars. The landing site is designated "Utopia." The craft is the second to safely land on the planet's surface—*Viking 1* had landed on July 3—and the first to send back closeup, color photographs. Both *Viking* probes were the first to carry experiments specifically designed to seek life on another planet, although the results of these experiments remain inconclusive.

1993 The *Sun Seeker*, piloted by Eric Raymond, completes the first solar-powered airplane flight across North America. The craft lands in Currituck, North Carolina, just a few miles from Kitty Hawk, which was the site of the historic 1903 Wright brothers flight. Their flight lasted only 93 seconds but consumed more fuel and created more pollution than the entire cross-country trip of the *Sun Seeker*.

1882 Electricity is first provided to a city from a central power station. The Edison Electric Illuminating Company in New York City begins service with one generator, producing enough power for 800 light bulbs. Fourteen months later the station was servicing 508 subscribers and 12,732 bulbs. In today's historic event, Thomas Edison begins the generator's operation with the flick of a switch in the Wall Street office of his primary financial backer.

1888 The name "Kodak" is registered and awarded patent No. 388,850 for the first roll-film camera not requiring a tripod or table for support. George Eastman of Rochester, New York, invented the 22-ounce, fixed-focus camera, which carries a film roll for 100 exposures. Each picture is round and 2½ inches in diameter.

1906 Microbiologist Max Delbrück is born in Berlin. He began his career as a nuclear physicist working with Otto Hahn—the discoverer of nuclear fission—but then turned to genetics and the study of bacteriophages (viruses that invade bacteria). Delbrück won the 1969 Nobel Prize for Physiology or Medicine for developing a new method for growing phages, for charting the enormously fast rate of phage reproduction within a bacterium, and for discovering that genetic material from different phage species could combine to form entirely new and different species.

1908 Stacy G. Carkhuff of the Firestone Tire and Rubber Company, Akron, Ohio, applies for a patent on the non-skid automobile tire.

1913 Biochemist Stanford Moore is born in Chicago. He shared the 1972 Nobel Prize for Chemistry for determining the structure of various proteins—most notably ribonuclease, which breaks-down RNA—using chromatography techniques.

1915 Despite opposition by the U.S. Forest Service, landowners, and various congressmen, Rocky Mountain National Park is dedicated in Colorado. The opposition was overcome mainly by a six-year crusade of articles and lectures by Enos Mills, "the John Muir of the Rockies," who first visited the Park area as a sickly youth, regained his health while hiking in its mountains, and eventually was buried there. The Park is 412 square miles in area, with the Continental Divide running down the middle.

1951 A message of peace is the subject of the first coast-to-coast television telecast across North America. At 10:30 PM Secretary of State Dean Acheson introduces President Harry Truman in the War Memorial Opera House in San Francisco; Truman then discusses the newly signed Japanese Peace Treaty. This is also the first transcontinental broadcast using coaxial cable.

1957 The Ford Motor Company begins selling the Edsel, an automobile soon to become a famous marketing failure.

1969 The Food and Drug Administration officially declares the birth control pill "safe."

1994 A 1993 study that declared cellular phones to be safe is now declared invalid by the primary researcher. Dr. Om Gandhi of the University of Utah conducted the original experiments, in which energy absorption by models of the human skull was measured when cellular phones were operating. Gandhi now says the phones should have been closer to the head, and more recent types of phones should be tested. Experts estimate it will be several years before definitive statements on the phones' safety can be made.

1964 Li-Li gives birth to Lin-Lin. It is just the second birth of a giant panda in captivity. It is 360 days since Li-Li birthed Ming-Ming, the first captive-born giant panda. Both events occurred in the Peking Zoo.

1667 Girolamo Saccheri, the first mathematician to almost discover non-Euclidean geometry more than a century before the actual discovery, is born in San Remo, Italy. In 1733 he published *Euclid Cleared of Every Flaw* after years of a futile search for a single error in Euclid's postulates; he was on the verge of a great breakthrough when he gave up.

1793 The Reign of Terror begins during the French Revolution, as the National Convention adopts harsh counterrevolutionary measures of sending thousands of citizens to the guillotine, including the great chemist Lavoisier.

1831 Charles Darwin first meets Captain Robert Fitzroy, in London. Fitzroy commands the HMS *Beagle*, and will soon be Darwin's cabinmate on the ship's five-year expedition (1831–1836), during which Darwin will visit the Galapagos Islands and develop the theory of evolution. Darwin almost missed the trip because his father opposed his participation, and because Fitzroy—who believed a person's character was revealed by facial features—had already offered the position to someone else, holding at first that the shape of Darwin's nose indicated a lack of energy and determination.

SEPTEMBER 5

1850 Physicist Eugen Goldstein, coiner of the term "cathode rays" and developer of cathode-ray tubes, which many scientists used to study atomic structure, is born in Gliwice, Prussia.

1857 From his country home in Down, England, 48-year-old Charles Darwin reveals the theory of evolution in a letter to Harvard botanist Asa Gray. The enthusiastic reaction of Gray and others, and the knowledge that Alfred Russel Wallace has independently hit on the same theory, convince Darwin that it is finally time to publish his theory after 20 years of indecision.

1885 Jake Gumper, owner of a service station in Fort Wayne, Indiana, buys the country's first gas pump. Sylvanus Bowser had invented and built the device in his barn. He was awarded a patent for it in 1887. It held one barrel of gas, and used marble valves and a wooden plunger.

1905 Arthur Koestler, author of *The Act of Creation* (1964, a scientific analysis of humor and creativity), is born in Budapest.

1923 The smoke screen is first demonstrated publicly, during a naval bombing exercise near Cape Hatteras, North Carolina. Invented by Thomas Buck Hine, the smoke screen was intended to hide the movement of troops and ships in battle.

1948 Physicist-chemist Richard Chace Tolman dies at 67 in Pasadena, California. It was Tolman who demonstrated that electrons are the charge-carrying entities in the flow of electricity; he was also able to measure the mass of the electron. During World War II, Tolman was the chief scientific advisor to Brigadier General Leslie R. Groves, overseer of the development of the atomic bomb. Tolman's younger brother Edward made great progess in the study of human behavior.

1977 *Voyager 1* is launched from Kennedy Space Center, destined to explore Jupiter and Saturn, then to pass out of the solar system with a number of audiovisual messages to any alien creatures that may encounter and recover the spacecraft.

1990 After 10 years and $537 million, a Congressional study of acid rain ends, concluding that acid rain is not a "crisis," only a "long-term problem," which presents hazards to human health, damages forests, reduces visibility, depletes soil nutrients, erodes buildings and statues, and kills aquatic life.

1991 Ms. Toni Conti, 52, begins a hunger strike in Albuquerque to protest a diving mule act at the New Mexico State Fair. Billed as the "World's Only High-Diving Mules," three of the animals are to plunge 40 feet into a 6-foot-deep pool of water, three times a day. "I want to attract their attention so they will know how people are reacting negatively to this cruel act," said Conti.

1889 Singing is first recorded. A cylinder is made of the Danish baritone Peter Schram on his 70th birthday, performing the role of Don Giovanni.

SEPTEMBER 6

1620 The Pilgrims depart Plymouth, England, for the New World.

1766 John Dalton is born the son of a weaver in Eaglesfield, Cumberland, England. He was a founder of physical science and the first to scientifically deduce the atomic theory—that all matter is composed of invisibly small atoms—from experimental results and data.

1776 History's first submarine attack occurs in New York Harbor, off Manhattan Island by the *Turtle*, a 7½-foot long submarine made of oak and built by David Bushnell of Saybrook, Connecticut (and so named because it looked like two turtle shells stuck together). It was powered by a hand-operated propeller, and its depth was controlled by taking in or forcing out water. Under cover of darkness, the one-man *Turtle*—the first submarine built for use in war—moves alongside Admiral Howe's flagship, the 64-gun *Eagle*; submarine pilot Ezra Lee attaches a time bomb to the hull, but the bomb drifts loose before it explodes, and no damage is done.

1810 The first colonists to reach the Pacific Coast depart New York City on the SS *Tonquin*, sponsored by John Jacob Astor and captained by Jacob Thorn. The ship anchored at Cape Disappointment, Washington State, the following April.

1819 A woodworking lathe is first patented in the United States, by Thomas Blanchard of Middlebury, Connecticut. His device was a profile lathe "for manufacturing gun stocks." The machine did the work of 13 men, and dramatically reduced the price of firearms.

1892 Sir Edward Victor Appleton is born the son of a millworker in Bradford, England. He received the 1947 Nobel Prize for Physics for discovering the "Appleton layer" in the ionosphere—a layer of ions and gas that lies approximately 150 miles above Earth's surface, and, as a constant reflector of radio waves, is important in radar and communications. The lower ionospheric layers only sporadically reflect radio waves.

1892 The gasoline-powered tractor was first manufactured in the United States by John Froelich, of Froelich, Iowa. On this date he ships the first one he sold to Langford, South Dakota. The device is a J.I. Case threshing machine combined with a Van Duzen gas engine mounted on wooden beams. The following year Froelich established the Waterloo Gasoline Traction Engine Company, which was later bought by the John Deere Plow Company.

1895 V-2 rocket designer Walter Dornberger is born in Giessen, Germany.

1902 Sir Frederick Augustus Abel, coinventor of the explosive cordite, dies at 75 in London (see July 17, 1827, for a biographical sketch).

1909 Robert Peary finally sends word that he had reached the North Pole five months ago. Peary's claim to be the first is still in dispute.

1952 Television is first broadcast in Canada, from Montreal.

1954 President Dwight Eisenhower breaks ground for the nation's first nuclear power plant devoted entirely to peaceful uses. Ike sends a remote signal from Denver for the groundbreaking, which occurs in Shippingport, Pennsylvania.

1990 Parents' smoking doubles their children's chances of getting cancer, according to a Yale study published in today's *New England Journal of Medicine*. The study estimates that 17% of all lung cancer cases in nonsmokers arise from secondhand smoke inhalation during childhood. "Here is another piece of evidence that smoke from other people's cigarettes is harmful to your health," said study director Dr. Dwight T. Janerich.

1992 The first person to receive a transplant of a nonhuman liver dies 71 days after the operation. The 35-year-old man, whose name was never made public, died as a result of bleeding within the skull, although the cause of the internal bleeding remains a mystery to doctors at the Pittsburgh Medical Center. The patient had received a baboon liver, which sparked protests by animal rights advocates.

1707 French naturalist Comte Georges-Louis Leclerc de Buffon is born to a wealthy family in Montbard. He is most remembered for his 44-volume *Histoire Naturelle* (1749–1812), the first work to join together a multitude of previously unconnected facts about natural history. One of the volumes, *Époques de la nature* (1778), was the first to present geological history as a series of stages.

1799 Dutch scientist Jan Ingenhousz, discoverer of photosynthesis, dies at 68 in Bowood Park, England. The son of a leather merchant, Ingenhousz became a doctor and achieved prominence by being one of the first to learn and advocate Jenner's method of smallpox vaccination. He inoculated the royal family of Empress Maria Theresa and became their personal physician. In 1779 he was elected to the Royal Society in England, where he conducted his famous plant physiology experiments. He built on Priestley's discovery that plants replenish some life-giving substance in air which animals deplete (oxygen). Ingenhousz showed that (1) plants must have sun to perform the replenishment (hence the name "photosynthesis"), (2) only the green parts of plants can perform this trick, and (3) plants also destroy oxygen, but much less than they replenish it.

SEPTEMBER 7

1829 Chemist Friedrich August Kekulé, one of the first to determine how atoms are combine to form molecules, is born in Darmstadt, Germany. He entered university as an architectural student, but switched to chemistry under the influence of Justus von Liebig. Kekulé's early interest and training in architecture may have helped him discover chemical structures.

1888 The first incubator for infants is used on a patient for the first time. At the State Emigrant Hospital on Ward's Island, New York City, Edith Eleanor McLean is placed in the chamber, weighing just 2 pounds 7 ounces at birth. Called a "hatching cradle," the incubator was built by Dr. William Champion Deming.

1908 Michael De Bakey, inventor and heart surgery pioneer, is born in Lake Charles, Louisiana.

1914 Physicist James Alfred Van Allen is born the son of a lawyer in Mount Pleasant, Iowa. V-2 rockets captured from Germany were important to Van Allen's early career, as he headed a program to use them to explore the upper atmosphere. In 1958 he discovered the Van Allen radiation belt—donut-shaped zones of high-energy particles around Earth.

1674 From his home in Delft, the Netherlands, Antonie van Leeuwenhoek writes to the Royal Society in London about "very little animalcules" he discovered with homemade microscopes. The announcement changed science forever. In the foreground of this painting are some cloth and ribbon, references to Leeuwenhoek's occupation as a haberdasher. It was his desire to inspect cloth closely that led Leeuwenhoek to invent his first microscopes.

1936 Boulder Dam—later Hoover Dam—begins operations. Located on the Colorado River, on the Nevada–Arizona border, the dam provides electric power, irrigation, and flood control. The dam created Lake Mead, which is the nation's largest reservoir.

1944 The infamous V-2 rocket is first used in combat by the Nazis against the Allies.

1967 The first successful biosatellite, *Biosatellite 2*, is launched from Cape Kennedy, Florida. It carries 13 experiments and a small zoo—including 10,000 gnats, 1000 flour beetles, 560 wasps, 120 frog eggs, 875 amoebas, 13,000 bacterial cells, 78 wheat seedlings, 10,000,000 bread mold spores, and 64 wild flowers—to study the effects of space on living organisms.

1992 A husky–retriever mix named Lady saves the lives of several children in woods outside Loveland, Colorado, when she attacks a group of rattlesnakes that the children had wandered into.

1994 At a Washington ceremony, Commerce Secretary Ron Brown has just finished a speech praising new communications technologies and the Information Superhighway when the sound system fails.

1157 English scholar Alexander Neckam is born in St. Albans, Hertfordshire. His textbook *De utensilibus*, ("On Instruments," c. 1180) was the first known written report in Western science about the magnetic compass as a navigational tool, although the Chinese had been using the device for two centuries. According to legend, Neckam and King Richard the Lion-Heart were born on the same night, and both were nursed by Neckam's mother.

1522 The first circumnavigation of Earth ends. The *Victoria*, under command of Juan Sebastián Elcano, reaches Seville, Spain, from where it had departed three years earlier, as part of a five-ship expedition led by Ferdinand Magellan. Magellan and four ships were lost during the voyage. Eighteen Europeans and four East Indians are on the *Victoria* when it anchors.

1565 The oldest town of European origin in North America is established by Spaniard Pedro Menéndez de Avilés at present-day St. Augustine, Florida.

SEPTEMBER 8

1848 Organic chemist Viktor Meyer is born into a wealthy family of textile merchants in Berlin (see biographical note on August 8, 1897).

1866 The first well-documented birth of sextuplets in the United States occurs in Chicago. The parents are James and Jennie Bushnell; Dr. James Edwards is the physician, and Priscilla Bancroft the midwife. Two of the infants died after eight months, but Norberto lived to 68 and Alberto, Alice, and Alincia survived past 70.

1915 History's first military tank has its maiden test run. The "No. 1 Lincoln," later modified and known as "Little Willie," was built by William Foster & Co. Ltd. of Lincoln, England.

1918 Sir Derek Harold Richard Barton is born in Gravesend, Kent, England. He became a 1969 Nobel laureate for helping establish "conformational analysis"—the study of the three-dimensional structure of molecules, and relating this structure to chemical properties—as an essential part of organic chemistry.

1944 A German V-2 rocket is first fired against London during World War II. Approximately 1230 of the missiles were eventually launched against that city.

1980 Willard F. Libby, inventor of radiocarbon dating, dies at 71 in Los Angeles. "Once you ask the question, where is the Carbon-14, and where does it go, it's like one, two, three, you have [radiocarbon] dating," Libby once said about his great innovation. The son of a farmer, Libby became a nuclear physicist, and was part of the atomic bomb project during World War II. In 1945 at the University of Chicago, Libby had his great insight about carbon dating. He was awarded the 1960 Nobel Prize for Chemistry. His technique has become standard in anthropology and fossil-dating; for example, it was used to demonstrate that the Dead Sea Scrolls were authentic and that the Piltdown man was a fake.

1994 Delaware becomes the first state to have video cameras installed on school buses. The cameras are an attempt to control unruly behavior. Not all buses will have real cameras in them; many buses will contain only "black box" camera housings. The mere threat of a camera has been found to restrict antisocial conduct.

1994 After days of emotional negotiations, and much debate over the wording of a statement on abortion, delegates in Cairo at the United Nations conference on population approve a 113-page plan to slow human population growth over the next 20 years.

1894 The genius Hermann von Helmholtz dies at 73 in Charlottenburg, Germany eight weeks after receiving a concussion. Best known for devising one of the bedrock principles of science—the law of conservation of energy—Helmholtz made a number of biological contributions, including the invention of the ophthalmoscope, the first measurements of nerve impulse speed, the first demonstration of heat and acid production by muscle contraction, and the experimental development of the Young–Helmholtz theory of color vision (that the eye responds only to three colors: red, green, and blue; the brain perceives many other colors by combining the nervous impulses from the eye). This theory is still controversial and continues to be researched today.

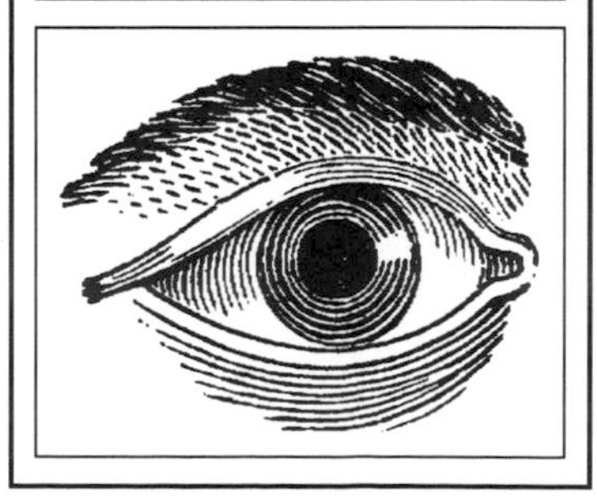

1492 Columbus departs the Canary Islands, his last stop before crossing the Atlantic to the New World.

1737 Luigi Galvani is born in Bologna, Italy, in the house in which he died 61 years later. A gifted physiologist, surgeon, and teacher, his name has become a household word—galvanized steel, "galvanized into action," galvanometer, galvanic skin response—through two seminal discoveries. Ironically, it was Galvani's assistant who made the first of these discoveries when he realized that electricity applied to nerves will cause muscle contraction; this finding was a first step in understanding how nerves and muscles work. The second discovery occurred during follow-up experiments; Galvani discovered that the contact of two dissimilar metals creates an electric current. This phenomenon later would greatly advance human technology when it was used to create the battery. Unfortunately Galvani did not realize its significance since he mistakenly assumed that the electricity arose from within the muscle rather than from the interaction of the metals. He staunchly defended this view throughout his life.

September 9

1789 Astronomer William Cranch Bond, the first to photograph the moon, is born into poverty in Portland, Maine.

1830 Charles Ferson Durant becomes the first American to make a balloon flight in the United States. He flies a homemade craft from Castle Garden, New York, to Perth Amboy, New Jersey. Durant was the country's first professional aeronaut, and the first to land a balloon on a ship.

1854 Darwin, 43, begins organizing his notes on species, which eventually develop into the theory of evolution.

1890 Social psychologist Kurt Lewin is born in Mogilno, Prussia.

1898 A milestone case in forensic medicine begins with the tragic disappearance of nine-year-old Else Langemeier from the small German village of Lechtingen. She and a classmate are found murdered in a nearby woods the next day. Police arrest a suspect, Ludwig Tessnow, who had been seen with the girls. In his shack clothes were found splattered with brown stains that Tessnow claimed were paint, not blood, but the police had no way of analyzing them until Dr. Paul Uhlenhuth soon published a chemical test that allowed him to identify dried blood and to differentiate blood of different species. The police chief read Uhlenhuth's report, and sent the Berlin researcher a sample of Tessnow's clothes to be analyzed. It was the first time in history that chemical analysis of blood stains led to a conviction.

1923 Virologist Daniel Carleton Gajdusek is born of immigrant parents in Yonkers, New York. In 1955 he began studying the fatal neural disease kuru, which is unique to one New Guinea tribe, the Fore. Tribesmen contracted the disease by ritually eating the brains of their deceased fellows. Gajdusek realized that the disorder was caused by a slow-acting virus, which uncovered an entirely new class of infectious disease, and its exploration won Gajdusek the 1976 Nobel Prize.

1934 A man-made object breaks the sound barrier for the first time. At Marine Park, Staten Island, New York, "Rocket No. 4" is launched by the American Rocket Society; it climbs 400 feet and then travels 1000 feet horizontally, reaching a speed of 700 mph.

1958 The Japanese Cabinet approves that country's first national water pollution control plan.

1963 The first birth in captivity of a giant panda takes place at dawn in the Peking Zoo. Ming-Ming is the offspring of Li-Li and Pi-Pi. The birth is kept secret for three months, for fear that the news would cause a huge wave of visitors that might stress and endanger the family. Li-Li, the mother, weighs 250 pounds at birthing, Ming-Ming weighs just 5 ounces.

1994 Wire services announce that artist Robert Wyland has agreed to paint a mural in a Mexico City zoo, El Nuevo Reino Aventura, in return for which the zoo will transfer one of their killer whales to a larger tank. "I've been saying for years that art can help save the whales," said Wyland. The whale to be moved is the 14-year-old star of *Free Willy*, a movie about a performing whale escaping captivity. Wyland's mural will depict the whale's life in the zoo.

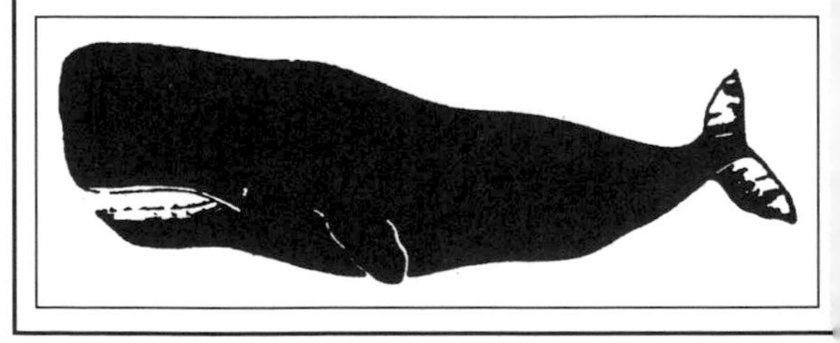

SEPTEMBER 10

1713 John Turberville Needham is born in London. He is famous for experiments that supported the theory of spontaneous generation, or that life could arise from nonliving sources. In these experiments he sealed boiled meat in jars, and found maggots on the meat days later. He interpreted this to show that life could arise spontaneously, but others later proved that Needham had not boiled the meat long enough to kill all of the live spores present in the meat.

1713 Gowan Knight, a pioneer in science, is baptized (his exact date of birth is unknown) in Corringham, England. He was a scientist and inventor whose work in magnetism led to innovations in the compass that are still used today. He developed a powerful technique for magnetizing metals in 1744, which replaced the crude and inaccurate needles then used in compasses. He introduced the common rhomboidal shape to needle design, and invented new means of suspending the needle. His improvements were readily accepted and used by England's Royal Navy.

1788 Archaeologist Jacques Boucher (de Crèvecoeur) de Perthes is born in Rethel, France, the son of a botanist who had influence with Napoleon. He was the first to discover Stone Age artifacts, and the first to theorize that man's prehistory could be measured by geological periods.

1846 Elias Howe patents the sewing machine.

1892 Physicist Arthur Holly Compton, who suggested the name "photon" to describe a unit of light, is born the son of a Presbyterian minister in Wooster, Ohio. He received the 1927 Nobel Prize for Physics for discovering the Compton effect (X-ray wavelength increases when struck with electrons).

1895 Melville Jean Herskovits, a pioneer in economic anthropology and the anthropological analysis of the "New World Negro," is born in Bellefontaine, Ohio.

1624 Thomas Sydenham, born today, is pictured at the bedside of a patient in London. Also pictured is Sydenham's friend and fellow physician, the philosopher John Locke, who often accompanied Sydenham on house calls. Sydenham became known as "the English Hippocrates" for his many advances in medical science, including the first differentiation between measles and scarlet fever (which he named), the first use of opiates to relieve pain and induce rest, and a textbook on epidemics that was standard until publication of Pasteur's germ theory of disease hundreds of years later.

1903 After three years of abusive treatment, Clifford Beers is released from an insane asylum and returns home to New Haven, Connecticut. He soon begins work on his autobiography, *A Mind That Found Itself.*

1913 The first coast-to-coast paved road in the United States is opened. The Lincoln Highway runs from New York City to San Francisco through 13 states, including Pennsylvania, Indiana, Nebraska, and Utah. It cost about $10 million and was opened under the direction of the Lincoln Highway Association, which disbanded in 1927. Neil Patterson made the first complete automobile run over the 3300-mile route.

1923 The first attempt to take motion pictures of a solar eclipse from an airplane occurs over Santa Catalina Island, California, but clouds spoil the event. The first usable images were not taken until April 28, 1930, by a crew flying 18,000 feet above Honey Lake, California. The eclipse lasted 1.5 seconds. The photographer for both flights was J.M.F. Haase of the U.S. Navy.

1936 The first World Speedway Championship is held in Wembley, London, Great Britain.

1977 Convicted murderer Hamida Djandoubi becomes the last to be executed with a guillotine.

1994 In the first atmospheric study of its kind, the space shuttle *Discovery* bounces bright green laser beam pulses off Tropical Storm Debby, to measure the exact height of the clouds.

1785 At sea off Massachusetts, oceanography sleuth Benjamin Franklin, nearly 80 years old, during his last voyage from Europe to the United States, conducts temperature and vegetation studies on the Gulf Stream. Begun when he was the colonial Deputy Postmaster and trying to explain why mail was slower when coming from Europe than going the other way, Franklin's study produced a wealth of information on the physical and biological properties of the Gulf Stream, including its first charting, which was one of the first charts of any of the world's ocean currents.

1831 Darwin, 22, sees HMS *Beagle* for the first time, when he and *Beagle* captain Robert Fitzroy travel from London to Plymouth Harbour to inspect the craft. This ship will take him on the history-making voyage during which he develops the theory of evolution.

1851 Sylvester Graham, inventor of the graham cracker, dies at 57 in Northampton, Massachusetts (see July 5, 1794, for a biographical sketch).

SEPTEMBER 11

1869 Thomas Graham, the father of colloidal chemistry, dies at 63 in London.

1877 Sir James Jeans, best known as the author of popular books on astronomy and as the first to propose that matter is continuously created through the universe, is born the son of a journalist in Ormskirk, England. As a child he was precocious, unhappy, and deeply interested in clocks. He made his first professional mark—using his exceptional mathematical ability and his analysis of spinning bodies—by destroying Laplace's "nebular hypothesis" that our sun and planets formed from a single gaseous cloud.

1884 Acoustical engineer Harvey Fletcher, the first to demonstrate stereophonic sound (in 1934) is born in Provo, Utah. His 1922 classic *Speech and Hearing* helped establish the field of psychoacoustics, the science of the relation between sound and sensations.

1935 Cosmonaut Gherman Stepanovich Titov is born near Barnaul, Russia. On August 6, 1961, in *Vostok 2*, he became the second human in space, the first to orbit Earth more than once, and the youngest ever in space. To date, no one younger has been in orbit.

1946 The first long-distance car-to-car telephone conversation occurs. A Texas reporter with the Houston *Post* calls a St. Louis reporter working for the *Globe Democrat.*

1947 For the first time in U.S. medical history, a major surgical operation occurs in one building and is televised in another. Surgeons at New York Hospital perform seven operations, which are simultaneously shown to delegates at the 33rd annual clinical congress of the American College of Surgeons in New York City's Waldorf-Astoria Hotel.

1969 Setting a record at the time for the largest land auction, 450,858 acres of the oil-bearing North Slope in Alaska is put up for sale in Anchorage. The event's largest bid is for 2560 acres, from a Hess–Getty oil consortium.

1991 The first major attack by African killer bees in the United States occurs in McAllen, Texas. Sixty-five-year-old Adan Garza is stung more than 300 times while clearing brush near a derelict shack in which the bees had a nest. Garza is hospitalized and listed as stable the next day. The bees earned their name because of their hyperaggressive nature, because they attack in swarms, and because they will pursue victims up to a mile.

1994 Rheagan Dickerson, the second smallest baby to survive in U.S. history, weighing just 12 ounces at birth, goes home after spending her first five months in All Saints Episcopal Hospital in Fort Worth, Texas. The smallest baby on record was a 9.9-ounce girl born at 27 weeks in 1989 at the Loyola University Medical Center in Chicago.

1522 Naturalist-physician Ulisse Aldrovandi is born in Bologna, Italy. He is especially remembered for his systematic and accurate descriptions and catalogues of animals, plants, and minerals. He was appointed inspector of drugs and pharmacies, and wrote the official pharmacopoeia *Antidotarii Bononiensis Epitome* in 1574, which became a model for subsequent works in the field.

1609 Henry Hudson, on the *Half Moon*, first sails into the river in New York that now bears his name.

1725 Astronomer Guillaume Le Gentil is born of impoverished noble parents in Coutances, France. In 1761 he traveled to Pondicherry, India, to observe a transit of Venus across the sun, but missed it because the outbreak of war prevented him from arriving on time. He stayed in India for eight more years for the next transit, but missed it again because of cloudy weather. When he returned to France his heirs had seized his property because he was assumed dead.

1793 A quarantine is imposed against a U.S. city for the first time. Yellow fever has broken out in Philadelphia, and Maryland Governor Thomas Sim Lee orders the cessation of all commerce between Philadelphia and Baltimore.

1811 Geologist James Hall is born in Hingham, Massachusetts. A charter member of the National Academy of Sciences, Hall is best known for theories on the formation of mountains.

SEPTEMBER 12

1818 Richard Jordan Gatling is born in Maney's Neck, North Carolina. He helped his father develop machines for sowing cottonseeds and thinning cotton plants, and later spent the rest of his life inventing various machines, the most famous of which was the first machine gun, the Gatling gun.

1897 Marie Curie gives birth in Paris to daughter Irene, destined to marry physicist Frédéric Joliot, and to share the 1935 Nobel Prize with him for their joint discovery of new radioactive elements produced artificially.

1915 The killer disease pellagra is conquered. For several months, 12 inmates of the state prison in Jackson, Mississippi, have been eating a protein-deficient diet designed by Dr. Joseph Goldberger. Just as the time allotted for the experiment is about to expire, one of the inmates is seen on this date with the telltale skin rash. Other inmates soon develop the same rash. Thus, the cause of the disease is proved to be dietary, not contagious. For the convict volunteers, the experiment means freedom, as each is granted a pardon although some of the sick men leave before they can be cured. For Goldberger, it is just the start of a struggle to win acceptance for his theory.

1940 The famous cave paintings of Lascaux are discovered by five school boys. Located near Montignac, France, the Grotte de Lascaux contains one of the most outstanding displays of prehistoric art known. Many animals, represented in a variety of styles, date back to 15,000 BC.

1992 Space shuttle *Endeavor* blasts off, carrying the first married couple in space, Mark Lee and Jan Davis; the first black woman in space, Mae Jemison; and the first Japanese citizen on a U.S. space mission, Mamoru Mohri.

1994 After a week of hard bargaining, delegates at the 182-nation U.N. Conference on Population in Cairo complete their "Program of Action," a 20-year plan for curbing the growth of human populations. The 100-plus-page document breaks new ground because it not only covers family planning, but also ties in economic development, empowerment of women, and environmental protection. "The world is never going to be the same after Cairo," said U.S. delegate Tim Wirth.

1994 General Norman Schwartzkopf declares himself prostate cancer posterboy. The war hero discusses his recent surgery in a newspaper interview; he recovered fully in just 3½ months because the cancer was caught early. "Look the urologist right in the eye and say, 'Bring me to my knees if you have to, but do a thorough exam,' because that way they'll save your life."

1941 Embryologist Hans Spemann dies at 72 in Freiburg im Breisgau, Germany. He tackled one of the great biological mysteries: How does an embryo develop from a single, invisible cell into a highly organized organism with trillions of cells? Spemann spent his life researching newt embryos. He discovered embryonic induction—in the early stages of life, embryo cells are highly plastic and their final form is determined by their location in the embryo. For example, if cells that would normally become skin are transplanted to an area of nerve development, these cells will become nerve. Spemann was the sole recipient of the 1935 Nobel Prize for Physiology or Medicine.

1851 Military surgeon Walter Reed is born the son of a minister in Belroi, Virginia. Reed led the research team that conquered yellow fever (see 1900 below).

1853 Bacteriologist Hans Christian Joachim Gram is born the son of a law professor in Copenhagen. Gram invented the Gram stain procedure, which remains a fundamental tool in identifying and classifying bacteria.

1886 Freud marries.

1886 Sir Robert Robinson, a 1947 Nobel laureate for Chemistry, is born the son of an inventor-manufacturer near Chesterfield, England. His research specialty was plant biology. He was the first to determine the chemical structure of several "alkaloids," a family of large, complex molecules that have dramatic effects on the body; nicotine, quinine, strychnine, morphine, and cocaine are all alkaloids.

SEPTEMBER 13

1887 Leopold Stephen Ružička, a 1939 Nobel laureate for Chemistry, is born in Vukovar, Croatia. He is famous for research on both sex hormones and perfume molecules.

1899 The first American automobile death occurs in New York City. H.H. Bliss, a 68-year-old real estate broker, is run over as he steps from a southbound streetcar on Central Park West and 74th Street. Bliss is taken to Roosevelt Hospital, where he died. Arthur Smith, the driver, is arrested and held on $1000 bail.

1900 Physician Jesse Lazear, 34, is bitten by a disease-carrying mosquito in Quemados, Cuba, during experiments that finally proved how yellow fever is transmitted. Lazear died two weeks later, becoming a famous martyr in the conquest of disease.

1912 Astronomer Horace Welcome Babcock is born in Pasadena, California. In 1951 with father Harold Delos Babcock, he invented the solar magnetograph, an instrument allowing detailed study of the sun's magnetic field. Several years later, Harold used the tool to discover that the sun periodically reverses its magnetic polarity.

1922 The highest shade temperature ever recorded on Earth is reached today in El Azizia, Libya. The ecological benchmark is 136.4°.

1946 George Washington Hill dies at 61 in Matapédia, Canada. He was a Madison Avenue wizard who made a fortune in marketing cigarettes. His great stroke of genius, which forever changed the cigarette industry, was targeting women specifically with cigarette ads.

1826 A rhinoceros is exhibited for the first time in the United States, at Peale's Museum and Gallery of the Fine Arts in New York City. Advertisements declare, "Its body and limbs are covered with a skin so hard and impervious that he fears neither the claws of the tiger nor the proboscis of the elephant. It will turn the edge of a scimitar and even resist the force of a musket ball."

1993 For the first time in history a major study establishes a link between profession and breast cancer. The federal Centers for Disease Control and Prevention releases an investigation of 2.9 million death certificates from 1979 to 1987: Librarians, nuns, and other professional women are considerably more likely to die of breast cancer than homemakers or women with less professional employment. Researchers emphasize that professional work does not necessarily cause breast cancer; rather, secondary factors, like the delay in childbearing, that often accompany a professional career may increase cancer likelihood.

1994 *Working Mother* magazine announces its selection of the best 100 U.S. companies to work for. On-site child care, flexible work schedules, rooms for breastfeeding, tuition reimbursement for workers' children, percentage of female executives and workers, and high pay for females were all factors that distinguished companies. "We've discovered that quality of life is a much greater asset in securing people than high salaries," said movie producer George Lucas whose company, Lucasfilm Ltd., has been on the list for four consecutive years.

1698 Charles Du Fay, an early experimenter in electricity, is born in Paris. Experiments led him to postulate the existence of two types of electrical fluid: "vitreous electricity" and "resinous electricity," depending on the objects that excited an electric charge. Later Benjamin Franklin renamed these "positive" and "negative."

1769 Explorer-naturalist Alexander von Humboldt is born the son of a military officer in Berlin. He is the founder of biogeography, the study of the geographical distribution of plants and animals. One of Earth's major ocean currents, the Humboldt Current off the west coast of South America, is named in his honor.

1804 Ornithologist John Gould is born in Lyme Regis, England, the son of the foreman of gardeners of Windsor Castle. Gould published many scientific papers and descriptions of new bird species, but he is most famous for large, lavishly illustrated, expensive books on birds for which his wife did most of the artwork. He produced over 40 such volumes.

SEPTEMBER 14

1883 Birth-control champion Margaret Higgins Sanger is born the sixth of eleven children in Corning, New York. She founded the country's first birth-control clinic in 1916 in Brooklyn, New York, and was an international leader in the field.

1901 When President William McKinley dies in Buffalo, New York, after having been shot eight days before. The presidency of Theodore Roosevelt begins, marking a time of unprecedented increase in protection for the environment.

1905 The oldest automobile race still run (the Royal Automobile Club Tourist Trophy) is first held, on the Isle of Man, Great Britain.

1956 The first prefrontal lobotomy in the United States is performed on a 63-year-old female by J.W. Watts and Walter Freeman in Washington, D.C., at the George Washington University Hospital.

1959 A space probe first hits the moon. Thirty-six hours after launch, the Soviet *Luna 2* crashes east of the Sea of Serenity. It is the first man-made object to reach a celestial body.

1961 Police in Columbus, Ohio, are summoned to the scene of a double murder in a McComb Street home where a woman and her lover have been shot. When the case reached court the following June, it included the first testimony in U.S. legal history by a hypnotized witness. The defendant was the woman's husband, Arthur C. Nebb. While hypnotized, he recounted the shootings as being accidental. The jury was convinced, and Nebb's crime was reduced to manslaughter.

1969 Possibly the heaviest starfish ever captured, a 13.2-pound *Thromidia catalai* is collected off Hot Amedee, New Caledonia, in the western Pacific.

1990 Gene therapy is first used on a human to treat an inherited disease. At the National Institutes of Health in Bethesda, Maryland, four-year-old Ashanthi Cutshall is injected with her own white blood cells, which had been removed so that their DNA could be altered to produce an enzyme that is critical to fighting infection. Ashanthi had been born with defective blood cells. Although regular reinjection of newly treated cells is necessary, the experiment is a huge success because it helped save her life and because it proved that gene therapy is possible.

1886 The typewriter ribbon is first patented in the United States. Patent No. 349,026 is awarded to George K. Anderson of Memphis, Tennessee. Although a typewriter was first patented in the United States in 1829, the first successful machine was not invented until 1873 by Christopher Latham Sholes. He coined the term "type-writer" to describe his device, which he also called a "literary piano." His first prototype was finished exactly 13 years and 2 days before Anderson's ribbon patent was issued.

1817 The Plumstock Rolling Mill, the first U.S. mill with facilities to roll and puddle iron, begins operation in Redstone Creek, Pennsylvania. Built and operated by Isaac Meason, it was destroyed by floods in 1824 and never rebuilt.

1885 Karen Horney (born Danielsen), one of the first great female psychoanalysts, is born in Hamburg, Germany, the daughter of a sea captain. She was an early rebel against strict Freudian theory; she suggested that environmental and social conditions—as opposed to the biological drives that Freud emphasized—influenced the development of personality and the formation of disorders or neuroses.

1910 Theodor Wulf, a Jesuit Father and amateur physicist, publishes in *Physikalische Zeitschrift* results of four days of observations he performed the previous spring atop the Eiffel Tower. This paper contains the first suggestion that Earth may be under constant bombardment by radiation from outer space from sources other than the sun.

SEPTEMBER 15

1916 Tanks enter battle for the first time, from the English "Heavy Section Machine Gun Corps," later called the "Royal Tank Corps," at the battle of Flers-Courcelette in France. Forty of the forty-nine Mark I Male tanks broke down before reaching the battleground.

1917 Dr. Félix d'Hérelle coins the term "bacteriophage"—meaning "eater of bacteria"—in a note to the French Academy of Sciences which discusses something extremely small that had been ruining his experiments on bacteria. Hérelle first saw these "*microbes invisibles*" as nuisances, and then as saviors that might rid the world of bacterial disease. This hypothesis was never proved, but many later scientists, including a number of Nobel laureates, used bacteriophages to explore the nature of viruses and the mysteries of heredity. Bacteriophage experiments first revealed that DNA stores genetic instructions.

1924 Embryologist Wilhelm Roux dies at 74 in Halle, Germany. He devoted his career to determining how organs and tissues are assigned their specific form after an egg is fertilized. Working with frog eggs, Roux destroyed one of the two initial subdivisions of a fertilized egg, which resulted in the development of just half an embryo. This led to Roux's "mosaic theory of development" that each embryo cell eventually forms only part of the whole organism. Later scientists were inspired by Roux's work, but proved his theory wrong; with different species, divided embryos could produce whole individuals. Nevertheless, Roux is known as a founder of experimental embryology, and in 1894 he established the first journal in this field.

1913 The first major goat show in the United States opens at Exhibition Park in Rochester, New York, as part of the sixth annual Rochester Industrial Exposition. Pedigreed goats are on display under the auspices of the Standard Milch Goat Breeders' Association of North America. Halfway through the show this organization changed its name to the New York Milch Goat Breeders' Association.

1928 Bryce Canyon National Park is established in Utah.

1929 Physicist Murray Gell-Mann, 1969 Nobel laureate who predicted the existence of quarks, is born the son of Austrian immigrants in New York City.

1980 A B-52 bomber containing thermonuclear weapons burns for three hours at an air base in Grand Forks, North Dakota.

1983 IBM announces the first 512K bit microchip. The following February four companies announced a chip twice as big.

1992 *Circulation*, a journal of the American Heart Association, reports a "provocative" Finnish study linking heart attacks with iron in the blood; each 1% increase in the blood of the iron-containing protein ferritin produces a 4% increase in heart attack risk. Only smoking is a stronger predictor of heart attack susceptibility. The findings support an old and controversial theory, originally developed from studies of menstruating women, that blood iron is a heavy risk factor. If verified, the findings could be bad news for the food-supplement industry and could call for changes elsewhere in nutrition.

1994 Astronauts on the space shuttle *Discovery* snatch the satellite "Spartan" from orbit, and during the next 1½ hours tuck it into a cargo bay. *Discovery* had set it adrift two days before, during which time it used two telescopes to study the sun's "solar wind." Solar wind contains charged particles traveling 1.5 million mph (over 416 miles per second) that can disrupt radio communications and electrical power on Earth.

1804 Physicist Joseph-Louis Gay-Lussac, 25, makes a solo balloon ascension to 23,018 feet. It is an altitude record that stands for over half a century, but more importantly Gay-Lussac took air samples and measurements that helped establish the sciences of meteorology and space biology.

1804 Inventor-engineer Squire Whipple is born in Hardwick, Massachusetts. In 1854 he published *Work on Bridge Building*, history's first set of scientific rules and theories on bridge construction.

1833 The nation's first interstate crime-fighting pact is signed in New York City by New Jersey and New York representatives. The pact was ratified the following February by both state legislatures, and then by Congress on June 28, 1834. The original signers were Benjamin Franklin Butler, Peter Augustus Jay, Henry Seymour, Theodore Frelinghuysen, James Parker, and Lucius Quintius Cincinnatus Elmer.

SEPTEMBER 16

1835 HMS *Beagle* reaches the Galapagos Islands, about 600 miles west of Ecuador. Aboard is naturalist Charles Darwin, 26. The ship departs the islands 34 days later, during which time the seeds for the theory of evolution were planted in Darwin's mind by his observations of the islands' unique fauna.

1853 Biochemist Albrecht Kossel, discoverer of the nucleic acids in DNA, is born the son of a merchant in Rostock, Germany. Kossel won the 1910 Nobel Prize in Medicine or Physiology.

1893 "Research is to see what everybody else has seen, and to think what nobody else has thought." Chemist Albert Szent-Györgyi is born in Budapest, Hungary. Best known as the codiscoverer of vitamin C—for which he won the 1937 Nobel Prize in Physiology or Medicine—Szent-Györgyi also did important cancer research and achieved critical insight into the way muscles work: He discovered and named the protein actin, determined that the interaction of actin and the protein myosin produces muscle contraction, and showed that ATP is the immediate energy source for muscle contraction.

1912 The first plant quarantine in U.S. history is issued against white-pine blister rust, a rust fungus that destroys white pine trees. The power of plant quarantine had been established 34 days earlier by a Congressional act against harmful pests and plant diseases "new to or not therefore widely prevalent or distributed within and throughout the United States." Blister rust as well as potato wart and the Mediterranean fruit fly were immediate targets of the act.

1953 *The Robe*, the first movie filmed in wide-screen CinemaScope, has its world premiere at the Roxy Theater in New York City.

1987 The Montreal Protocol, a treaty to save Earth's ozone layer by reducing harmful chemical emissions, is signed by two dozen nations.

1991 Archaeologists have located the exact site of Thomas Jefferson's birth near Shadwell, Virginia, according to a report today by the Associated Press. Fire destroyed the home in 1770, along with Jefferson's early writings and papers. When the site was rediscovered in 1941 by architect Fiske Kimball, haste and faulty archaeology led researchers to place the home some 100 feet from its actual location. A cellar still remains and scientists hope that excavations will reveal new details of Jefferson's background, including whether or not he was born into wealth.

1994 Michigan State researchers announce in *Science* that they have isolated, cloned, and reinserted a gene that controls plant growth. The gene iaglu seems universally to activate a particular hormone, IAA, which leads to enhanced growth of all parts of a plant. An excited USDA scientist said, "This is just the first step [to exploiting the gene to grow more and bigger crops]." Scientists see the gene as replacing chemical sprays that are expensive and environmentally harmful.

1994 The Exxon Corporation is ordered by a federal jury to pay $5 billion in punitive damages to fishermen, native Alaskans, property owners, and others in the aftermath of the 1989 *Exxon Valdez* oil spill.

1725 Geologist Nicolas Desmarest, the first to suggest that valleys were formed by streams that ran through them, is born the son of a school-teacher in Soulaines, France. Desmarest also discovered that basalt rock is volcanic in origin, ending the theory that all rocks came from sedimentation of ancient oceans.

1677 Stephen Hales, a lifelong clergyman and the founder of plant physiology, is born in Bekesbourne, England. Hales was the first to accurately measure several key functions of plant life, including transpiration—the regular loss of water as it moves through the plant—and the pressure of sap movement. He also made the first measurements of human blood pressure.

1764 Astonomer John Goodricke, a martyr to his science, is born the son of an English diplomat in Gröningen, the Netherlands. Goodricke was deaf and mute from birth. At age 17 he became the first to discover a regular pattern in the fluctuation in brightness of a "variable star." Goodricke hypothesized that another star was periodically eclipsing the view of the variable star. This guess was confirmed a century later. At just 21 Goodricke died of exposure, the result of years of research and observations in the cold night air.

1778 The first treaty between the United States and native Americans is signed at Fort Pitt (now Pittsburgh) by Andrew and Thomas Lewis (for the United States) and by Captain White Eyes, Captain Pipe, and Captain John Kill Buck (for the Delaware Indian Nation). The treaty pledges perpetual peace and friendship, mutual forgiveness of offenses, and an Indian representative in Congress "on certain conditions."

1842 Darwin, 43, moves himself, and his wife and two children into Down House in the countryside outside Down, Kent, England. It was there he wrote *On the Origin of Species.*

1857 Pioneering rocket scientist Konstantin Tsiolkovsky is born the son of a forester in Izhevskoye, Russia. He built Russia's first wind tunnel, was the first to suggest the possibility of space stations, and foresaw the need for liquid fuel. The Russian government planned the launch of *Sputnik I*—and hence the launch of the Space Age—to coincide with the 100th anniversary of Tsiolkovsky's birth, but technical reasons delayed the actual launch 29 days. "Mankind will not remain tied to earth forever" reads the tombstone on his grave.

1993 The meteor that hit Earth 65 million years ago and possibly led to the extinction of the dinosaurs may have been much bigger than previously thought. Virgil L. Sharpton (of the Planetary and Lunar Institute in Houston) reports in *Science* that the Chicxulub crater in Mexico's Yucatan Peninsula and the Gulf of Mexico seems to be 186 miles wide, not 110. The force of a meteor creating such a hole "would be several times larger than the explosion at one point of the complete global nuclear arsenal.... We are talking about the equivalent of blasting out the state of Connecticut to a depth of about 10 kilometers (6.2 miles)." The resultant dust and chemical storm, tidal waves, and wildfires all worked to doom the dinosaurs, according to the "impact theory" of extinction.

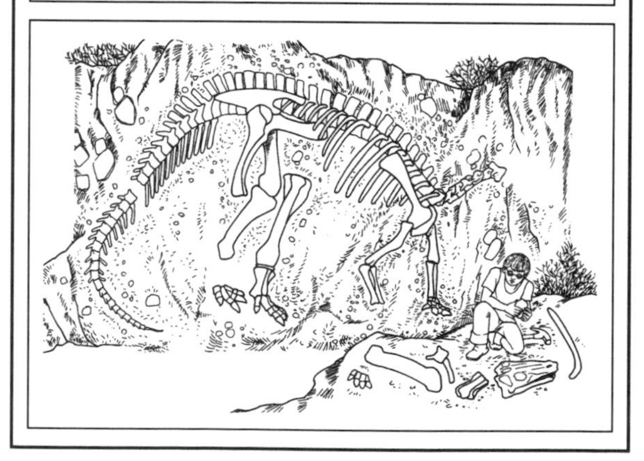

1908 Fort Myer, Virginia, is the site of the nation's first airplane fatality. Thomas Etholen Selfridge of the U.S. Army dies of head injuries in a crash resulting from a propeller blade hitting a guy wire that had come loose during the flight. Orville Wright receives multiple hip and leg fractures in the accident.

1922 A sound-on-film "talking" movie is first presented in public, at the Alhambra Theater in Berlin, Germany. It was produced by the "Tri-ergon" process.

1959 "Life is precious to the old person. He is not interested merely in thoughts of yesterday's good life and tomorrow's path to the grave. He does not want his later years to be a sentence of solitary confinement in society. Nor does he want them to be a death watch."—Dr. David Allman, *The Right to Be Useful*, an address to the AMA Conference on Aging in Boston.

1962 NASA announces the selection of nine new astronauts; included is Neil Armstrong, the first man on the moon.

1984 Sierra Sneith becomes the first child born of a heart transplantee. Sierra is born at Stanford University in California; her mother Betsy received a new heart in February 1980.

1994 Twenty-one-year-old Heather Whitestone is named the 1995 Miss America. She is deaf, save 5% hearing in her left ear, making her the first handicapped woman to wear the crown.

SEPTEMBER 18

1752 French mathematician Adrien-Marie Legendre is born in Paris. He was the first to describe the "method of least squares." His work on elliptical integrals was critical to the development of mathematical physics.

1819 Physicist Jean-Bernard-Léon Foucault, the first to demonstrate experimentally that Earth rotates on its axis, is born the son of a bookseller-publisher in Paris. Foucault was a sickly child, and was privately educated. He obtained a medical degree, but failed in that profession because, like Darwin, he could not stand the sight of blood. Instead, he turned to physics research. He produced the most accurate measurements of light's speed up to that time, and devised the Foucault pendulum, a free-swinging weight on a very long tether, with which Earth's rotation was finally demonstrated in 1851.

1840 Naturalist-traveler-educator Constantine Rafinesque dies at 56 in Philadelphia. Well known and controversial in his lifetime, Rafinesque wrote several books and more than 950 articles on natural history, religion, banking, and literature; he made important contributions to ichthyology and botany, and set the stage for Darwin's theory of evolution with his original notion that each variety of a species is a "deviant" that may become a permanent species through time and reproduction.

1851 The *New York Times* is first published.

1895 The first chiropractic adjustment in U.S. history is performed in Davenport, Iowa. Daniel David Palmer cracks the back of Harvey Lillard. Five years later in the same city Palmer established the nation's first chiropractic school.

1900 Sir James George Frasier completes and dates the introduction to *The Golden Bough*, a classic anthropological analysis of magic and religion in humans.

1907 Physicist Edwin Mattison McMillan, discoverer of the element neptunium, is born the son of a physician in Redondo Beach, California. McMillan shared the 1951 Nobel Prize with Glenn Seaborg, who also discovered "transuranium" elements (those elements with an atomic number greater than uranuim).

1908 Viktor Amazaspovich Ambartsumian, founder of theoretical astrophysics and famous for theories on stellar evolution, is born the son of a literature professor in Tbilisi, Russia.

1915 The Narragansett Speedway, the nation's first automobile racetrack with an asphalt top, opens in Cranston, Rhode Island. Two world records are broken during the first day of racing.

1944 In a secret meeting in Hyde Park, Churchill convinces Roosevelt that atomic weapons should be developed without Russia's collaboration—even at the risk of an international arms race—and that the Nobel laureate Niels Bohr should not be trusted for suggesting such a joint venture.

1965 The bird with the largest wingspan on record is caught in the Tasman Sea by crew members of the Antarctic research ship USNS *Eltanin*. It is a male "wandering albatross" with a wingspan of 11 feet 11 inches.

1975 Newspaper heiress Patty Hearst is captured in San Francisco by the FBI after 19 months under the control of the Symbionese Liberation Army. It is the end of a chapter in one of history's most famous cases of brainwashing.

1992 Having been hunted to near extinction, the population of Florida black bears may be as low as 400 individuals, and occurring only in a few, isolated pockets as the area of their natural habitat has reduced in size. On this date, the state game commission votes to ban hunting of the regal beast in the Osceola National Forest until 1996, hoping the population will rebound by that time.

1993 Two state-of-the-art telescopes are dedicated on Arizona's Mount Graham International Observatory. The $8 million Hertz telescope, built in conjunction with Germany's Max Planck Institute, will look at radio waves from space. A $3 million device, built in conjunction with the Vatican, which has been conducting continuous astronomy research for the past 400 years, will focus on infrared emanations. Added to these two will be a huge binocular telescope. The high, dry location is perfect for star-watching, but several groups have protested the project. The Observatory construction has desecrated part of the holy ground of the San Carlos Apaches, and according to environmentalists, imperiled the habitat of Earth's 300 remaining Mount Graham red squirrels.

1749 French astronomer Jean-Baptiste Delambre is born into poverty in Amiens. Mentor Joseph Lalande introduced him to astronomy. After several years of observations, Delambre prepared charts for the movements of the sun, Jupiter, Saturn and the newly discovered Uranus. From 1792 to 1799—in the midst of revolution and war—Delambre and Pierre Méchain measured the arc of the meridian across France; from this they calculated the size of Earth, and it was this measurement that formed the basis of the metric system.

1863 Ernst Haeckel, coiner of the word "ecology," addresses a congress of scientists in Stettin, Germany, about man's evolution according to the newly published theories of Charles Darwin. Ironically, the scientists react to this theory—soon to become one of the most important and influential in the history of science—with criticism and indignation.

September 19

1864 Geneticist Karl Correns is born the son of an artist in Munich, Germany. Correns discovered genetic linkage, cytoplasmic inheritance, and the fundamental laws of genetics. The monk Mendel had independently discovered these same laws decades before, but his research was buried in obscurity until Correns and others rediscovered the laws at the turn of the century. Ironically, Correns married the niece of botanist Karl Wilhelm von Nägeli, whose poor advice and lack of insight caused Mendel to abandon further publication and further research.

1892 "If you can make forestry profitable at Biltmore within the next ten years, I shall consider you the wisest forester and financier of the age," writes chief of the U.S. Forestry Division, Dr. Bernhard Fernow, to 27-year-old Gifford Pinchot, who eventually succeeds in turning a profit within two years while installing ecologically sound forestry at the North Carolina estate of the Vanderbilts. Pinchot was history's first U.S.-born silviculturalist; he founded the Yale School of Forestry, was the first chief of the U.S. Forest Service, and was twice the governor of Pennsylvania in (1923 and 1931). In 1898, the nation's first undergraduate school of forestry was established within Cornell University in Ithaca, New York. German-born and -educated Dr. Bernhard Eduard Fernow is the first director and dean of the New York State College of Forestry. Governor Frank Swett Black of New York signed the legislation creating the school the previous April. The first Forest Engineer degree was awarded in 1900 to Ralph Clement Bryant.

1876 The carpet sweeper is patented. Melville R. Bissell of Grand Rapids, Michigan, receives patent No. 182,346 for his "broom-action" machine which adjusts to different surfaces simply by varying pressure on the handle. The idea for a carpet sweeper is not new, but Bissell's device is. He founded the successful Bissell Carpet Sweeper Company in Grand Rapids around his idea.

1915 Dr. Elizabeth Stern is born in Cobalt, Canada. She was the first to report a link between a specific virus and a specific cancer (herpes simplex virus and cervical cancer), discovered a link between prolonged use of "the pill" and cervical cancer, and also discovered that a normal cell goes through 250 distinct stages before reaching advanced cancer, thus allowing early diagnosis.

1928 Mickey Mouse makes his screen debut, when the animated cartoon *Steamboat Willie*, the first animated cartoon talking picture, opens at the Colony Theater in New York City.

1957 The United States conducts its first underground nuclear test, in the Nevada desert.

1991 A human from the Bronze Age is discovered by hikers in the Alps on the Italian–Austrian border. "He was very nourished, strong and well dressed, truly an elegant chap," observed archaeologist Konrad Spindler (University of Innsbruck). The body, clothing, and accessories, including bow, arrows, bronze axe, and straw-lined boots, are all very well preserved; it is the first time a Bronze Age man—about 4000 years old—has been discovered in a natural setting, rather than in a grave, in which objects surrounding the corpse would have been carefully selected.

1992 The world's largest beach cleanup occurs, when 200,000 volunteers in 34 states and 25 nations comb coastlines to pick up and catalogue other people's rubbish. In the United States the Center for Marine Conservation sponsors the event. Not only unsightly, the garbage harms wildlife. Plastic bags and balloons can kill sea turtles that injest them because of their resemblance to jellyfish.

1519 The first circumnavigation of Earth begins, as Ferdinand Magellen leads a party of five ships out of Seville harbor in Spain. Only one of the ships made it back to Seville three years later.

1546 England's Royal College of Physicians receives its Grant of Arms in London.

1842 Physicist-chemist Sir James Dewar, inventor of the Dewar flask (better known as the "thermos bottle"), is born the son of an innkeeper in Kincardine, Fife, Scotland. Although he researched and wrote on a wide range of topics, his most important work was on very low temperatures. In 1891 he was able to produce large quantities of liquid oxygen, and in the following year he invented the thermos, a container with two walls that are separated by a vacuum, the inner wall of which is silvered, which reduces heat loss through conduction, convection, and radiation, to preserve the gas in a liquid state. Dewar also collaborated with Sir Frederick Abel to produce cordite, the first practical smokeless explosive.

SEPTEMBER 20

1848 The American Association for the Advancement of Science is organized in Philadelphia. William Charles Redfield is the first president. The AAAS was the country's first national scientific society, and it remains the most important. It publishes the prestigious *Science* magazine weekly.

1859 The electric stove is patented by George B. Simpson of Washington, D.C.

1860 Pasteur, 37, arrives in Chamonix in the French Alps, to take samples of air from a relatively pure environment. He later compares these samples with germ-rich city air. Pasteur's classic series of experiments destroyed the theory of spontaneous generation and established the germ theory of disease.

1951 The country's first shopping mall opens in Seattle. Also on this date, a monkey and eleven mice are sent to an altitude of 236,000 feet in an Aerobee rocket. It is one of the first journeys into the upper atmosphere by mammals. Unlike the previous five U.S. rockets carrying mammals into space, the recovery parachute worked on today's flight.

1963 President John F. Kennedy proposes a joint Russian–U.S. expedition to the moon, in a U.N. address.

1965 The U.S. Forest Service announces that 10% of the trees in Alaska's Tongass National Forest will be sold for "clear cutting" by private timber companies. An area larger than Rhode Island will be stripped to the ground. This is the largest sale of trees in Forest Service history.

1970 Earth's northernmost volcano, Beeren Berg on the island of Jan Mayen in the Greenland Sea, erupts. The island's 39 inhabitants (all male) are evacuated.

1971 Psychologist B.F. Skinner appears on the cover of *Time* magazine (see March 20, 1904).

1973 East and West Germany—political enemies during the Cold War—unite for the sake of environmental health. An agreement that specifies how common waters are to be managed, and outlines joint action in the event of environmental emergency is signed in Berlin.

1975 Gail A. Cobb becomes the first female police officer in the United States to be killed in the line of duty. She is murdered by a robbery suspect in an underground garage at 20th and L Streets in Washington, D.C.

1992 A project to rescue 400 elephants from the drought-stricken Gona Re Zhou national park in Zimbabwe is announced by the U.S. Fish and Wildlife Service, which is donating $200,000 to the effort. The imperiled pachyderms are being shot with tranquilizer darts and trucked to farmland.

1993 Actress Marlee Matlin gives a sign-language rendition of *The Star-Spangled Banner* at a Burbank, California, ceremony announcing two postage stamps that honor the deaf.

1878 "I aimed at the public's heart and by accident I hit it in the stomach," stated Upton Sinclair, born on this day in Baltimore, of his reformist novel *The Jungle*. Intended to anger the public over the treatment of meat packers, the book stirred greater anger over the treatment of the meat, and thereby led to the passage of food inspection laws.

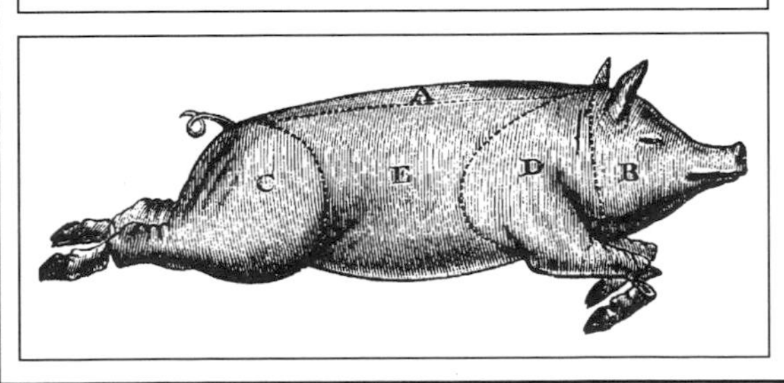

1756 John Loudon McAdam, inventor of the macadam road surface, is born in Ayr, Scotland. After making a small fortune during the American Revolution, McAdam made a small fortune in New York City by selling war prizes. In 1783 he returned to Ayrshire, the county of his birth, where he took a position as road trustee, and at his own expense conducted experiments to improve road construction. He suggested elevating the roads and covering them with several grades of crushed rock. Later he refined and implemented his new road design as surveyor general of the Bristol (England) roads. Many countries soon adopted his design.

1832 French physicist Louis-Paul Cailletet, famed as one of the first to liquefy several gases, including oxygen, hydrogen, and nitrogen, is born the son of a metallurgist in Châtillon-sur-Seine.

1853 Dutch physicist Heike Kamerlingh Onnes, the first to liquefy helium, is born the son of a prosperous manufacturer in Groningen. In 1911, during his work with extremely low temperatures, Onnes found that some metals lost all electrical resistance when supercooled. This discovery of superconductivity won Onnes the 1913 Nobel Prize for Physics.

SEPTEMBER 21

1860 Louis Pasteur, 38, already famous for developing pasteurization, hikes from Chamonix into the Alps with a guide, several mules, and 33 sealed flasks carefully packed in boxes. Over the next two days he allows mountain air to enter the flasks thus mixing the air with a nutrient broth already in them, and reseals each flask. The experiment is a success. The pristine alpine air produces no microbial life whereas urban air, sampled in Paris, generates dense colonies of microbes. The results of this simple experiment disprove the doctrine of spontaneous generation and lay the foundation for Pasteur's "germ theory of disease," which Isaac Asimov called "the greatest single medical discovery of all time."

1866 "Chance favors only those who know how to court her," said Charles J.H. Nicolle, born on this date in Rouen, France. He "courted" diseases and won the 1928 Nobel Prize for Physiology or Medicine for discovering in 1909 that typhus is transmitted by body lice. He also did important work on measles, diphtheria, brucellosis, rinderpest, and tuberculosis.

1895 The Duryea Motor Wagon Company of Springfield, Massachusetts, becomes the first automobile company to incorporate in the United States.

1926 Donald A(rthur) Glaser, inventor of the bubble chamber (a device for observing the behavior of subatomic particles) is born in Cleveland, Ohio. He was one of history's youngest Nobel laureates, receiving the 1960 Nobel Prize for Physics at age 34. Then he promptly announced that he was switching research interests from nuclear physics to molecular biology.

1982 A striped bass weighing 78 pounds 8 ounces is caught from a jetty in Atlantic City, New Jersey, during a fishing contest. The animal sets the record as the heaviest fish ever caught with rod and reel, and with the $250,000 prize money, becomes the most valuable fish ever caught.

1994 Scientists at Chicago's Argonne National Laboratory unveil a specially designed "superchip" for use in the Human Genome Project, a collaboration of 350 laboratories around the world that has one mission: to locate and describe each of the 100,000 genes in a human cell. Although only one inch square, the chip is expected to cut years from the Project and reduce the cost by a factor of ten, from $3 billion to $300 million.

1994 Premature babies who receive care that is tailored to their bodies and personalities do much better, even after 10 or 12 years, according to a study in the *Journal of the American Medical Association.*

1893 Frank Duryea test-drives the "horseless carriage" that he and brother Charles built in Springfield, Massachusetts. This vehicle is believed to be the country's first gasoline-powered automobile.

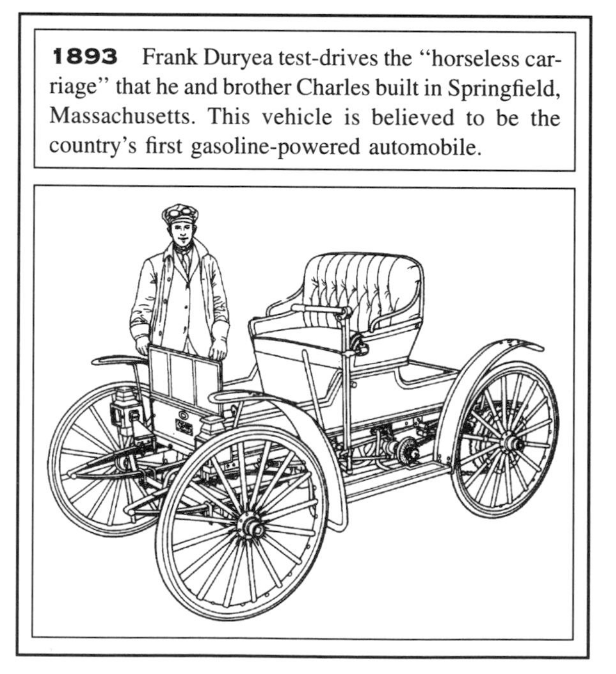

SEPTEMBER 22

1656 The first all-woman jury in American history hears a horrible case in Patuxent, Maryland: Judith Catchpole is tried for the murder of her own child. The jury is composed of seven married and four single women; they find Catchpole not guilty because they are not convinced that she ever had a child in the first place.

1711 English astronomer Thomas Wright (the first to describe the Milky Way as a flattened, rotating galaxy) is born the son of a carpenter in Byers Green, near Durham. Wright suffered throughout life with a horrible speech impediment, which, as a child, kept him from formal education. When he discovered astronomy and feverishly began studying it, his father burned his books as being a waste of time.

1715 The first to survey and geologically map France, Jean-Étienne Guettard is born in Étampes.

1791 Physicist/chemist Michael Faraday is born in Newington, England, one of ten children of a poor journeyman blacksmith.

1901 Surgeon Charles Branton Huggins is born in Halifax, Nova Scotia. In 1941 he showed that prostate cancer in males could be treated with female sex hormones. This was the first case of a major cancer being controlled solely with chemicals, and it won Huggins the 1966 Nobel Prize.

1932 Beneath the ocean off Nonsuch Island, Bermuda, zoologist William Beebe and adventurer–financier Otis Barton descend in their specially built, leak-plagued "bathysphere" to a depth of 2200 feet, further into the ocean depths than humans had gone before. The travelers make a number of discoveries: animal life is plentiful in the darkness at such great depths, many unknown species are present, and a variety of abyssal organisms produce their own luminescence.

1932 Psychiatrist Karen Horney and her daughter reach America, the psychoanalytic movement having been crushed in Germany and Austria by the Nazis.

1947 The first airplane flight across the Atlantic without a pilot takes off from Stephensville, Newfoundland. Ten hours and 15 minutes later the U.S. Army C-54 four-engine "Skymaster" landed safely at Brise Norton, four miles outside London. The robot-controlled flight carried 14 people over the 2400 miles.

1949 Russia explodes its first atomic bomb.

1955 A male Kodiak bear in the Cheyenne Mountain Zoological Park (Colorado Springs, Colorado) sets a record as history's largest-ever living terrestrial carnivore (1670 pounds). He dies on the same day.

1956 Explaining in a *New York Times* interview why a diet high in fruits and vegetables (commonly called "rabbit food") is good for executives, endocrinologist Dr. Aurelia Potter observes that, "Middle-aged rabbits don't have a paunch, do have their their own teeth, and haven't lost their romantic appeal."

1991 The Huntington Library of San Marino, California, announces that its microfilmed photographs of the Dead Sea Scrolls will finally be open to a much wider field of scholars than previously allowed. The 800 animal-skin scrolls, are the oldest-known copies of the Old Testament, and hold invaluable information on the turbulent period that spawned Christianity and modern Judaism. Access has previously been denied to all but a very few scholar–editors authorized by the Israeli government. This restriction has accounted for just 100 of the scrolls being translated and published.

1992 The National Cancer Institute announces a study showing farmers have elevated risks of a variety of cancers, and suggest the use of pesticides as the cause.

1692 The last execution takes place in the famous Salem, Massachusetts, witchhunts. In all, 19 "witches" were hanged and one was pressed to death. Two were readmitted to The First Church in Salem on the 300th anniversary of their executions. "It's supposed to be the start of a healing process," said one church member during the readmittance ceremony in 1992.

1819 Armand-Hippolyte Fizeau, the first to measure the speed of light by a terrestrial method, is born in Paris. In 1849 he created a spinning device, with which he measured the velocity of light within 5% of today's recognized value.

1835 The governor of the Galapagos Islands tells Darwin that he can tell which Galapagos island any tortoise comes from, simply by looking at the shape of the tortoise's shell. It is a dramatic demonstration of species formation, and spurs Darwin to explain how nature could produce such variation.

1846 The planet Neptune is discovered by German astronomer Johann Gottfried Galle, 34. Émile Gallé, recognized as a pioneer in glass manufacturing, dies exactly 58 years later (1904) in Nancy, France.

1850 German biologist Richard von Hertwig is born in Friedberg. He is best known for developing the germ-layer theory that all organs and tissues develop from just three layers of cells in the embryo.

SEPTEMBER 23

1852 William Steward Halsted, an innovator in scientific surgery and a founder of Johns Hopkins University as the first U.S. surgical school, is born in New York City.

1882 Chemist Friedrich Wöhler dies at 82 in Göttingen, Germany. The son of a veterinary, Wöhler was a gynecologist until one of his professors, Leopold Gmelin, persuaded him to pursue chemistry. His work changed science and philosophy dramatically. In 1828, by accident, he created urea in the lab while working on an unrelated project. Never before had man synthesized a chemical made by the body of a living thing. This work destroyed the distinction between organic and inorganic chemicals, and it vanquished the doctrine of Vitalism, which held that only the bodies of living things had the ability to produce organic molecules.

1885 Professor Edmund Beecher Wilson of Bryn Mawr begins the first general biology course at a U.S. college. The course contained five lectures a week and eight hours of lab work. It first covered the structure of familiar plants and animals, then one-celled organisms, then a progression of increasingly complex creatures. The course ended with a study of chicken embryos.

1632 Galileo is summoned by letter to appear before the Inquisition.

1791 German astronomer Johann Franz Encke, dicoverer of Encke's Comet and Encke's Division, is born the son of a minister in Hamburg. Encke's Comet has the shortest orbital period of any known comet, and Encke's Division is found in the outermost ring of Saturn.

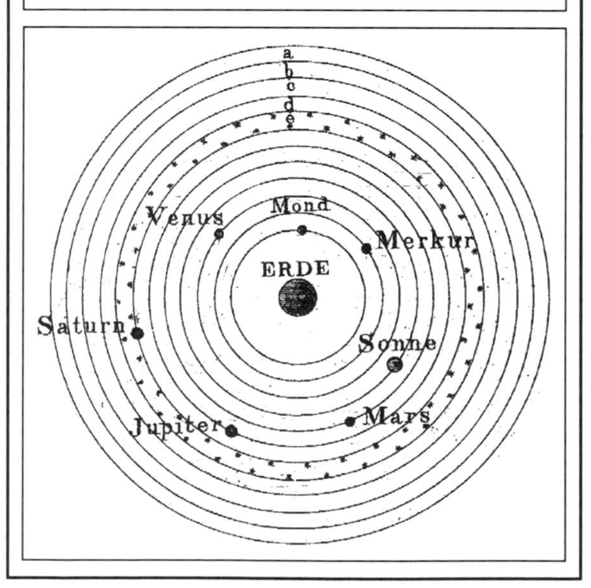

1907 One of Earth's last wild passenger pigeons, is shot at St. Vincent, Quebec.

1934 A radio broadcast is heard in both Arctic and Antarctic regions for the first time in history. Admiral Richard Byrd, near the South Pole at "Little America," picks up a radio broadcast to Labrador in the North Pole by the New York Coffee House over W2XAF, the shortwave station of the General Electric Company in Schenectady, New York.

1938 A time capsule, to be opened in 6939, is buried at the World's Fair in New York City.

1939 Sigmund Freud dies in London of throat and mouth cancer after years of heavy cigar smoking.

1993 Just one day after the exclusive Ocean Reef Club in Florida agreed to release two young dolphins into the wild, another captive dolphin dies in captivity "from chronic illness." The victim is a 42-year-old female named Lady. The two freed dolphins, Bogie and Bacall, will first go to a holding lagoon before final release to the open ocean. The Dolphin Alliance of Melbourne, Florida, negotiated their release.

SEPTEMBER 24

1501 Geronimo Cardano, the first physician to clinically describe typhus fever, and a world-class mathematician, also renowned as a gambler, cheat, and braggart, is born in Pavia, Italy, the son of a lawyer-mathematician who was friendly with Leonardo da Vinci. Cardano wrote the *Ars magna*, which is now a milestone in the history of algebra.

1657 America's first autopsy is performed in Maryland on a slave allegedly murdered by his master. For "dissecting and viewing the corpse" the surgeon is paid "one hogshead of tobacco."

1852 History's first dirigible flight occurs. French inventor Henri Giffard takes off in Paris, and flies his steam-powered balloon for 17.3 miles at an average speed of 4.3 mph.

1858 The cornerstone for the world's first hospital for alcoholics is laid in Binghamton, New York. The United States Inebriate Asylum, later named the New York State Inebriate Asylum, was organized by Dr. James Edward Turner "for the reformation of the poor and destitute inebriates."

1870 Chemist Georges Claude, inventor of the neon light, is born in Paris.

1895 Physiologist André Frédéric Cournand is born in Paris. He shared the 1956 Nobel Prize for Physiology or Medicine for developing heart catheterization in which a tube is run through part of the circulatory system to study normal and diseased heart functioning.

1898 Pathologist Howard Walter Florey is born in Adelaide, Australia. He shared the 1945 Nobel Prize for Physiology or Medicine for his part in bringing the first antibiotic, penicillin, to practical use.

1905 Molecular biologist Severo Ochoa, the first to discover an enzyme that allows cells to construct the genetic material RNA, is born the youngest son of a lawyer in Luarca, Spain. Ochoa used this enzyme—polynucleotide phosphorylase—to create the first synthetic RNA. He shared the 1959 Nobel Prize for Physiology or Medicine with biochemist Arthur Kornberg, who created the first synthetic DNA.

1907 Physicist John Ray Dunning is born in Shelby, Nebraska. He was at the 1939 Washington physics conference in which Niels Bohr announced that Hahn and Strassmann may have split the atom; physicists all over the world rushed to duplicate the feat, but Dunning was the first to succeed. Dunning also confirmed Bohr's speculation that uranium-235, not uranium-238, was the material being split. Dunning then developed a gas-diffusion method of separating these two isotopes; his method was the first, and still most popular method of separation.

1936 AND 1991 Two giants in the field of imaginary animals undergo passages. In 1936, Jim Henson, creator of the Muppets, is born in Greenville, Mississippi. In 1991, Theodore Seuss Geisel, creator of the "Dr. Seuss" books, dies in his home at 87 in La Jolla, California.

1979 Bernard Haemmerle patents eyeglasses for horses.

1992 A Baylor researcher and an English doctor announce in the *New England Journal of Medicine* a test that screens for genetic disease shortly after the sperm and egg unite. The presence of cystic fibrosis, muscular dystrophy, Tay–Sachs, and hemophilia can all be detected just days after conception. In its first use, the test allowed a couple with a high chance of producing a cystic fibrosis child to select a normal embryo, which eventually became healthy, 7-pound Chloe O'Brien in Burnley, England.

1993 Talk show hostess Sarah Purcell is injected with a used syringe on national television. Physician Edward Gilbert is demonstrating flu shots when, apparently distracted by the lights and the pressure of television, he injects Purcell with the same needle that he just stuck into co-host Gary Collins. Purcell and Collins are advised to get blood tests for AIDS and hepatitis. Gilbert is visibly shaken and has to be comforted by Collins and the show's medical reporter, Dr. Art Ulene.

1960 Having survived civil unrest, the rigors of jungle life, and threats by the animals on which he intruded constantly, George Schaller leaves Albert National Park, the Congo, after completing the first scientific study of the ecology and behavior of gorillas in their natural habitat.

1492 The lookout on the *Pinta* in Christopher Columbus's first trip to the New World screams that he sees land. Unfortunately, it was a false alarm.

1493 Columbus departs Cadiz, Spain, with a flotilla of 17 ships on his second trip to the New World.

1513 Spanish explorer Vasco Núñez de Balboa "discovers" the Pacific Ocean.

1534 Pope Clement VII dies from eating "death cap" mushrooms (*Amanita phalloides*), the most poisonous fungus on Earth.

1644 Dutch astronomer Olaus Roemer is born the son of a shipowner in Århus. In the 1670s he made the first measurements of the speed of light that were not contaminated by human reaction time. Roemer performed his measurements by precisely calculating when Jupiter should eclipse its satellites, and then comparing these predictions with the times when the eclipses were actually observed on Earth.

SEPTEMBER 25

1750 Abraham Gottlob Werner, founder of the Neptunist school of geology, which maintained that all rocks were created as sediment through the action of water, is born in Wehrau, Silesia (now Osiecznica, Poland), the son of an ironworks inspector.

1843 Geologist Thomas Chrowder Chamberlin, one of the first to see that the Earth has undergone not just one but several Ice Ages, is born the son of a farmer in Mattoon, Illinois. He also proposed the planetesimal hypothesis—that our planets were formed around material that was ripped from the sun and another star when the two passed very close to each other.

1846 Wladimir Peter Köppen, a strong force in the development of climatology known especially for delineating and mapping Earth's climatic regions, is born the son of a geographer-statistician in St. Petersburg, Russia.

1866 Thomas Hunt Morgan, named for his uncle John Hunt Morgan (who led "Morgan's Raiders" farther north than any other Confederate force went during the Civil War), is born in Lexington, Kentucky. Although he believed neither Darwin's theory of evolution nor Mendel's theory of genetics at the start of his career, Morgan eventually verified and tied them both together with his "chromosome theory of heredity," based on his experiments with the fruit fly. He received the Nobel Prize for Physiology or Medicine in 1933.

1877 The centrifugal cream separator is patented in the United States. Containers of milk are spun at great speed, forcing cream and heavy milk to separate from skim milk. Wilhelm Lefeldt and Carl Lentsch of Schoeningen, Germany, receive patent No. 195,515 for an "improvement in centrifugal machines for creaming milk."

1924 Professor Jesse James Galloway of Columbia University, New York City, begins the first course in micropaleontology in the United States. It covered the fundamentals of paleontology, classification of ancient plants and animals, and the use of the microscope to identify ancient organisms.

1990 The Bolivian government announces that control of 1.9 million acres of rain forest will be turned over to Chimane Indians, and that new logging restrictions will soon be in place. Mahogany wood has previously been widely exported, and its exploitation has led to serious degradation of the Chimane lands. Today's announcement comes after 800 Chimanes marched for 35 days over the Andes from their home in the lowlands to the country's capital of La Paz in protest of the logging.

1992 A judge in Orlando, Florida, rules in favor of 12-year-old Gregory Kingsley, seeking a divorce from his biological parents.

1992 The first U.S. rocket to Mars in 17 years blasts off from Cape Canaveral. The *Mars Observer* is lifted at 1:05 PM atop a Titan 3 rocket; the mission is expected to last several years, during which scientists hope to learn if life ever existed on the Red Planet. The international mission includes 11 Russian scientists, and once in orbit, the U.S. spacecraft will transmit data back to Earth with a French–Russian radio.

SEPTEMBER 26

1754 French chemist Joseph-Louis Proust is born the son of an apothecary in Angers. Although he was the first to study the sugar in grapes, now called glucose, and one of the first to experiment with rideable balloons, Proust is most famous for proving that chemicals combine in fixed and definite proportions, regardless of their source or concentration. This is called the "law of definite proportions," sometimes called "Proust's law," and it was a major step in establishing atomism, or that all matter is composed of atoms.

1772 The first law to license medical practitioners in America is passed by New Jersey. The act creates a licensing board of two Supreme Court judges and a third member they appoint. Strict fines will be applied to those who practice medicine without a license; exempted from licensing are those who bleed patients, those who do not charge for their services, and those who pull teeth.

1774 John Chapman, (a.k.a. Johnny Appleseed) is born in Leominster, Massachusetts. Beginning as a professional nurseryman who sold seedlings for profit, Chapman became a character of myth through real-life seed-planting treks into the wilderness and through his personal characteristics, including his eccentric and bedraggled appearance, his outdoor life-style and knowledge, and his comraderie with animals and Indians.

1849 Ivan Pavlov, who advanced psychology while studying digestion in dogs, is born in Ryazan, Russia. He came from a long line of priests, and originally studied to follow this heritage, but in theological seminary he read Darwin's *On the Origin of Species*, which changed his interests to natural science research. During his digestion studies, he accidentally discovered that stimuli associated with food—for example, the sounds of assistants bringing food—would produce saliva flow just as the real food would. He measured the strength of this effect by counting the rate and quantity of saliva flow. This led to his development of "Pavlovian responses" and conditioned reflexes, which provided great insights into learning, and gave science powerful tools to examine the perceptions and psychology of nonhumans. Pavlov received the Nobel Prize for Physiology or Medicine in 1904.

1871 Cement is patented. David Oliver Saylor of Allentown, Pennsylvania, receives patent No. 119,413 for his "portland" cement mixture of magnesium-clay with limestone-clay. English cement outsold Saylor's mixture until 1897, when portland cement became more popular.

1886 Physiologist A. V. Hill, whose studies of heat created during muscle contraction led to an understanding of muscle chemistry and to a Nobel Prize in 1922, is born in Bristol, England.

1887 Aeronautical/military engineer Sir Barnes Wallis is born in Great Britain. During World War II he created several flying machines and bombs, the most famous of which was the rotating, bouncing bomb that skipped over water after being dropped from an airplane; when the bomb reached a dam, it sank and exploded at the base of the retaining wall. Such bombs were very effective, and earned the name "dambuster" bombs.

1900 Dr. Jesse Lazear, 36, dies of yellow fever in Quemados, Cuba, during his research efforts to find the cause and the cure for the disease.

1936 Ruth Harkness departs Shanghai for the Chinese interior, in an expedition that will produce the first capture of a living giant panda by westerners.

1950 The moon appears blue, to viewers in Great Britain. The illusion is caused by sulfur particles in the upper atmosphere, released during an enormous forest fire in northern Canada.

1991 The first Biosphere experiment begins at 8:16 AM when four men and four women walk through air-lock doors into a 3.15-acre steel and glass enclosed dome in the desert outside Oracle, Arizona. As a prototype space colony, the enclosure is intended to be totally self-sufficient, and contains a rain forest, ocean, coral reef, savanna, and a farm. The experiment ended 3 years later after having successfully remained self-sufficient, with a few exceptions: one biospherian had to leave for surgery and oxygen had to be pumped in.

1992 CBS-TV debuts a series of 55-second commercials for good nutrition, squeezed in among the usual Saturday morning ads for fast foods and sugar-laden treats for young cartoon watchers. Willie Munchright, a claymation figure, stars in the ads. The series is called "What's on Your Plate."

1507 Naturalist-physician Guillaume Rondelet is born in Montpellier, France. His major contribution to science was his *Book of Marine Fish* (1554–1555), in which he described in detail some 250 marine animals, most of which were also pictured. It was the most ambitious work of its kind at that time. Rondelet defined fish as "whatsoever passeth through the paths of the sea," and so included whales, seals, snails, and dolphins.

1818 German chemist Hermann Kolbe, the second person to synthesize an organic molecule from inorganic ingredients, is born the eldest of 15 children of a minister in Elliehausen. He created acetic acid, commonly produced in many living organisms, in a series of lab experiments from 1843 to 1845. It established the validity of Wöhler's historic 1828 synthesis of urea—the first man-made reproduction of an organic compound—and forever destroyed strict boundaries between the world of living, or organic, compounds and nonliving, or inorganic, compounds. Kolbe also introduced the term "synthesis" into chemical usage.

1825 Railroad transportation is born. George Stephenson's locomotive *Active* (later renamed *Locomotion*) pulls 38 cars (carrying 450 people) from Darlington to Stockton, England, at an average speed of 15 mph. It is the first passenger trip on a railroad, and finally opens the possibility of land travel at speeds faster than a galloping horse.

1852 Sir William Willcocks, designer of the first Aswan Dam on the Nile, is born in India.

1854 History's first major disaster involving an ocean liner in the Atlantic occurs. The steamship *Arctic* sinks with 300 passengers aboard.

1918 Sir Martin Ryle, a cowinner of the first Nobel Prize for astronomy observations, is born the son of a physician in Brighton, England. Ryle developed telescope systems capable of precisely locating sources of very weak radio signals in outer space. He and others have used the systems to examine the most distant known galaxies. Ryle shared the 1976 Nobel Prize for Physics with Sir Antony Hewish.

1922 Radar is first used to detect an object. Dr. Albert Hoyt Taylor and Leo C. Young of the Naval Aircraft Radio Laboratory begin a series of experiments that determine that radio equipment on two ships can detect a vessel moving between them in fog, darkness, and smoke screen. They also discover that tall buildings reflect radio signals.

1962 The book that spawned the ecology movement, *Silent Spring* by Rachel Carson, is published by Houghton Mifflin in Boston. Even before publication, the book created a storm of controversy; the publisher had been threatened with legal action, and advance sales had reached 40,000 copies.

1979 The U.S. Congress approves formation of the Department of Education, the 13th Cabinet agency.

1981 France's superfast railroad train, TGV Sud-Est, is brought into public service. It soon is making the Paris–Lyon run at an average speed of 132 mph. In 1990 the TGV Atlantic set the all-time speed record for a passenger train at 320.2 mph in the Courtalain–Tours run.

1991 "The mother of us all," the one human female from whom all people are descended, lived in sub-Sahara Africa between 166,000 and 249,000 years ago; these are the conclusions in *Science* in a study of mitochondrial DNA patterns throughout the world by Mark Stoneking, a Penn State anthropologist. Most controversial is the age estimation by Stoneking, which is based on the assumption that the mutation rate has been constant since the birth of man. Other scientists have attacked the analysis as a "layer cake of assumptions" and "a house of cards."

1993 Ex-Beatle Paul McCartney performs a sold-out concert in Norway, a nation that recently resumed commercial whaling. Seventy percent of Norwegians support whaling by fishermen in small villages, but McCartney uses his concert to make an antiwhaling statement. "I come from Liverpool and we used to sell slaves to the United States, so when the Norwegians defend the killing of whales with it being an old tradition, I don't buy it."

1542 Europeans first land on the west coast of America. The naval expedition of Juan Rodriguez Cabrillo (which had left Navidad, Mexico, 3 months before) lands at an area now known as Ballast Point, San Diego.

1698 Mathematician-cosmologist Pierre-Louis Moreau de Maupertuis is born in Saint-Malo, France.

1808 Arnold Henry Guyot, the geologist for whom flat-topped underwater volcanoes, guyots, are named, is born in Boudevilliers, Switzerland.

1838 Darwin begins reading *Essay on Population* by social philosopher Thomas Malthus. It provided the critical insight Darwin needed to complete his theory of evolution. (Malthus maintained that the human population grows faster than the food supply, and that only disease, war, or starvation would slow it down. Darwin immediately thought that this applied to all forms of life, that the weakest individuals would die first, and that those organisms that adapted to times of starvation would survive; this is the essence of "natural selection.")

1852 Henri Moissan, the first to isolate Earth's most active element, fluorine, is born the son of a poor railway employee in Paris. His feat won him the 1906 Nobel Prize for Chemistry. Prior to his work, many noted chemists had tried and failed to isolate flourine; some of those chemists were badly injured in the attempt, and others died. Moissan died prematurely at 54, shortly after winning the Noble Prize, as a result of repeated exposure to fluorine compounds.

1854 Zoologist Adam Sedgwick, known for studying the wormlike organism *Peripatus*—an evolutionary link between worms and insects—is born in Norwich, England, a grandnephew of the famous geologist of the same name.

1869 French paper manufacturer Aristide Berges becomes the first to convert the mechanical energy of a waterfall into electrical energy to run equipment. His paper plant is in the French Alps. In 1889 Berges exhibited his innovations at the Paris World's Fair; his brochures were the first to use the term "hydroelectric power."

1908 Otto Hahn first meets Lise Meitner, in Berlin. Their 30-year collaboration led to the first nuclear fission and the atomic bomb.

1924 Man's first flight around the world ends in Seattle. The journey began in Seattle on April 6, 1924, by three U.S. Army planes, the *Chicago* (Lt. Lowell Herbert Smith, pilot), the *Boston* (Lt. Leigh Wade, pilot), and the *New Orleans* (Lt. Erik Henning Nelson, pilot). The *Boston* did not complete the 175-day, 26,103-mile trip because it was forced down near the Faroe Islands in the North Atlantic. The other two made 57 hops, landing in 21 countries, 25 states, and one U.S. territory. The trip also was the first air crossing of the China Sea and the first transatlantic flight via Iceland and Greenland.

1895 At 4:40 PM Louis Pasteur dies in his sleep at home, surrounded by his family. Pasteur examined a soup of microbes that developed in one of his famous swan-neck flasks, in a series of experiments that proved that organisms do not arise spontaneously.

1954 George Harrison Shull, the father of hybrid corn, dies at 80 in Princeton, New Jersey, after an illustrious career in genetics, during which he founded the journal *Genetics* (1916) and developed breeds of corn with 25 to 50% better yields than previously available.

1994 The government of Ecuador announces new measures to protect the Galapagos Islands from overfishing and from damage by the estimated 49,000 tourists who annually visit the archipelago 600 miles off the country's coast in the Pacific.

1994 The South Korean carmaker Daewoo announces plans to market a car powered by natural gas in North America starting in 1998.

1511 Physician Michael Servetus, the first to describe the circulation of blood between heart and lungs, is born the son of a notary in Villanueva de Sixena, Spain. The idea that blood circulated clashed with conventional wisdom of the time. Servetus's religious views were also unorthodox and he was burned at the stake at age 42.

1787 This date marks the beginning of the U.S. Army Medical Corps. Richard Allison is appointed surgeon to the army. As Congress created new regiments, a surgeon was appointed to each. In 1798, the system changed when a single physician general was appointed for the entire army. James Craik of Virginia was the first to hold this post.

1829 One of the great establishments in forensic science, Scotland Yard in London, begins operations.

1895 Parapsychologist-ESP researcher Joseph Banks Rhine is born in Waterloo, Pennsylvania.

SEPTEMBER 29

1898 The infamous Trofim Denisovich Lysenko is born in Karlovka, Ukraine. His name will forever be linked with bad science, falsified data, rabid defense of an outdated theory, and cruel suppression of his opponents. Lysenko's view that upbringing and environment changes an organism's genes—an idea that was abandoned by most biologists since Darwin—was supported by Stalin. Lysenko's power grew and he ordered aberrant agricultural procedures that were soon adopted nationwide.

1901 Nuclear physicist Enrico Fermi is born in Rome. Fermi was in charge of the Manhattan Project when, on December 2, 1942, at 3:45 PM in a squash court at the University of Chicago, it produced history's first nuclear chain reaction. This event began the Atomic Age.

1993 Miller Brewing Co. announces that it will no longer produce clear beer. "We had a tremendous initial trial, but repeat business was not necessarily as good," said spokesman Eric Kraus. The company was trying to exploit a current marketing fad in "clean," totally clear products, including deodorant and Pepsi. The odd-looking beer sold well at first, but sales dropped when the novelty wore off.

1915 A transcontinental radio telephone is first demonstrated in the United States. Speech is transmitted from New York City to Mare Island, at San Francisco, a distance of 2500 miles. A call goes through to Honolulu later that night.

1920 Chemist Peter Dennis Mitchell is born in Mitcham, England. He received the 1978 Nobel Prize for discovering how enzymes are arranged on mitochondria (the cell's energy-processors). His "chemiosmotic theory" explains how this arrangement facilitates energy conversions in the cell.

1927 Willem Einthoven, the developer of the electrocardiogram (EKG or ECG), dies at 67 in Leiden, the Netherlands. Einthoven combined the discoveries of Galvani (see September 9, 1737) with his own interests in physics and medicine, and invented the string galvanometer in 1903. This device reacts to very small electric currents under the skin, and it allowed Einthoven to determine normal and abnormal heart rhythms. Others soon used it to track brain waves, thus producing the first electroencephalograms (EEGs). Einthoven won the 1924 Nobel Prize for Physiology or Medicine.

1943 Nuclear physicist and Nobel laureate Niels Bohr learns in Copenhagen that the Nazis have ordered his arrest and his removal to Germany. That night he sneaks by motorboat to Sweden, where he helps evacuate Jews from Denmark.

1992 An oil rig in the Gulf of Mexico blows out and starts spouting oil; a nuclear power plant in Fukushima, Japan, shuts down when human error causes a core-meltdown mechanism to activate; and the Turkey Point nuclear plant in Dade County, Florida, begins operating again after a hurricane-caused shutdown, but it ceases operation again two days later because evacuation plans are not yet finished.

1994 Tennessee health officials announce results of a review of Elvis Presley's medical records: Although there is no mention of drugs in the original coroner's report, and heart disease is listed as the cause of death, "There is no basis to conclude that any person willfully and knowingly made false statements on the death certificate of Elvis Presley." The new report does not say heart disease was the *best* diagnosis for the King's demise, just that there is no reason to accuse the medical examiner at the time, Jerry Francisco, of false reports.

1604 Astronomer Johannes Kepler, 32, first observes the flaring of a supernova. The temporary bright light boosts Kepler's belief that the heavens are not immutable, and spurs his efforts to understand nature, science, and religion. At the time he was studying data on planetary motion collected by Tycho Brahe. Eventually Kepler derived a set of laws of planetary motion based on Brahe's work. Brahe had died in 1601, but Kepler was in Brahe's observatory in Prague when he saw the supernova on this date.

1630 The first execution in colonial America takes place in Plymouth, Massachusetts. John Billington, an original signer of the Mayflower Compact, is hanged. He had been "arraigned, and both by grand and petie jurie found guilty of willful murder, by plaine and notorious evidence, and was for the same accordingly executed. This, as it was ye first execution amongst them, so was it a matter of great sadness unto them. He way-laid a young man, one John Newcomin (about a former quarele) and shote him with a gune, whereof he dyed."

SEPTEMBER 30

1846 Late in the afternoon, having put it off as long as possible, music teacher Eben Frost enters the Boston dental office of Dr. William Morton with a horrid toothache. Frost asks to be hypnotized for the extraction procedure. Morton had been investigating ether, and was about to have one of his teeth pulled by an assistant to test ether's effects when Frost knocked on his door. It was Frost who benefited from the first published use of anesthesia in surgery.

1882 German Physicist Hans Geiger, inventor of the Geiger counter for detecting radioactivity, is born the son of a philology professor in Neustadt an der Haardt.

1902 The first rayon patent is issued to Harry Mork, Arthur Little, and William Walker. Patent No. 709,922, and a second patent issued the next month, cover the spinning of artificial silk from the man-made chemical cellulose acetate. The name "rayon" was not adopted until 1924 to replace "artificial silk."

1905 Physicist Sir Nevill Francis Mott, a pioneer in studying the electromagnetic properties of noncrystalline solids, now widely used in computers, tape recorders, and solar-energy devices, is born in Leeds, England. He shared the 1977 Nobel Prize with his research assistant and with J.H. Van Vleck.

1929 A television broadcast service is first launched, by TV inventor John Logie Baird using a BBC transmitter in England.

1939 French chemist Jean-Marie Lehn is born in Rosheim. His 1987 Nobel Prize was awarded for synthesizing biologically active molecules that mimic the actions of chemicals found naturally in the bodies of living things.

1954 The submarine *Nautilus* is commissioned in Groton, Connecticut. It is history's first nuclear-powered ship.

1959 Ross Granville Harrison, the first to successfully grow isolated animal tissue in the lab, dies in New Haven, Connecticut, at 89. During his first year as a Yale professor (1907) Harrison was able to cultivate sections of tadpole; this and subsequent innovations have been widely applied to fighting disease and to organ transplantation and grafting.

1993 A report in the *New England Journal of Medicine* confirms that obesity is a heavy social and economic burden in modern society. Women are more adversely affected than men. When compared with the general population, overweight women: are 20% less likely to get married, have $6710 lower household incomes, are 10% more likely to live in poverty, and average 4 months less schooling. "I don't think this will come as news to obese people," said researcher William Dietz. "They are freely discriminated against." This weight-based bias is now called "sizism."

1992 The number of printed pages stored by the U.S. government reaches 5.2 billion. In 1992 the bureaucracy used an estimated 1380 tons of ink, and recycled 24 tons of high-quality paper—every day. The Government Printing Office is a block-long building which used 88,485,994 pounds of paper in 1993; it employs 4900 people, making it one of the world's largest printing houses, and it still has to contract three-quarters of its work to private printers at a cost of $600 million annually. During the ecological battles to save the forests of the northern spotted owl, the government has printed multiple copies of seven different reports; the first report in 1990 was 1152 pages long, of which 15,000 copies were made.

1846 Charles Darwin, 37, begins an 8-year study of barnacles. He amassed a collection of 10,000 barnacles in his home. His monographs on living and fossil barnacles, published in 1851 and 1854, still remain the standard works of reference in the field.

1847 Maria Mitchell, 29, discovers a comet from her homemade observatory in Nantucket, Massachusetts. The event brings her international recognition, a series of positions as a professional astronomer—in 1865 she became the first astronomy professor of Vassar College—and her election as the first female member of the American Academy of Arts and Sciences.

1848 The nation's first school for the retarded, the Massachusetts School for the Idiotic and Feeble-Minded Youth, opens its doors to students. The state legislature originally appropriated $2500 a year for three years. The school survives to this day as the Walter E. Fernald State School, named for its first resident supervisor.

1869 The prestamped postcard is first issued. It was conceived by Emmanuel Herrman of the Neustadt Military Academy in Vienna.

1902 Otto Hahn, 23, destined to discover nuclear fission, begins his first job as a professional chemist, in the laboratory of Professor Theodor Zincke of the University of Marburg, Germany.

1904 Nuclear physicist Otto Robert Frisch is born in Vienna. His fame in linked to Otto Hahn's; Hahn and his partner Fritz Strassmann split the atom in the 1938, but did not realize completely what they had done. They communicated their results to Frisch and his aunt Lise Meitner, who were the first to understand that fission had been produced and that enormous amounts of energy were released.

1908 Henry Ford introduces the Model T automobile. Each car costs $825.

1913 The nation's first monument to a bird is unveiled. Designed by Mahonri Young (a grandson of Brigham Young), the Salt Lake City monument celebrates sea gulls for devouring the crickets and grasshoppers that destroyed Mormon wheat fields in May 1848.

1915 Psychologist-educational theorist Jerome S. Bruner is born in New York City.

1890 Home of "the noblest forests, the loftiest granite domes, the deepest ice-sculptured canyons, and snowy mountains soaring into the sky," in the words of John Muir, the Yosemite Valley in California is established as a National Park.

1956 Tsung-Dao Lee and Chen Ning Yang publish "Question of Parity Conservation in Weak Interactions" in the journal *The Physical Review*. The article "aroused only mild interest among nuclear physicists" at the time, but eventually led to the overthrow of a major tenet of modern physics—the conservation of parity—and it garnered the Nobel Prize for the authors.

1971 History's largest amusement park opens. It is Disney World in central Florida, and covers 28,000 acres of previously tropical wilderness.

1972 Legendary anthropologist Louis Seymour Bazett Leakey dies at 69 in London. Born the son of missionary parents in Kabete, Kenya, Leakey was the first white baby ever seen in Kikuyuland. His many fossil discoveries in East Africa proved that man is far older than previously thought, and that man's evolution was centered in Africa, not Asia. He also hired Jane Goodall and Dian Fossey to perform historic research on the behavior of man's closest animal relatives, the great apes.

1993 To protest commuter traffic gridlock, Portugal's Socialist Party sponsors a race between a Ferrari 348 TS coupe and a burro. The course is the 1.6 miles of highway from Lisbon to the suburb of Odivelas. The burro wins by four minutes.

1994 The state of Florida begins collecting a 1.5% gross receipts tax on dry cleaners and laundry businesses to create a slush fund for cleaning up toxic spills of dry cleaning chemicals. Some 100 spills have been recorded in the state, and a number of drinking water wells have been polluted.

1608 The first known refracting telescope is offered by lens grinder-glasses maker Hans Lippershey to the Dutch government for use on the battlefield. The principle of aligning two lenses to produce great magnification was apparently a chance discovery by a Lippershey apprentice.

1746 Swedish mineralogist Peter Jacob Hjelm, discoverer of the element molybdenum, is born in Sunnerbo Härad.

1832 Botanist Julius von Sachs is born in Breslau, Germany. He showed that plants, like animals, respond to environmental stimuli, and he determined the process by which water travels from the ground up the stem to the leaves. His most important work was on plant nutrition (he discovered the role of chlorophyll and determined the location of chlorophyll in plant cells) and plant metabolism (he discovered that plants, like animals, also respire).

1836 HMS *Beagle* docks in Falmouth, England, after its 5-year voyage around Earth with its young naturalist Charles Darwin.

OCTOBER 2

1906 Space scientist, rocket engineer, and science popularizer Willy Ley is born in Berlin, Germany, from which he escaped when the Nazis rose to power.

1907 Scottish biochemist Sir Alexander Todd is born in Glasgow. He won the 1957 Nobel Prize in Chemistry for synthesizing nucleotides, the building blocks of DNA.

1917 Christian René de Duve, discoverer of lysozymes, the organelles that handle digestion within cells, is born in Thames Ditton, England. He shared the 1974 Nobel with George Palade and Albert Claude, who also made fundamental discoveries about how cells work.

1940 Automaker Freelan O. Stanley dies at 91 in Boston. In 1897 he and twin brother Francis began manufacturing the "Stanley steamer," history's best-known steam-driven car. The Stanley Motor Company continued production through World War I. In 1906 one of their cars was clocked at 127 mph, faster than any vehicle had ever gone before. Freelan outlived Francis by 22 years.

1948 Mary Leakey discovers a shattered skull of *Proconsul africanus* outside Rusinga, Kenya; it is the first skull of a fossilized ape ever found. The skull dates back 20 million years. Reconstruction of the skull was tedious. "Many of the pieces were about the size of a matchhead." Thirty-six pieces alone comprised one square inch of the jaw.

1956 The first nuclear-powered clock is unveiled at the Overseas Press Club in New York City. The "Atomicron," made by the National Company Inc., is 84 × 22 × 18 inches, and costs $50,000. Its timekeeping mechanism is driven by the cesium atom, which oscillates at a never-changing 9,192,830 megacycles per second.

1963 E.I. du Pont de Nemours and Company of Wilmington, Delaware, exhibits history's first artificial leather for shoes at a press conference. Shoes with upper sections made of "Corfam" appear in stores in 20 cities on January 27, 1964.

1968 North Cascades National Park (Washington State) and Redwood National Park (California) are established.

1985 Rock Hudson dies of AIDS in his home in Beverly Hills, California. The death of the beloved movie star was a wake-up call for many to the prevalence and danger of the disease.

1994 A 70-year-old mystery of the Russian czars is finally settled. Germany's *Der Spiegel* and London's *Sunday Times* publish DNA tests on Anna Anderson Manahan, a woman who claimed for 60 years to be Anastasia, the youngest daughter of Czar Nicholas II. He was Russia's last czar; he and his family (with the possible exception of Anastasia) were shot by Bolsheviks on the night of July 16–17, 1918, ending 300 years of Romanov rule. The DNA tests prove that Manahan, who died in 1984, was not related to Nicholas. The DNA was taken from a blood sample that has been kept in Heidelberg for 43 years, and the Romanov DNA was taken from a pit near Sverdlovsk, where the skeletons were finally discovered in July 1991.

1994 Cynthia Silveira gives birth to Hope, the first of two twins, in Good Samaritan Hospital in San Jose, California. The second twin Hailey is born eight days later, each twin came from a separate uterus, their mother being one in 50,000 women who has two uteri. The odds of having twin pregnancies in both uteri at the same time are astronomically small.

1226 St. Francis of Assisi dies at just 44 in Assisi, Italy. Founder of the Franciscan order, he was renowned for his respect for nonhuman animals, which stemmed from his belief that all of nature, not especially and not particularly man, manifests God.

1632 The first tax on tobacco, and apparently the first ban on tobacco use, in colonial America is authorized in Boston by the Massachusetts Court of Assistants and General Court. "No person shall take any tobacco publicly, under pain of punishment; also that everyone shall pay 1d. for every time he is convicted of taking tobacco in any place, and that any Assistant shall have power to receive evidence and give order for the levying of it, as also to give order for the levying of the officer's charge. This order to begin the tenth of November next."

1803 John Gorrie, inventor of the refrigerator, is born in Charleston, South Carolina.

1838 Darwin finishes reading *Essay on Population* by social scientist Thomas Malthus. The book gave Darwin the final insight with which he constructed the theory of evolution.

OCTOBER 3

1844 Sir Patrick Manson, the father of tropical medicine, is born the son of a bank manager in Aberdeen, Scotland. He was the first to discover that an insect could harbor a parasite that later infects man. Many diseases have since been shown to follow this model. Manson made his discovery in the late 1870s, working with mosquitoes that transmit the worms causing elephantiasis. In 1899 he founded the world-famous London School of Tropical Medicine. He was knighted in 1903.

1854 Army surgeon William Crawford Gorgas is born the son of a Confederate soldier in Mobile, Alabama. Gorgas's fame lies with his war against the disease-carrying mosquito. Gorgas was sent to Panama in 1904 to lead the eradication effort against mosquitoes, and was so successful that malaria and yellow fever were eradicated. This medical victory, more than any engineering feat, enabled completion of the Panama Canal in 1914.

1931 The first nonstop flight across the Pacific is begun by Major Clyde Pangborn and Hugh Herndon in a Bellanca cabin monoplane in Sabishiro, Japan. They landed in Wenatchee, Washington, 41 hours 13 minutes later.

1942 The V-2 rocket is first tested-fired, at Peenemünde, Germany. Nearly 4300 V-2s were launched from Germany against England from 1944 to 1945. Later, captured V-2s were the first space rockets used by the United States.

1948 The *Albatross* expedition ends in Göteborg, Sweden, where it began 15 months before. It was the first large-scale exploration of the ocean floor with modern oceanography instruments.

1955 The Mickey Mouse Club (on ABC-TV) and Captain Kangaroo (on CBS-TV) debut.

1969 The country's first outpatient vasectomy clinic opens at the Margaret Sanger Research Bureau, Inc. in New York City, under the direction of Dr. Aquiles Jose Sobrero.

1989 The record depth for a breath-held dive is set at 351 feet off Elba, Italy, by Ms. Angelina Bandini.

1992 Scottish scientists report in *The Lancet* that breast-fed babies have different brain chemistry than bottle-fed children. A fatty acid, docosahexaenoic acid, was more prevalent in postmortem examinations of the brains of the breast-fed group. The full significance of this chemical, and of the findings as a whole, is uncertain, but it supports evidence of the superiority of mother's milk to cow's milk.

1994 Universal Press Syndicate announces the retirement of Gary Larson, one of history's great science cartoonists. His drawings have been appearing for 15 years, and now are seen as *The Far Side* in nearly 1900 newspapers. Larson, 44, explains his retirement as "simple fatigue, and a fear that if I continue for many more years my work will begin to suffer or at the very least ease into the Graveyard of Mediocre Cartoons."

1995 O.J. Simpson is found not guilty of murder in California, in a blockbuster case about domestic violence and sociological conflict between races, in which DNA testing was given a major role during testimony.

OCTOBER 4

1582 This day is declared to be October 15 by Pope Gregory XIII, under advisement of an astronomical conference in Rome. The event marks the advent of the Gregorian calendar, used universally today.

1716 James Lind, the physician who determined the cause and cure of scurvy to be dietary (specifically, the lack of ascorbic acid), is born the son of a merchant in Edinburgh, Scotland. For these discoveries and for advocating many reforms in the treatment of sailors, Lind has been called "the father of naval hygiene."

1858 Inventor-physicist Michael Pupin is born the son of illiterate parents in Idvor, Austria- Hungary. At 15 he reached the United States penniless, but graduated from Columbia University 10 years later. Among his creations were the fluoroscope (a fluorescent screen permitting direct observation and easy photographing of X rays) and an improved method of transmitting signals over long stretches of wire. The Bell Telephone Company bought the innovation in 1901, making long-distance telephone calls practical.

1915 Dinosaur National Monument (Utah–Colorado), home of the world's most concentrated deposit of petrified bones of dinosaurs, crocodiles, and turtles, is established.

1932 *Woven Dreams*, the first antivivisection play to be seen in the United States, debuts in Philadelphia. Nina Halvey wrote the drama, which won the International Humanitarian Prize of 1931. The American Anti-Vivisection Society sponsored the showings. Antivivisectionists object to the practice of dissecting or cutting into live animals.

1946 Gifford Pinchot, the first U.S.-born silviculturist, a scientific manager of forests, dies at 81 in New York City (see September 19, 1892 for details of his life).

1951 History's second shopping mall opens (in Framingham, Massachusetts), 14 days after the first mall opened in Seattle.

1955 The first solar-powered telephone conversation (commercial) takes place in Americus, Georgia, over a distance of about 14 miles. "Hello, Gene. This is George Mathews. How many bales of cotton do I have in your warehouse?" are the first words. Mathew's phone is powered by a square-yard battery containing 432 silicon cells; it was built by the Bell Telephone Laboratories.

1957 The Space Age begins. Russia first launches a man-made object (*Sputnik I*) into orbit.

1958 The first transatlantic passenger jetliner service begins. BOAC flies one jet from New York to London, and a second jet in the opposite direction.

1959 The Russian *Cosmic Rocket III* is launched, destined to send back the first photographs of the dark side of the moon.

1970 The poster girl for the War on Drugs, rock 'n roll singer Janis Joplin, 27, is found dead in her Hollywood hotel room.

1971 The first wheeled vehicle on the moon, *Lunokhod 1*, breaks down. It had been roving the moon under remote command from Earth since the previous November 17. It had gone a total of 6.54 miles in the Mare Imbrium, and was able to navigate gradients up to 30°.

1991 In Madrid, Spain, 26 nations sign the Antarctic Treaty, prohibiting mineral exploration and mining on that continent for 50 years.

1994 A genetic basis for Alzheimer's has been discovered, announce Indiana University School of Medicine researchers in *Science*. A study of four generations of one afflicted Indiana family shows a universal defect in one gene on chromosome 21. The defect results in production of a substance called amyloid, which is found in Alzheimer's sufferers.

1708 The *Boston News-Letter* prints America's first advertisement for a patent medicine: "DAFFY'S Elixir Salutis, very good, at four shillings and sixpence per half pint bottle."

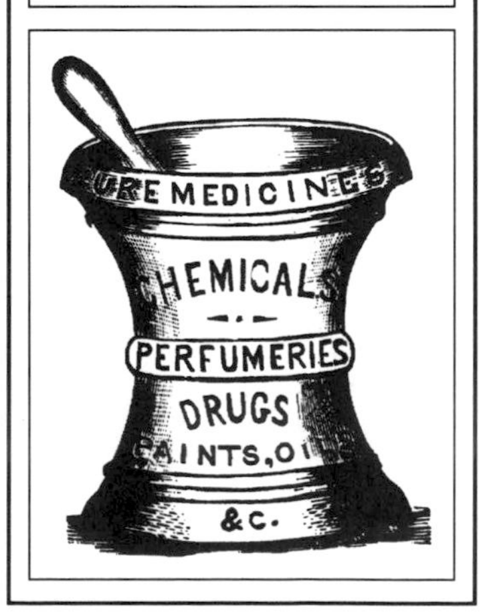

1853 Antioch College opens in Yellow Springs, Ohio, immediately becoming the nation's first liberal arts institute to offer courses in physiology, hygiene, and didactics, the science of teaching, and the first U.S. college to employ a female professor. Rebecca Mann Pennell taught physical geography, natural history, drawing, didactics, and civil history. Legendary educator Horace Mann is the college's first president.

1879 Pathologist Francis Peyton Rous, the first to discover that some cancers are caused by viruses, is born in Baltimore, Maryland. He published his discovery in 1911, but was not awarded his Nobel Prize until 1966, partly because of the fact that neither he nor the medical establishment believed in a cancer-virus link for many years; this interval of 55 years set a record. At the age of 87, Rous set another record as the oldest recipient of the Nobel Prize. He was still actively researching in his 90s.

1882 Robert Hutchings Goddard, the father of modern rocketry, is born the only child of a bookkeeper-salesman-machine-shop owner in Worcester, Massachusetts.

OCTOBER 5

1889 Physicist Dick Coster, codiscoverer of the element hafnium, is born in Amsterdam.

1902 Raymond Albert Kroc, founder of the McDonald's empire, is born in Chicago. Kroc was an ambulance driver at age 15 in World War I, and then held a number of postwar jobs, including jazz pianist, real estate salesman, and paper cup merchant. While selling blenders in 1954, Kroc received an order from a restaurant owned by brothers Maurice and Richard McDonald in San Bernardino, California that employed an assembly line to market burgers, fries, and milk shakes. Impressed by the exceptional number of blenders needed to keep up with the orders, Kroc decided to establish a chain of drive-in restaurants based on this model, and agreed to pay the McDonalds 0.5% of the gross receipts. The first McDonald's as we know them opened on April 15, 1955, in Des Plaines, Illinois. The rest(aurant) is history.

1921 Biochemist Mahlon Bush Hoagland, discoverer of transfer RNA, a critical molecule in the functioning of DNA, is born in Boston.

1930 Laura Ingalls departs Roosevelt Field in New York on the first transcontinental flight piloted by a woman.

1931 The first transpacific nonstop flight is completed in Wenatchee, Washington, by Clyde Pangborn and Hugh Herndon.

1947 In the first televised White House address, President Truman asks Americans to refrain from eating meat several days a week, so that the extra grain can be sent to starving peoples in Europe.

1966 A malfunction in the sodium cooling system at the Enrico Fermi demonstration nuclear reactor near Detroit causes partial meltdown of the core. The incident is more embarrassing than dangerous, as all radiation is contained.

1974 The first verified walk around the world ends in Waseca, Minnesota, where it began over four years before. David Kunst began the trek on June 10, 1970 with his brother John (who was killed in 1972 by Afghan bandits). Kunst wore out 21 pairs of shoes along the way.

1991 In Larsen Bay, Alaska, the bones of 756 Kodiak Island natives are reburied, 60 years after they had been removed without permission by Smithsonian Institution researchers. "It marks a change of heart, of sensitivity, between scientists who at one time were insensitive to natives, and the scientists of today," says Frank Talbot, director of Smithsonian's Museum of Natural History, at the reburial ceremony.

1992 Wire services announce that Mattel, Inc. will recall "Teen Talk Barbie" dolls and exchange them for mute models. The American Association of University Women and others had protested that Barbie's proclamations, like "Math class is tough," were sexist and could discourage girls from pursuing technical careers.

1854 The country's first baby show is held in Springfield, Ohio. It was announced largely as a joke, but parents took it very seriously and 127 babies were entered. The winner (the 10-month-old daughter of William Ronemus of Vienna, Ohio) was awarded a silver plate service and a $300 tray.

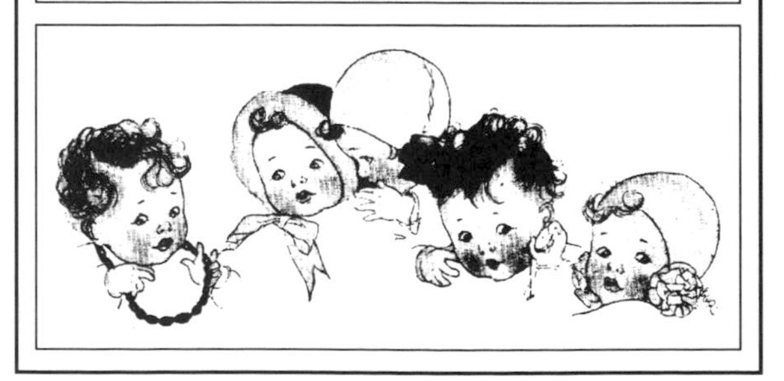

1732 Astronomer Nevil Maskelyne, the first to accurately measure time to a tenth of a second, and inventor of navigational methods and guides that were standards for over a century, is born in London.

1783 François Magendie is born in Bordeaux, France. He is credited with founding the science of nutrition, during classic studies during the Napoleonic Wars, in which he discovered that proteins are critical to life. He is also credited as the founder of experimental pharmacology because he was the first to test the effects of various chemicals on the human system; he introduced morphine, strychnine, and compounds with iodine and bromine into medical practice. Among his other accomplishments is the Bell–Magendie law—the back of the spinal cord is sensory, or carries messages from the sense organs to the brain, while the front is motor, or carries messages from the brain to the muscles.

OCTOBER 6

1807 The wild experimenter Sir Humphry Davy first produces and discovers potassium. He is only 18, working in his own lab in the recently founded Royal Institution in London. In a chance experiment, Davy passes an electric current through molten potash and lavender flames burst forth as potassium is released and contacts the air. "Capital experiment!" writes Davy in his notebook. Several days later he becomes the first to produce sodium using a similar apparatus. Through his fabled career, Davy was known for having chemistry experiments explode, and for producing gases that he inhaled recklessly.

1825 A giant is first exhibited as a theater attraction in the United States. "Just arrived from Ireland" and "conspicuous for the masculine beauty of his form and his surprising strength," Patrick Magee can be seen from 7 AM to 10 pm at 13 Park, Park Exchange, New York City, for 25 cents.

1846 Inventor George Westinghouse is born in Central Bridge, New York, where his father ran a machine shop and farm-implement factory. Westinghouse made a fortune from inventing the air-powered brake, and was mainly responsible for acceptance of alternating current as the world's standard in electric power transmission.

1866 Reginald Fessenden, inventor of AM radio and the first to broadcast voice and music over long distances, is born the son of a minister in Milton, Quebec, Canada.

1903 E.T.S. Walton, codeveloper of the first nuclear particle accelerator, known as the Cockcroft–Walton generator, is born the son of a priest in Dungarvan, Ireland. He shared the 1951 Nobel Prize with Sir John Douglas Cockcroft of England.

1914 Anthropologist-adventurer Thor Heyerdahl is born in Larvik, Norway. He led both the *Kon-Tiki* (1947) and *Ra* (1969) transoceanic expeditions to prove theories of migration by ancient man. His theories were never accepted.

1927 *The Jazz Singer* opens. It is history's first major talking movie. It begins an era in both entertainment and technology.

1939 Harvey Cushing, a pioneer in surgery on brain tumors, dies of a brain tumor at 70 in New Haven, Connecticut.

1943 To escape possible imprisonment by the Nazis, nuclear physicist and Nobel laureate Niels Bohr takes an airplane from Sweden (where he had been aiding Jews escape Hitler) to England. The plane was tiny, and Bohr almost died of lack of oxygen.

1993 A 5-year-old fan of MTV's "Beavis and Butt-head" cartoon show sets fire to his family's mobile home in Moraine, Ohio. One person dies as a result. It is the second such incident in Ohio. The network vows to "reexamine issues regarding Beavis and Butt-head."

1994 Results from "America's most comprehensive survey of sexual behavior" are announced by researchers with the University of Chicago and the State University of New York at Stony Brook. Researchers are surprised to find that the sex lives of Americans are conventional and based in marriage. "Unlike what we're led to believe by watching movies and reading novels, most people have few partners and have rather infrequent sex," said coauthor Robert Michael.

1783 The first U.S. patent application for a self-winding clock is filed by Benjamin Hanks of Litchfield, Connecticut, for a "clock or machine that winds itself up by help of the air and will continue to do so without any other aid or assistance."

1806 Carbon paper is patented by the Englishman R. Wedgewood.

1833 England's George Stephenson files the first patent on a high-speed steam locomotive.

1855 François Magendie dies one day after his 72nd birthday in Seine-et-Oise, France (see October 6, 1783).

1856 A folding machine for books and newspapers is first patented in the United States. Cyrus Chamber, Jr. of Kennett Square, Pennsylvania, receives patent No. 15,842 for his right-angle three-fold machine, which was installed in the Bible printing house of Jasper Harding & Son in Philadelphia.

1885 Nuclear physicist Niels Bohr is born the son of a physiology professor in Copenhagen. Both Bohr and his son, Aage Bohr, were to receive Nobel Prizes.

1905 *Collier's* magazine begins a series of articles entitled "The Great American Fraud" by freelancer Samuel Hopkins Adams, which exposed rampant fraud and deceit in the marketing of medicines.

1919 The oldest airline still in service, KLM, or Koninklijke Luchtvaart-Maatschappij, in the Netherlands, is established.

1927 Psychiatrist-mystic R.D. Laing is born in Glasgow, Scotland.

1931 The newly invented process of infrared photography is first used to photograph a large number of people. A picture is taken of 50 visitors to the Eastman Kodak Research Laboratory in Rochester, New York, in a totally dark room that is flooded with invisible infrared rays.

1935 The people of Denmark present one-half gram of radium to physicist Niels Bohr in honor of his 50th birthday.

1957 The first large nuclear accident occurs, in England's Windscale plutonium reactor near Liverpool. A fire scatters radioactive material over the countryside. Authorities seize all milk and growing foodstuffs within 400 miles of the plant. In 1983 the British government estimated that 39 people died of cancer as a result of the mishap.

1959 The dark side of the moon is photographed for the first time, by Russia's *Lunik III* spaceprobe.

1822 Zoologist Rudolf Leuckart, founder of the science of parasitology, by being the first to describe the complicated life cycles of several parasites, is born the son of a printing plant owner in Helmstedt, Germany. Leuckart first made his mark by demonstrating that jellyfish and starfish are not related, even though they share a similar circular body plan.

1960 Senator John Kennedy and Richard M. Nixon engage in their second televised debate. Kennedy's narrow victory in the coming presidential election has been attributed to his success and his more appealing appearance in the televised debates; it is a historic lesson in the power of television.

1963 President John Kennedy signs a nuclear test ban treaty with Russia and England.

1992 The Philippine government moves against dog-eating. Rep. Salvador Escudero, an ex-vegetarian, files a bill that would outlaw mistreatment or killing of stolen dogs and cats. Although uncommon in cities, the stealing of dogs for purposes of making hors d' oeuvres in beer halls is common in the countryside. Escudero defended his legislation as a means to prevent humans from catching canine-borne diseases.

1994 In the first medical conference of its kind, 60 doctors, nurses, and lightning victims gather at Tampa General Hospital for a Lightning Strike–Electrical Shock Conference. "There's an amazing amount of ignorance in the health care profession regarding this issue," said Penny Ackerly, 40, who was hit by lightning at a Miami toll booth. Coincidentally, today is the 17th anniversary of the greatest nonlethal electric shock on record; in 1977 Harry F. McGrew received and survived a blast of 340,000 volts from a high-tension line in Huntington Canyon, Utah.

1754 One of the earliest expeditions to hunt Stellar's sea cow, a relative of the manatee, reaches Bering Island off Alaska. The sea cow was hunted into extinction 26 years after discovery.

1850 Henri-Louis Le Châtelier is born in Paris, the son of France's inspector general of mines. Among his achievements were invention of the thermoelectric couple for measuring high temperature, which consists of two metals stuck together that generate an electric current when heated, and Le Châtelier's principle, which states that if one factor changes in an equilibrium, other factors rearrange so as to minimize the change.

1871 The first major forest fire in U.S. history occurs near Peshtigo, Wisconsin, north of Green Bay. The blaze reaches a width of 8–10 miles, and destroys over a million acres of trees. The area was in the midst of a three-month drought.

1883 Biochemist Otto Warburg is born the son of a physics professor in Freiburg im Breisgau, Germany. He explored how cells use oxygen, and the chemicals involved in this process, and won the 1931 Nobel Prize for Physiology or Medicine for these studies. He was offered a second Nobel Prize in 1944, but was forced to refuse it by the Nazis because he was Jewish. Warburg's international standing kept him from imprisonment.

1901 Australian physicist Marcus Laurence Oliphant is born in Adelaide. In 1934 he created tritium, the only radioactive form of hydrogen, which paved the way for others to create the hydrogen bomb and to advance nuclear fusion technology. Oliphant also proposed the design of the first proton synchrotrons, the most powerful particle accelerators available.

1906 The hot permanent wave for hair is first demonstrated, by a German hairdresser named Nestle working in London. It is the first "perm" method on record. The hair was soaked in ammonia, pinned in place, then heated intensely with an iron. Nestle took his method to the United States, where it achieved great success under the name "permanent wave."

1909 An earthquake rips through the Kulpa Valley in central Europe. The event leads geophysicist Andrija Mohorovicic, 51, to discover the boundary between Earth's crust and mantle—now called the Mohorovicic discontinuity—after carefully analyzing the earthquake's outward movement to measuring stations in the area.

1917 Biochemist Rodney Robert Porter, the first to determine the four-chain structure of antibodies, is born in Newton-le-Willows, Lancaster, England. He shared the 1972 Nobel Prize with Gerald M. Edelman who determined the same structure by different methods.

1929 Franklin W. Stahl is born in Boston. In 1958 with Mathew Meselson, using bacteria, he was the first to determine that DNA copies itself in the cells of living creatures. This process of "replication" ensures that genetic blueprints are passed from cell to cell and from parent to offspring.

1945 President Truman announces that the secrets of the atomic bomb will be shared only with Canada and England.

1968 The first international astronaut rescue treaty is ratified by a 68-0 vote of the U.S. Senate. On December 3, President Lyndon Johnson publicly announced that the "Agreement on the Rescue of Astronauts, the Return of Astronauts and the Return of Objects Launched into Outer Space" was in effect. Signatories are the United States, the United Kingdom, and Russia.

1994 Washington State implements the nation's toughest-ever ban on smoking. The habit, prohibited in stores, state offices, and most restaurant spaces, now cannot occur in private offices or associated areas.

1994 The Associated Press reports that repairmen working on Egypt's 4600-year-old Sphinx have found a new, mysterious passage leading deep into the body. The evidence suggests that the tunnel was created in pharaonic times. As yet, no treasure has emerged from the passageway. In 1987 another unknown entrance was discovered in the Sphinx; it yielded several shoes and a newspaper from the early 1900s.

1000 Norwegian explorer Leif Erickson enters the Western Hemisphere, landing at present-day Nova Scotia or Newfoundland (some historians believe that the Icelander Bjarni Herjulfsson beat Erickson to the New World by 14 years). Like the rediscovery of the Americas by Christopher Columbus 400 years later, Erickson's discovery was an accident. Erickson was trying to sail from Greenland back to Norway when a storm blew him off course.

1253 Scholar Robert Grosseteste dies at about 78 in Buckden, England. He introduced direct translations of Aristotle and other ancient Greeks to European Christendom. He is also known as Roger Bacon's teacher.

1676 Antonie van Leeuwenhoek writes a second letter from his home in Delft, the Netherlands, to London's Royal Society, describing his monumental discovery of one-celled animals. The event marks the birth of microbiology. The first letter (in 1674) stirred little interest, but this time the response is great, partly because Leeuwenhoek reports that a simple solution of water and pepper contains huge amounts of tiny organisms, and are therefore unknowingly ingested by humans.

OCTOBER 9

1780 The first U.S. astronomy expedition to view a solar eclipse sets out from Harvard College in Cambridge, Massachusetts. Four professors and six students sail in a boat provided by the Commonwealth of Massachusetts to Penobscot Bay, Maine. Although at war with the United States, the British commander at Penobscot allows the group to land and observe the eclipse on October 27, 1780. It lasted from 11:11 AM to 1:50 PM.

1843 Gregor Mendel, destined to discover the laws of genetics, becomes a novice at the monastery in Brünn, Austria-Hungary.

1852 Biochemist Emil H. Fischer is born in Euskirchen, Prussia, the son of a successful merchant who unsuccessfully urged his son to enter business. Instead, Fischer was influenced by the great chemist August Kekulé. Fischer is famous for his work with (1) sugars—he showed that the most common sugars, like glucose and fructose, are made of six carbon atoms, with differences in properties corresponding to differences in the arrangements of the carbons, (2) "purines"—a class of compounds he named and whose structure he determined which are now known to be a major part of DNA, and (3) proteins—he showed how amino acids are linked to form protein, and he devised methods to link amino acids in the lab. Fischer was the sole recipient of the 1902 Nobel Prize for Chemistry.

1879 Physicist Max von Laue, the first to diffract X rays with crystals—a technique that revealed much about the structure of the crystals, the X rays, and other substances, including DNA—is born the son of a soldier in Pfaffendorf, Germany. He won the 1914 Nobel Prize.

1855 Isaac M. Singer patents the sewing machine motor.

1890 The first flight by an airplane is made in Armainvilliers, France, by Clément Ader in his craft *Eole*. It is not a sustained flight, which was the great achievement of the Wright brothers 13 years later, but a "hop" of 164 feet.

1936 Boulder Dam (later called Hoover Dam) begins transmitting electricity to Los Angeles.

1937 Jimmy Angel crashes his plane into a mountain near Angel Falls, Venezuela, the world's tallest waterfall, at a total drop of 3212 feet, which he had discovered in 1935.

1946 The electric blanket is first manufactured. The Simmons Company of Petersburg, Virginia, produces and markets the device for $39.50.

1987 Laboratoires Roussel applies in Paris for the first license to manufacture and market the controversial female contraceptive pill RU486.

1992 Indian scientists in New Delhi announce that they have produced and tested a birth-control vaccine. Not only does one shot last an entire year, but there are no apparent side effects and the vaccine may work to block fertilization, preventing objections by antiabortionists.

1731 Physicist-chemist Henry Cavendish is born into an aristocratic English family in Nice, France. He made many fundamental discoveries, including the composition of air, the nature and properties of hydrogen, the specific heat of several substances, and the structure of water.

1810 The now-defunct *Berlin Evening News* publishes an article entitled "Preliminary Thoughts About Mortar Mail," suggesting that mail could be carried by rocket.

1816 English physician Sir John Simon is born in London. His efforts to improve urban hygiene helped modernize public health.

1861 Explorer-oceanographer-humanitarian Fridtjof Nansen is born in Store-Frøen, Norway.

1865 The ivory-colored billiard ball is patented. John Wesley Hyatt receives patent No. 50,359 for the ball, which also won a $10,000 prize from the firm of Phelan and Collender in New York City for the best ivory-substitute ball. Hyatt eventually received four different patents on the composition and manufacture of billiard balls; one of these patents (in 1869) was for celluloid, the first synthetic plastic.

1881 Charles Darwin's last book is published. *The Formation of Vegetable Mould, through the Action of Worms, with Observations on Their Habits* is published in London by John Murray. It is Darwin's 21st book over a 42-year span, and appears 6 months before his death.

1886 The tuxedo coat makes its debut, in Tuxedo Park, New York. Griswold Lorillard is said to have introduced the tailless dress coat from England, worn on this date with a scarlet satin vest.

1913 Engineers dynamite the Gamboa Dam during construction of the Panama Canal, thereby uniting the Pacific and Atlantic Oceans.

1916 Physician Jean Dausset, destined for the 1980 Nobel Prize for exploring the link between genetics and immunological reactions, is born in Toulouse, France.

1930 The cyclotron—a new type of atom smasher—is first described in an article in *Science* by E.O. Lawrence, later a Nobel laureate for inventing the device, and coauthor N.E. Edlefson.

1931 L.S.B. "Louis" Leakey, 28, leads a small party, including E. Vivian Fuchs, later one of the great explorers of Antarctica, into Olduvai Gorge, Tanzania, for the first time. Leakey became famous for his many discoveries of the remains of human ancestors in the renowned Olduvai Gorge.

1971 London Bridge is rededicated at Lake Havasu City, Arizona. The Bridge remains the largest antique ever sold. It was purchased in 1968 by McCulloch Oil Corporation of Los Angeles and transported, stone by stone, to Arizona at a cost of $7.2 million (three times the cost of its puchase price of $2,469,600).

1980 One of the world's largest telescopes is completed in the New Mexico desert. The VLA, or Very Large Array radio telescope, consists of 27 mobile antennas riding on 13-mile-long arms.

1992 The nation's first mass screening for depression is held in 100 hospitals and medical schools. As many as 10 million Americans, mostly female, are regularly depressed, but only one in three is properly treated.

1994 Cynthia Silveira gives birth to Hailey, the second of unusual twins that developed in two separate uteri in the mother (see October 2, 1994).

1802 Geologist-theologian-science popularizer Hugh Miller is born in Cromarty, Scotland. Considered one of the finest science writers of the 1800s, Miller worked as a bank accountant and newspaper editor before his *The Old Red Sandstone* appeared in 1841; it described 408- to 360-million- year-old fossil discoveries he had made in Scottish Devonian strata. It established his reputation as a science writer who appealed to a lay audience. Miller's other great book, *Footprints of the Creator* (1849), which focused specifically on fossil fish, argued that their perfection in structure disproved Darwin's theory of evolution. Miller's writings are largely responsible for why the Devonian period is known as the "Age of Fishes."

1755 Mineralogist Don Fausto D'Elhuyar, codiscoverer of the element tungsten, is born in Logroño, Spain. He and older brother Juan accidentally discovered the metal while working on a mineral called wolframite in 1782; tungsten was originally called wolfram, and its chemical symbol remains W.

1758 German astronomer Heinrich Wilhelm Olbers is born the 8th of 16 children of a minister in Arbergen. He discovered the minor planets Pallas and Vesta and five comets, including Olbers' Comet. He also proposed Olbers' paradox: given that the universe is flooded with light from an infinite number of stars, why is the sky dark at night?

1811 The first steam-powered ferryboat, the *Juliana*, is put into service between New York City and Hoboken, New Jersey.

1881 Roll film for cameras is patented. David H. Houston of Cambria, Wisconsin, receives patent No. 248,179 for "photographic apparatus" involving a camera and a "roll of sensitized paper or any other suitable tissue, such as gelatine that may be discovered, and an empty reel, upon which the sensitized band is wound as rapidly as it has been acted upon by the light." The device anticipated motion pictures because its stated purpose was "to facilitate taking a number of photographic views successively in a short time."

OCTOBER 11

1881 Lewis Fry Richardson, physicist, psychologist, and the first to apply mathematical techniques to accurately predict the weather, is born in Newcastle upon Tyne, England.

1887 Dorr Eugene Felt of Chicago receives patent No. 371,496 for the first adding machine in the United States that was accurate and error-free. Felt worked with Robert Tarrant to produce the "comptometer," and until 1902 it was the only multiple-order, key-operated machine on the market.

1939 President Franklin Roosevelt receives and reads a letter from Albert Einstein stating that an atomic bomb is both possible and advisable, in light of nuclear fission's discovery in Nazi Germany. Roosevelt is convinced, and he orders funding for the Manhattan Project ($6000 was initially allocated in February 1940). After 5 years and at a cost of $2 billion, the atomic bomb had been built and exploded.

1958 The second U.S. moonshot, *Pioneer 1*, is launched. It is a little more successful than the first attempt (an unnamed rocket that blew up shortly after takeoff the previous August), but it does not attain enough speed to leave Earth's gravity and burns up in the atmosphere.

1962 Erich Tschermak, a codiscoverer of the laws of genetics, dies at 90 in the city of his birth, Vienna. The son of a mineralogy professor, Tschermak became a botanist who began a series of breeding experiments with peas in 1898. Two years later, while writing the results, he encountered an obscure reference to the work of Gregor Mendel, who 33 years before had described the very same results and laws Tschermak was writing about. The same year as Tschermak'a discovery, Hugo de Vries and Karl Erich Correns also reported Mendel's work and similar genetic experiments of their own.

1802 The first parachute patent is issued to Frenchman Jacques Garnerin.

1968 *Apollo 7* is launched. It is the first three-man U.S. spaceflight—carrying astronauts Schirra, Eisele, and Cunningham—and the first manned flight in the *Apollo* series of launches that successfully lands a man on the moon.

1982 The English warship *Mary Rose* is raised to the surface off Portsmouth, England, having been sunk by the French over 400 years before. Most of the frame is intact.

1984 *Challenger* astronaut Kathy Sullivan becomes the first woman to take an untethered walk in space.

1992 The first pig-to-human organ transplant takes place in Los Angeles, when doctors insert the pig liver into Susan Fowler, 26, as a stopgap measure until a suitable human liver can be located. Fowler shows signs of improvement at first, but dies the next day of acute liver failure just as a human liver becomes available.

1773 America's first state-run hospital for the insane opens in Williamsburg, Virginia. Zachariah Mallory of Hanover County was the first patient at the Publick Hospital for Persons of Insane and Disordered Minds. The name was later changed to the Eastern State Hospital.

1820 John James Audubon, 35, leaves his wife and a failed business career in Cincinnati, and sets off down the Ohio River to New Orleans to devote himself to painting. It is the beginning of his career as a renowned naturalist.

1860 Inventor Elmer Ambrose Sperry is born the son of a farmer in Cortland, New York. The most famous of his 400 patents is the gyroscopic compass—the first essential improvement in compasses in 1000 years—which developed from a toy.

1862 Cytologist Theodor Boveri, the first to show that chromosomes exist in cells all of the time, rather than forming just during cell division, is born in Bamberg, Germany, the son of a doctor.

OCTOBER 12

1865 Sir Arthur Harden, winner of the 1929 Nobel Prize for Chemistry, and the first to discover a coenzyme (a trace molecule essential to enzyme action), is born the son of a businessman in Manchester, England.

1881 Physicist Albert A. Michelson, 30, begins the first of a series of 23 measurements of the speed of light. It was the greatest passion of his professional life, and his measurements were the best for many years. The last of the calculations was made over 40 years later.

1915 Legendary nurse Edith Cavell is executed at age 49 by German troops in occupied Belgium.

1928 The iron lung is first used in a hospital, on a young girl with polio-induced breathing failure at Children's Hospital in Boston. The prototype for the machine was invented by Harvard's Philip Drinker and Louis Agassiz Shaw, using two household vacuum cleaners, hand-operated valves, and a crude iron tank. Consolidated Gas Company of New York gave $7000 to Harvard to manufacture a second model, which was produced by Warren E. Collins Inc. of Boston.

1928 The nation's first institute devoted to medical teaching, treatment, and research is dedicated in New York City. The Columbia-Presbyterian Medical Center, at 168th Street west of Broadway, opened on March 6, 1928, with 100,000 books and a capacity for 600 students. Its first dean of faculty was William Darrach.

1964 Russia launches a *Voskhod* capsule with three astronauts aboard. It is the first time any ship has carried more than one person into space.

1965 Chemist Paul Hermann Müller, who discovered that DDT could be an exceptional insecticide, dies at 66 in Basel, Switzerland (see January 12, 1899).

1992 On this Columbus Day, the Brazilian science journal *Goeldiana* reports the discovery of a new treasure from the New World: A previously unknown monkey species has been found in the Amazonian rain forest. The "Maues marmoset" (named after the river where it was discovered) is pocket-sized, koala-faced, and striped like a zebra. It is the third new monkey species found in Brazil in the last two years. The country is now known to have 68 primate species, more than one-quarter of all of the world's primates.

1992 NASA scientists celebrate Columbus Day by launching a new expedition—the search for life beyond Earth. Powerful radio telescopes at Goldstone, California, in the Mojave Desert, and on Mount Arecibo, Puerto Rico, are connected to huge antennas and computerized analyzers; they will be looking for signals from aliens and other life in the universe. "Columbus set off across the terrestrial ocean, and now we're standing on the shores of the cosmic ocean, searching for intelligent life out there," said project manager Michael Klein.

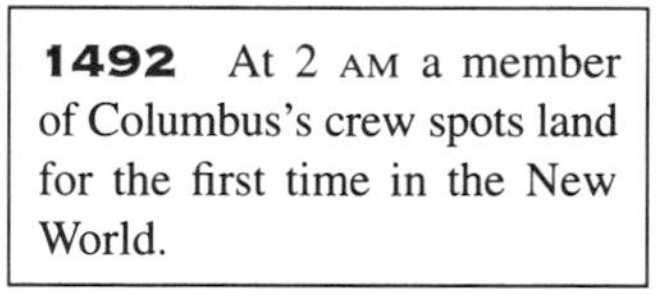
1492 At 2 AM a member of Columbus's crew spots land for the first time in the New World.

OCTOBER 13

1796 Swedish anatomist Anders Adolf Retzius is born in Stockholm, the son of a natural history professor. Retzius's greatest contribution was in anthropology, in which he made extensive skull measurements in an effort to distinguish one race from another. Although the effort was eventually futile in distinguishing different races, it did lay the groundwork for others (using different physical features) to scientifically distinguishthe races.

1805 The *Investigator* expedition returns to England after five years exploring the geography and natural history of Australia. The chief naturalist on board was famous Scottish botanist Robert Brown.

1860 The nation's first aerial photograph is taken by two men in a balloon tethered 1200 feet above Boston. Samuel Archer King navigated the balloon and James Wallace Black operated the camera. They take eight exposures, but only one is usable. Wet plates are used, and each is processed right in the balloon.

1901 Ecologist Orlando Park is born in Elizabethtown, New Jersey. His *Principles of Animal Ecology* helped establish this new branch of science. His fieldwork largely centered on the pselapids, a family of small beetles that often live with ants.

1908 Biophysicist Robley Cook Williams is born in Santa Rosa, California. Perhaps his greatest contribution came from his hobby: astronomy. Williams noticed that lunar mountains became highly visible when struck by sunlight at oblique angles; he believed that tiny objects sprayed with a metal film at such angles might enhance their visibility when viewed through the electron microscope. He was right. In collaboration with Ralph Wyckoff, Williams developed techniques that permitted the first three-dimensional pictures of very small objects.

1940 Visual long-distance communication between deaf persons is first accomplished orally and visually. Bertha O'Donnell and Adele Costa "talk" to each other in sign language over a distance of 8 miles through a two-way television hookup in New York City.

1945 Chocolate pioneer Milton S. Hershey dies at 88 in Hershey, Pennsylvania. After a rural school education, Hershey went to work as an apprentice confectioner in Lancaster, Pennsylvania. At age 19 he established his own candy store in Philadelphia, but after it failed, he moved it to New York City, only to fail again. Hershey then returned to Lancaster where he finally achieved success; he founded a caramel company which he sold for $1 million in 1900. He then went to work perfecting a chocolate bar; and by 1903 he had built a factory near his hometown, Hockersville, Pennsylvania, which became the world's largest chocolate-making plant. The town of Hershey developed around the plant. In his later years Hershey became a philanthropist, especially aiding the cause of orphan boys.

1821 Rudolph Virchow, statesman, anthropologist, and a founder of modern pathology, is born the son of a small merchant in Schivelbein, Prussia.

1962 The Edward Albee play *Who's Afraid of Virginia Woolf?* opens on Broadway in New York City. Uta Hagen stars as Martha and Arthur Hill as George. As well as providing great entertainment, the drama is a pioneering analysis of psychology, including the study of game-playing, role-playing, and conflict resolution.

1986 Dutch scientists at the German Alfred Wegener Institute measure visibility in the Weddell Sea off Antarctica, establishing it as the clearest seawater on Earth. A 1-foot Secchi Disk can be seen to a depth of 262 feet.

1992 The United States joins 53 other nations, as President George Bush signs a treaty to help minimize man's effects on Earth's climate.

1993 Clones of human embryos have been created in a laboratory, report George Washington University biologists at a joint meeting of the American Fertility Society and the Canadian Fertility and Andrology Society.

OCTOBER 14

1066 The Battle of Hastings, which changed the government and culture of England forever, is fought. Historians have attributed William's victory over Harold II to one small technological advantage: the stirrup.

1788 Astronomer-geodesist Sir Edward Sabine, noted for studies of Earth's shape and its magnetism, is born in Dublin, Ireland. In 1852 he discovered that sunspots and Earth's magnetism fluctuate together; this was the first link known between the sun and Earth, other than the sun's radiation of light and its gravity.

1827 Anthropologist John Ferguson McLennan is born in Inverness, Scotland. He developed influential theories on cultural evolution, kinship, and the origins of religion, and introduced the terms "endogamy" and "exogamy" to describe different marriage patterns in different cultures.

1834 Henry Blair of Glenross, Maryland, obtains a patent on a corn planting device, becoming the first African American in history to receive a patent. In 1836, he obtained a patent for a cotton seed planter.

1863 Alfred Nobel is granted his first patent; it is a Swedish patent for preparing nitroglycerine. During his lifetime he obtained 355 patents.

1885 In Villers-Farley, France, 14-year-old J.B. Jupille kills a rabid dog with his bare hands, thereby protecting five other young shepherds. Jupille is bitten and develops rabies, but he is saved when he is taken to Louis Pasteur's laboratory and becomes the second person to be treated with Pasteur's experimental vaccine.

1891 Zoologist Sir James Gray is born in London. As an experimenter and a longtime editor of the *Journal of Experimental Biology* (1923–1954) Gray helped expand scientific zoology from the narrow field of comparative anatomy to physiology and function. This progress was essential in understanding how animals are built and how they function. Gray's research focused on the movement mechanisms in cells and in whole animals. He pioneered the application of engineering principles to study biology. Gray wrote a number of popular books on cells and animal movement, and was knighted in 1954.

1947 Chuck Yeager becomes the first human to fly faster than sound, breaking the sound barrier in the rocket-powered plane *XS-1* over Murac, California.

1968 The first live telecast from a manned US spacecraft is beamed back to Earth from *Apollo 7*.

1980 The genetics research company Genentech first offers shares of stock for public sale. Within minutes the price per share jumps from $35 to $89. This is one of the greatest leaps in value in Wall Street history.

1993 An Australian study in the *New England Journal of Medicine* provides the first scientific proof that sunscreen prevents skin cancer. The 588 volunteers were randomly assigned to get either SPF (sun protection factor) 17 sunscreen or a similar placebo cream. By the end of the summer those volunteers with the real sunscreen averaged a decrease in keratoses, or precancerous growths, while those without real protection showed an increase in such keratoses. Dr. Darrell Rigel of New York University Medical School commented, "It's the first time we have been able to definitely show that sunscreen lowers the risk of getting skin cancer later in life."

1994 Twenty-seven-year-old Fyona Campbell becomes the first woman to walk around the world. It took her 11 years. During the 19,586-mile trek, Campbell wore out 100 pairs of shoes, was nearly attacked in Morocco, and had to be rescued by the French Foreign Legion in Zaire. She ends her journey in John o'Groat's, Scotland. Near tears, hugging her father, Campbell says, "I don't think my walking days are over. But my walking days by myself are, thank God."

1920 The first college radio station in U.S. history goes on the air. WRUC of Union College in Schenectady, New York, begins a series of weekly 30-minute broadcasts of vocal and instrumental records. The programs could usually be heard over a radius of 50 miles, or more if weather conditions were right.

1608 Evangelista Torricelli is born in Faenza, Italy. He invented the barometer, and was the first to create a sustained vacuum. These feats came from experiments suggested by Galileo, for whom Torricelli served as secretary and companion for the last three months of his life. Torricelli died at just 39 of typhoid fever.

1783 Man's first test flight occurs when France's Jean-François Pilatre de Rozier ascends in a hot-air balloon tethered to land in Paris. A month later de Rozier and a partner used the same balloon and no tether for the first true flight above land.

1789 John Morgan, the nation's first professor of medicine and founder of its first medical school, dies at 54 in Philadelphia. Educated in Europe, Morgan returned to North America in 1765 to establish a medical training institute within the College of Philadelphia, now the University of Pennsylvania, where he was appointed its first professor. He advocated a liberal education for trainee doctors, and he thought that surgery, pharmacy, and medicine should be taught as individual subjects. His ideas, too advanced for their time, were widely criticized by other colonial doctors, and were eventually abandoned. Morgan was appointed surgeon general of the colonial army during the Revolution. He attempted reforms that again were attacked by colleagues, and he was forced from office by a coalition of enemies. Although later exonerated of wrongdoing by George Washington, Morgan never recovered completely; he died in poverty, alone, ten years later.

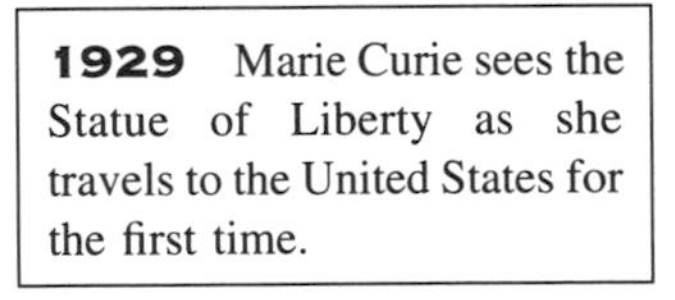

OCTOBER 15

1827 Charles Darwin, 18, is admitted as a student to Christ College, Cambridge. He had already disappointed his family by quitting his medical training in Edinburgh, and his career at Cambridge was undistinguished.

1878 The first electric company in U.S. history incorporates. The Edison Electric Light Company in New York City immediately sells 3000 shares for $100 to finance the invention of the incandescent lamp by Thomas Alva Edison. Two years later another Edison company formed to furnish electric lighting to New York City.

1880 Marie Stopes is born in Edinburgh, Scotland. She was a crusading advocate of birth control, who braved violent opposition to found the United Kingdom's first clinic for contraception education in 1921.

1900 The first building with scientifically designed acoustics, Boston Symphony Hall, opens.

1910 Cell biologist Torbjörn Caspersson is born in Motala, Sweden. He pioneered the use of ultraviolet microscopy in studying nucleic acids, and explored the role of RNA in protein synthesis.

1926 Social philosopher Michel Foucault is born in Poitiers, France. He is best remembered for his studies of the rules that control human societies. His first major interest was the history of mental illness and its therapy, during which he developed the idea of societal "principles of exclusion" (for example, distinctions between the sane and insane, the haves and have-nots). His *Madness and Civilization* (1965) dealt with classification of insanity in the 1600s. Foucault then turned his attention to the history of modern prisons and wrote *Discipline and Punishment: The Birth of the Prison* (1977). Finally he examined sexuality in his *History of Sexuality* (1978), which was a study of Western attitudes toward sex from the ancient Greeks to the present.

1929 Marie Curie sees the Statue of Liberty as she travels to the United States for the first time.

1990 A 100-pound mushroom is found in the New Forest, Great Britain. It is the heaviest edible fungus on record. It is an *L. sulphureus*, nicknamed the "chicken of the woods."

1992 Harvard astronomer Brian Marsden announces the disturbing results of some calculations on the comet Swift-Tuttle: He realizes that there is an appreciable chance it will hit Earth on August 14, 2126. He adds that because the comet crosses Earth's orbit, "sooner or later, it will hit us. Over a million years, there's an excellent chance of being hit."

1994 The environmental group Greenpeace launches a campaign to reduce the use of products made with chlorine, which has been linked to breast cancer.

1708 Albrecht von Haller, the father of experimental physiology, is born the son of a lawyer in Bern, Switzerland. Poor health forced him into quiet activities as a child, and he soon displayed exceptional intellect; he started writing on scholarly topics at age 8, and by 10 he had written a Greek dictionary.

1793 John Hunter, surgeon and the founder of pathological anatomy in England, dies at 65 in London. An early advocate of experimental biology, his death was hastened by venereal disease, which he gave himself to demonstrate, incorrectly, that syphilis and gonorrhea are the same disease. As with many surgeons in the 1700s, Hunter never graduated from college, learning his craft instead as an apprentice and by dissecting cadavers during lectures by his famous brother William Hunter. John held various posts, eventually becoming the physician extraordinary to King George III in 1776. Among his students was the discoverer of vaccination, Edward Jenner.

1843 One of science history's great "flashes of genius" strikes mathematician William Rowan Hamilton, 38, as he walks along Royal Canal to Dublin, Ireland. In an instant he sees the solution to a problem he has been working on for 10 years, that of algebraically dissecting three-dimensional space. He has invented the theory of quaternions, a landmark in the development of algebra. Hamilton etches the fundamental formula of quaternions into the stonework of Brougham Bridge, before he can get to his desk.

1848 Instruction begins at America's first college specializing in homeopathic medicine, in which a disease is treated by giving patients herbs and other substances that produce symptoms of that disease. The Homeopathic Medical College of Pennsylvania graduated its first class of six students on March 19, 1849. Dr. Walter Williamson was the first dean of the Philadelphia school.

1914 The first blood transfusion of World War I takes place in Biarritz, France. Corporal Henri Legrain, 45th Infantry, is brought from the front after suffering massive blood loss; he is saved by a transfusion from a Breton soldier, Isodore Colas. The technique is still experimental, in that blood groupings in humans had only recently been discovered.

1916 Margaret Sanger opens the world's first birth control clinic in the poverty-ridden Brownsville section of Brooklyn, New York City. A circular announcing the event was printed in English, Yiddish, and Italian.

1927 The first remnant of the human ancestor Peking man is discovered on Dragon-Bone Hill, outside Peking, by Swedish paleontologist Birgir Bohlin. It is a tooth. The 6-month expedition was due to end in just 3 days.

1951 A motion picture taken inside the heart of a living creature is first shown, at a clinical session of the New York Academy of Medicine at Montefiore Hospital, New York City. The 9½-minute color film shows the opening and closing of the mitral valve of a dog's heart. It was made by Drs. Elliott Hurwitt and Adrian Kantrowitz and photographer Anatol Herskovitz.

1964 China becomes the fifth nation to explode the atomic bomb.

1982 Halley's Comet is sighted for the 33rd time on record, by astronomers using the 200-inch Hale Telescope on Mount Palomar, California.

1991 Arlette Schweitzer, 42, leaves St. Luke's Midland Hospital in Aberdeen, South Dakota, after giving birth to her own grandchildren. She is the first American to be implanted with fertilized eggs from her daughter. She gives birth to twins, a boy and a girl. The youngsters are their mother's brother and sister, and their own aunt and uncle.

1846 Surgical anesthesia is first demonstrated publicly. In a theater full of doctors at Massachusetts General Hospital in Boston, dentist William T. G. Morton removes a tumor from the jaw of a man unconscious from ether inhalation. "Gentlemen, this is no humbug," is the famous appraisal of Dr. John Collins Warren after the operation.

1757 Scientific jack-of-all-trades René-Antoine Réaumur dies at 74 in Saint-Julien-Terroux, France (see February 28, 1683).

1833 Paul Bert, founder of aerospace medicine, is born the son of a lawyer in Auxerre, France. His research into the effects of pressure on the body enabled travel into outer space and into the ocean depths.

1834 James Bogardus of New York City receives the first U.S. patent for a dry gas meter. His "gasometer" used the filling and emptying of a bellows to indicate the quantity of gas being used.

1886 Pathologist Ernest William Goodpasture is born in Montgomery County, Tennessee. In 1931 he developed a technique for growing viruses in fertile chicken eggs, which enabled others to create vaccines for many diseases.

1902 On Mount Sabinio in the Congo, a family of mountain gorillas (*Gorilla gorilla beringei*) is first seen by a westerner, Oscar von Beringe. He immediately shoots and kills two gorillas.

OCTOBER 17

1903 Samuel Pierpont Langley, the first to build an unmanned machine that successfully flew, attempts his first manned flight with his assistant, Charles M. Manly as the pilot. It is unsuccessful. He tries again 2 months later, but crashes into the Potomac River. Nine days later the Wright brothers make their historic flight.

1933 Albert Einstein arrives in the United States, a refugee from the Nazis.

1949 A long-distance telephone number is dialed for the first time. In New York City, Mark Sullivan, president of the Pacific Telephone and Telegraph Company, dials Oakland, California, and talks with Keith S. McHugh, president of the New York Telephone Company, and Dr. Oliver E. Buckley, president of the Bell Telephone Laboratories. The call goes through in about a minute.

1958 Six Yugoslavian scientists are accidentally irradiated at the Nuclear Institute of Vinca. They are taken to the Pierre Curie Hospital in Paris, where they undergo the first *mass* bone marrow transplant in history (although the technique was first used on humans in 1957). Five of the six patients fully recover in 4 months.

1973 Arab oil-producing nations announce cutbacks in oil exports to Western nations and Japan. Days later an all-out embargo is declared. The embargo lasts until the following March.

1986 *Pioneer 10* crosses the orbit of Pluto and becomes the first man-made object in history to travel so far from Earth; it is 3.67 billion miles away at that point.

1991 An experimental program to rehabilitate drug dealers by teaching them farming is a success, according to an Associated Press report. Inmates at the San Francisco County jail in San Bruno run the 12-acre farm, which now supplies seven city soup kitchens, feeding 3000 people a day. "Now they're giving back and starting to feel better about people," said program founder Cathy Marcum about the prisoners.

1991 The *New England Journal of Medicine* reports that a very common bacterium has been linked to a very common and deadly cancer. The microbe, *Helicobacter pylori*, had already made science history in the 1970s when it was proved to cause ulcers; now a Stanford study has shown that persons infected with *Helicobacter* at an early age have a drastically elevated risk of developing stomach cancer later in life. It is still unknown why people with ulcers somehow seem protected against stomach cancer; and why many more people are infected with *Helicobacter* than actually develop cancer.

1820 Astronomer-mathematician Édouard Albert Roche is born in Montpellier, France. His best-known contribution, the Roche limit, is a calculation of the nearest a satellite can get to the body it orbits before it is destroyed by tidal forces. This limit is about 2½ times the diameter of the primary body. Because the rings of Saturn lie entirely within the Roche limit, it is guessed that the rings are composed of debris from a demolished moon.

1799 Chemist Christian Friedrich Schönbein, discoverer of ozone, is born of poor parents in Metzingen, Germany. Self-taught, he also discovered the explosive guncotton. This discovery resulted from an accident in his kitchen in 1845; although strictly forbidden by his wife to work in this room, Schönbein was experimenting with nitric and sulfuric acids there one day when his wife was not home, and accidentally spilled some. In haste he reached for the first thing at hand to wipe up the liquids. This was his wife's apron, which he then hung over the stove to dry before she came home. When dried, the apron evaporated in a smokeless flash; by chance Schönbein had produced nitrocellulose, the first smokeless explosive.

1842 Telegraph cable is first laid. Samuel Morse supervises the laying of an insulated copper line from the Battery to Governors Island in New York Harbor. The next day signals were transmitted perfectly until a ship's anchor snagged on the line and destroyed 200 feet of the cable. The following year, another historic inventor, Samuel Colt, again laid a cable across New York Harbor. This time the line was heavily protected, with cotton yarn, beeswax, and asphalt, all set in lead pipe. The line connected New York City with Coney Island and Fire Island.

OCTOBER 18

1861 One of the first U.S. physicians to achieve renown in Europe, Dr. James Marion Sims (the founder of gynecology in the United States) demonstrates a new fistula operation in front of a host of medical elite in the Hôtel Voltaire in Paris. The patient, a 40-year-old woman who had been suffering for 20 years with a fistula—an abnormal passage from an organ to the skin or to another organ—recovers completely, which, in Sims's words, "created a *furore* among the profession in regard to the curability of an affection which they had until now supposed to be totally incurable."

1870 Sandblasting is patented. Benjamin Chew Tilghman of Philadelphia receives patent No. 108,408 for "cutting and engraving stone, metal, glass, etc ... with sand as a projectile."

1896 History's first cartoon strip ("The Yellow Kid") debuts in the *New York Journal*.

1931 Thomas Edison dies at 84 in West Orange, New Jersey.

1962 The international team of James Watson, Francis Crick, and Maurice Wilkins are named winners of the Nobel Prize for Physiology or Medicine for determining the structure of DNA.

1962 *Ranger 5* is launched. Electrical failure causes the craft to miss its target, the moon, by 450 miles, causing it to fall into orbit around the sun. This negative result was typical of the *Ranger* series of moonshots; it was not until *Ranger 7* (June, 1964) that a U.S. spaceship reached the moon with functional photographic equipment.

1967 The first telescope-carrying space observatory is launched. It is NASA's Orbiting Solar Observatory 0504.

1969 The artificial sweeteners, cyclamates, are banned as cancer-causing by the federal government.

1989 Space shuttle *Atlantis* is launched on a five-day mission, which will include deployment of the Galileo space probe bound for Jupiter.

1990 The *Sacramento Bee* reports today that hundreds of birds were killed on purpose after the *Exxon Valdez* oil spill. The U.S. Department of Justice and the U.S. Fish and Wildlife Service allegedly paid $600,000 to Ecological Consulting Inc. of Portland, Oregon, to conduct a study in which some 350 birds, including auklets, eiders, and scoters, were shot, fitted with radio transmitters, smeared with oil and tossed back into the sea. The idea was to follow the corpses' movement and thereby, somehow, estimate the number of bird deaths in the actual oil spill. The data became part of the legal case against Exxon.

1993 Splashing "hot sauce" on raw oysters may save your life! LSU researchers report at a meeting of the American Society for Microbiology that various toppings can kill harmful and possibly fatal microbes. "Some of the findings were a little astonishing to us. We had no idea these condiments would be so powerful," said researcher Kenneth Aldridge. The scientists acknowledge that the germs produced in test tubes in the lab, differ from the naturally occurring germs living in raw oysters.

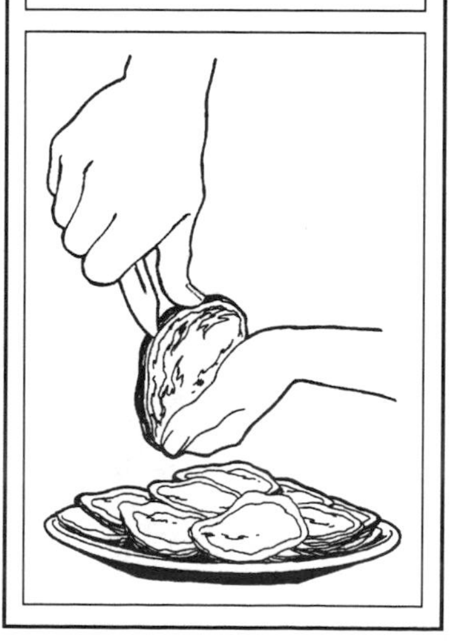

1856 Biologist Edmund Beecher Wilson, the first to note the existence of X and Y chromosomes and their correlation with the sex of an individual, is born the son of a lawyer in Geneva, Illinois.

1868 Louis Pasteur, 46, suffers a massive stroke on the way to work; his life is in danger for a week.

1909 Cesare Lombroso, pioneer in scientific criminology, dies at 73 in Turin, Italy. In 1876 Lombroso was a psychiatry professor at the University of Pavia when he published *The Criminal Man*, which proposed the notion of the "atavistic criminal," a wrongdoer by birth who was a throwback to an early stage in human evolution. The book also suggested that physical characteristics could identify a criminal in some cases. Although these views are now discounted, Lombroso was important because his work produced (1) more humane and constructive treatment of criminals and (2) more scientific analysis of criminal behavior.

OCTOBER 19

1910 Astrophysicist Subrahmanyan Chandrasekhar, who determined the origin of black holes, is born in Lahore, Pakistan. He was a 1983 Nobel laureate for calculating what happens when stars exhaust their energy and collapse. His best-known computation (performed in 1928 on a ship from India to England) is the "Chandrasekhar limit": any star with a mass more than 1.44 times that of our sun will eventually collapse into a neutron star or, if especially massive, into a black hole; whereas stars smaller than the 1.44 figure will become white dwarfs.

1926 The semiautomatic rifle is patented by John C. Garand of Somerset, Maryland.

1941 Grandpa's Knob, Vermont, becomes the nation's first community to receive electrical power from a wind turbine. Palmer Putnam invented the turbine, the output of which was phased into the Central Vermont Public Service Corporation's system for two hours on this date. The wind blew at 28 mph, and power output reached 800 kilowatts.

1950 An illegal television station is closed by the FCC for the first time in broadcast history. Using a 90-foot tower on Whittemore Mountain, the Tube Division of Sylvania Electric Products in Emporium, Pennsylvania, had been televising programs from WJAC-TV (Johnstown, Pennsylvania) without a license.

1871 Physiologist Walter Bradford Cannon is born in Prairie du Chien, Wisconsin. In Cannon's hometown, William Beaumont had conducted a classic series of experiments on digestion in the early 1800s. Then in the early 1900s Cannon also conducted important digestion experiments. He devised a "bismuth meal," which allowed him to X-ray the movement of food through the digestive tract. This was the first use of X rays for physiological research of any kind. Cannon is also famous for studying how the body responds to stress.

1959 *The Miracle Worker*, William Gibson's play about the life of Helen Keller, opens in New York City.

1977 The asteroid *Chiron* is discovered between Saturn and Uranus by Charles Kowal at the Hale Observatory in California. At the time it was the most distant of the known asteroids ("minor planets").

1977 The French–English supersonic jet *Concorde* flies into New York City for the first time.

1990 U.S. Fish and Wildlife spokesman Craig Rieben acknowledges a backlog of 601 species that deserve immediate protection, but are not yet classified as "endangered." Another 3000 species are probably endangered, but "are receiving little or no protection." At least 34 U.S. species became extinct in the 1980s.

1993 The star of *Free Willy*, a 14-year-old killer whale named Keiko, will receive improved medical and living facilities in its Mexico City enclosure. This is the announcement of the Alliance of Marine Mammal Parks and Aquariums, a Washington-based organization of some 30 exhibition parks, which negotiated an agreement with Keiko's current home, Reino Aventura. The behemoth will receive immediate care for its skin infection, other health problems will be investigated, new filtration and cooling equipment will be provided, and a new home will be considered (eventually Keiko did get a new home, in a tank in Oregon).

1616 Anatomist-mathematician Thomas Bartholin is born in Copenhagen. In 1652 he became the first to fully describe the human lymphatic system. He was also one of the earliest defenders of Harvey's discovery of the circulation of blood. From 1670 to 1680 he was physician to Denmark's King Christian V. His brother Erasmus discovered double refraction of light, and his father Caspar was a famous anatomist.

1632 "If you would see his monument, look about you" are the words on the tomb of Christopher Wren (born on this date in East Knoyle, England). Wren is buried in London's St. Paul's Cathedral, which he had built. Wren was the greatest English architect of his time, and he was also a noted astronomer and mathematician. He was a founding member of the Royal Society, and served as its president. His speculations on gravity lay the groundwork for Newton's discoveries.

1813 British forces bomb the city of Danzig with Congreve rockets—the most famous black-powder rockets—during the Napoleonic Wars. It is a great victory for the missiles. Congreve rockets were also used in 1814 in the British attack on Fort McHenry, near Baltimore, about which Francis Scott Key wrote, "the rockets' red glare, the bombs bursting in air" in the *Star Spangled Banner*.

OCTOBER 20

1846 The first U.S. college to give equal rights to women, Mount Union College in Alliance, Ohio, is founded. The Reverend Orville Nelson Hartshorn established the Methodist Episcopal school as "Mount Union Seminary." Females are granted full degrees and allowed to stand on the platform at commencement, which was highly unusual at the time.

1891 Sir James Chadwick, discoverer of the neutron, is born in Manchester, England. Not only was the discovery (1932) of academic interest because it helped explain the structure of atoms, but it also had tremendous practical importance. Having large mass but no electric charge, the neutron is especially effective in smashing atoms, and the first nuclear fission and the first atomic bombs used neutrons. Chadwick received the 1935 Nobel Prize for Physics.

1902 For months, the number of cases of beriberi had been climbing steadily among the inmates of Bilibad Prison in Manila, the Philippines, ever since the well-intentioned U.S. administrators had changed the prisoners' diet to white polished rice. On this day, the diet is changed back to "dirty," unpolished brown rice. The results are startling; beriberi cases fall immediately. These events provide an important clue in the discovery of vitamins.

1936 Anne Sullivan (teacher of the blind and deaf Helen Keller) dies at 70 in Forest Hills, New York.

1961 The country's first vending machine to dispense live flowers is installed in Grand Central Station, New York City, by the Automated Flowers Company of Greenwich, Connecticut. Manufactured in Denmark, the machine stands 6 feet tall, 3 feet wide, and 2 feet deep; it uses regular wall current to run its refrigerator, and requires no outside plumbing.

1983 The length of the meter is redefined, for the first time since the meter was created nearly 200 years before. The international body *Conférence Générale des Poids et Mesures* (CGPM) decides that the meter will now be equal to the distance that light travels in a vacuum in 1/299,792,458 of a second. The old definition was 1/10,000,000 of the distance from the North Pole to the Equator.

1984 History's largest artificial environment for marine life, the Monterey Bay Aquarium in California, opens. It houses over 6500 marine animals from at least 525 species.

1993 Astronauts on the space shuttle *Columbia* perform a variety of medical experiments. Blood samples are taken to analyze why bones deteriorate in space, and stress tests are administered on exercise bicycles. The astronauts also face an unexpected problem when their toilet springs a leak.

1994 Rutgers scientists report at a news conference that the world's fastest-growing worms are found on a volcanic ridge 1½ miles deep in the Pacific ocean.

1520 The Strait of Magellan is first navigated. Ferdinand Magellan, 40, leads the expedition around the southern tip of South America. One of the original five ships had already abandoned the expedition, the other ships were nearly out of water, and the crews were eating rat-fouled food and leather.

1639 The first medical law in America is passed by Virginia. It is an "act to compel physicians and surgeons to declare on oath the value of their medicines."

1660 Georg Ernst Stahl, the first to scientifically explain burning, is born in Ansbach, Germany, the son of a minister. Stahl was a famous lecturer and a physician (to royalty), and one of the first to realize that burning and rusting were related processes. He developed the phlogiston theory that held that combustible substances were rich in a hypothesized substance (phlogiston), which was lost during burning. Eventually disproved, this theory was important for over a century because it was the first systematic account of a chemical transformation.

OCTOBER 21

1833 Swedish inventor-philanthropist Alfred Bernhard Nobel is born in Stockholm. The family fortunes changed in 1842 when the Russian government bought an underwater mine that the elder Nobel invented, and moved him and his family to St. Petersburg to supervise its manufacture. Alfred studied in the United States from 1850 to 1854; on returning to Russia he found his father working on the powerful but uncontrollable explosive nitroglycerine. Alfred continued this work, and discovered (by accident) a way to control the substance. In 1866 Nobel was experimenting on a barge in the middle of a lake (so as to minimize the hazards to others) when he found that a cask of nitroglycerine had leaked and mixed with packing material, composed of "diatomaceous earth" (silicone-rich skeletons of millions of tiny sea creatures called diatoms). Experiments revealed that the combination was extremely stable, and could only be ignited with a detonating cap. Nobel called the combination "dynamite." His invention brought him fame and wealth. Alfred remained a bachelor, and in his will established the Nobel Prizes with his fortune.

1849 A tattooed man is first exhibited in the United States, at the Franklin Theater, Chatham Square, New York City. Advertisements in the New York *Herald* read, "The manager has at an enormous expense engaged Mr. J.F. O'Connell, the wonderful 'Tattooed Man' who will go through a variety of performances peculiar to himself, and perfectly original."

1879 Thomas Edison invents the first successful incandescent light bulb, at Menlo Park, New Jersey. In 1931 the anniversary of this event was celebrated with Edison's funeral and burial, on a hillside overlooking West Orange, New Jersey (where Edison moved his "invention factory" in 1887), now aglow with electric light.

1877 Bacteriologist Oswald Avery, a founder of immunochemistry, and the first to show that DNA (not protein) carries genetic instructions, is born the son of a clergyman in Halifax, Nova Scotia.

1907 The human ancestor Heidelberg man is discovered in a commercial sandpit near Mauer, Germany, when a laborer strikes a skull with his shovel, splitting the priceless relic in half.

1959 Frank Lloyd Wright's Guggenheim Museum opens in New York City, six months after Wright had died.

1967 The era of antiship missiles begins, when the Israeli destroyer *Eilat* is attacked by Egyptian sentry boats firing Russian-made Styx missiles.

1976 President Gerald Ford signs the Resource Conservation and Recovery Act, which establishes federal standards on toxic waste disposal.

1991 Balloons fly over Mount Everest for the first time. Two British crews (each consisting of a cameraman and a pilot) achieve the feat in the hot-air ships *Star Flyer 1* and *Star Flyer 2*.

1993 A cream that removes fat from women's thighs is announced by the Associated Press, reporting on a meeting of the North American Association for the Study of Obesity. Safer than liposuction and easier than exercise, the cream causes a media avalanche. Scientists reported remarkable results in the two studies, but these studies involving just a few women, have not yet been duplicated by others. The active ingredient in the cream is aminophylline, an asthma medicine that can be purchased over the counter.

1511 German mathematician Erasmus Reinhold, the first to prepare planetary tables based on Copernicus's highly controversial theory that Earth orbits the sun, is born in Saalfeld.

1783 Naturalist-traveler-educator Constantine Rafinesque is born in Gelata, Turkey. Well known and controversial in his lifetime, Rafinesque wrote several books and more than 950 articles on natural history, religion, banking, and literature; he made important contributions to ichthyology and botany, and he set the stage for Darwin's theory of evolution with his original notion that each variety of a species is a "deviant" that may become a permanent species through time and reproduction.

1797 The first true parachute jump by a human is made by inventor André-Jacques Garnerin, who ascends a half-mile above Paris in a balloon. He then cuts the cord holding the basket to the balloon and lands safely.

1825 Charles Darwin, 16, enters Edinburgh University for a brief and futile attempt to become a doctor.

1877 Frederick William Twort, discoverer of bacteriophages (tiny organisms that invade bacterial cells), is born the son of a physician in Camberley, England. Bacteriophages have since been extremely important in experiments on genetics and cell function.

1881 Physicist Clinton Joseph Davisson is born the son of a paperhanger in Bloomington, Illinois. He shared the 1937 Nobel Prize for Physics for discovering that electrons can be diffracted (just like light), thereby confirming de Broglie's theory that electrons behave like waves and like particles. Davisson made his discovery by accident. In 1925 a vacuum tube fell to the floor and shattered in his lab, causing a nickel target in the tube to develop a film; Davisson heated the nickel to remove the film, which crystallized the nickel into an object that diffracted electrons.

1884 George Washington Hill is born in Philadelphia. He was a Madison Avenue wizard who made a fortune in the marketing of cigarettes. His great stroke of genius, which forever changed the cigarette industry, was targeting women specifically with cigarette ads.

1903 George W. Beadle, geneticist, Noble laureate, and president of the University of Chicago, is born on a farm in Wahoo, Nebraska.

1905 Karl Jansky, discoverer of radio waves from outer space, is born in Norman, Oklahoma.

1934 History's first instantaneous production of a phonograph record (without the intermediate step of creating a wax master) occurs. Professor Jean Picard and his wife record and describe their balloon flight from Ford Airport in Detroit. The couple spends the next eight hours in the air, including a two-hour expedition into the stratosphere. The flight ends in a treetop in Cadiz, Ohio.

1938 Inventor Chester Carlson produces history's first xerographic image.

1939 A professional football game is televised for the first time. The NBC station W2XBS televises the 23–14 defeat of the Philadelphia Eagles by the Brooklyn Dodgers in Ebbets Field in Brooklyn.

1975 The Russian *Venera 9* becomes the first man made object to orbit Venus.

1992 The origin of AIDS remains a mystery; one theory is that humans first contracted the disease during the testing of a monkey-derived polio vaccine in the Congo in 1957. This theory was proposed earlier this year in *Rolling Stone* magazine, and subsequently investigated by a team of specialists who announce on this date "with almost complete certainty" that the vaccine trial was not the AIDS genesis. The most telling evidence against the theory is the case of a British sailor who had been in Africa at that time, but who returned to England with AIDS before the trial began.

1843 Stephen Moulton Babcock, the father of scientific dairying, is born near Bridgewater, New York. He was an agricultural chemist who studied both in the United States and Germany, finally settling for good at the University of Wisconsin. He developed the Babcock test (introduced in 1890), which was a simple method of gauging the amount of butterfat in milk. The test helped detect, regulate, and prevent milk adulteration, spurred development of dairy methods, and improved the production of cheese and butter.

October 23

4004 BC Earth is created at 9:00 A.M on this day according to long-esteemed calculations published in 1650 by Archbishop Ussher of Armagh, Ireland. It is often incorrectly reported that Ussher made the calculations, when it was actually an anonymous cleric who produced them from a detailed, creative study of the Bible.

1873 Physicist-engineer William Coolidge is born a distant cousin of President Calvin Coolidge in Hudson, Massachusetts. Coolidge's improvements to tungsten filaments were essential to the modern electric light bulb and the modern X-ray tube.

1893 Soviet cosmologist-astronomer Ernst Julius Öpik is born in Port Kunda, Estonia. His research focused on meteors and what happens to them as they enter our atmosphere; his work has been used to design satellites that will return to Earth from space.

1896 Sigmund Freud's father (Jakob) dies. Freud, 40, is in shock at first, then overcome with long-repressed feelings and emotions toward his father and toward his earliest familial experiences. The incident strengthens Freud's attempt to understand the mind, and directly leads to Freud's development of dream interpretation (begun in earnest, on himself at first, in July 1897).

1905 Physicist Felix Bloch, developer of NMR (nuclear magnetic resonance) technology/theory, is born in Zurich, Switzerland. Bloch shared the 1952 Nobel Prize for his work, which has helped explain magnetism within atoms and electrical conduction, and which has led to new tools for diagnosing disease.

1752 Nicholas Appert, inventor of the bouillon cube and of food preservation by using hermetically sealed containers, is born in Châlons-sur-Marne, France. His food preservation invention is the basis for modern canning methods, and took 14 years of experimentation to develop.

1908 Physicist Ilya Mikhaylovich Frank, winner of the 1958 Nobel Prize for explaining Cherenkov radiation, is born in St. Petersburg, Russia (see July 8, 1895).

1910 Blanche S. Scott becomes the first woman to pilot a public, solo airplane flight. She reaches an altitude of 12 feet at Driving Park in Fort Wayne, Indiana.

1913 Edwin Klebs, one of the first to link bacteria to disease, dies at 79 in Bern, Switzerland. He was able to establish this link in endocarditis, tuberculosis, and syphilis. In 1878, he was the first to transmit syphilis from man to monkey, thus providing science with a valuable nonhuman subject with which to study the disease. In 1884 with Friedrich Löffler he discovered the killer diphtheria bacillus, now known as the Klebs–Löffler bacillus.

1936 Senator Arthur Harry Moore dedicates the country's first old age colony, at Roosevelt Park in Millville, New Jersey. It rents seven houses for couples (at a cost of $7 a month), six houses for singles ($5 a month), and a community house. Millville donated land that had been taken over for taxes; the structures were built by the WPA. The plan was the brainchild of Effie Morrison, deputy director of the Cumberland County Welfare Board.

1956 A video recording on magnetic tape is televised coast to coast for the first time. The Jonathan Winters show is aired from 7:30 to 7:45 PM by WRCA-TV, New York City. The magnetic tape process, developed by RCA for the National Broadcasting Company, allowed immediate playback and erasing, both impossible with film.

1992 *Science* reports yet another gene flaw is linked to Alzheimer's disease. University of Washington researchers found that an abnormality in one arm of chromosome 14 predisposes a person to develop one form of the incurable disease that annually cripples some 4 million Americans. Other researchers have recently reported that flaws in chromosomes 19 and 21 are also linked to Alzheimer's, which reaffirms that the disease is caused by a variety of processes.

1993 "Drive-through" flu shots become available, at the University of Kansas Medical Center in Kansas City.

OCTOBER 24

1601 Tycho Brahe, the last and probably the greatest naked-eye astronomer, dies at 54 in Prague (see December 14, 1546).

1632 Antonie van Leeuwenhoek, the first to see bacteria and one-celled animals, is born the son of a basketmaker in Delft, the Netherlands. Leeuwenhoek was a linen draper-haberdasher by trade, but his passion was grinding lenses, originally begun so that he could inspect cloth in detail. His great skill at this allowed him to make exceptional microscopes.

1804 Physicist Wilhelm Eduard Weber is born in Wittenberg, Germany. He is best known for studies of magnetism (in partnership with friend Carl Friedrich Gauss), and the basic unit of magnetism is now called the weber in his honor.

1831 Charles Darwin, 22, joins HMS *Beagle* at Plymouth Harbour, but the actual start of the historic trip is repeatedly delayed until late December. "These two months at Plymouth were the most miserable which I ever spent," he later wrote.

1836 Matches are patented. Alonzo Dwight Phillips of Springfield, Massachusetts, receives patent No. 68 for "manufacturing of friction matches." The head of Phillips's match contains chalk, phosphorus, glue, and brimstone. It replaces the cumbersome, dangerous practice of scraping bits of sulfur through sandpaper.

1854 Dutch physical chemist Hendrik Roozeboom is born in Alkmaar. He is known for popularizing and experimenting with the phase rule (derived by theorist Josiah Willard Gibbs) that describes how physical conditions affect the three phases, or states, of matter—gas, solid, and liquid.

1861 A telegraph message is first sent across North America, from Stephen Johnson Field (chief justice of the California Supreme Court) to Abraham Lincoln.

1901 Ms. Anna Edson Taylor becomes the first person to go over Niagara Falls in a barrel. Protected by cushions and a leather harness, she drops over Horseshoe Falls on the Canadian side in a barrel 4½ feet in height and 3 feet in diameter. She survived the ride, but died in poverty in 1921.

1939 The age of synthetics dawns. Nylon stockings are first sold to the public, in Wilmington, Delaware.

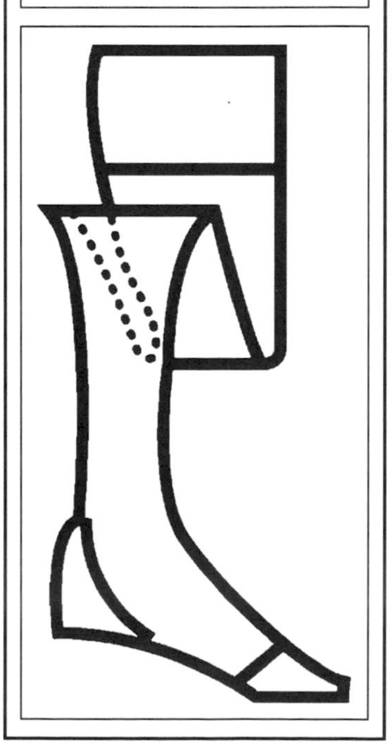

1944 One of the two largest battleships ever built, the Japanese *Musashi*, is sunk in the Philippine Sea. Its sister ship *Yamato* was sunk the following April.

1956 President Dwight Eisenhower concludes in a Washington speech that the testing of nuclear weapons is safe. "The continuance of the present rate of H-bomb testing, by the most sober and responsible scientific judgement ... does not imperil the health of humanity."

1991 Dr. Jack Kevorkian is released from police custody in Pontiac, Michigan, after questioning about the double physician-assisted suicide that he conducted the night before.

1993 "If society would look at this [the cloning of human embryos] askance and not wish scientists to proceed, then it's better to have that ethical debate currently now, before anyone would do the real thing," says Dr. Robert J. Stillman, director of George Washington University's *in vitro* fertilization program, during a CNN interview. "The real thing" has almost happened; his colleagues announced just 11 days before that they successfully cloned human embryos that were abnormal and could not develop to viable individuals. Biomedical ethicist Arthur Caplan calls human cloning "staggering, in terms of its ethical implications." After all, if a cloned embryo produces a "good" child, dozens more just like it could be produced if the clones were properly stored in a refrigerator.

1789 Amateur astronomer Samuel Heinrich Schwabe, the first to sketch Jupiter's Great Red Spot in detail, and the first to see that the number of sunspots cycles regularly, is born in Dessau, Germany.

1826 Philippe Pinel, the man who "unchained the insane," dies at 81 in Paris. Prior to his work, insanity was commonly thought to be the result of demoniac possession, and those afflicted were often locked up, fettered, purged, bled, steamed, and blistered. When Pinel became the chief physician of Bicêtre (the Paris asylum for men) in 1792, his first bold reform was to remove the chains from inmates—some of whom had been chained for 40 years—and to institute therapy involving counseling, friendly contact, and meaningful activity. In 1794 he became the director of the female facility of Salpêtrière and instituted the same changes.

1870 The Averill Chemical Paint Company of New York City registers a "trade-mark for liquid paint." It is the first official trademark in U.S. history. It was filed under an act of July 8, 1870. After 121 registrations, the act was declared unconstitutional. The Lanham Act of 1946 (U.S. Code, Title 15, Chap. 22) established the current system of trademark protection.

1875 Gilbert Newton Lewis, the first to describe chemical bonds as electron-sharing and electron-transfer, is born the precocious son of a lawyer in West Newton, Massachusetts. Lewis was also the first to isolate deuterium—the heavy isotope of hydrogen, which later became a major component in the first atomic weapons.

1877 Astronomer Henry Norris Russell is born the son of a Presbyterian minister in Oyster Bay, New York. He is remembered for his method of calculating distances to binary stars and for developing the Hertzsprung–Russell diagram, which describes the relationship between brightness and color in stars.

1881 An air-brush painting device is first patented. Leslie L. Curtis of Cape Elizabeth, Maine, receives patent No. 248,579 for an "atomizer for coloring pictures."

1888 Richard Evelyn Byrd, explorer of both the North and South Poles, is born in Winchester, Virginia.

1904 Sword-swallowing champion Alex Linton is born in Ireland. He later emigrated to Sarasota, Florida. His great claim to fame was "swallowing" four 27-inch blades—all at the same time.

1929 Albert B. Fall, ex-Secretary of the Interior, is convicted of accepting a $100,000 bribe in connection with the Elk Hills Naval Oil Reserve in California.

1935 Astronaut Russell L. Schweickart is born in Bailey's Corner, New Jersey. During the March 1969 flight of *Apollo 9*, Schweickart and James McDivitt gave the Lunar Landing Module its first test in outer space; the pair flew the Module for 100 miles, and Schweickart took a 40-minute spacewalk from it.

1975 The earliest programmable electronic computer, the 1500-valve *Colossus* (developed in the early 1940s by British intelligence officers to break the German coding machine *Enigma*), is declassified.

1984 Power begins flowing from the Itaipu power plant on the Paraná River on the Brazil–Paraguay border. It remains the largest power plant on Earth, eventually reaching an output of 13,320 megawatts.

1984 *Komsomolskaya Pravda* reports the discovery of more than 1800 stone artifacts frozen in the permafrost near Yakutskaya, Siberia. Estimated to be about 2 million years old, the tools are some of the oldest human implements on record.

1990 For the first time in history, in Stanford, California, a lung is transplanted from a living donor. The risk of tissue rejection is reduced because the donor is living, and because the donor is the mother of the recipient.

1992 Brain surgery was performed 4200 years ago. The Antiquities Authority of Israel announces in Jerusalem that archaeologists recently found an 8-year-old child from the Bronze Age with an oblong matchbox-sized hole in the top of its skull. The operation probably was done with metal scrapers, either to relieve pressure or to drain an infection. Such operations, called trephining, are known to have occurred as early as 8000 BC, but this is the first case from the Eastern Mediterranean involving a child.

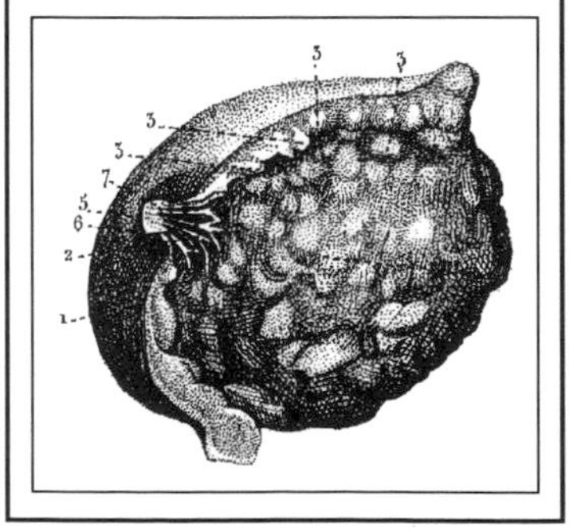

1812 *Medical Inquiries and Observations Upon the Diseases of the Mind* by Benjamin Rush, M.D., is copyrighted. Published by Kimber and Richardson of Philadelphia, it is the first U.S. book on psychiatry.

1825 The Erie Canal opens in upstate New York, joining the Hudson River with Lake Erie. It is the first canal of significance in North America.

1858 The washing machine is patented. Hamilton Erastus Smith of Philadelphia receives patent No. 21,909 for a manual, rotary motion, crank-operated device consisting of a wooden tub with a perforated, moving cylinder in the center. Five years later Smith received a patent on the first washing machine with a self-reversing action.

1863 Soccer, one of the oldest and most popular games in history, becomes standardized, with the formation of the Football Association in England.

1877 For the first time, a broken kneecap is repaired by opening the patient's skin and wiring the bone pieces together. The operation's greatest significance is that the surgeon is Joseph Lister, who is using his new antiseptic techniques for the first time in a major city (London). The operation was never attempted before because of the high risk of infection before Lister's innovations. The operation on this day convinces many in the medical establishment that cleanliness prevents infection.

1898 French inventor Eugene Ducretet sends a wireless telegraph message through the air between the Eiffel Tower and the Pantheon in Paris, a distance of 4 km. It sets a new distance record for the telegraph.

1956 A brief, dramatic event on the moon is photographed by U.S. astronomer Dinsmoor Alter. Russian and British astronomers also detect the event, a sudden cloud formation. One observer called it a "volcanic eruption," but others saw it as a simple release of gas.

1970 Astronomer Marcel Minnaert dies at 77 in Utrecht, the Netherlands. He began his professional life as a botanist, but a desire to understand the nature of light and its effects on plants led him to study physics. He made his name by using the newly invented microphotometer to analyze the intensities of light waves coming from the sun. His monumental *Photometric Atlas of the Solar Spectrum* (1940) is still a standard reference on the sun's energy. He is also known for *Light and Color in the Open Air* (1954), which analyzed how weather conditions affect light.

1977 History's last case of endemic smallpox is recorded in Somalia.

1977 The experimental space shuttle *Enterprise* makes a safe but bumpy landing at Edwards Air Force Base in California after its maiden test flight. England's Prince Charles witnesses the adventure.

1984 "Baby Fae," an anonymous 12-day-old newborn with a fatal heart defect, is given a walnut-sized baboon heart at the Loma Linda University Medical Center, 60 miles from Los Angeles. It is history's first transplant of an animal organ into an infant.

1993 Christopher J. Burnette, 25, is sentenced in Williamsport, Pennsylvania, to one year in prison for falsely reporting that he found a hypodermic needle in a can of Diet Pepsi. Burnette was the first person charged in a strange wave of false reports that swept across the country the previous summer.

1994 Britain's leading Roman Catholic private school will not allow its pupils immunizations against measles, mumps, and rubella, announces Father Leo Chamberlain (of Ampleforth College in North Yorkshire) in a BBC interview. Chamberlain charges that the vaccine was developed from an aborted fetus. Health authorities deny that fetuses were used in the current vaccine (but admit that vaccines from the 1960s did use such tissue).

1785 The mule first arrives in the United States. Two "jacks" are shipped to Boston as a gift to George Washington from Charles III of Spain. Although exporting full-blooded jacks was illegal, King Charles sent the animals after hearing of Washington's interest in them.

1449 Ulugh Beg was the greatest Mongolian scientist in history. He produced star charts far superior to anything that existed at the time, and founded a university and famous observatory in Samarkand. On this day he is assassinated by his son at age 55.

1728 Explorer-oceanographer James Cook is born in Marton-in-Cleveland, England (see February 14, 1779).

1780 A party of four professors and six students from Harvard "College" view a solar eclipse from Penobscot Bay, Maine. It is the first U.S. astronomy expedition to record an eclipse. The eclipse lasts from 11:11 to 1:50 PM. Although Penobscot Bay was in the hands of the British and the two countries were at war, the British commander allowed the scientists to land, view the eclipse, and depart.

1788 Antoine Lavoisier (the founder of modern chemistry) and his wife Marie Paulze are almost killed in an accident at a munitions factory that Lavoisier was inspecting in Essonne, France.

OCTOBER 27

1806 Eminent Swiss botanist Alphonse Pyrame de Candolle, famous for studying how plants are distributed geographically, is born the son of eminent Swiss botanist Augustin Pyrame de Candolle.

1827 French chemist Marcellin Berthelot is born the son of a physician in Paris. He coined the common terms "endothermic" and "exothermic," and was the first to synthesize organic compounds not found in nature. His prolific writing—1600 papers and books on chemistry—had great influence in science in the late 1800s.

1859 German physicist Gustav Robert Kirchhoff announces invention of the spectroscope, a device that reveals the chemical composition of substances by their emission and absorption of light. Spectroscopy greatly advanced chemical analysis.

1904 The first rapid-transit subway opens in New York City. It is the IRT (Interborough Rapid Transit) running from the Brooklyn Bridge to Grand Central Station to Times Square to 145th Street. It is also the first time aluminum has been used in subway cars.

1920 The lord mayor of Cork, Ireland, Terence James MacSwiney, dies on day 76 of a hunger strike protesting his imprisonment by the British. Meticulous records of his temperature, heart rate, blood pressure, and other aspects of his physical condition have been kept by prison doctors, thus providing a very accurate and complete description of starvation.

1938 The du Pont company announces the name "nylon" for its new synthetic yarn.

1991 The national Sunday magazine *Parade* releases a poll of the top ten issues that frustrate Americans. The top three are the economy, government fraud/incompetence, and health care. Crime, abortion, taxes, government waste, roads, education decline, and free trade are the remaining seven. Environmental destruction is not on the list.

1992 President George Bush takes time out from campaigning in Paducah, Kentucky, to sign five acts: the "Native American Languages Act of 1992" (to preserve Indian languages), the "International Dolphin Conservation Act of 1992" (to preserve marine mammals), the "Battered Women's Testimony Act of 1992" (authorizing a study on the admissibility of expert testimony in domestic violence cases), the "Telecommunications Authorization Act of 1992" (formally establishing the responsibilities of the National Telecommunications and Information Administration), and a bill establishing the Brown v. Board of Education National Historic Site in Topeka, Kansas (celebrating the historic 1954 Supreme Court ruling that outlawed school segregation).

1858 Theodore Roosevelt is born in New York City. His presidency brought unprecedented protection for the environment and the health of citizens.

900 Alfred the Great, known primarily as a great warrior who stopped the Danes' conquest of Britain, dies at 51 in Winchester, England. He was also a scholar who (like Charlemagne) tried to bring education to his subjects during the Dark Ages. He had important books translated from Latin into Anglo-Saxon, and translated some of them himself.

1636 Harvard College is founded in Massachusetts.

1793 Eli Whitney applies for a patent on his cotton gin.

1799 The first aeronautical patent in U.S. history is awarded to Moses McFarland of Massachusetts for a "federal balloon."

1845 Polish physicist Zygmunt Florenty von Wroblewski, known for creating techniques to liquefy gases, is born in Grodno, Russia. He was the first to liquefy appreciable amounts of oxygen, nitrogen, carbon monoxide, and hydrogen. His scientific career began slowly, however, because as an undergraduate he took part in a Polish rebellion against Russia and was sent to Siberia for several years.

1848 The hollow brick is patented by French brothers Paul and Henri-Jules Borie. They used an extrusion procedure to produce blocks that were strong and self-supporting, yet cheaper and lighter than previous bricks.

1867 Hans Driesch is born in Bad Kreuznach, Germany. He made a number of important contributions to embryology: He discovered that when the first two cells of an embryo are separated, each will produce a whole individual; he merged two embryos to produce a giant individual; and he speculated, correctly, that all of an organism's genetic information is held in each cell. Because he could find no physical explanation for some of his experimental results, he became history's last great scientist to defend "vitalism," the belief that life cannot be explained by chemical and physical laws.

1886 President Grover Cleveland accepts the Statue of Liberty as a gift from France. Its sculptor, Frédéric Auguste Bartholdi, attends the ceremony in New York Harbor. At a height of 152 feet, it remains the nation's tallest statue. (The tallest statue in the world is of Buddha, 394 feet high, in Tokyo.)

1892 The first animated cartoon is shown in Paris, in a process invented by Emile Reynaud in which "luminous pantomimes" dance across a cloth screen. Edison's kinetoscope predates this showing, but kinetoscope pictures could only be viewed by one person at a time, rather than being projected on a screen for an audience.

1914 Two historic biologists are born. Jonas Salk, creator of the Salk vaccine for polio, is born in New York City, the son of a Polish-Jewish garment worker. In Liverpool, England, R.L.M. Synge is born the son of a stockbroker. In 1952 he shared the Nobel Prize for Chemistry with Archer J.P. Martin for inventing paper chromatography, a simple means of determining components of complex substances.

1919 Congress overrides President Woodrow Wilson's veto and enacts the Volstead Act, which provides for enforcement of Prohibition against the sale of alcoholic beverages.

1929 A child is born in midair for the first time in U.S. history. A daughter is born to Mr. and Mrs. T.W. Evans in a transport plane flying over Miami, Florida.

1965 Designed in 1947 by architect Eero Saarinen, the stainless-steel "Gateway to the West" arch in St. Louis, Missouri, is completed. It commemorates human expansion westward after the Louisiana Purchase in 1803. It spans 630 feet, and at 630 feet high is Earth's tallest man made monument.

1993 Chinese rocket scientists predict that a two-ton chunk of "space junk"—a piece of a satellite that had split in half ten days before—will continue orbiting Earth for at least six months. The space junk hits the Pacific at 17,000 mph later that afternoon.

1991 The City Council of Simi Valley, California, allows full pensions for its police dogs. When the canines retire (usually at age 9) they can now be purchased for $1 by their handlers, and the City will pay for all food and veterinary bills.

1766 A fox-hunting club meets for the first time in America. The Gloucester Fox Hunting Club, comprising residents of Philadelphia and Gloucester County, New Jersey, holds an organizational meeting in Philadelphia. Twenty-nine dog owners attend. The club began hunting activities on January 1, 1767, and disbanded in 1818.

1791 John Elliotson, a pioneer in medical practice and education, is born in London. He was one of the first to urge use of the stethoscope and he pushed for lecturing as part of clinical training. Forced to resign his teaching position in 1838 at University College Hospital, London, because of his interest in hypnosis, he started his own hospital in 1849.

1831 Paleontologist Othniel Charles Marsh is born the son of a shoe manufacturer in Lockport, New York. He was raised by a rich uncle, George Peabody, whom he persuaded to endow the Peabody Museum of Natural History at Yale University (where Marsh graduated in 1860). Among the 500 extinct species that Marsh discovered was the pterodactyl and a famous sequence of ancestral horses that helped support Darwinism in its earliest days. He is also remembered for his legendary and hate-filled rivalry with fellow bone-hunter Edward Cope.

OCTOBER 29

1920 Baruj Benacerraf, discoverer of a genetic basis for immune responsiveness, is born in Caracas, Venezuela. The discovery was especially important in studying autoimmune reactions, in which an organism produces antibodies that attack the organism itself. Benacerraf won the 1980 Nobel Prize for Physiology or Medicine.

1928 German airship *Graf Zeppelin* begins a nonstop flight of 3967.1 miles, still the record distance for a dirigible.

1928 Physiologist-author Cleveland P. Hickman is born in Greencastle, Indiana.

1947 A forest fire in the United States is attacked for the first time by rain from artificially seeded clouds. Seeders from the General Electric Company in Schenectady, New York, fly to a blaze in Concord, New Hampshire, and spread dry ice through cumulus clouds. Rain is artificially induced, although natural rain begins falling shortly thereafter, making it impossible to gauge exactly the efficacy of the artificial seeding. The work is part of Project Cirrus, a weather research program run jointly by the U.S. Army Signal Corps and the Office of Naval Research.

1963 The moon comes alive for 20 minutes. Three red spots are sighted briefly near the lunar crater Aristarchus by James A. Greenacre and Edward Barr through a large telescope at Lowell Observatory in Flagstaff, Arizona. The cause and composition of these spots remain mysteries.

1964 The Star of India and other gems are stolen from the American Museum of Natural History in New York City. The Star and most of the other gems are eventually found, and three thieves were later taken into custody.

1966 The National Organization for Women (NOW) is founded.

1979 Antinuclear protestors try unsuccessfully to shutdown the New York Stock Exchange; it is the 50th anniversary of Black Tuesday (1929) when stock prices plummeted and the Great Depression began.

1991 U.S. space probe *Galileo* encounter its first asteroid, Gaspra.

1992 "Artificial vision for the blinded person is an achievable goal, hopefully before the end of the decade," says Dr. Murray Goldstein, director of the National Institute of Neurological Disorders and Stroke at a Los Angeles press conference. He has just announced results of an experiment in which tiny electrical stimulators in the brain of a blind woman allowed her to "see" colors and shapes.

1994 Puppeteer Shari Lewis (famous for decades as the operator of the cuddly puppet "Lamb Chop") confesses in a *T.V. Guide* interview that she has a passion for dining on rack of lamb. Part of her enjoyment is shocking waiters "at every opportunity" when she orders lamb. Lewis uses the rest of the interview to denounce television violence directed at kids.

1618 Sir Walter Raleigh is beheaded for treason in London. He was instrumental in the European colonization of North America, and in the introduction of tobacco and potatoes to Ireland.

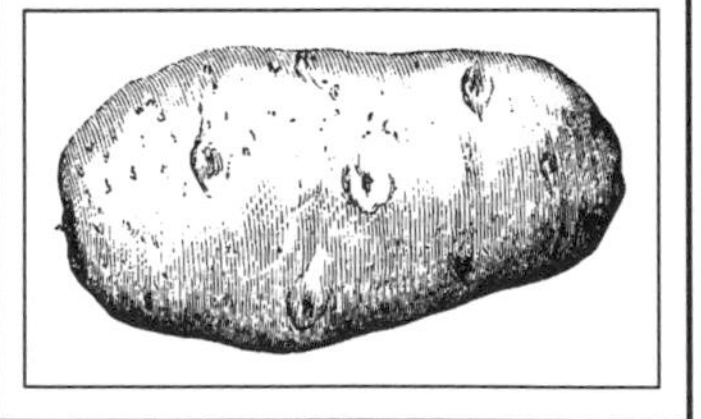

OCTOBER 30

1794 The first commercial use of ball bearings in the United States occurs in Lancaster, Pennsylvania. The ball bearings are attached to a weather vane atop the steeple of the Evangelical Lutheran Church of the Holy Trinity. The base of the weather vane is 195 feet above ground.

1817 Hermann Franz Moritz Kopp, a pioneer in physical organic chemistry, is born the son of a physician in Hanau, Germany. He first achieved fame with his four-volume history of chemistry (1843–1847). In the lab he was the first to carefully measure boiling points of organic substances, and was able to relate these points and other physical properties with chemical structure, or specifically, the length of carbon atom chains.

1895 Two Nobel laureates are born. Dickinson Woodruff Richards, developer of the heart catheter and winner of the 1956 Nobel Prize, is born in Orange, New Jersey. Gerhard Domagk, discoverer of the antibiotic properties of Prontosil (the first sulfa drug), is born the son of a teacher in Lagow, Germany. He won the 1939 Nobel Prize (which the Nazis forbid him from accepting). The use of antibiotics grew dramatically when Domagk saved his daughter's life with Prontosil, and shortly thereafter U.S. doctors saved the life of President Franklin Roosevelt's son with the same "miracle" substance.

1920 Physiologist Sir Frederick Banting reads a research article by Moses Baron describing how tying off ducts from the pancreas causes that organ to degenerate. This suggests an experiment to Banting—or as he later wrote, "the idea presented itself"—that led to his discovering the cause of diabetes. Banting received the Nobel Prize in 1923.

1925 A human face appears on television for the first time. In England, using equipment invented by John Logie Baird, the face of 15-year-old William Taynton, who apparently demanded a bribe of two shillings sixpence to appear, is broadcast on a television screen.

1928 Microbiologist Daniel Nathans is born in Wilmington, Delaware. He studied enzymes that can break DNA chains at specific points. This paved the way for recombinant DNA techniques in which DNA can be taken apart and put back together according to the biologist's designs. Nathans won the Nobel Prize in 1971.

1938 Earth is attacked by Martians! Thousands of radio listeners hear and believe *The War of the Worlds*, starring Orson Welles on CBS. The use of fake news bulletins and simulated on-scene reports is so realistic that panic ensues in the New York–New Jersey area. It is a classic in broadcast history.

1961 The most powerful man-made explosion in history occurs. A 57-megaton thermonuclear device is detonated today in the Novaya Zemlya region of Russia. The resulting shock wave circles Earth three times, the first lap taking 36 hours 27 minutes.

1963 "Every time you scientists make a major invention, we politicians have to invent a new institution to cope with it—and almost invariably, these days, it must be an international institution," says President John Kennedy, addressing the National Academy of Sciences, quoted in the *Wall Street Journal*.

1985 The largest crew ever on a space mission—eight crew members—blasts off in space shuttle *Challenger*. It is the 22nd shuttle flight, and lasts just over a week. Watching the launch is schoolteacher Christa McAuliffe, destined to perish in the ill-fated *Challenger* mission the following January.

1989 Earth's largest known frog is the African giant frog or the "goliath frog"; on this date in Cameroon a 34½-inch specimen is weighed at 8 pounds 1 ounce.

1992 *Science* reports a University of Pennsylvania study that confirms pediatric folklore: Babies do grow in spurts, sometimes as much as a half-inch in one day. Just before such a spurt, a child is restless and irritable.

1993 History's first dissections in space are performed aboard the orbiting *Columbia* by Dr. Martin Fettman (the first U.S. veterinary in space) and astronaut-physician M. Rhea Seddon. Six male rats are beheaded and dissected. "This is really a milestone in space life sciences research," said a NASA spokesman.

1888 The ballpoint pen is patented. J.J. Loud of Weymouth, Massachusetts, receives patent No. 392,046 for a marking device with a spheroid point able to revolve in all directions.

1620 John Evelyn, author of one of the earliest and greatest books on scientific forestry, is born the son of a wealthy landowner in Wotton, England. His *Sylva, or a Discourse of Forest-trees, and the Propagation of Timber* (1664) went through ten editions over the next 200 years. It was just one of the 30 books he wrote on a range of topics, including fine arts, history, stamp collecting, and religion.

1692 Archaeologist Anne-Claude-Philippe de Tubières, Comte de Caylus is born in Paris.

1802 Benoit Fourneyron, inventor of the water turbine, is born the son of a mathematician in Saint- Étienne, France.

1815 Mathematician Karl Weierstrass, the father of modern analysis, is born the son of a city official in Ostenfelde, Germany.

1831 Physiologist Karl von Voit, the first to accurately measure human energy requirements, is born the son of an architect in Amberg, Germany. His research forms cornerstones in our understanding of both metabolism and nutrition.

1835 Adolph von Baeyer, the 1905 Nobel laureate for Chemistry, is born the son of a Prussian general in Berlin. Baeyer's numerous achievements included the synthesis of indigo and the discovery of barbituric acid, the basis of all sedative-hypnotic drugs called barbiturates. It is said that Baeyer named the acid after a woman "Barbara" whom he was dating at the time of the discovery.

1883 The World Woman's Christian Temperance Union is organized at a four-day convention in Detroit. This follows the establishment nine years earlier of the first national women's temperance society in the United States, the National Woman's Christian Temperance Union.

1930 Astronaut Michael Collins, a member of the first manned mission to the moon, is born in Rome, Italy.

1952 An atomic fusion bomb is first detonated. Nicknamed "Mike" (but known as a thermonuclear bomb, hydrogen bomb, or H-bomb), the device explodes at 19:14:59.4 G.C.T. from a tower in the Elugelab Atoll in the Marshall Islands in the heart of the Pacific Ocean. Its massive energy comes from the fusing together of hydrogen atoms—the same process used by the sun.

1956 An airplane lands at the South Pole for the first time. Rear Admiral George John Dufek flies the transport plane *Que Sera Sera* to the Pole, where he lands, disembarks, and erects a flag. He is the first American to set foot on the South Pole.

1987 History's first five-organ transplant is conducted at Children's Hospital in Pittsburgh, Pennsylvania. Tabatha Foster (age 3½ years), never having been able to eat solid food is given a new liver, pancreas, small intestine, large intestine, and stomach segments.

1991 "We're accusing the cigarette companies of selling cancerous addiction and then affecting innocent people," states a Miami attorney in the first-ever suit filed on behalf of victims of secondhand smoke.

1992 Pope John Paul II formally proclaims that the Catholic Church erred 400 years ago in condemning Galileo for his belief that Earth is not the center of the universe.

1994 President Bill Clinton signs the California Desert Protection Act.

1994 Manic-depression seems attributable to an inherited defect in chromosome 21, announces Dr. Miron Baron (director of psychogenetics, New York State Psychiatric Institute) at a New York press conference. Workers at Thomas Jefferson University in Philadelphia had reported in June that abnormalities on chromosome 18 also predispose a person to manic-depression.

1828 Physicist-chemist Sir Joseph Swan is born in Sunderland, England. He produced an electric light bulb (long before Edison) but the vacuum in his bulb was too weak to make the light practical. He also created the first dry photographic plates (beating George Eastman by 15 years). Edison and Eastman became rich and famous by building on the foundations laid years earlier by Swan.

NOVEMBER 1

1755 The largest earthquake of the eighteenth century savages Lisbon, Portugal. Its fire and tidal wave destroy almost all of the city, ending decades of economic growth based on colonial exploitation of Brazil.

1772 "A week ago I discovered that sulfur on being heated gained weight. It is the same with phosphorus," reports Antoine Lavoisier, 29, in a simple note to the Secretary of the French Academy of Sciences. The discovery advanced and changed chemistry for all time.

1781 The first state medical society in U.S. history incorporates in Boston.

1828 Meteorologist-geophysicist Balfour Stewart is born in Edinburgh.

1833 In Plattsburg, Missouri, Army surgeon William Beaumont finishes the last of 238 experiments on a trapper, whose stomach had been accidentally shot open eight years earlier. The hole in the stomach never closed, allowing Beaumont to observe and study digestion directly.

1870 The U.S. Weather Bureau (originally a branch of the War Department) files its first weather report, based on observations telegraphed in from 24 locations throughout the country.

1873 Barbed wire is manufactured for the first time, in inventor Joseph Farwell Glidden's factory.

1880 Alfred Wegener, the first to propose the modern theory of continental drift, is born the son of an orphanage director in Berlin, Germany.

1888 The last breed of wild horse to become extinct in nature was Przhevalsky's horse; its discoverer, Nikolay Przhevalsky, dies today of typhus at 49 in Karakol, Russia (now renamed Przhevalsk in his honor).

1911 Donald William Kerst, inventor of the betratron, is born in Galena, Illinois. His device, which appeared in 1940, was the first to accelerate electrons ("beta particles") fast enough to smash atoms.

1912 "Rays of very great penetrating power are entering our atmosphere from above," was the conclusion of Victor Hess in today's issue of *Physikalische Zeitschrift*. It is history's first report by a professional physicist that Earth is constantly bombarded by radiation from outer space.

1918 The worst subway disaster in U.S. history occurs in Brooklyn, New York, when a BRT train derails; 97 die at the scene, 5 die later, and the BRT goes bankrupt in December.

1919 Sir Hermann Bondi, a formulator of the "steady state" or "continuous creation" theory of the universe, is born in Vienna. Collaborating with Fred Hoyle and Thomas Gold in 1948, Bondi developed the theory (and argued it mathematically) that the universe is the same everywhere and for all time.

1939 An animal created through artificial insemination is exhibited for the first time. It is a rabbit, shown by Dr. Gregory Pincus to the New York Academy of Medicine.

1941 *Under the Sea-Wind* is published. It is Rachel Carson's first book. She later became famous for her ecology classic, *Silent Spring*.

1951 The United States explodes history's first hydrogen bomb, in the skies over Eniwetok in the Marshall Islands.

1979 A collision causes the tanker *Burmah Agate* to dump 10,700,000 gallons of oil into Galveston Bay, an inlet of the Gulf of Mexico.

1990 "An historic step in cleaning up the sea" is taken by 43 nations in London, when they agree to gradually eliminate dumping of industrial waste at sea by 1995.

1993 Russian and U.S. space officials sign a pact in Washington specifying that NASA will send ten space shuttle missions up to the Soviet space station *Mir*, and that the two superpowers will jointly build a space station to be in service by October 2001.

1677 Physicist Robert Hooke cancels a demonstration of protozoa to the Royal Society in London. It would have been the first time a group of scientists saw one-celled animals.

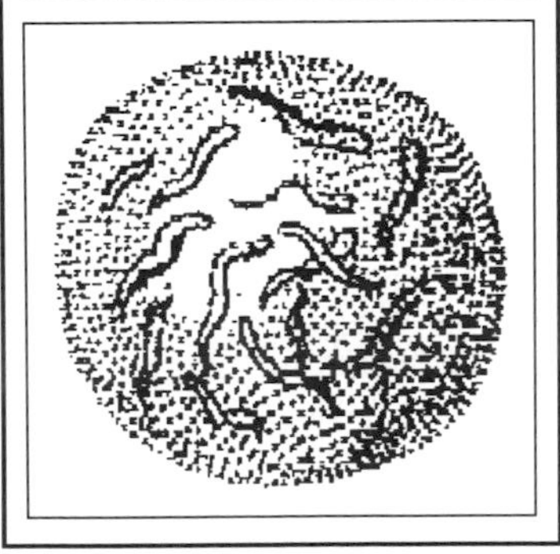

1570 One of the great floods in North Sea history swamps the Frisian Islands off Germany, killing 41,000 inhabitants. Huge winds combined with a huge tide cause the deluge.

1734 Daniel Boone is born in Berks County, Pennsylvania. As a courageous frontiersman who helped open the West to European settlement, he inadvertently contributed to ecological destruction in the United States.

1793 Nikolay Lobachevsky, "the Copernicus of geometry," is born in Nizhni Novgorod, Russia.

1815 Mathematician George Boole is born in Lincoln, England, the son of a poor shoemaker. Boole created "symbolic logic," the mathematical analysis of cognition, now known as Boolean algebra.

1878 The Atlantic giant squid is a triple record-breaker: It is the largest invertebrate, the largest mollusk, and the animal with the biggest eyes. On this day, the largest squid ever seen runs aground in Thimble Thickle Bay, Canada. It weighs 2.2 tons, and has 38-foot-long arms.

NOVEMBER 2

1885 Astronomer Harlow Shapley, whose work led to the first close estimate of the size of our galaxy, is born the son of a farmer in Nashville, Missouri.

1897 Jacob Bjerknes, a founder of modern meteorology, is born in Stockholm the son of Vilhelm Bjerknes. The father–son team established a network of weather stations throughout Norway and researched how air masses of different temperatures—which they named "fronts"—interact.

1906 Bengt Edlén, known for estimating temperatures on the sun, is born in Gusum, Sweden.

1920 The first commercial broadcast in radio history hits the airwaves of Pittsburgh, Pennsylvania.

1931 Synthetic rubber is produced on a commercial scale for the first time. The du Pont company in Wilmington, Delaware, begins mass-producing "Du Prene," which was first used to make hosing to carry oil.

1933 Treatment of epilepsy by elevating the patient's skull cap is demonstrated for the first time by neurosurgeon Karl Ney who presents the procedure in New York City.

1947 The airplane with the largest wingspan in history, and the nation's first plane with eight engines, makes its one and only flight. The wooden *Spruce Goose*, built for $40 million, was flown for about one minute by Howard Hughes, off the coast of Long Beach, California. It reaches a height of 70 feet.

1954 Four-way split-screen television is seen for the first time as reporters from New York City, Washington, Chicago, and Los Angeles appear on the same screen simultaneously.

1904 A prisoner in a federal penitentiary is fingerprinted for the first time, at Leavenworth, Kansas.

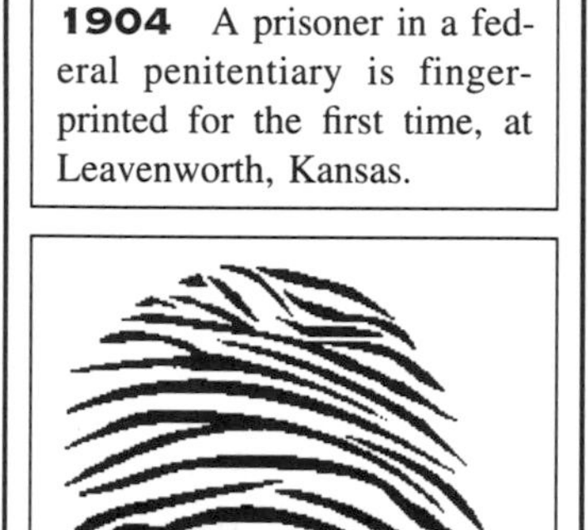

1957 The country's first titanium mill opens in Toronto, Ohio.

1963 The first light-emitting postage stamp is issued in the United States. Impregnated with phosphorescent material, the stamp glows briefly after exposure to ultraviolet light.

1994 Mayor Carty Finkbeiner of Toledo, Ohio, offers a unique solution to the problem of noise from the airport: move deaf people into the area. The idea, presented at a staff meeting, draws immediate and widespread opposition. "That's like saying let the blind work at night because they can't see," said Dave Wielinski, chair of the handicapped-rights group, Barrier Free Toledo. "It would be an insult to the deaf community."

1664 Today marks the debut of *Microphagia*. Robert Hooke shows an advanced copy of his book to the Royal Society in London. It was an immediate bestseller, and is now a classic. It was history's first treatise on cells and microbiology and it coined the word "cell" in a biological context.

1749 Daniel Rutherford, discoverer of nitrogen and creator of the first maximum–minimum thermometer, is born in Edinburgh, son of a medical professor and uncle to Sir Walter Scott.

1839 One of the original drug wars, the first Opium War between China and Britain, erupts when two English frigates clash with a fleet of junks off the Chinese coast.

1854 Chemist Jokichi Takamine is born in Takaoka, Japan, in the year Commodore Perry opened that country to the West. Takamine moved to New Jersey, where he discovered adrenaline. In 1912 he coaxed the mayor of Tokyo to send a stand of cherry trees to Washington, where they have bloomed ever since.

NOVEMBER 3

1892 The first successful automatic telephone exchange system opens in La Porte, Indiana. Almon B. Strowger invented the device in the 1890s. The first such exchange appeared, but failed, in 1879.

1896 Shortly after the most important discoveries of his life—that iodine is found naturally in animals and that the thyroid is iodine-rich—chemist Eugen Baumann dies at 49 in Freiburg, Germany.

1900 The nation's first car show is held in Madison Square Garden, New York City. Fifty-one exhibitors are at the week-long exposition, which includes hill-climbing, steering, and braking contests.

1930 The first vehicular tunnel between the United States and a foreign country opens to traffic under the Detroit River between Detroit, Michigan, and Windsor, Ontario.

1952 Frozen bread appears in stores for the first time; it is produced by Arnold Bakers of Port Chester, New York.

1953 The nation's first coast-to-coast broadcast of a live TV show in color is filmed at the Colonial Theatre, New York City, by WNBT-TV and transmitted to Burbank, California. Nanette Fabray stars.

1953 A podiatry section is established in the U.S. Navy. The first naval foot doctor was Ensign Richard Stewart Gilbert of New York City.

1955 An animal- or human-infecting virus has been crystallized for the first time, announce scientists at a meeting of the National Academy of Sciences at the California Institute of Technology in Pasadena. Dr. Carlton Everett Schwerdt led the team at the University of California at Berkeley that crystallized the poliomyelitis virus.

1957 Russia launches *Sputnik II*, the second man-made object to orbit Earth. Laika the dog is aboard, becoming the first animal in space. Only several days of oxygen are provided, and Laika dies.

1958 Russian astronomer Nikolay Kozyrev obtains a spectrogram of a momentary cloud formation on the moon. He calls it a "volcanic eruption"; others have called it a simple release of carbon dioxide.

1963 Valentina Tereshkova, the first woman in space, marries a fellow cosmonaut.

1964 Votes are counted electronically for the first time in a U.S. presidential election.

1982 The worst road accident in history occurs in Afghanistan. A gas tanker explodes inside Salang Tunnel, killing 176 people.

1994 Godzilla celebrates its 40th birthday. "Hey, I'm young for a monster!" proclaims the mutant reptile at the celebration at Tokyo's Toho Co., the studio that created the movie monster during the Eisenhower administration.

1863 The first U.S. patent for the preparation of yeast is issued to J.T. Alden of Cincinnati.

1472 Columbus first writes about potatoes in his journal. He and his men have encountered a vegetable that looked "like carrots and tasted like chestnuts."

1741 Bering Island off Kamchatka is discovered by the ill-fated crew led by Vitus Bering. His ship wrecked on the island shortly after its discovery, then scurvy killed him and many of his crew members. Also, Steller's sea cow (a relative of the manatee) was first seen there, and lived on the island until it was hunted to extinction.

1837 Mining engineer-philanthropist James Douglas is born in Quebec. A coinventor of the Hunt–Douglas process for extracting copper from ore, Douglas established the first commercial electrolytic copper plant, in Phoenixville, Pennsylvania. His wealth and mining activities opened the U.S. Southwest.

1862 Richard Jordan Gatling patents the "Gatling gun," the first rapid-fire machine gun. It shot 250 bullets per minute and was manufactured in Indianapolis.

NOVEMBER 4

1873 The dental gold crown is patented by Dr. John Beers of San Francisco, who receives patent No. 144,182.

1873 Anthony Iske of Lancaster, Pennsylvania, receives the nation's first patent for a food-slicing machine. Patent No. 144,206 describes his "machine for slicing dried beef."

1879 The Ritty brothers, John and James, patent the cash register. James "Jake" Ritty, a businessman from Dayton, Ohio, thought of the idea for the machine during an ocean voyage to Europe, when he happened to observe a device that automatically counted the revolutions of the ship's propeller.

1846 Benjamin F. Palmer of New Hampshire receives the nation's first patent for an artificial leg.

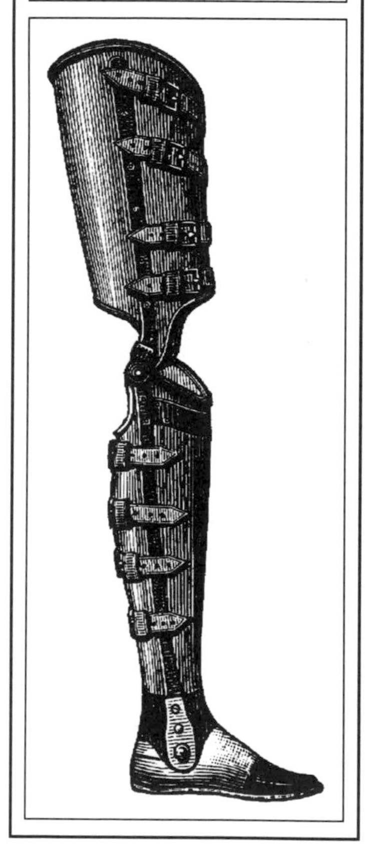

1899 *The Interpretation of Dreams* by Sigmund Freud is published.

1911 Marie Curie receives notification that she will be receiving her second Nobel Prize. None before her and few since have attained the honor of receiving two Nobel Prizes. The news is announced amid the scandal over her affair with a mathematician at the Sorbonne.

1922 The staircase to King Tutankhamen's tomb is discovered in the desert at Luxor, Egypt, by a worker in the 200-man party of amateur archaeologist Howard Carter. After three weeks of excavation, Carter stood at the tomb's door, into which he drilled a small hole. "At first I could see nothing," he wrote, "but slowly from the mist, strange animals, statues and gold."

1927 Captain Hawthorne Gray of the U.S. Army Air Corps becomes the first U.S. aviator to die from lack of oxygen. He ascends to 40,000 feet in a balloon from Scott Field in Belleville, Illinois, but suffocates when he is unable to open a reserve oxygen cylinder.

1930 The first attempted airplane trip across the Atlantic since Lindbergh takes place. The plane is the enormous German Dornier Do X. Bad weather forces it to ditch in the ocean off the coast of France.

1939 An air-conditioned automobile is exhibited in the United States for the first time. The Packard Motor Car Company of Detroit puts the vehicle on display at the 40th Automobile Show in Chicago.

1943 The X-10 nuclear reactor "goes critical" at the Oak Ridge National Laboratory in Tennessee. It is history's first reactor to produce large quantities of radioisotopes for medicine and research.

1965 A woman drives a car faster than 300 mph for the first time. Lee Ann Roberts Breedlove (wife of car racer Craig Breedlove) pilots the jet-engined *Spirit of America* at Utah's Bonneville Salt Flats.

1994 *Science* reports the discovery in the Mongolian desert of an 80-million-year-old egg.

1639 A post office is established in colonial America. The General Court of Massachusetts in Boston orders that the house of one Richard Fairbanks be used for "all letters which are brought from beyond the seas, or are to be sent thither." Fairbanks was allowed to charge one penny for each letter handled; he was accountable to the authorities who governed the colony.

1879 Math/physics genius James Clerk Maxwell dies of cancer at 47 in Cambridge, England (see November 13, 1831).

1891 Two days before she turns 24, Marie Curie enrolls at the Sorbonne. She has been out of school for five years, she is in a foreign country, and she barely has enough money to survive, fainting from hunger on at least one occasion in the classroom. Yet she eventually graduates top of her class.

1892 The fiery J.B.S. Haldane is born on Guy Fawkes Day in Oxford, England. Like his famous father, John, Haldane made significant discoveries about animal respiration and is also remembered for his genetic theories. In 1957 he emigrated to India in protest of British colonial policies.

NOVEMBER 5

1895 The first automobile patent in U.S. history is No. 549,160, awarded to attorney George Baldwin Selden of Rochester, New York, for combining an internal-combustion hydrocarbon engine with a road vehicle.

1906 Marie Curie delivers her first lecture at the Sorbonne. She is the first female physics teacher in the school's history.

1944 Transplant surgery advanced tremendously because of the innovations of Alexis Carrel, who dies on this day at 81 in Paris. He became skilled in surgery, but quit the field in 1904, going to Canada to become a cattle rancher; in 1906, he returned to medicine as a researcher at New York City's Rockefeller Institute. He is remembered for his technique of delicately suturing blood vessels together, and for his methods of keeping tissues and organs alive outside of the body. In one famous experiment he kept a section of chicken heart alive for over 34 years—much longer than the life span of the chicken. Carrel received the 1912 Nobel Prize in Physiology or Medicine.

1975 The most famous alleged alien abduction in history occurs when seven loggers claim to encounter a UFO in northeastern Arizona. One of the group, Travis Walton, claims to have been lifted into the sky by a beam of light, and returned to Earth five days later. The 1993 movie *Fire in the Sky* recounts the story.

1992 Wire services report the discovery of 5000-year-old beer. This is the earliest known evidence that ancient man brewed and drank the beverage. Beer residue was found in hatch marks in a double-handled jar taken from the Zagros Mountains in Iran. This area was occupied by the ancient Sumerians whose symbol for beer was a jar with hatch marks. It is guessed that the grooves were intended to help remove bitter chemicals from the barley beverage.

1994 Former president Ronald Reagan, 83, announces that he has Alzheimer's disease. A handwritten letter to his "fellow Americans" is intended to spur awareness of the incurable brain affliction that is the fourth leading cause of death among U.S. adults. "I now begin the journey that will lead me into the sunset of my life. I know that for America there will always be a bright dawn ahead."

1492 Columbus first writes about corn in his journal. "[The sailors] found a great quantity of the grain that the Indians called maize, which was well tasted, bak'd and dry'd and made into flour." Europeans went to the New World in search of gold—and they discovered a golden grain.

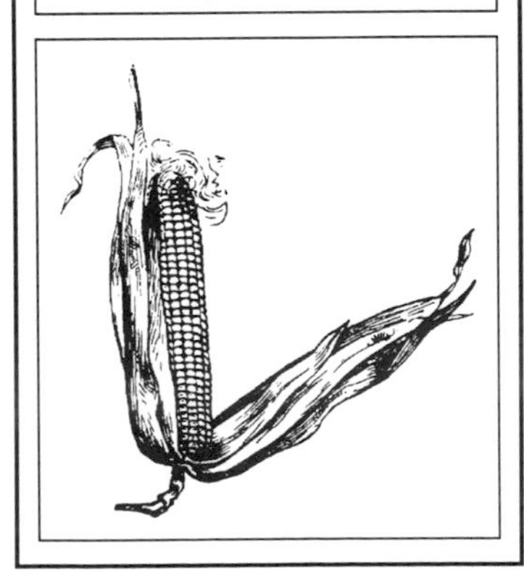

1777 Botanist Bernard de Jussieu, one of the first to adopt Linnaeus's system for naming and grouping organisms, dies at 78 in Paris. His nephew, Antoine Laurent, and two brothers, Joseph and Antoine, were also famous plant scientists.

1822 Claude Berthollet dies at 73 near Paris. He developed the modern system of chemical nomenclature with Lavoisier, and conducted the experiments that laid the foundation for the law of mass action, that the rate of a chemical reaction depends on the concentration of the chemicals.

1835 Cesare Lombroso, pioneer in scientific criminology, is born in Verona, Italy.

1848 Naturalist-writer Richard Jefferies is born on a farm near Swindon, England. He is remembered for detailed, yet poetic descriptions of country life. *Bevis: The Story of a Boy* was one of his best-known works. Unappreciated during his own life, Jefferies died in poverty at 38.

NOVEMBER 6

1848 Classes begin at the Homeopathic Medical College of Pennsylvania, Philadelphia. It is the first U.S. college devoted to homeopathy—in which herbs are used to produce a disease's symptoms in order to cure that disease. Homeopathy was popular in the United Staates and Europe in the mid-1800s, then it floundered, and is gaining popularity again in the late twentieth century.

1851 Charles Henry Dow (creator of the industrial tool, the Dow Jones averages) is born in Sterling, Connecticut.

1865 Sir William Boog Leishman is born the son of a professor in Glasgow, Scotland. An army doctor in the tropics, he developed a vaccine for typhoid fever, and discovered the cause of leishmaniasis—a disease or infection caused by a parasitic protozoan.

1880 The cause of malaria is discovered. In Constantine, Algeria, French parasitologist Alphonse Laveran finds the protozoan *Plasmodium vivax* in the blood of an afflicted human. It is the first time that a one-celled animal, rather than a bacterium, was shown to cause a human disease.

1919 Einstein is right! British astronomers announce in London that expeditions to Africa and Brazil have proved that gravity bends light, as Einstein's theory of relativity predicts. At today's announcement, Nobel laureate J.J. Thomson describes the theory of relativity as "one of the highest achievements of human thought."

1928 The nation's first flashing electric sign is installed. It is enormous. It is the "Motogram," installed on the New York Times building in New York City, to report election returns. At 360 feet long and 5 feet high, the contraption has 14,800 bulbs which flash 21,925,664 times an hour.

1935 Henry Fairfield Osborn dies at 78 in Garrison, New York. He advanced the art of museum display, popularized paleontology, and introduced the evolutionary concept of adaptive radiation (which is the evolution from a primitive, nonspecific ancestor into a group of new species, each of which is adapted to a specific niche).

1947 *Meet The Press*, the most durable television show in history, first airs.

1967 *The Phil Donahue Show* debuts in Dayton, Ohio. It is the first talk show with audience participation, thereby creating a new genre in television history. The format was an accidental creation; the audience was there to see a variety show that was canceled. The crowd stayed to watch Donahue interview Madalyn Murray O'Hair, and soon the audience was also asking questions.

1978 The FBI arrests Stanley Mark Rifkin in Carlsbad, California, ending history's greatest cyberfraud. Rifkin allegedly stole $10.2 million from a Los Angeles bank through its own computers.

1991 History's worst oil fire disaster is over. Kuwait's emir (Sheik Jaber al-Ahmed al-Sabah) pushes a lever at the Burgan oil field, which finally caps the last of 732 burning wells set on fire by Iraqi troops during their recent military occupation of Kuwait. The firefighting involved 10 nations and cost $1.5 billion.

NOVEMBER 7

1818 Emil Du Bois-Reymond, the founder of modern electrophysiology, is born in Berlin. After studying electric fish as a college student, Du Bois-Reymond invented an apparatus that allowed him to detect and measure electric currents moving along nerves. He discovered that a "wave of relative negativity" along both nerves and muscles was the key physiological event in neural communication and muscle contraction.

1867 Marie Curie is born in Warsaw, the daughter of a physics teacher and a girls' school principal.

1874 The elephant is first used to symbolize a political party in the United States. *Harper's Weekly* in New York City publishes a Thomas Nast cartoon about the chances of Ulysses Grant seeking a third term; the elephant symbolizes the Republican Party.

1876 The nation's first cigarette-making machine is patented by Albert Hook of New York City. The "Hook Machine" produced a continuous cigarette of undetermined length, which was then sliced into separate cigarettes. The device was put into commercial use in 1882.

1878 Nuclear physicist Lise Meitner is born the daughter of a lawyer in Vienna. Inspired by the discoveries of Marie Curie, Meitner entered science against strong antifemale prejudice. She collaborated with Otto Hahn during 30 years of research, and was the first to understand the mechanism and significance of Hahn's indirect discovery of nuclear fission. She was not named in Hahn's 1944 Nobel Prize.

1885 Canada is spanned by a transcontinental railway; the last spike is driven at Craigellachie.

1888 Sir Chandrasekhara Raman, the first Asian to win a Nobel Prize in science, is born the son of a physicist in Tiruchchirappalli, India. His work on light scattering revealed particulate properties of light.

1903 Konrad Lorenz, a cowinner of the first Nobel Prize for animal behavior research, is born the son of a surgeon in Vienna. Lorenz was the first to describe the form of early learning called "imprinting" (in 1935), which has been demonstrated in birds and some mammals, and may occur in humans. His Nobel Prize was awarded in 1972.

1913 Alfred Russel Wallace, who independently of Darwin created the theory of evolution, dies at 90 in Broadstone, England (see January 8, 1823).

1940 The 2800-foot center span of the Tacoma Narrows Bridge collapses and falls into Puget Sound, Washington, during a windstorm.

1946 Coin-operated television is first exhibited, in Asbury Park, New Jersey. It is called "Tradio-Vision."

1964 Hans Euler-Chelpin dies at 91 in Stockholm, where he had received the 1923 Nobel Prize in Chemistry for determining the structure of enzymes that ferment sugar. He is also known for his research into vitamins.

1967 *Surveyor 6* blasts off from Cape Kennedy for the moon, where 10 days later it launches itself 10 feet off the surface; this is the first time a satellite was launched from a heavenly body outside Earth.

1967 President Johnson signs a bill establishing the Corporation for Public Broadcasting.

1974 An African bush elephant is shot to death in Mucusso, Angola. Weighing 13 tons and measuring over 13 feet at the shoulder, it is the largest elephant on record.

1991 A bloodmobile for dogs begins taking donations in Philadelphia. It is apparently the world's first.

1991 Basketball superstar Earvin "Magic" Johnson, 32, announces in California that he has tested positive for the AIDS virus and has retired from professional sports. "This is one of those things you think can't happen to you, but it can."

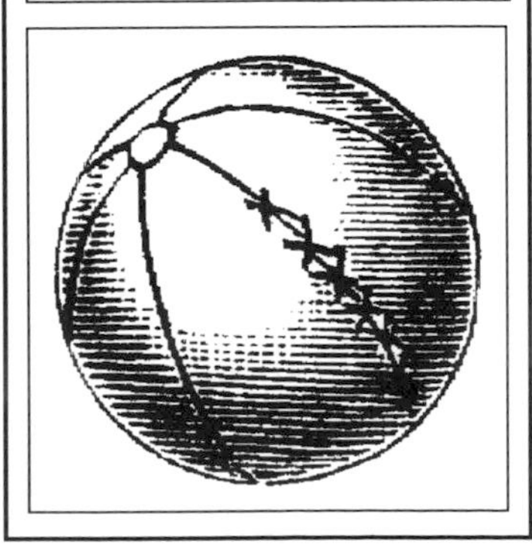

1656 Astronomer Edmund Halley, of Halley's Comet fame, is born the son of a successful merchant in Haggerston, England.

1703 John Wallis, mathematical prodigy, historian, and innovator, dies at 86 in Oxford, England.

1793 The Louvre Museum, Paris, first opens its doors to the public.

1854 Physicist Johannes Rydberg is born in Halmstad, Sweden. He developed a mathematical equation to describe the spectral lines of various elements when heated to glowing, thereby providing Niels Bohr with important clues to the structure of the atom, which Bohr announced in 1913.

1858 George Peacock dies at 67 in Ely, England. He was a leader of the British mathematicians who established algebra as an abstract science, thus laying the foundation for the design of modern computers.

NOVEMBER 8

1880 Edwin Drake, the first to drill an oil well, dies in poverty in Bethlehem, Pennsylvania, at age 61.

1884 Hermann Rorschach is born the son of an art teacher in Zurich. It was here that Carl Jung influenced him to enter psychiatry, and where he developed the famous test that bears his name—the Rorschach test is an inkblot test of personality and intelligence in which the patient interprets a variety of inkblots.

1895 In Würzburg, Germany, Wilhelm Konrad Roentgen discovers X rays—by accident (see February 10, 1923).

1904 The electric power plug is patented by Harvey Hubbell of Bridgeport, Connecticut.

1910 A bug zapper is first patented in the United States. The "insect electrocutor" is patent No. 974,785, granted to William M. Frost of Spokane, Washington.

1922 Christiaan Barnard, the first to successfully perform a human heart transplant, is born in Beaufort, South Africa. The 1967 operation involved implanting the heart of a black man into a white man's body.

1934 James Mark Baldwin, heavily influential in advancing psychology from the field of philosophy to science, dies at 73 in Paris. His *Handbook of Psychology* (two volumes, 1889–1891) was the first treatment in English of the emerging science; his *Mental Development in the Child and the Race* (1895) and *Social and Ethical Interpretations in Mental Development* (1897) treated mental phenomena for the first time from the perspective of biology and evolution. In 1894 he collaborated with James McKeen Cattell to found *Psychological Review*, which spawned *Psychological Index* and *Psychological Bulletin*, which is still in circulation today.

1805 The Lewis and Clark expedition reaches the Pacific Coast, at the mouth of the Columbia River. It was the first recorded trip by man across North America, and foreshadowed the end of the Native American way of ecologically sound living.

1969 Astronomer Vesto Slipher dies three days before his 94th birthday in Flagstaff, Arizona. He obtained the first practical photos of Mars, the first absorption spectra of Jupiter and Saturn, and the first evidence of "the redshift" (which Hubble later used to prove that the universe is expanding).

1990 *Nature* presents evidence for a theory that seeds of life arrived on Earth from outer space in a dust storm deposited by the disintegration of a giant comet. Amino acids have recently been discovered in ancient soils in Denmark above and below a layer of iridium, which apparently was created when a comet fragment hit Earth.

1991 The day after basketball superstar Earvin "Magic" Johnson's shocking announcement that he has the AIDS virus, President George Bush announces that he will fight harder against AIDS. Also, the FOX television network announces it will be the first to air condom commercials and stocks in condom companies rise.

1994 It is election day in the United States, but five citizens are too far away to vote. The astronauts on the space shuttle *Atlantis* are orbiting Earth at a distance of 160 miles. This is the first time that astronauts have been in space during a presidential election.

NOVEMBER 9

1801 Gail Borden, father of the modern preserved food industry, is born in Norwich, New York.

1825 Limelight is first used practically. Inventor Thomas Drummond heats a small ball of lime to glowing in front of a reflector on Scotland's Slieve Snaght; the light is seen on Divis Mountain over 66 miles away. The discovery changes lighting techniques forever in lighthouses and theaters.

1833 Sally Louisa Tompkins, humanitarian and hospital administrator and founder, is born in Mathews County, Virginia. She was the only female commissioned in the Confederate army, as a result of her selfless service to wounded soldiers.

1864 Microbiologist Dmitry I. Ivanovsky, discoverer of viruses, is born in Nizy, Russia.

1874 Albert Francis Blakeslee is born in Geneseo, New York. Best known for developing a way of changing the normal number of chromosomes in living cells (1937), Blakeslee first achieved prominence as a Harvard graduate student when he discovered sexuality in a lower fungus.

1891 Theodor Geiger, the first sociology professor in Danish history, is born in Munich, Germany, from which he fled when the Nazis came to power.

1897 Chemist Ronald Norrish is born in Cambridge, England. He shared the 1967 Nobel Prize in chemistry for his research into extremely fast chemical reactions.

1911 A patent application is filed for the first neon advertising sign. George Claude of Paris eventually received patent No. 1,125,476 in 1915, and a sign using his design was erected in July 1923 on the Cosmopolitan Theater in New York City.

1915 Soon to be famous as a founder of abstract algebra (which today is one of mathematics's largest and most active branches), 33-year-old Emmy Noether delivers her inaugural lecture at the University of Göttingen, despite a 1908 Prussian law prohibiting women from teaching at universities.

1934 Carl Sagan, astronomer, ecologist, and explorer of extraterrestrial life, is born in New York City.

1936 "The most famous animal of the twentieth century," a baby panda named Su-Lin, is captured by the hunting party of Ruth Harkness near Tsaopo, China. It will become the first live panda seen in the Western world. It dies tragically in a Chicago zoo on April 1, 1938.

1952 Chaim Weizmann, one of the first to harvest a material produced by a microbe, dies in Rehovoth, Israel, at 77. In 1911 he was able to culture bacteria to produce acetone for use in explosives. He was also Israel's first president, and one of the very few scientists to head a country.

1965 The biggest blackout in U.S. history darkens the Northeast and parts of Canada for 13 hours.

1967 Brian Latasa receives 230,000 volts from a defective high-tension line in Los Angeles, setting a record for a nonlethal electric shock.

1967 Taller than the Statue of Liberty, the massive Saturn V rocket is given its first test flight at Cape Kennedy. It is a success and it becomes the rocket that the United States will use to power the Apollo moon missions.

1974 Scientists at Stanford University begin an experiment that ultimately leads to the discovery of the "charmed quark" subatomic particle and to the Nobel Prize for the team's leader, Burton Richter.

1991 An international team of scientists in Culham, England, produces "a significant amount of power" from nuclear fusion for the first time in history. About 1.7 megawatts of electric power is produced during the two-second reaction.

1994 DNA tests on a beef help identify cattle rustlers, announces a spokeswoman for the police department of Brevard County, Florida. The uncooked beef was found to have the same genetic makeup as a purebred Angus cow that had been shot, butchered, and illegally sold several months before.

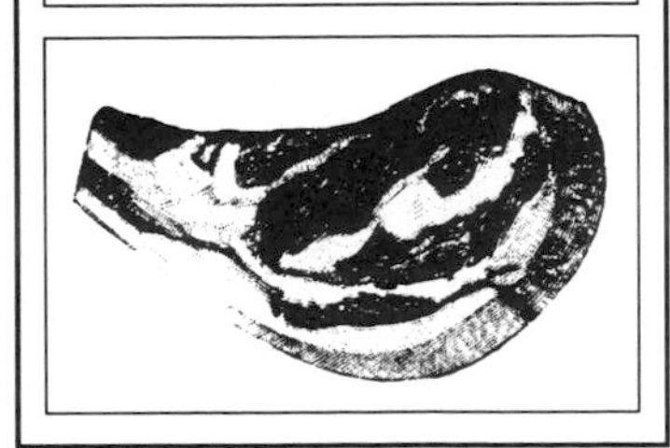

1619 René Descartes, 23, has three "visionary dreams" in the Bavarian town of Ulm. Although it is St. Martin's Eve, a day of celebration and drinking, Descartes claimed he drank no alcohol for three months prior to these dreams. He apparently spent the day meditating in a heated room before going to sleep. The dreams changed the direction of his life, and eventually of all mathematics and science. The dreams led Descartes to doubt all knowledge and concepts of reality, and from this he developed a method of precise inquiry that foreshadowed "the scientific method." Descartes described this inquiry technique in the 1637 classic *Discourse on Method.*

1764 Spanish-Mexican mineralogist, Andrés Del Rio, discoverer of vanadium, is born in Madrid.

1799 Joseph Black, discoverer of carbon dioxide and bicarbonates, dies at 71 in Edinburgh.

1851 Francis Maitland Balfour, a founder of modern embryology, is born in Edinburgh, Scotland (see biographical note on July 19, 1882).

NOVEMBER 10

1852 Paleontologist Gideon Mantell dies at 62 in London. He discovered four of the five dinosaur genera known during his time. He had a successful medical practice and marriage until he became obsessed with fossils in 1822 when his wife found some iguanodon teeth during a walk in the countryside.

1855 Alexandre Darracq, one of the first to mass-produce cars, is born in Bordeaux, France.

1861 Self-taught astronomer Robert Innes is born in Edinburgh. He discovered 1628 binary stars, and in 1915 discovered Proxima Centauri, the star nearest Earth (excluding our sun).

1871 "Dr. Livingstone, I presume?" Perhaps history's most famous quotation, spoken by *New York Herald* journalist Henry M. Stanley when he finally (after nearly a year in Africa) reaches Scottish explorer-missionary David Livingstone at Ujiji on Lake Tanganyika. "Yes, and I feel thankful that I am here to welcome you," replies the malnourished and ill Livingstone, who at the time was searching for the source of the Nile River.

1801 Dueling is outlawed by a U.S. state for the first time. Tennessee Governor Archibald Roan signs the "act to prevent the evil practice of dueling," recently passed by the state's General Assembly.

1899 Dr. Rudolph Matas uses spinal anesthesia on a patient in New Orleans. The following year he publishes his method in the *Journal of the Louisiana State Medical Society*; it is the nation's first report on spinal anesthesia.

1916 Geneticist Walter S. Sutton dies at 39 of a ruptured appendix. He was the first to suggest (in 1902) that chromosomes exist in pairs, and that they are the hereditary units theorized by Mendel.

1918 Ernst Otto Fischer is born the day before World War I ends, in München-Solln, Germany. He shared the 1973 Nobel Prize in Chemistry for determining the structure of ferrocene, an organic metal.

1934 Wilhelm His dies at 70 in Wiesental, Switzerland. Son of the renowned anatomist of the same name, His discovered in 1893 the bundle of His, which controls contraction of the heart muscle.

1951 Direct-dial, coast-to-coast telephoning begins in North America, in New Jersey.

1975 The *Edmund Fitzgerald* and its crew of 29 disappear in a storm on Lake Superior. At the time, it was the largest freshwater ship on Earth. "We are holding our own," were the captain's last words.

1977 The world's first test-tube baby, Louise Brown, is conceived in England.

1992 Two scientists announce in La Jolla, California, that they have developed a "molecular knife" that impairs reproduction in the AIDS virus by slicing through its nuclear material, RNA. The team of Flossie Wong-Staal and Arnold Hempel obtained an enzyme called a "hairpin ribosome" from a potato virus.

1993 A Harvard heart specialist announces that getting out of bed in the morning is much more likely to kill you than having sex.

1620 The *Mayflower* anchors off Massachusetts, and the Mayflower Compact is signed.

1675 Thomas Willis, history's first great epidemiologist, dies of pneumonia at 54 in London.

1729 Louis Bougainville is born the son of a notary in Paris. To avoid becoming a notary, Louis joined the French army, and then the French navy, in which he led the first French expedition around the world, making important discoveries, including Bougainville Island, in the Pacific.

1851 The nation's first telescope patent is awarded to Alvan Clark of Cambridge, Massachusetts.

1851 Jacques Bertillon, a major innovator in the application of statistics and quantitative methods to the social sciences, is born in Valmondois, France, the son of the head of the Paris Bureau of Vital Statistics. The "Bertillon classification" of causes of death, invented by Jacques, became a standard in many nations around the world.

NOVEMBER 11

1875 Astronomer Vesto Slipher is born in Mulberry, Indiana (see November 8, 1969, for a biographical note).

1886 Paul Bert, the founder of modern aerospace medicine, dies at 53 in Hanoi. As a physiologist, he discovered the causes of illness induced by traveling very high into the atmosphere (low pressure causes oxygen deprivation) and deep into the ocean (high pressure causes decompression sickness when divers rise to the surface too quickly).

1925 The discovery of cosmic rays by Robert Andrews Millikan (California Institute of Technology) is publicly announced at a meeting of the National Academy of Sciences in Madison, Wisconsin.

1927 Wilhelm Johannsen, coiner of the word "gene," dies in Copenhagen, where he was born 70 years before. He also coined the terms "phenotype" and "genotype."

1933 The first electrical contract in U.S. history, whereby a city agrees to purchase power from the federal government, is signed in Tupelo, Mississippi.

1936 A weather map appears on British television for the first time.

1973 Biochemist and Nobel laureate Artturi Virtanen dies at 78 in his place of birth, Helsinki. He discovered that acidifying cattle fodder prevents rotting without reducing its nutritional value.

1974 Discovery of the "charmed quark" subatomic particle is announced by two research teams.

1982 Space shuttle *Columbia* is launched with a crew of four, the largest crew at the time to voyage into space.

1986 The record high jump for a dog is set in Newton, Great Britain, when the German shepherd "Duke" scales a ribbed wall 11 feet 9 inches high.

1990 History's first heart–liver transplant patient, 13-year-old Stormie Jones, dies of sudden heart rejection in Pittsburgh. Dr. Thomas Starzl, her physician, vows never to operate again.

1994 *Science* reports that simple eye drops may be a new, powerful tool with which to diagnose Alzheimer's. Researchers at Brigham and Women's Hospital and Harvard tested a hunch that "tropicamide" (a chemical commonly found in eye drops) might cause extra dilation in Alzheimer's victims.

1527 Tycho Brahe observes the flaring of a new star from his observatory in Scania, Denmark. His report of the event established the existence of "novas," challenged the Aristotelian notion of a fixed universe, and helped win support for Copernicus's theory that the sun is the center of the solar system.

1614 Galileo, 50, reports to a high-ranking church official in Padua, Italy, that he has created a magnifying tool by looking through the wrong end of the telescope that he invented. "With this tube I have seen flies which look as big as lambs." It is one of history's first compound microscopes.

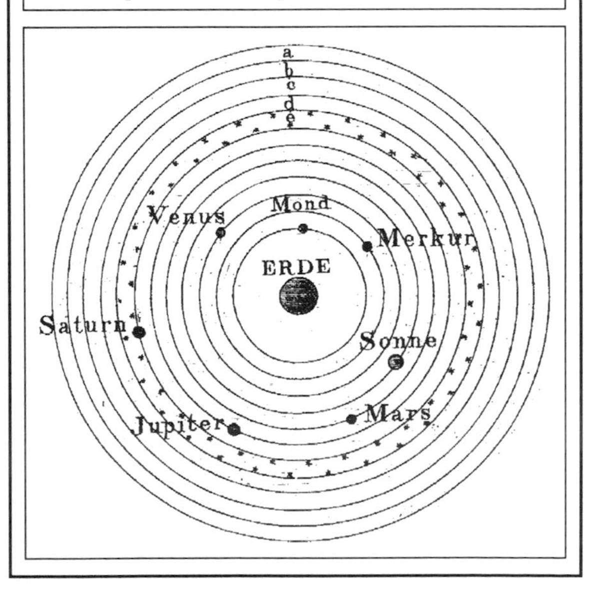

1746 Physicist Jacques Charles is born in Beaugency, France. He is remembered for Charles' law that explains the relationship between pressure and volume of a gas, and for inventing the hydrogen balloon. Legendary tales as a balloonist saved him from the guillotine.

1783 "Water, besides oxygen, contains another element," reports the great chemist Antoine Lavoisier, 40, to the French Academy of Sciences in Paris. He later named the element "hydrogen."

1799 The first known record of shooting stars in U.S. history is written. The meteoric display is seen and recorded by Andrew Ellicot off the Florida Keys.

1842 John Strutt, a.k.a. Lord Rayleigh, is born in Terling Place, England. He received the 1904 Nobel Prize for Physics for isolating argon, and also made fundamental discoveries about wave propagation.

NOVEMBER 12

1871 Erich Tschermak is born the son of a geologist in Vienna. By 1900, he, Correns, and DeVries independently rediscovered the laws of genetics discovered by Mendel in the 1860s, which remained hidden for many years.

1891 Seth Nicholson is born the son of a geologist in Springfield, Illinois. He is known for measuring temperatures of various planetary objects and for discovering four satellites of Jupiter.

1901 The first Nobel Prize for Physics is presented to Wilhelm Roentgen, the discoverer of X rays.

1911 At the end of the first airplane trip across North America, C.P. Rogers crashes at Compton, California. He is badly injured, but a month later resumes his flight to the Pacific coast.

1912 The body of Robert Falcon Scott is found in the Antarctic after his tragic attempt to reach the South Pole.

1916 Percival Lowell dies at 61 in Flagstaff, Arizona. Although a respected diplomat, traveler, and author, he is remembered as an astronomer. He successfully predicted the discovery of Pluto and unsuccessfully argued for years that Mars contained life. His sister was the poet Amy Lowell.

1941 The nation's first heredity clinic opens within the University of Michigan at Ann Arbor. Its purpose was to gather data on human heredity and to provide family counseling.

1941 The nation's first female pilot to test standard production aircraft, Alma Heflin, makes her first test flight, for the Piper Aircraft Corporation of Lock Haven, Pennsylvania.

1966 The first photograph of a solar eclipse taken from the atmosphere is snapped from satellite *Gemini XII*.

1970 History's most lethal natural disaster begins in the Ganges delta islands of Bangladesh; it is a circular storm that kills an estimated 1 million people by the time it ends the next day.

1933 The Loch Ness monster is photographed for the first time, by a local factory worker.

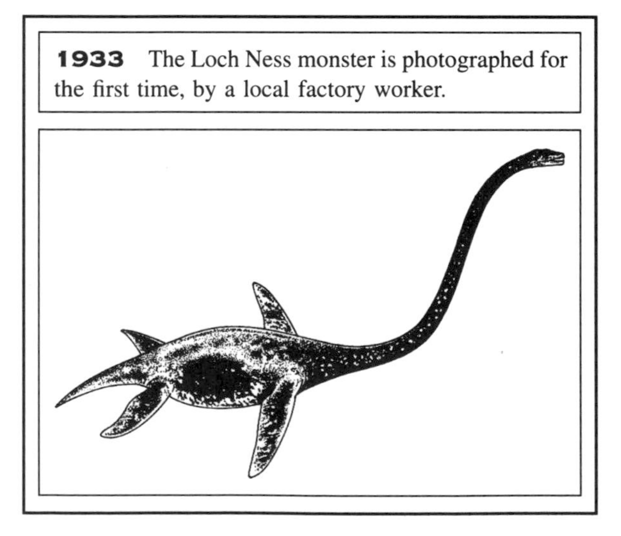

1975 Leo Wiener opens the country's first hotel for dogs, "The Kennelworth," in New York City.

1980 Space probe *Voyager I* comes within 77,000 miles of Saturn, sailing beneath its rings.

1984 The first salvage operation in space takes place. Astronauts on the shuttle *Discovery* haul a Palapa B-2 satellite into its cargo bay for transport back to Earth.

1992 People who wear disposable contact lenses are 14 times more likely to develop eye ulcers than those who wear nondisposable lenses, according to research released on this date in Chicago. The scientists blame misuse, not the lenses themselves.

1994 Wire services report help caffeine addicts: A Portland, Oregon, group and 12-step program.

NOVEMBER 13

1460 Henry the Navigator dies in debt in Vila do Infante (near Sagres, Portugal) at 66. His name was a misnomer, as he did little exploring himself; instead he financed many voyages, thereby helping launch the great age of exploration by Europeans.

1749 Benjamin Franklin is appointed first president of the first academy in American history (the Academy and College of Philadelphia), and in 1789 he writes in a letter, "In this world nothing can be said to be certain, except death and taxes."

1802 French botanist André Michaux dies of fever in Tamatave, Madagascar, at 56, during one of the several plant-gathering expeditions around the world that made him famous as an explorer and scientist.

1831 James Clerk Maxwell is born an only son in Edinburgh. Maxwell published his first scientific article at age 14. He later achieved enormous insights into the physical world, best known of which was his electromagnetic theory which stated that light, electricity, and magnetism are different forms of the same type of energy.

1848 Albert Grimaldi is born the Prince of Monaco in Paris, and becomes a major patron of the sciences, especially oceanography. He helped develop equipment and institutions in this budding science.

1862 Mary Henrietta Kingsley is born into a secluded life in London. At age 30, she traveled to Africa to complete a book on native religion and law begun by her deceased father. She became a famous traveler, adventuress, and specimen collector.

1868 Albert Smith Bickmore, the "father" and first director of the American Museum of Natural History, writes to automobile magnate W.E. Dodge about financing the purchase of France's Verreaux Collection of animal specimens. It soon became one of the four original collections of the New York Museum.

1893 Edward Adelbert Doisy, the first to crystallize a female sex hormone, estrone, is born in Hume, Illinois.

1902 G.H.R. von Koenigswald, investigator of the humanoid "Java man," is born in Berlin.

1913 Dr. Daniel H. Williams is first African American admitted to the American College of Surgeons.

1927 The Holland Tunnel opens to traffic between New Jersey and New York City.

1957 Gordon Gould has a notary witness his lab notebook, calculations in which created the laser.

1961 At noon, history's greatest gas fire begins at Gassi Touil in the Algerian Sahara. Flames rise 450 feet. It continues until April.

1971 A man-made object orbits another planet for the first time. The unmanned *Mariner 9* enters orbit around Mars at 7:33 EST, to photograph its surface and to collect data on its atmosphere and chemistry.

1974 Karen Silkwood, nuclear technician and union activist, dies in a mysterious car crash after reporting safety violations at the Kerr-McGee Cimarron plutonium plant near Crescent, Oklahoma.

1989 The most violent quasar outburst ever observed is seen by a joint U.S.–Japanese team. The total energy released in 3 minutes is estimated to be equal to that released by the sun in 340,000 years.

1992 A "window to the seventeenth century" opens. Anthropologists in St. Marys City, Maryland, use a crane to hoist a third coffin out of a 300-year-old crypt in Chapel Field. The recovered skeletons are "without peer" the best-preserved remains of colonial Americans known to science. DNA tests are possible.

1993 The medical journal *Lancet* reports (contrary to popular belief) that chronic drinking does *not* kill brain cells—it disconnects them.

1946 Artificial snow from a natural cloud is produced for the first time. Vincent Joseph Schaefer of the General Electric Company sprinkles dry ice pellets for several miles as he flies 14,000 feet above Mount Greylock in Massachusetts. Snow develops and falls an estimated 3000 feet.

1716 Gottfried Leibniz, philosopher and coinventor of calculus, dies at 70 in Hannover, Germany.

1732 Louis Timothee, a young French immigrant, is hired by the Library Company of Philadelphia, becoming the first paid librarian in U.S. history.

1765 Steamboat inventor Robert Fulton is born the son of a farmer in Little Briton (now Fulton), Pennsylvania.

1771 Marie-François-Xavier Bichat, the first to use the word "tissue" in a biological context, is born in Thoirette, France, the son of a doctor. Without use of a microscope, and without knowledge of cells as the basic unit of life, Bichat identified 21 different types of tissue in the human body. His *General Anatomy* (1801) is considered the birth of histology.

NOVEMBER 14

1797 Geology giant Sir Charles Lyell is born near the Grampian Mountains in Scotland. He is largely responsible for the acceptance of the theory that Earth's features and living organisms are the result of very slow and gradual natural processes. He paved the way for Darwinism.

1807 Chemist Auguste Laurent is born the son of a wine merchant in La Folie, France. He discovered various compounds, but is most famous for his acrimonious debate with Berzelius over the true nature of chemical structure. Laurent was correct, but his beliefs are not proven until after his death.

1829 Louis Vauquelin, discoverer of the elements chromium (1797) and beryllium (1798), dies a 66-year-old bachelor in his place of birth, Saint-André-d'Hébertôt, France.

1851 *Moby Dick* is first published in the United States.

1863 Leo Baekeland is born in Ghent, Belgium. Already wealthy from other chemical inventions, he created Bakelite (announced in 1909); it was the first thermosetting plastic, and became the cornerstone of the modern plastics industry.

1888 The Pasteur Institute is dedicated before a dignitary-filled assemblage in Paris.

1891 Sir Frederick Grant Banting, the first Canadian to win a Nobel Prize, is born the son of a farmer in Alliston, Ontario. He isolated and discovered insulin in 1922.

1921 In the summer, Dr. Frederick G. Banting and his young assistant Charles H. Best began experiments at the University of Toronto that eventually determined the cause of diabetes. Shown behind Banting is the unsung hero in the story, the dog; all of the Banting and Best experiments were performed on canines. See November 14, 1891, for Banting's birth; see also references to Best and J. J. R. Macleod for other details of the story.

1922 The BBC begins broadcasting in England.

1933 Astronaut Fred W. Haise is born in Biloxi, Mississippi. His first flight in space was *Apollo XIII*.

1935 A life-size photograph of an entire human appears in a U.S. newspaper for the first time. A picture of Larry Quinn, a 14-pound 7-ounce baby who is 21½ inches tall, appears in the *Call-Bulletin* of San Francisco, where Quinn was born two days before at St. Mary's Help Hospital.

1938 Hans Gram, inventor of microbiology's classic "Gram stain technique," dies at 85 in Copenhagen.

1969 One month after announcing that cyclamates cause cancer and chromosome damage, and are not safe for human consumption, the U.S. Food and Drug Administration decides that cyclamates can still be sold in packaged foods for seven more months.

1969 *Apollo XII* is launched to the moon from Cape Kennedy. It is the first time that a president (Richard Nixon) attends the launch of a manned spaceflight.

NOVEMBER 15

1280 The scholar Count Albert von Bollstädt dies at 87 in Cologne, Germany. Best known as Albertus Magnus (Albert the Great) because of his great learning in a range of fields, from botany to chemistry to geology, he correctly theorized that spots on the moon are surface features and that the Milky Way is composed of many stars; his description of arsenic was so accurate that he is sometimes given credit for its discovery. Magnus was accused of wizardry, but his position in the church as a bishop saved him from prosecution.

1630 Astronomer Johannes Kepler, alleged never to have taken a bath in his life, dies at 58 in Bavaria.

1672 Franciscus Sylvius dies at 58 in Leiden, the Netherlands. A founder of the "iatrochemical school of medicine"—which held that all living processes are based on chemical reactions— he was instrumental in shifting medicine from mysticism to science. He is credited with establishing the first university chemistry laboratory, discovering the brain's Sylvian fissure, and inventing gin.

1677 Robert Hooke demonstrates the existence of protozoa to the Royal Society in London.

1738 William Herschel, the first to discover a planet beyond Earth, is born the son of a musician in Hannover, Germany. He discovered Uranus in 1781.

1835 Some of Darwin's writings are first presented to science. They are excerpts from letters about geology written aboard the *Beagle* and sent to Professor John Henslow, who reads them on this day to a meeting of the Cambridge Philosophical Society.

1849 The first major poultry show in U.S. history—"Grand Show of Domestic Poultry and Convention of Fowl Breeders and Fanciers" opens at the Public Gardens, Boston.

1871 Erich Tschermak, one of the scientists who discovered the laws of genetics, is born in Vienna.

1874 August Krogh is born the son of a brewer in Grenå, Denmark. In 1920 he won the Nobel Prize for Physiology or Medicine for discovering how the body regulates blood flow to the capillaries.

1888 Oceanographer H.U. Sverdrup is born in Sogndal, Norway. He is known for his physical, chemical, and biological studies of the oceans, and especially for explaining equatorial countercurrents.

1896 Niagara Falls provides electric power to an entire city—Buffalo, New York—for the first time.

1919 Chemist Alfred Werner dies at 52 in Zurich. He won the 1913 Nobel Prize for his "coordination theory" of chemical structure that described the distribution and bonding of atoms in molecules.

1928 Geologist Thomas Chamberlain dies at 85 in Chicago. He was one of the first to realize that Earth has undergone several Ice Ages, not just one. He also studied the formation of planets.

1959 Charles Thomson Rees Wilson, Nobel laureate and inventor of the nuclear "cloud chamber," dies at 90 in Carlops, Scotland.

1967 Elmer McCollum, discoverer of fat-soluble vitamins, dies at 88 in Baltimore.

1974 The pressure-sensitive, self-sticking postage stamp debuts, in New York City. The Dove of Peace is the first image on this stamp.

1978 Anthropologist Margaret Mead dies at 76 in New York City.

1990 President George Bush signs the Clean Air Act of 1990, the first update of the Clean Air Act of 1977.

1993 The World Health Organization warns that we are losing the war against tuberculosis.

1839 William Murdock, inventor of the gas light, dies at 85 in Birmingham, England.

1620 Corn is first encountered by Pilgrims, at a site now called Corn Hill in Provincetown, Massachusetts. Miles Standish was one of the leaders of the party of 16 that made the discovery.

1676 The first prison warden in U.S. history is hired in Nantucket, Massachusetts. William Bunker agrees to tend the new jail for one year for a salary of "foeur pounds, halfe in wheat, the other in graine."

1717 Mathematician Jean Le Rond d'Alembert is born in Paris. He was the illegitimate son of an aristocrat who abandoned him at the church of Jean-Le-Rond. He was found and raised by a glazier.

1814 After 113 years 124 days on Earth, Pierre Joubert dies, setting the Canadian record for human longevity.

1837 Horace Mann begins writing his first annual report as Secretary for the Massachusetts Board of Education. The 12 documents he produced in this post transformed public education worldwide.

NOVEMBER 16

1869 The Suez Canal opens, joining the Mediterranean and Red Seas. It is the world's longest canal for large ships. Of the 1.5 million workers, 120,000 died during its construction.

1875 A dental mallet is first patented in the United States. Dr. William Bonwill receives patent No. 170,045 for "electro-magnetic dental pluggers," a vibrating device for filling cavities with gold.

1901 An automobile exceeds the speed of 1-mile a minute for the first time in the United States. It happens in Brooklyn. A.C. Bostwick drives a 40-hp Winton during an Automobile Club race on Ocean Parkway. His record is broken several minutes later by H. Fournier, whose record then falls to Foxhall Keene.

1933 The world's highest waterfalls are first sighted by U.S. pilot Jimmie Angel, who records sighting them in Venezuela in a branch of the Carrao River. The falls are now called Salto Angel.

1934 Karl Linde, inventor of the first practical refrigerator, dies at 92 in Munich. His research into the cooling of gases enabled him to be the first to produce liquid oxygen and nitrogen in industrial quantities.

1945 The discoveries of elements 95 (americium) and 96 (curium) are announced to the public. They are discovered by two teams led by Glenn T. Seaborg.

1954 Geneticist A.F. Blakeslee dies at 80 in Northampton, Massachusetts (see November 9, 1874).

1955 A motorboat exceeds the speed of 200 mph for the first time in the United States. Donald Campbell in the *Bluebird* covers the 1-km course on Lake Mead, Nevada, at an average speed of 216.2 mph.

1955 A dugong—a close relative of the endangered manatee—first arrives in the United States. It is put on display in San Francisco's Steinhart Aquarium until December 27, when it dies.

1966 The greatest meteor shower in recorded history begins. The Leonid meteors are visible over the western United States and the eastern USSR, at a rate of 2300 meteors per minute at the peak.

1969 Moon rocks are put on public display, in New York City.

1973 The Alaska Pipeline act is signed into law by President Richard Nixon.

1973 *Skylab IV* blasts into outer space from Cape Canaveral; it returns 84 days 1 hour 15 minutes 31 seconds later, making it the longest U.S. spaceflight to date.

1974 The world's largest radio telescope dish, covering 18½ acres, is rededicated at Arecibo, Puerto Rico.

NOVEMBER 17

1597 Henry Gellibrand, the first to discover that Earth's magnetic field is constantly changing in strength and orientation (a mystery not yet explained), is born in London.

1790 August Möbius, creator of the paradox in fluidity, the Möbius strip, is born in Schulpforte, Germany, the son of a dancing teacher.

1791 The world learns that breathing and burning are essentially the same. A famous memo by Armand Ségun and Antoine Lavoisier describing the chemical similarities in the two processes is presented to the French Academy of Sciences today.

1865 Astronomer John Plaskett is born in Hickson, Canada. Although a high school dropout, he found a job in an Ottawa observatory, where he built an enormous telescope of his own design, and discovered "Plaskett's twins"—which for half a century were the most massive stars known.

1902 Nobel Prize-winning physicist Eugene Wigner is born the son of a businessman in Budapest. He developed several important theories about nuclear structure, and helped build the atomic bomb.

1908 Ecologist Thomas Park is born in Danville, Illinois. He is best known for studies in beetle population studies, which have since been applied to human biology by others. He wrote *Principles of Animal Ecology* (1949), which analyzed the ecology and evolution of animals with principles first derived from plants.

1913 The nation's first course for dental hygienists starts. Thirty-three women begin the course created by Dr. Alfred Civilion Fones, founder of the Fones Clinic in Bridgeport, Connecticut. Twenty-seven of the women graduated on June 5, 1914. Fones later published the nation's first textbook for dental hygienists, *Mouth Hygiene*.

1921 The first report of Rhodesian man is published in London. This Neanderthal remains man's oldest known relative for several years, until bones of other relatives are found in Java, China, and Olduvai Gorge.

1925 Rock Hudson is born in Winnetka, Illinois. He shook the world when it was revealed that he was homosexual and that he had AIDS, at a time when neither condition was socially acceptable.

1926 Inventor-naturalist-explorer Carl Akeley dies at 62 in Virunga National Park, Zaire, where he is buried. His taxidermy innovations created museum displays of unprecedented realism. The Akeley camera provided the first footage of gorillas in the wild, the Akeley African Hall is in the American Museum of Natural History, and Akeley helped establish the first wildlife sanctuary in central Africa.

1940 Raymond Pearl, a founder of biometry, or the application of statistics to biology and medicine, dies at 61 in Hershey, Pennsylvania. He founded the department of medical statistics at Johns Hopkins, founded several biology journals, and wrote more than 700 articles and books.

1952 Mrs. Margaret Rice dies of breast cancer in Chicago, having neglected real therapy while using the infamous Ruth B. Drown's "Radio Therapeutic Instrument," a worthless "miracle" machine. Drown was eventually fined $1000 for false advertising claims, but she continued treating up to 35,000 patients with several devices in her Hollywood offices.

1970 "The first extraterrestrial robot," Russia's *Lunakhod 1*, lands in the Sea of Rains on the moon.

1980 The modern turkey-plucking record is set in Dublin by Vincent Pilkington: one bird in 1.5 minutes.

1980 Despite bitter faculty debate and promise of great profit, Harvard's president Derek Bok announces that the school will *not* become a partner in a gene-engineering company.

1889 Direct rail service between Chicago and the West Coast is begun by the Union Pacific RR Co.

1789 Artist Louis Daguerre is born in Cormeilles, outside Paris. He is known for inventing the first form of photography, the daguerreotype.

1810 Asa Gray, botanist and early champion of Darwin, is born in Samquoit, New York.

1812 Russian troops overrun Napoleon's forces at Krasnoy. Left for dead on the frozen battlefield, Jean-Victor Poncelet, 24, recovers and the following spring he invents projective geometry in a prison camp.

1820 Antarctica is discovered, by Captain Nathaniel Brown Palmer on the *Hero*, a 44-ton sloop with a crew of just six. They had sailed from Connecticut the previous July. The Russian Fabian von Bellingshausen and the Englishman Edward Bransfield also claimed the first sighting of the continent.

1832 Adolf Erich Nordenskiöld, the first to successfully navigate the Northeast Passage (from Norway to the Pacific across the Asian Arctic), is born into an aristocratic Swedish family in Helsinki, Finland.

1839 Physicist August Kundt is born in Schwerin, Germany. He is remembered as Roentgen's mentor, but he also developed an ingenious method of measuring the velocity of sound in gases and solids (by sending the sound through a tube lined with fine powder).

1874 The country's first national women's temperance society is organized in Cleveland.

1897 P.M.S. Blackett, the first to prove that atomic bombardment causes one element to change into another, is born in London. He received the 1948 Nobel Prize for physics for his (atom) smashing research.

1901 George Horace Gallup, creator of the Gallup Poll, is born in Jefferson, Iowa. He taught journalism at Drake University and Northwestern University until 1932 when he was hired by a New York City advertising firm to do public-opinion surveys. He changed social science forever.

1906 What happens when light hits the eye? George Wald, born on this date in New York City, won the 1967 Nobel Prize for Physiology or Medicine for determining the chemical reaction that occurs in the cells of the retina.

1913 The first airplane loop-the-loop over U.S. soil is performed by Lincoln Beachey in San Diego.

1923 Alan Shepard, the first American in space (May 1961) and the fifth man on the moon (as commander of *Apollo 14* in 1971), is born in East Derry, New Hampshire.

1928 *Steamboat Willie* premieres in New York City's Colony Theater. It is Mickey Mouse's debut, Walt Disney's breakthrough, and history's first animated cartoon with synchronized sound.

1938 After unwittingly spreading disease and death to hundreds while working as a cook, "Typhoid Mary" Mallon dies in New York City of a stroke.

1932 Psychiatrist Karen Horney delivers her first paper in the United States, at a conference in Chicago.

1961 The science journal *Nature* carries a report by George Claus and Bartholomew Nagy about meteors that brought algalike structures to Earth. The shapes, materials, and chemistry of these extraterrestrial forms resemble Earth-generated fossils.

1963 The push-button telephone goes into public service.

1978 The greatest mass suicide in modern history takes place in the People's Temple religious cult, Jonestown, Guyana.

1992 Superman is killed, in issue No. 75 of the Superman comic book. At 4 million copies, it is the largest seller in the history of DC comics. Unfortunately, the killer is a "superlunatic, called Doomsday, dressed in a straitjacket and escaped from a cosmic insane asylum." The event upsets mental health workers because it "is stereotyping at its worst, pitting the ultimate evil against the ultimate good—and promising to make life difficult for the mentally ill into the next generation."

NOVEMBER 19

1711 The stormy life of chemist-poet Mikhail Lomonosov begins in Denisovka (now called Lomonosov), Russia. The founder of Russian science, his life was tormented by drink and his temper.

1722 Leopold Auenbrugger is born in Graz, Vienna. As the son of an innkeeper, he often observed wine barrels being tapped to determine how full they were. When he became a doctor, he invented the still-standard diagnostic technique of percussion, in which the body's surface is tapped to determine the state of the organs underneath.

1840 Embryologist Aleksandr Kobalevsky, the first to find a common pattern of development for all "higher" animals, is born in Shustyanka, Russia.

1872 The first U.S. adding machine to print totals and subtotals is patented. Edmund D. Barbour of Boston receives patent No. 133,188 for his "calculating machine." It was never successfully marketed.

1887 Biochemist James B. Sumner is born in Canton, Massachusetts. At 17 a hunting accident led to the amputation of his left arm, which caused some teachers to advise against a career in chemistry. Sumner went on to become an excellent athlete (especially in tennis) and a Nobel Prize-winning chemist. He investigated the nature of enzymes, the body's little chemical processing plants. Before his work, many thought enzymes were proteins, but experimental evidence both supported and contradicted this position. In 1926 Sumner extracted an enzyme from jack beans that transforms urea into ammonia and carbon dioxide; he called it urease. He was then able to crystallize it. This was the first time any enzyme had been crystallized, and it finally proved that enzymes are proteins. Sumner shared the 1946 Nobel Prize for Chemistry with Wendell M. Stanley, the first to crystallize a virus.

1918 Hendrik Hulst, the pen-and-paper astronomer, is born in Utrecht, the Netherlands. During Nazi occupation of his country, when science research was impossible, Hulst predicted that the universe should contain 21-centimeter radio waves produced by infrequent behavior of hydrogen atoms. Such waves were later discovered by others, and they permitted extensive mapping of the Milky Way.

1945 Harry Truman delivers the first presidential address to Congress devoted entirely to health care.

1954 An automatic toll machine goes into operation for the first time in U.S. roadway history. Two exact-change machines are put in service at the Union Toll Plaza near Newark, New Jersey.

1959 Ford announces that it is halting production of the Edsel.

1969 *Apollo 12* astronauts Alan Bean and "Pete" Conrad make man's second landing on the moon. They become the first to stay there for a day, and the first to retrieve an object—a piece of 1967's *Surveyor 3*.

1971 The water storage unit of a nuclear reactor in Monticello, Minnesota, spills over, dumping 50,000 gallons of radioactive waste into the Mississippi River. Some of it enters the St. Paul water system.

1973 Barbara Ringer is appointed as the first female Register of Copyrights in U.S. history.

1992 Stuntman Jim Mouth smokes 154 cigarettes—all at the same time. The feat is performed in Los Angeles during the year's "Great American Smokeout" to publicize the harm and ugliness of the habit.

1993 Two teams publish independent reports on the discovery of a key mechanism in the spread of cancer. Both found that protein p53 regulates a gene that in turn regulates cell growth; an abnormal p53 can lead to the abnormal cell profusion that is cancer.

1912 Biologist George Palade is born in Iasi, Romania. His innovations in microscopy allowed him to discover several cell structures, best known of which are ribosomes. He won the 1974 Nobel Prize.

1915 Earl Sutherland, who isolated and discovered cyclic AMP, is born in Burlingame, Kansas. His feat occurred in 1956, his Nobel Prize in 1971.

1519 Magellan's expedition, the first to circumnavigate Earth, crosses the Equator. The North Star is no longer visible, which disrupts navigation and causes panic among the sailors.

1602 Physicist Otto von Guericke is born in Magdeburg, Germany. Inspired by philosophic debate over the impossibility of a vacuum, he built history's first. To create this first vacuum (in 1650), he built the first air pump. He also built the first electrical generator, and with it became the first to see electrically created light (when he ran electricity through a ball of sulfur).

1620 Peregrine White is born aboard the *Mayflower* in Cape Cod harbor, becoming the first child of English descent born in the New World.

1873 Physicist William Coblentz is born the son of a farmer in North Lima, Ohio. He is famous for measurements of infrared radiation from stars, planets, and nebulae.

NOVEMBER 20

1886 Ethologist Karl von Frisch is born in Vienna. Frisch opened the eyes of science to the sensory and communication capacities of "simple" animals. He showed that fish can distinguish colors and brightness, and that they can distinguish sounds with greater acuity than humans; he also discovered bees' famous "waggle dance," in which individuals communicate the direction and distance of food to hive-mates by a series of "dancing" movements. He shared the 1973 Nobel Prize in Physiology or Medicine with fellow ethologists Niko Tinbergen and Konrad Lorenz.

1888 An employees' time clock is patented for the first time in the United States. The device was invented by Willard L. Bundy of Auburn, New York, who receives patent No. 393,205. The Bundy Manufacturing company produced the machines, and later became a branch of IBM.

1567 Pope Pius V prohibits bullfighting: Any princes who permit it will be excommunicated, and those who die in the ring will not receive a Christian burial.

1889 Edwin Powell Hubble, the founder of extragalactic astronomy, is born the son of a lawyer in Marshfield, Missouri. He obtained a law degree, but became bored with it and returned to his love of the stars. He is best known for providing the first evidence that the universe is expanding.

1930 Pioneering psychiatrist Karen Horney delivers her famous paper "The distrust between the sexes" to a Berlin medical conference during her split with Sigmund Freud.

1948 Believed to be extinct, the flightless takahe bird of New Zealand is photographed for the first time.

1966 The world's largest time capsule is sealed and buried (until AD 2866) in Rosamond, California.

1967 A few minutes past 11 AM, the U.S. Department of Commerce Census Clock passes 200 million.

1969 Charles ("Pete") Conrad and A.L. Bean depart the moon after becoming the first humans to spend an entire day on its surface. Their lunar module *Intrepid* had landed on the moon's Ocean of Storms the previous day. They rejoin *Apollo XII*, piloted by Richard Gordon, eventually splashing down in Earth's Pacific Ocean, 400 miles from Samoa, on November 24.

1969 The Nixon administration announces a ban on residential DDT use, en route to a total ban of the insecticide.

1980 The *Solar Challenger* flies for the first time with no source of power other than the sun. The following July it became the first solar airship to cross the English Channel.

1991 It is announced today in New York City that the Philip Morris tobacco company has awarded a research grant of $1.2 million to James E. Woods, an EPA advisor on health. "It sounds like the fox guarding the henhouse to me," quips a member of Americans for Non-Smokers' Rights in response.

1783 Man's first flight: Pilâtre de Rozier and the Marquis d'Arlandes rise above Paris in a balloon.

1785 Dr. William Beaumont is born the son of a farmer in Lebanon, Connecticut. He earned medical immortality for eight years' research on digestion in Alexis St. Martin, a trapper whose stomach was accidentally shot open and never closed completely. St. Martin lived to be 82, Beaumont 67.

1818 Lewis Henry Morgan, a father of scientific anthropology, is born near Aurora, New York. A lawyer by profession, Morgan developed an interest in the culture and plight of the Native Americans and became a lifelong advocate for their rights. Through this interest he discovered that the Seneca had a system of designating relatives that was completely different from the Anglo-American system, but which was identical to that of other Native American tribes. His publication in 1871 of his findings was history's first major thesis on comparative kinship. In later works Morgan developed the first scientific theory of the origin and evolution of civilization, positing that changes in social organization followed changes in food production, and that man progressed from "savagery" to "barbarism" to "civilization." Most of this theory is accepted today.

NOVEMBER 21

1824 Hieronymus Richter, codiscoverer of the element indium, is born in Dresden, Germany.

1846 Oliver Wendell Holmes coins the word "anesthesia" in a letter to William Thomas Green Morton, the surgeon who first publicly demonstrated the pain-killing effects of ether.

1867 Vladimir Ipatieff is born the son of a Moscow architect. Ipatieff became one of history's great hydrocarbon chemists. He discovered the structure of isoprene (the basic molecule in rubber), developed ways to catalytically control breakdown reactions, and produced high-octane gas from low-octane fuels.

1871 The nation's first cigar lighter patent is No. 121,049, issued on this date to Moses F. Gale of New York City. Oscar Hammerstein was to receive the first U.S. patent for a practical cigar rolling machine 12 years later.

1877 Edison announces his invention of a "talking machine," the phonograph.

1891 Alfred Sturtevant, the first to map the position of genes on a chromosome, is born in Jacksonville, Illinois. He invented gene-mapping by studying fruit flies.

1922 The first U.S. cruise ship to circumnavigate the globe, the Cunard liner *Laconia*, departs New York City with 440 passengers. It completed its around-the-world voyage 130 days later. Meanwhile, back in Washington, Rebecca Latimer Felton—the first woman to occupy a seat in the U.S. Senate—completes her second and last day on the job. She had been appointed by Georgia Governor Thomas Hardwick to temporarily fill the seat vacated by the death of Thomas Watson.

1953 One of the great hoaxes in science history ends. The British Museum declares that the Piltdown man is a "perfectly executed and carefully prepared fraud."

1963 Robert Stroud (a.k.a. the Birdman of Alcatraz) dies at 73 in federal prison in Springfield, Missouri.

1968 The Love Canal environmental debacle begins with the birth of Sheri Schroeder in Niagara Falls; her physical problems at birth cause her mother Karen to mount a famous antipollution campaign.

1994 DNA tests are used to vindicate pop idol Michael Jackson, as Superior Court Judge Glenn Ritchey, Jr. dismisses a paternity suit by Michelle Flowers, 31, who claimed her 10-year-old son was fathered by Jackson.

1694 Voltaire is born the son of a minor bureaucrat in Paris. A great author, critic, and philosopher, Voltaire's main contribution to science was his popularization of Newton's works and his advocacy of rational analysis.

1787 Rasmus Rask, a founder of the science of comparative linguistics, is born in Braendekilde, Denmark. His study of the sources and similarities of European languages led Jacob Grimm to formulate Grimm's law, a cornerstone in linguistics that describes how consonants evolve from one language to another.

1820 The oldest Irish person on record, the Hon. Katherine Plunket, is born. She lives for 111 years 327 days.

1842 Mount Saint Helens erupts; it is the first accurately dated volcanic eruption in North American history.

1899 The Appalachian National Park Association is formed in Asheville, North Carolina. It is the first organization dedicated to protecting forests in the eastern United States.

1904 Louis Néel, 1970 Nobel laureate and developer of solid-state physics, is born in Lyon, France.

1904 A direct-current electric motor is first patented in the United States. Mathias Pfatischer of Philadelphia is awarded patent No. 775,310 for a "variable speed motor."

NOVEMBER 22

1906 SOS is adopted as the universal radio distress signal, by delegates to the International Radio Telegraphic Convention in Berlin. The new letters replace the original distress call of CQD, established in 1904, which was popularly interpreted to mean "Come quick, danger," but which actually meant "Stop sending and listen; danger."

1917 Sir Andrew Fielding Huxley, grandson of famed biologist T.H. Huxley, is born in London. He shared the 1963 Nobel Prize in Physiology or Medicine for determining with A.L. Hodgkin the chemistry of nerve conduction; the pair discovered the all-important "sodium pump" that allows nerves to repeatedly carry impulses along their length.

1922 The New York Philharmonic Orchestra is first broadcast on radio; regular Sunday concerts soon become a cultural fixture.

1927 The snowmobile is patented. Carl Eliason of Sayner, Wisconsin, receives patent No. 1,650,334 for his "vehicle for snow travel."

1941 Kurt Koffka, explorer of the mind, dies at 55 in Northampton, Massachusetts. In collaboration with Wolfgang Köhler and Max Wertheimer, Koffka founded Gestalt psychology in the early 1900s. Their doctrine stressed wholeness; they maintained that people perceive their environment as a whole (rather than responding to its parts separately) and that mental phenomena, such as thoughts and feelings, occur as wholes (rather than being merely the sum of the individual, underlying neural activities).

1942 The first African American in space, Guion S. Bluford, is born in Philadelphia.

1910 The steel-shafted golf club is patented by Arthur F. Knight of Schenectady, New York.

1953 A color television program appears coast to coast for the first time in the United States. It is the *Colgate Comedy Hour*, starring Donald O'Connor and Ralph Bellamy.

1957 Dr. William Menninger observes in the *New York Times*, "Mental health problems do not affect three or four out of every five persons, but one out of one."

1977 Regular SST (supersonic transport) service between New York and Europe begins on a trial basis, with the landing of the *Concorde* at JFK airport.

1991 The longest slot-car track in history (958 feet) is built at the Mallory Park Circuit, Great Britain.

1993 Nine-year-old Tarah Schaeffer makes her debut on *Sesame Street*. Afflicted with a disease called osteogenesis imperfecta (or "brittle bones"), Tarah will be the first regularly appearing character in a wheelchair.

NOVEMBER 23

1221 Alfonso X is born in Burgos, Spain. The Alphonsus crater on the moon is named for him; he was an enlightened monarch who established schools, fostered the sciences, and oversaw preparation of the "Alfonsine Tables"—the most accurate planetary charts for the next three centuries.

1553 Prospero Alpini, the first botanist to note that plants come in two sexes, is born in Venice, Italy. The discovery came during studies of the date palm while practicing medicine in Egypt. He died on this date in 1616.

1616 Mathematician John Wallis is born the son of a rector in Ashford, England. A child prodigy who reportedly calculated the square root of a 53-digit number in his head to 17 places, Wallis achieved prominence in his 20s by deciphering enemy codes during the English Civil War. His *The Arithmetic of Infinitesimals* (1655) foreshadowed Newton's invention of calculus (Newton acknowledged that calculus developed from a study of Wallis's work). Wallis was also the first to use fractions and negative numbers as exponents, or powers of a given number, and in 1656 he was the first to use a sideways "8" to symbolize infinity. He was also the first to suggest the law of conservation of momentum (1668), and he cofounded the Royal Society with Robert Boyle in 1663.

1654 Blaise Pascal's "night of fire": the famous 31-year-old mathematician-physicist suffers a near-death experience when his carriage horses run wild. He interprets this as a sign of divine displeasure, abandons research completely, and spends the last seven years of his life in meditation and religious writing (see June 19, 1623).

1837 Johannes van der Waals, the first to postulate the existence of van der Waals forces, is born the son of a carpenter in Leiden, the Netherlands. His worked helped explain how molecules are held together.

1869 Inventor Valdemar Poulsen is born in Copenhagen. He produced the first device that generated continuous radio waves and the first workable device to translate sound waves into magnetic variations in metal. Both innovations are critical to modern communications.

1887 Henry Moseley is born in Weymouth, England, the son of a naturalist who sailed on the historic voyage of the *Challenger* that began modern oceanography. Moseley's life was one of genius with a tragic end at age 27. He was the first to demonstrate that an element's properties depend on its atomic number (not its atomic weight), and he established the relationship between charge and atomic number.

1921 The Sheppard-Towner Act becomes law, despite opposition of the American Medical Association. The new law provides federal subsidies to states that establish infancy and maternity welfare programs.

1948 The zoom lens (for television cameras) is patented. Dr. Frank Back of New York City receives patent No. 2,454,686 for a "varifocal lens for cameras."

1965 France puts its first satellite into orbit.

1969 The first news conference from outer space is televised, starting at 7:30 PM New York time. Aboard *Apollo XII* 108,000 miles from Earth and traveling 3670 mph, Pete Conrad, Alan Bean, and Richard Gordon answer questions that are submitted in writing by reporters to Marine Colonel Gerald P. Carr in the Manned Spacecraft Center in Houston. Carr then reads the questions to the astronauts.

1992 Russian engineers announce in an ABC-TV interview that a sunken nuclear submarine is leaking radioactive cesium into the ocean about 150 miles from Norway and near rich fishing grounds.

1992 Animal rights activist Marilee Geyer has two turkeys in her home in Scotts Valley, California, for Thanksgiving dinner and she stuffs them with all of the bird feed they can eat. It is a protest against cruelty of "farm factory" methods. The two "toms" were saved by the group Farm Sanctuary from being abandoned for dead by a New York poultry farm.

1922 Financier Lord Carnarvon and his daughter arrive in Luxor, summoned by archaeologist Howard Carter who has already discovered the staircase to King Tut's tomb. During his stay in Egypt, Carnarvon nicks himself shaving; soon the wound becomes infected and kills him, initiating the legendary "King Tut's curse."

1639 A transit of Venus across the face of the sun is first observed. The feat is accomplished by self-taught astronomer Jeremiah Horrocks, 22, who had predicted it. He is the curate at Hoole, England, and rushes outside between church services to witness the event.

1871 The National Rifle Association (NRA) is organized and chartered in New York City. The first officers among the 35 members were Ambrose Burnside (president), William Church (vice president), and George Wood Wingate (secretary). Its first shooting meet was held on April 23, 1873, at Creedmoor, Long Island, New York, with various National Guard and Army units competing.

1873 Barbed wire is first patented, by Joseph Glidden of DeKalb, Illinois.

1876 Architect-city planner Walter Burley Griffin is born in Maywood, Illinois. A disciple of Frank Lloyd Wright, Griffin is best known for planning the design of the Australian capital, Canberra.

NOVEMBER 24

1892 Missionary Mary Bazett arrives in Kenya with her two sisters from England. Eleven years later she gives birth to the first white baby seen in Kikuyuland, Louis Seymour Bazett Leakey, one of history's greatest anthropologists (see August 7, 1903).

1903 The automobile electric self-starter is patented by Clyde Coleman of New York City.

1908 The first national forest in the southern United States is established by proclamation of President Theodore Roosevelt. It is the Ocala National Forest, in central Florida.

1916 Inventor of the first fully automatic machine gun, Hiram Maxim, dies at 76 in London.

1926 Nuclear physicist Tsung-Dao Lee is born in Shanghai. In 1957 he and partner Chen Ning Yang became the first of Chinese birth to win Nobel Prizes, for proving that parity conservation (which relates to the structure of atoms) does not hold true for nuclear physics.

1957 "Being a good psychoanalyst, in short, has the same disadvantage as being a good parent: the children desert one as they grow up." —Dr. Morton Hunt, quoted today in a *New York Times* article.

1963 A grim "first" in television history: An actual murder is telecast. It is Jack Ruby's attack on Lee Harvey Oswald.

1965 An enormous toad (the largest toad known to Western science) is trapped at Miraflores Vaupes, Colombia. She is 9.37 inches long and weighs nearly 3 pounds. She is taken to New York City, where she died in 1967.

1859 A small, green book is published in London by John Murray. Five hundred copies are first printed, to sell for 15 shillings each. It is *On the Origin of Species* by Charles Darwin; one of the most influential books in history, it forced man to rethink his place in the universe.

1969 Man's second trip to the moon ends. *Apollo 12* splashes down in the Pacific Ocean.

1971 "D.B. Cooper" hijacks a Northwest Airlines 727 over Washington State, and escapes with $200,000. It is the first successful skyjacking in U.S. history, but it may have been a hollow victory for "Cooper," who made his getaway by parachuting into the mountains at night, during a raging thunderstorm with winds up to 200 mph. He has never been heard from since.

1987 The first United States–USSR treaty to ban an entire class of nuclear weapons is signed.

1992 In West Palm Beach, Florida, a baby girl is delivered at home—by a fourth grader. Nine-year-old Daniel Dresbach was watching *Lethal Weapon II* when his pregnant mother went into labor. He called the doctor, kept her spirits up, and coached her breathing (using instructions from the Lamaze classes which he attended with her). When paramedics arrived the only thing left to do was cut the umbilical cord.

1994 Wire services report that an endangered harpy eagle—native to the fast-disappearing rain forests of South and Central America—has hatched in captivity for the first time in U.S. history, at the San Diego zoo. The chick is the first of its species to be artificially incubated.

NOVEMBER 25

1715 An English patent is first granted to a resident of America. It goes to "Thomas Masters, Planter of Pennsylvania, for an invention found out by Sibylla his wife for cleaning and curing the Indian Corn growing in several colonies in America."

1735 The world's heaviest bell, the Tsar Kolokol, is cast by brothers I.F. and M.I. Motorin in Moscow. It is 22 feet high and weighs 222.6 tons. It cracked in a 1737 fire, but still is housed in the Kremlin.

1814 Julius Mayer is born the son of an apothecary in Heilbronn, Germany. He was the first to state the law of conservation of energy, and the first to suggest that solar radiation was kinetic energy converted into radiant energy; unfortunately, Joule and Helmholtz are credited for the first idea, and Kelvin and Helmholtz for the second.

1884 A machine for producing evaporated milk is first patented. John B. Meyenberg of St. Louis receives patent No. 308,421 for an "apparatus for preserving milk." The following Valentine Day he founded the Helvetia Milk Condensing Co. in Highland, Illinois.

1887 Botanist-geneticist Nikolay Vavilov is born the son of a shoe factory owner in Moscow. He established over 400 research institutes, and imported an immense collection of plants into Russia, where he did important research into their heredity and immunity. When Vavilov adopted and practiced Mendelian theory, which clashed with Stalin's notorious pseudo-geneticist Lysenko, Vavilov was imprisoned and spent his last years in a prison camp.

1887 Zoologist E. Newton Harvey is born in Philadelphia. He is known for discovering why fireflies glow. In the early 1900s, using chemical extracts from glowing shrimp, he determined the chemical pathway that is responsible for bioluminescence in many species, including fireflies, squid, bacteria, fungus, jellyfish, and worms.

1912 The American College of Surgeons incorporates in Springfield, Illinois.

1953 Five conservatives and a socialist introduce a bill to the English Parliament to reprimand the British Museum for believing the "Piltdown man" hoax. The Piltdown skull was discovered in 1912 in a Surrey gravel pit; it was a combination of an orangutan jaw and a human skull, skillfully doctored to look millions of years old. Many scientists believed it to be a missing link between human and monkey evolution. It was the most famous hoax in anthropology history, and set back paleontology theory for decades.

1968 Author Upton Sinclair dies at 90 in Bound Brook, New Jersey. His reformist novel *The Jungle*, intended to anger the public over the treatment of meat packers, instead stirred greater anger over the treatment of the meat, and thereby led to the passage of food inspection laws.

1978 The longest coma in history comes to an end. Elaine Esposito succumbs in Tarpon Springs, Florida, having spent the last 37 years 111 days in a coma that began during an appendectomy at age 6.

1984 William Schroeder becomes the second human to successfully receive an artificial heart. In Louisville, Kentucky, he is given Jarvik-7, and survives for 620 days.

1992 In Copenhagen, 74 nations agree to phase out ozone-depleting chemicals faster than previously established by the historic 1987 Montreal Protocol. The move comes after recent data indicated that Earth's ozone layer is sicker than previously realized.

1993 History's greatest tree-planting spree begins. Three hundred residents of Walsall, Great Britain, plant 1774 trees in 17 hours 20 minutes (over a 6-day period).

1994 A young female pygmy whale is found close to death in shallow water near Longport, New Jersey. Eventually, "Inky" is nursed back to health and released into the wild, thus becoming the first of her species to recover from such poor condition. Her rehabilitation was initially ineffective, until researchers sent a tiny camera down her throat and found her stomach clogged with plastic.

1917 The longest blimp flight in history is over. The German dirigible *Zeppelin L59* docks in Yambol, Bulgaria, after flying to Khartoum and back, spanning a distance of 4500 miles, in four days.

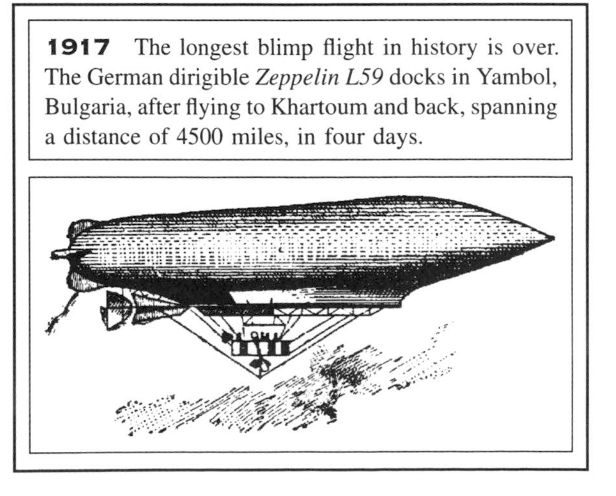

1817 Charles Adolphe Wurtz, the Sorbonne's first professor of organic chemistry and a major force in making Paris a world center in chemistry research, is born the son of a minister in Wolfisheim, France.

1832 The first streetcar in U.S. history goes into public service, in New York City. The horse-drawn car was designed and built by John Stephenson of Philadelphia, and is named the *John Mason*, after the prominent banker who organized the New York and Harlem Railroad.

1837 John Newlands, who joined music theory and chemistry theory (and suffered tremendous ridicule for it), is born the son of a minister in London. After fighting in Garibaldi's army to liberate Italy, he became a professional chemist. He independently noticed that every eighth chemical element shared common properties, and so published his "law of octaves." His discovery has long outlived him and his critics.

1867 A U.S. patent for a refrigerated railroad car is first issued, to J.B. Sutherland of Detroit.

NOVEMBER 26

1894 Mathematician Norbert Wiener, originator of the term and concept of "cybernetics," or giving inanimate objects such as computers the ability to "think," is born in Columbia, Missouri, to a renowned linguist who drove his prodigy son "mercilessly."

1895 Astronomer Bertil Lindblad is born the son of an army officer in Örebro, Sweden. He was the first to offer substantial proof that the Milky Way galaxy rotates, and he improved estimates of the absolute brightness of stars.

1898 Karl Ziegler, Nobel laureate (1963) and pioneer in plastics chemistry, is born the son of a minister in Helsa, Germany. His greatest discovery came in 1953 when he found that adding trace metals to a certain preparation led to production of a very strong form of polyethylene; this happened by accident, when metal from previous experiments contaminated his new work.

1915 W. Atlee Burpee, college dropout and founder of the world's largest mail-order seed company, dies at 57 in Doylestown, Pennsylvania. He established a seed business with two partners in 1876 (at age 18), but left in 1878, eventually to establish seed farms in Bucks County, Pennsylvania, Gloucester County, New Jersey, and Santa Barbara County, California. Much of his success came from the new hybrids of flowers and vegetables he created.

1919 "Can you see anything?" asks Carnarvon. "Yes, wonderful things," replies Howard Carter, on peering into King Tut's tomb for the first time.

1716 A lion is first exhibited in America. It is put on display in Boston, by "Captain Arthur Savage, at his house in Brattle Street, where is to be shewn by William Nichols, a Lyon of Barbary, with many other rarities, the like never before in America."

1966 The world's first major tidal power station opens in the Golfe de St. Malo, Brittany, France. About a half-mile long, the *Usine marèmotrice de la Rance* harnesses the rise and fall of tidal waters to produce 544,000,000 kilowatt-hours of electricity annually.

1967 Using high-speed recording instruments for the first time, Cambridge student Jocelyn Bell detects a "bit of scruff" in the skies. She had noticed it once before, in October. The "scruff" turns out to be history's first pulsar, and its discovery wins the Nobel Prize, awarded to her boss, Antony Hewish.

1991 The nation's largest school system (in New York City) begins handing out condoms on demand.

1992 The Associated Press reports that Greece is returning to ancient power sources: the wind and the sun. "This is one of the few places in the world where we are trying to take advantage of both elements," said an engineer at the country's first alternative-power plant in Kithnos Island. It draws energy from solar panels and windmills. The cost to generate a kilowatt of power there is \$0.175, in contrast to \$2.36 on the nearby island of Anti-Kithira, which relies on the burning of fossil fuels.

1994 A final round of drug tests at the Asian Games in Tokyo prove that the record-setting Chinese athletes have been training on testosterone. Swedish swimming coach Hans Chrunak, who saw the Chinese women win 12 of 16 swimming golds in the World Championships the previous month, described the female swimmers as looking like tractors but racing like Ferraris.

1676 The first major fire in the recorded history of North America occurs in Boston. It "burned down to the ground 46 dwelling houses, besides other buildings, meeting houses, etc." The second major fire in American history also occurred in Boston, in August 1679; 150 buildings were destroyed, and damage was set at about $1 million.

1701 Anders Celsius, who devised the world's most popular scale of temperature, is born in Uppsala, Sweden, into a family of esteemed scientists. Celsius became an astronomer, and was the first to link the aurora borealis (the northern lights) with magnetic phenomena in Earth's atmosphere. In 1742 he announced the centigrade ("hundred steps") temperature scale. Initially, 100 degrees was freezing.

1822 The oldest horse on record dies at 62 in Great Britain.

1839 The American Statistical Association is organized in Boston.

NOVEMBER 27

1843 Englishman Alexander Bain applies for a patent on his "Method for transmitting copies over a distance, by means of electricity." Although never commercialized, it is the forerunner of television.

1848 Physicist Henry Augustus Rowland is born the son of a minister is Honesdale, Pennsylvania. His most important feat was creating a grating with 15,000 lines per inch, thereby opening a new era in the spectrographic analysis of light, especially the light and energy from outer space.

1857 Sir Charles Scott Sherrington is born in London. At the urging of his stepfather, Sherrington obtained a medical degree (Cambridge, 1885), but spent his life as a physiology professor in London, then Liverpool, and finally Oxford. Among his contributions during 50 years of research on animal nervous systems are Sherrington's law (when one muscle group is activated, muscles opposing this action are automatically inhibited), classification of sense receptors (as extero-, intero-, or proprioceptive), and coining of the terms "neuron" and "synapse." He received the 1932 Nobel Prize in Medicine or Physiology.

1874 Chaim Weizsmann is born in Motol', Russia. He was one of the first to harvest a material produced by a microbe: He was able to culture bacteria to produce acetone for use in explosives (1911). He was also Israel's first president, and one of the very few scientists to head a country.

1876 Theophile Schloesing and Achille Müntz begin an experiment at the French National Institute of Agronomy that proves that nitrification of plants is accomplished by bacteria.

1895 The Nobel Prizes are created. A year before his death, Alfred Nobel signs his last will in the Swedish Club in Paris. Believing that "large inherited wealth [is] a misfortune which merely serves to dull men's facilities," Nobel specifies that most of his huge fortune (resulting mainly from his invention of dynamite) will establish the Nobel Prizes, intended to enhance world peace and the welfare of mankind.

1903 Chemist Lars Onsager is born in Oslo. He won the 1968 Nobel Prize for his theory of irreversible chemical processes, used in producing U-235 for the A-bomb.

1951 A rocket intercepts an airplane for the first time, over White Sands Proving Grounds, New Mexico.

1963 A red cloud formation is seen on the moon by several scientists at several observatories. It lasts for 75 minutes, and its cause and chemistry have never been completely explained.

1886 A new cook takes charge in the kitchen of the Dutch military hospital in Djakarta, Indonesia, and immediately refuses to "allow military rice to be taken for civilian chickens." The chickens, housed in the laboratory of Christiaan Eijkman, had been getting ill mysteriously. When the fine, polished "military rice" was not fed to them, as ordered by the cook, they regained health on eating the unpolished (with the husks still attached) rice. This was the clue that led to the discovery of vitamins.

1520 Magellan reaches the Pacific, through the strait that now bears his name at the tip of South America.

1793 As friends and relatives are imprisoned daily, Lavoisier surrenders to the Reign of Terror government in Paris. The founder of modern chemistry never sees freedom again.

1858 Metallurgist Sir Robert Hadfield is born the son of a steel manufacturer in Sheffield, England. Hadfield transformed his father's industry by creating superhard steel (by adding large amounts of manganese). This was the first great alloy steel, and led to development of materials like stainless steel.

1887 Alfred Nobel is granted a patent for smokeless gunpowder; it was a great step in munitions technology.

1895 The first "real" car race in U.S. history occurs in the snow from Chicago to Waukegan, Illinois. Of the 80 cars originally entered, only 6 start. James F. Duryea wins, with a speed of 7½ mph. His brother Charles built the car.

November 28

1908 Social anthropologist Claude Lévi-Strauss is born in Brussels. He became a leading advocate of structuralism, which influenced not only social science, but also philosophy, literature, religion, and film.

1922 Skywriting is first exhibited in U.S. skies. "Hello, U.S.A. Call Vanderbilt 7200," appears at noon over Times Square in New York City. The pilot is Captain Cyril Turner, Royal Air Force, who writes the message a half-mile above ground. The visible vapors are created when oil hits the hot exhaust pipe at the plane's rear. The message brings 47,000 calls to the written number over the next 2½ hours.

1929 The first airplane flight over the South Pole takes off from Little America, Antarctica (see November 29, 1929).

1938 Psychologist William McDougall dies at 67 in Durham, North Carolina. In the early 1900s he was a major force in bringing Darwinian and biological perspectives to human psychology, especially social behavior. At the time, psychology was still firmly rooted in philosophy. McDougall's *Physiological Psychology* (1905) and *Introduction to Social Psychology* (1908) were especially influential.

1942 The infamous and unsolved Coconut Grove fire occurs in a Boston nightclub; nearly 500 perish.

1954 Nuclear physicist Enrico Fermi dies of stomach cancer at age 53 in Chicago. Fermi led the Manhattan Project when, on December 2, 1942, at 3:45 PM in a squash court at the University of Chicago, it produced history's first nuclear chain reaction. This began the Atomic Age.

1963 Cape Canaveral is renamed Cape Kennedy until 1973, when it is changed back to its original name.

1837 John Wesley Hyatt, inventor of the first synthetic plastic, celluloid (1869), is born in Starkey, New York. Although he received over 200 patents, this was Hyatt's greatest feat, which he produced in efforts to create a better billiard ball (and to collect the $10,000 prize offered for a substitute ivory).

1964 *Mariner 4* blasts off from Cape Kennedy, destined to provide the first closeup photos of Mars.

1983 The $1 billion *Spacelab* is launched into space aboard the shuttle *Columbia*.

1994 Round-the-clock shifts are instituted in the battle to clean up one of the world's worst oil spills. An old pipeline recently burst near Usinsk, 1000 miles northeast of Moscow, spilling an estimated 80 million gallons onto the tundra. This dwarfs the 1989 *Exxon Valdez* spill of 11 million gallons.

1994 The Associated Press reviews two Boston studies showing that men who feel anxious are six times more likely to suffer sudden cardiac death than calmer individuals. In comparison, smoking doubles the likelihood of cardiac arrest.

NOVEMBER 29

1627 The great naturalist John Ray is born the son of a blacksmith in Black Notley, England. As part of a partnership intending to catalogue all living things on Earth, he produced an encyclopedia of 18,600 plants in the late 1600s. Ray established the species as the fundamental unit of classification.

1762 Pierre Latreille, the founder of modern entomology, is born in Brive-la-Gaillarde, France. An ordained priest, he created the first detailed classifications of insects and crustaceans, and published a massive 14-volume opus in the early 1800s.

1803 Physicist Christian Doppler is born in Salzburg, Austria, the son of a master mason. He was the first to explain, quantify, and test the Doppler effect—the apparent change in pitch of the sound (or light) caused by an object's movement. Many speed-measuring devices are based on this effect.

1849 Electrical engineer Sir John Fleming is born the son of a minister in Lancaster, England. Among his many contributions was the invention of the vacuum tube, making radio communication possible.

1874 Egas Moniz, the father of psychosurgery, is born in Avanca, Portugal. He shared the 1949 Nobel Prize for inventing the frontal lobotomy.

1929 The South Pole is crossed by airplane for the first time. Richard Byrd commands the flight in a Ford trimotor plane, crewed by Bernt Balchen (pilot), Ashley McKinley (photographer), and Harold June (radio operator). They cross the Pole at about 8:55 AM (New York time), drop a U.S. flag, and return to base at Little America, Antarctica, by 5:10 PM.

1951 History's first underground nuclear bomb test occurs at Frenchman Flat, Nevada.

1961 The United States sends an animal into orbit around Earth for the first time. Enos the chimp (a 37.5-pounds 5½-year-old male) blasts off from Cape Canaveral in a Mercury satellite on an Atlas rocket. Over the next 3 hours 21 minutes, Enos orbits Earth twice at 17,500 mph. He lands in the ocean near Puerto Rico, and is retrieved by the USS *Stormes* after 1 hour 25 minutes.

1963 "The process of living is the process of reacting to stress." —Dr. Stanley J. Sarnoff, physiologist, speaking to the National Institutes of Health, quoted in *Time* magazine.

1965 *Unsafe At Any Speed* by Ralph Nader is published. It improved automobile safety forever.

1973 The highest bird flight in history is recorded when a Ruppell's vulture collides with a commercial airliner at 37,000 feet over Abidjan, Ivory Coast.

1984 The second man to receive an artificial heart, William J. Schroeder, rises from his hospital bed for the first time with his new organ and takes "the Coors cure," a can of cold beer.

1989 The first acknowledged military tracking of a UFO occurs over Eupen, Belgium.

1994 After two extensive searches with the Hubble Space Telescope, two astronomers announce that most of the universe is still missing. For years scientists have estimated that much more matter must exist than has been detected, according to various calculations and theories. Approximately 95% of what should be out there has not been found. "Our results increase the mystery of the missing mass."

1825 Neurologist Jean-Martin Charcot is born in Paris. His father was a successful carriage-maker who was able to offer Jean-Martin the choice of a career in medicine or in art. Charcot chose the former. He revolutionized the treatment of neuroemotional disorders by his careful attention to symptoms and by his use of hypnotism. Freud's interest in the subconscious began while a student of Charcot's. "To take from neurology all the discoveries made by Charcot would be to render it unrecognizable," wrote renowned neurologist Joseph Babinski.

1694 Italian physician Marcello Malpighi—the biologist who founded the field of microscopic anatomy—dies at 66 in Rome. He made many discoveries that are now taken for granted. He was the first to see red blood cells and to suggest that they determine the color of blood, he was the first to see the small blood vessels that connect arteries with veins (this discovery was crucial in establishing that blood does circulate), and he discovered taste buds. His extensive research on the fine structure of many human organs earned him the title "the first histologist." He was also the object of much jealousy; in the last decade of his life, his home, library, and laboratory were destroyed by unknown assailants. He spent the remainder of his days as personal physician to Pope Innocent XII.

1761 Chemist Smithson Tennant is born the son a of clergyman in Selby, England. He discovered the elements osmium and iridium, and was the first to demonstrate that diamonds are all carbon.

1819 Cyrus Field, the industrialist who laid the first cable across the Atlantic, is born in Stockbridge, Massachusetts.

NOVEMBER 30

1858 Sir Jagadis Chandra Bose, the first world-famous scientist from India, is born the son of a civil servant in Mymensingh. He applied his genius as a physicist to make instruments that provided the most accurate and sensitive records of the growth and movement of plants.

1866 Robert Broom, the son of a textile designer, is born in Paisley, Scotland. His fame came in determining man's evolutionary past; he found samples of the hominid *Australopithecus* that confirmed Raymond Dart's contention that the remains are truly the long-sought "missing link" between the evolution of the ape and the human.

1869 Physicist Nils Dalén is born in Stenstorp, Sweden. He won the 1912 Nobel Prize for inventing a sun-controlled gas-supply valve that advanced the design of buoys and unmanned lighthouses.

1915 Henry Taube, the 1983 Nobel laureate in Chemistry, is born in Neudorf, Canada. His work explained the result and configuration of inorganic metallic compounds dissolved in water.

1924 A Radio Facsimile Transmission (in which photographs are transmitted as radio waves) is first sent across the Atlantic as a public demonstration. A variety of images are sent from the Marconi offices in London, to New York City, including photos of Calvin Coolidge, the Prince of Wales, and the Chinese proverb "One picture is worth ten thousand words."

1993 The Brady bill, intended to curb handgun violence, is signed into law by President Clinton.

1936 Site of the first public display of a reconstructed dinosaur, the famed Crystal Palace in London, is destroyed by fire. It had been built for the International Exhibition of 1851.

1954 The only meteorite known to have hit a person in U.S. history crashes into the home of Mrs. Elizabeth Hodges in Sylacauga, Alabama. The sulfide meteorite (7 inches long, 8½ pounds) hits her on the arm. She is declared healthy by a doctor, but is later hospitalized because of the publicity.

1971 Japan's greatest oil spill occurs when a tanker breaks in half and spills 6,258,000 gallons of oil off the coast.

1974 Donald C. Johanson discovers the remains of the hominid "Lucy" in Hadar, Ethiopia. Dating back 3.5 million years, she is the oldest known ancestor related to the human. Her name comes from the Beatles's song that plays throughout this night at the expedition's celebration.

1990 Norman Cousins, a founder of the holistic health movement, dies at 75 of a heart attack in Los Angeles. His pioneering bestseller *Anatomy of an Illness as Perceived by the Patient* (1979) described his recovery from a life-threatening form of arthritis through mind–body therapy, consisting mainly of humor, vitamin C, and positive thinking.

1743 German chemist Martin Heinrich Klaproth, discoverer of uranium (1789), zirconium (1789), and cerium (1803), is born the son of a tailor in Wernigerode. Eight years later his family was impoverished by fire, but Klaproth regained wealth in 1780 by marrying the rich niece of eminent chemist Andreas Marggraf. Klaproth helped discover several other elements, but always was unselfish and meticulous in giving credit to others. He was also careful about reporting exact measurements (which is not always true with scientists in all ages), and which is partly why Klaproth is sometimes called "the father of analytic chemistry."

1783 The birth of space biology and the birth of gas ballooning occur on the same flight, when Jacques Alexandre Charles and the Robert brothers ascend from the Jardin de Tuileries, Paris, in a craft built by Charles. It is the first hydrogen-filled balloon, the first balloon equipped with up–down controls, the first time a barometer is used to measure altitude, and the first time physiological measurements are taken on a human body at high altitude.

1845 The *Water Cure Journal* is first published, in New York City. Sixteen pages, costing $1 and edited by Drs. Joel Shew and T.D. Pierson, the *Journal* is the nation's first magazine devoted to "proper explanation of hydropathy, or water cure, including bathing in its various forms, attention to diet, drink, air, exercise, cleanliness, and clothing, as affecting bodily and mental health."

1913 A drive-in automobile service station first opens in the United States, at the corner of Baum Boulevard and St. Clair Street, Pittsburgh, Pennsylvania. Over 30 gallons of gasoline is sold on opening day. The station is open all night, and offers free crankcase service.

1936 A hydroponicum—a means for growing plants without soil—is first patented in the United States. Ernest Walfrid Brundin and Frank Farrington Lyon of Montebello, California, obtain patent No. 2,062,755 for a "system of water culture." In October 1937, they established the Chemi-Culture Company based on their technique.

1947 Mathematician Godfrey Harold Hardy dies at 70 in Cambridge, England. Although he solved many problems in prime number theory, he is most famous for discovering the Hardy–Weinberg law in biology. In 1908 he and Wilhelm Weinberg independently developed a simple mathematical formula showing that in a large population the proportion of dominant and recessive genes tends to remain constant. This law is a cornerstone of population genetics, and explains why even very rare and very harmful genes tend to stay in a population.

1959 The Antarctic Treaty is signed. Twelve nations agree in Washington, D.C. to establish Antarctica as a demilitarized zone; no maneuvers, bases, or weapons may be brought there. The continent is set aside as a scientific preserve. Claims of territorial sovereignty are not affected.

1960 The 1.4-million-year-old resting place of a missing link, Chellean man, is first spotted by paleontologist Louis Leakey. The find occurs in Olduvai Gorge, Tanzania, which Leakey established as one of history's greatest sources of humanoid fossils.

1990 A breakthrough, literally, occurs in the "Chunnel"—the tunnel under the English Channel between England and France. French and English workers finally meet each other when a passage large enough to walk through is dug connecting the excavations from both countries.

1994 Annabelle Goodwin, 77, is discharged in good health from a hospital in Seligman, Arizona, after spending two weeks in a van that was stranded in the wilderness. It snowed and the night temperatures fell below freezing. Mrs. Goodwin had several blankets, and survived on food and water, which she and her husband had packed. In mid-November they had run out of gas, and Mr. Goodwin (76, suffering from Alzheimer's disease) set out on foot for help. He was later found about a mile from the van.

1994 The Associated Press announces that history's first test of an oral AIDS vaccine has begun in California. Researchers at San Francisco General Hospital have started enrolling patients in a new trial of a compound that had proved successful in a study of 24 persons treated by injection.

1904 A patent application for the first steam-operated cloth-pressing machine is filed by Adon J. Hoffman of Seattle, Washington.

1881 German physicist Heinrich Barkhausen is born the son of a judge in Bremen. In 1919 he discovered the Barkhausen effect: As iron is subjected to an increasingly strong magnetic field, its magnetization proceeds in jumps, not smoothly; these jumps can be heard when magnified through a loudspeaker.

1885 George Richards Minot is born the son of a physician in Boston. He obtained a medical degree from Harvard, and in 1915 he began service at Massachusetts General and Peter Bent Brigham hospitals, as his father, uncle, and grandfather had done before him. His accomplishments surpassed theirs, culminating in the 1934 Nobel Prize for Physiology or Medicine. Minot teamed with William Murphy in 1924 to discover a cure for pernicious anemia, which was invariably fatal before Minot's work. Minot suspected the disease stemmed from a vitamin shortage, and believed that liver (known to be high in vitamins) might help. It was that simple: To be cured, patients only had to eat a half pound of raw liver every day.

DECEMBER 2

1929 A skull of Peking man, a missing link between humans and their nonhuman ancestors, is discovered 40 miles from Peking, just outside the village of Choukoutien, China. It is found by Dr. W.C. Pei after he was lowered into a cave on Dragon-bone Hill. Anthropologist Roy Chapman Andrews wrote, "There it was, the skull of an individual who had lived half a million years ago. It was one of the most important discoveries in the whole history of human evolution. He could not have been very impressive when he was alive, but dead and fossilized, he was awe-inspiring."

1934 Glass is poured at the Corning Glass Works in Corning, New York, for history's first telescope with a lens of 200 inches in diameter. The 2700°F glass is gradually cooled, a degree or two a day over the next 11 months. The lens weighed 20 tons when it was shipped on March 26, 1936, to the California Institute of Technology where it was ground and polished for 11 years. Finished on October 3, 1947, it was incorporated into the Hale telescope at the Mount Palomar Observatory in San Diego.

1942 A nuclear chain reaction is demonstrated, conducted, and produced by scientists for the first time, at 3:45 PM in a squash court at the University of Chicago, under the direction of Enrico Fermi. Although the reaction is the same as used in the impending Nagasaki and Hiroshima bombs, in today's experiment only enough power to light a small flashlight is produced.

1957 The reactor at the nation's first nuclear power plant devoted to peaceful uses (in Shippingport, Pennsylvania) reaches criticality for the first time. By December 23 it achieved its full rated capacity of 60,000 net kilowatts (calculated to provide for the household needs of 250,000 people). President Dwight Eisenhower dedicated the plant on May 26, 1958.

1970 The Environmental Protection Agency begins operation, under its first director William Ruckelshaus.

1984 The worst industrial/chemical accident in history (eventually to claim the lives of more than 4000) begins at the Union Carbide methylisocyanate plant in Bhopal, India.

1993 Space shuttle *Endeavor*, "the most expensive repair truck in history," blasts off from Cape Canaveral. Its mission is to repair the fault-plagued Hubble Space Telescope. Hubble was launched in April 1990 with great fanfare and promise of shedding light on some of man's most profound questions, including the age of the universe, the existence of life outside Earth, and the nature of "black holes." But problems such as a misshapen mirror that blurred its images, broken gyroscopes, a faulty computer, and unstable solar panels, arose quickly and made the Hubble a $3 billion embarrassment. Today's 11-day mission ended in success; it fixed the telescope and proved that complex engineering tasks could be performed by humans in space.

1994 The "master switch" that determines the sex of a human embryo has been found, reports a University of Chicago team in *Science*. At conception, all embryos begin developing female structures; at about day 35, the SRY gene (on the male Y chromosome) goes into action—triggering a gene (MIS) that dissolves the female structures. Embryos that undergo this process become male; those that do not, remain female. The scientists have high hopes that "[if] we can understand the general switches involved in sex determination, then we could possibly relate that to other basic processes, such as how organs differentiate or how cancer arises," said team leader Michael Weiss.

DECEMBER 3

1616 Mathematician John Wallis is born the son of a rector in Ashford, England. Wallis's work directly helped Newton invent calculus. He was the first to use fractions and negative numbers as exponents and the first to use a sideways "8" to represent infinity, he wrote one of the first histories of mathematics, and he was a cofounder of the Royal Society.

1838 Meteorologist Cleveland Abbe, father of the U.S. Weather Bureau, is born in New York City.

1857 Carl Koller, inventor of local anesthesia, is born in Schüttenhofen, Bavaria (now Susice, Czechoslovakia). In the early 1880s Koller was an intern at the Vienna General Hospital when a colleague named Sigmund Freud asked him to research the bodily effects of cocaine, as a possible cure for morphine addiction. The work suggested to Koller that the substance might deaden pain in very small areas (thus eliminating the dangers of general anesthesia). In 1884 he studied cocaine in animal experiments, and then performed the first surgery on a human under local anesthesia during an eye operation.

1873 History's heaviest wooden ship is launched in Toulon, France. The vessel is the *Richelieu*, 333 feet long and weighing 9548 tons.

1900 Biochemist Richard Kuhn is born in Vienna. He shared the 1938 Nobel Prize for isolating and determining the structure of many carotenoids and vitamins. To extract pure vitamin B_6, Kuhn processed 14,000 gallons of skim milk. He was forbidden by Hitler to accept the Prize because in 1935 the Nobel Prize for Peace had been awarded to Carl von Ossietzky, who was in a Nazi concentration camp at the time.

1910 A neon tube is first put to practical use. Recently invented by French physicist Georges Claude, tubes are used on this date to illuminate sections of the Grand Palais in Paris. Two years later, neon tubes are used in an advertising sign for the first time, at a hairdresser's shop on the Boulevard Montmartre in Paris.

1922 *Toll of the Sea*, history's first successful movie in Technicolor, is released at the Rialto Theater, New York City. Technicolor was invented by Dr. Herbert Thomas Kalmus, the president of the Technicolor Motion Picture Corporation.

1967 Dr. Christiaan Barnard performs the first human-to-human heart transplant in Capetown, South Africa. The patient, a white male, lives for 18 days after receiving the heart of a black male.

1968 President Lyndon Johnson publicly announces that the United States has entered into the "Agreement on the Rescue of Astronauts, the Return of Astronauts and the Return of Objects Launched into Outer Space" with Russia and the United Kingdom. It is history's first astronaut-rescue treaty.

1982 Recipient of the world's first permanent artificial heart, Barney Clark is removed from a respirator, one day after the heart operation in Salt Lake City, Utah.

1992 "The world's most expensive drug"—a product called Ceredase, used to treat the rare genetic illness Gaucher's disease and costing about $380,000 annually—can be given effectively in smaller, more frequent doses, according to a report in the *New England Journal of Medicine*. The new decreased dose would cut the total intake of the medicine by 75%, and the cost for a year's treatment would drop to $88,000. Scientists realize the possibility for the dosage reduction after studying how most of the drug is quickly lost from the bloodstream.

1993 There are causes of lung cancer other than smoking: eating a high-fat diet. The *Journal of the National Cancer Institute* reports today that fatty diets—already linked to cancers of the colon, prostate, and breast—are now linked to respiratory cancer. The study involved some 1500 nonsmoking women from Missouri. Those with high-fat diets had a 6-fold higher risk of lung cancer than those with the lowest fat consumption. The risk of adenocarcinoma was enhanced 11-fold.

1903 Hungarian-U.S. mathematician John von Neumann, a founder of "game theory" and an innovator in computer science, quantum physics, and meteorology, is born a prodigy son of a Jewish banker in Budapest. It is said that by age 6, he could divide two eight-digit numbers in his head.

1680 Anatomist-mathematician Thomas Bartholin dies at 64 in Copenhagen. In 1652 he became the first to fully describe the human lymphatic system. He was also one of the earliest defenders of Harvey's discovery of the circulation of blood. From 1670 to 1680 he was physician to Denmark's King Christian V. His brother Erasmus discovered double refraction of light, and his father Caspar was a famous anatomist.

1798 Luigi Galvani dies in Bologna, Italy, in the house in which he was born 61 years earlier (see September 9, 1737).

1893 Physicist John Tyndall, who explained why the sky is blue, dies at 73 in Hindhead, England. His research centered on the movement of heat and light through gases. He is known for the Tyndall effect—a light beam cannot be seen from the side when it moves through a clear liquid or gas, but when dust and other particles are in the medium, the light beam can be seen. The scattering of sunlight by atmospheric dust, which is especially intense for blue wavelengths, accounts for the overall blue appearance of the sky.

DECEMBER 4

1906 Physician Robert Wallace Wilkins, who introduced reserpine to Western medicine, is born in Chattanooga, Tennessee. The drug had been used for centuries in India to calm the mentally ill; Wilkins began using it in the 1950s to treat high blood pressure, then realized and reported that it has excellent pacifying qualities without causing drowsiness. Reserpine was the first in a new class of drug, known as the tranquilizer.

1908 Alfred Day Hershey is born in Lansing, Michigan. He was awarded the Nobel Prize in 1969 for genetic research. In 1952 with Martha Chase he performed the critical experiments showing that DNA, rather than protein, carries the hereditary blueprints.

1945 Thomas Hunt Morgan, named for his uncle John Hunt Morgan (who led Morgan's Raiders farther north than any other Confederate force went during the Civil War), dies at 79 in Pasadena, California. Although he did not completely believe either Darwin's or Mendel's theories at the start of his career, Morgan eventually verified and tied them both together with his "chromosome theory of heredity," based on his experiments with the humble fruit fly. He received the Nobel Prize for Physiology or Medicine in 1933.

1952 The worst smog disaster in history begins in London. After five days 3500 to 4000 people die of smog-related causes.

1991 David Baltimore is one of the nation's most distinguished scientists and winner of the 1975 Nobel Prize for Physiology or Medicine for research into the fundamental biology of viruses. Today he resigns as president of Rockefeller University after five years of prolonged allegations of fraud in a paper that he coauthored. Although never accused of fraud himself, Baltimore has been severely criticized for his casual attitude toward the investigation and for his decidedly noncasual "stonewalling defense" of the research. He eventually abandoned this defense, apologized to the scientist who alleged that fraud occurred, and resigned the presidency of Rockefeller University, one of the world's most prestigious research establishments. He continued to work at the New York City institute, returning to the AIDS research he abandoned when he became president two years before.

1993 *Endeavor* astronauts capture the faulty Hubble Space Telescope for repairs.

1131 Omar Khayyám, Persian mathematician, astronomer, and poet, dies at 83 in the town of his birth, Nishapur. He wrote the principal book on algebra in his time, and reformed the Muslim calendar so it coincided with observed astronomical data. The same change was made in the Gregorian calendar five centuries later.

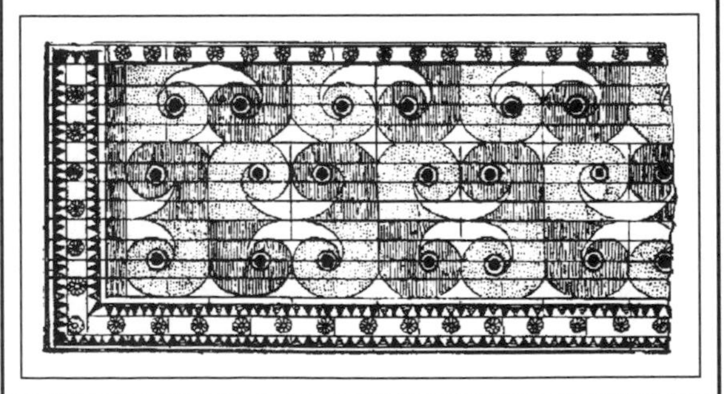

DECEMBER 5

1537 Vesalius receives his medical degree at age 23 in Padua, Italy, and is appointed a professor of surgery the next day. He became the first to teach human anatomy using human bodies, marking a turning point in medicine, ethics, and human thought.

1855 Clinton Hart Merriam, cofounder of both the National Geographic Society and the U.S. Fish and Wildlife Service, is born in New York City. In his early 20s, Merriam began traveling throughout the western United States as a naturalist—eventually exploring and collecting specimens in 48 states and Bermuda. Later he was placed in charge of the U.S. Biological Survey, which became the Fish and Wildlife Service. Merriam was also an anthropologist, collecting data on the distribution, mythology, and language of 157 tribes of Pacific coast Native Americans.

1868 The country's first bicycle school is opened in New York City at 932 Broadway by the Pearsall brothers. Instruction in the riding of "velocipedes" is provided.

1876 Darwin publishes *Effects of Cross and Self Fertilization in the Vegetable Kingdom*. Through 12 years of experimenting on 57 species of plants, Darwin discovered that the offspring of cross-fertilized plants were much healthier, bigger, and more vigorous than plants that had fertilized themselves. The findings are the foundation of understanding of phenomena like hybrid vigor and incest taboo.

1901 Cinema pioneer Walt Disney is born in Chicago.

1901 Werner Karl Heisenberg, author of the uncertainty principle in nuclear physics, for which he won the 1932 Nobel Prize, is born the son of a humanities professor in Würzburg, Germany.

1932 Einstein is granted a visa, permitting him to emigrate from Germany to the United States. Shortly thereafter, Nazi storm troopers ransack his summer home outside Berlin and confiscate his sailboat.

1932 Physicist Sheldon Lee Glashow, winner of the 1979 Nobel Prize for his theory that connected electromagnetism with weak nuclear interactions, is born in New York City.

1935 The first large-scale, commercial hydroponicum—a system for growing plants without soil—is established in Montebello, California, by Ernest Walfrid Brundin and Frank Farrington Lyon. They patent their "system of water culture" the following year, and formed the Chemi-Culture Company in 1937.

1941 All of the existing bones of Peking man—an extinct human ancestor and evolutionary missing link—are taken aboard a train by a detachment of nine U.S. marines, in an attempt to smuggle the bones out of China before the invading Japanese army seizes them. Two days later Pearl Harbor erupts. In the confusion, the bones are lost and they have never been found. Only plaster casts and sketches remain.

1978 The space probe *Pioneer Venus 1* begins transmitting pictures and data about Venus from its orbit around the planet.

1991 According to international wire services, the mayor of Sunol, California, has resigned to spend more time with his family. The mayor is an 11-year-old dog. The Labrador–Rottweiler mix named Bosco had been elected 10 years before by an unofficial poll in a bar in this unincorporated, 400-person town 40 miles from San Francisco. Because it is unincorporated, state-sanctioned elections are not held. Bosco's victory made international news, however, causing the *People's Daily* in communist China to observe, "Western 'democracy' has reached such a peak of perfection that not only can one talk of democracy between people, but between dogs and people."

1993 Astronauts from the shuttle *Endeavor* complete the first of five spacewalks to repair the $3 billion Hubble Space Telescope.

1933 Prohibition ends at 5:32 PM, when Utah becomes the 36th state to ratify the 21st Amendment.

1742 French chemist Nicolas Leblanc, discoverer of the Leblanc process (for making soda ash, sodium bicarbonate, from common salt), is born in Issoudun, the son of an ironworks director who soon dies and leaves Nicolas an orphan. Soda ash was used in paper, soap, glass, and porcelain, and in 1775 the French Academy of Sciences offered a prize for a new method of producing it cheaply from salt. Leblanc developed the method and won the prize after years of work; but he never actually received the prize money, nor was his patent protected. The National Assembly confiscated both his patent and his factory, and Leblanc ended his own life in 1806.

1778 French chemist-physicist Joseph-Louis Gay-Lussac is born the son of a judge in Saint-Léonard. Gay-Lussac was just 24 when he discovered that all gases expand by the same, precise amount for any given temperature increase (Gay-Lussac's law); it was an extremely important finding, not only for industrial purposes, but as a clue to the nature of matter. Gay-Lussac then explored the nature of the atmosphere in several daring balloon flights during which he recorded much data, and which have earned him recognition as a founder of meteorology. His other achievements include discovering that gases combine in simple and specific proportions, being the first to isolate boron, and coining the terms "pipette" and "burette."

DECEMBER 6

1830 The country's first national astronomical observatory is established by the Navy in Washington, D.C. Lieutenant Louis Malesherbes Goldsborough was the first officer in charge, and the first technical instrument was a 30-inch transit built by Richard Patten of New York.

1863 Chemist Charles Martin Hall is born the son of a minister in Thompson, Ohio. One of his early chemistry teachers made the chance remark that anyone discovering a cheap method of producing aluminum would become rich and famous. At age 22 Martin did, through the use of electrolysis. His process made the metal an industrial mainstay.

1868 Comparative linguist August Schleicher dies at 47 in Jena, Prussia. The first to show the development of languages as a family tree, he was influenced by Darwin's theory of evolution, and attempted to explain the development of languages in similar terms. He thought of a language as a living organism, exhibiting development, maturity, and demise, and therefore amenable to study by techniques of natural science. He created a classification system based on botanical classification, which led him to the family tree diagram. His specialty was the Indo-European languages, but his methods and insights were applicable to all languages.

1917 Two munitions ships collide in Halifax Harbor, Nova Scotia, causing a blast that destroys much of the city's north side and kills approximately 2000. It remains the worst nonnuclear explosion on record.

1920 Sir George Porter is born in Stainforth, England. He shared the 1967 Nobel Prize for Chemistry for studies on flash photolysis, a technique for observing the stages of extremely fast chemical reactions.

1929 The first-discovered skull of the human ancestor Peking man reaches Peking from Chou-k'ou-tien cave outside the city where it was recently unearthed (see December 5, 1941).

1947 About 2300 square miles of precious wetlands is saved from human exploitation when President Harry Truman dedicates the Everglades National Park in Florida. None of the preserve is more than eight feet above sea level. Senator Spessard Holland and journalist John D. Pennekamp (whose name was given to the country's first underwater national park) led the fight to preserve the land.

1957 The nation's first attempt to send a satellite into orbit ends in a fiery explosion on the launch pad at Cape Canaveral.

1993 Spacewalking astronauts from the shuttle *Endeavor* attack the greatest problem of the $3 billion bus-sized Hubble Space Telescope: the faulty mirror system. A misshapen mirror, installed before takeoff, has caused all images to blur. On this day, Story Musgrave and Jeffrey Hoffman begin installing a new camera with optics that correct for the mirror's problems. These and a host of other repairs during the 11-day *Endeavor* mission allowed Hubble to eventually gather images and information on some of the oldest and faintest objects in the universe.

1994 An all-terrain wheelchair is demonstrated at Florida A&M in Tallahassee. Designed and exhibited by Fred Davis, the "Mobile Aquatic Rehabilitation Vehicle for Exercise and Leisure" is made to travel over rough and smooth terrain, and water. The demonstration takes place in the college's pool.

1786 Johann von Charpentier, one of the first to propose glaciers as agents of geological change, is born in Freiberg, Saxony. At first he followed his father as a mining engineer, but in 1818 his interest turned to studying glaciers when a nearby lake was dammed by a glacier, which ruptured and killed many shortly thereafter.

1810 Theodor Schwann, who founded modern biology by identifying cells as the basic unit of life, is born in Neuss, Prussia. As a research assistant to Johannes Müller, Schwann, 24, made the major discovery that digestion does not occur through the action of acid, but through acid plus another substance (which he named "pepsin"); it was the first time that an enzyme was prepared from animal tissue, and was a milestone in biochemistry. In 1838 Matthias Schleiden hypothesized that all plants are composed of cells; Schwann knew Schleiden and his work, and in 1839 he extended Schleiden's theory to encompass animals, stating that *all* living things are composed of cells. Schwann called this doctrine "the cell theory." He was also the first to identify eggs as single cells that, once fertilized, eventually develop into whole organisms through repeated splitting. He also coined the term "metabolism" to describe chemical processes in living things, and he discovered "Schwann cells"—the wrapping around nerve fibers that enables impulses to travel very fast.

DECEMBER 7

1872 The HMS *Challenger*, a 2309-ton corvette with "a penchant for rolling like a barrel," departs the Thames River, England, on history's first round-the-world scientific voyage. Mountains of data are collected in the next 3½ years, including the discovery of 4717 new species. The event marks the birth of modern oceanography.

1887 The nation's first cancer hospital begins accepting patients. The New York Cancer Hospital opens at 106th Street and Central Park West in New York City. James Hunter, Clement Cleveland, and William Bull are the first attending surgeons. In 1916 the hospital name became the Memorial Hospital for the Treatment of Cancer and Allied Diseases; it is now part of the Memorial Sloan-Kettering Cancer Center.

1909 "Bakelite," the first completely synthetic plastic, is patented by Dr. Leo Baekeland of Yonkers, New York, who created the material by taming the reaction between phenol and formaldehyde. Two patents are granted on this date: No. 942,699 for "an improvement in methods of making insoluble condensation products of phenol–formaldehyde" and No. 942,700 for "a condensation product of phenol and formaldehyde and a method of making the same."

1936 *Streptococcus* meningitis (heretofore 99% fatal) is treated for the first time with a sulfa drug. A six-year-old girl is saved by Dr. Francis Schwenkter in Baltimore.

1970 Linus Pauling publishes *Vitamin C and the Common Cold.* Pauling was one of the very few to win two Nobel Prizes, but this book remains highly controversial because it advocates taking very large quantities of vitamin C.

1971 A dike at a chemical plant ruptures in Fort Meade, Florida, sending 2 billion gallons of toxic sludge into the Peace River.

1972 *Apollo 17* is launched. It is the only night launch of the *Apollo* program, and the craft sets the duration record for time on the moon's surface (74 hours 59 minutes). Two hundred fifty-three pounds of moon rocks was collected during the mission.

1993 Energy Secretary Hazel O'Leary begins a new era of openness in the U.S. nuclear program by acknowledging at a Washington press conference that since the 1940s hundreds of secret weapons tests have been conducted, and some 800 radiation experiments have been conducted on humans. "It is apparent that informed consent could not have taken place," says O'Leary about the case of 18 civilians, all now deceased, who were injected with highly radioactive plutonium.

1994 The Ukrainian wrestling team visits Jacksonville, Florida, to scout the city as a possible training site for the 1996 Olympics in nearby Atlanta. When visitng the zoo, a mature chimpanzee named Jackson showers the group with feces. A zoo official speculates that Jackson saw the burly athletes as a territorial threat, and explains that hurling waste is a form of chimpanzee defense, which caused the wrestlers to explode with laughter.

1982 Lethal injection is used for the first time as punishment in the United States. Charlie Brooks Junior is executed in Huntsville, Texas.

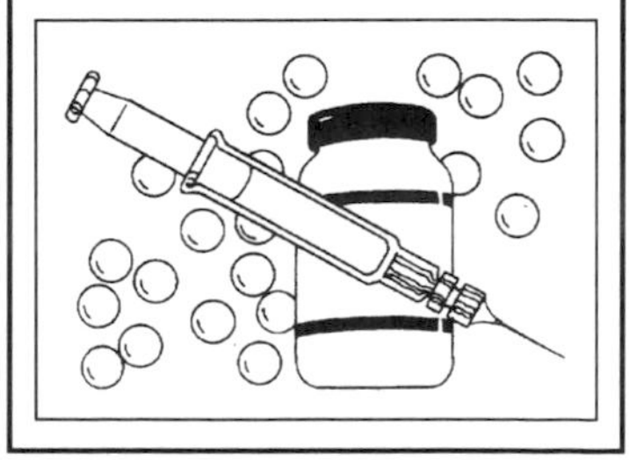

1765 Inventor Eli Whitney is born the son of a farmer in Westboro, Massachusetts. He is famous for creating the cotton gin (which ironically was a major cause of the American Civil War because it made cotton such a valuable crop). Whitney was also responsible for introducing interchangeable parts and division of labor into his factories (in which he was making muskets). His innovations have earned him the title "grandfather of mass production."

1792 The first known cremation in U.S. history was that of Henry Laurens, who dies at 68 on this date. Born in Charleston, South Carolina, Laurens distinguished himself as a patriot during the Revolution, and was involved in peace negotiations between England and America. His last will read, "I solemnly enjoin it upon my son as an indispensable duty that, as soon as he conveniently can after my decease, he cause my body to be wrapped in twelve yards of tow cloth, and burnt until it is entirely consumed, and then, collecting my ashes, deposit them wherever he may see proper."

DECEMBER 8

1862 Pasteur is elected to the French Academy of Sciences. His membership had been denied the first time.

1864 George Boole dies at 50 of pneumonia, which he caught by keeping a lecture date even though he was soaking wet after a two-mile walk through a downpour. He created "symbolic logic," the mathematical analysis of logic. This branch of science is now called Boolean algebra. Although largely ignored during his life, his work eventually formed the basis for information theory, and is now incorporated into the design of computers. His great written work was *Laws of Thought* (1854).

1959 "Smellovision" makes its public debut in the United States, at the DeMille Theatre in New York City. The motion picture *Behind the Great Wall* (a travelogue of China, including a tiger hunt, a May Day parade, and fishing with cormorants) includes odors that are forced through ceiling vents into the audience, using the Aromarama process.

1987 Reagan and Gorbachev sign a treaty under which the intermediate-range nuclear weapons of both countries will be destroyed.

1858 An electric light is first used in a public installation. It is an arc lamp—in which electricity is continuously sent across a gap between two conductors—in the lighthouse at South Foreland, England.

1992 NASA scientists in the Mojave Desert take radar photographs of "the most irregularly shaped object we've yet seen in the solar system." It is the jagged and battered asteroid Toutatis, about 4 miles wide and apparently comprised of two bodies held together by gravity. The pictures are by far the clearest ever of an "Earth-approaching" or "near-Earth" asteroid; the images are 100 times more detailed than the 1991 pictures of the asteroid Gaspra, taken by the Jupiter-bound spacecraft *Galileo*. Sixty-five million years ago a 6- to 9-mile-wide asteroid crashed into Mexico's Yucatan Peninsula. Some believe that the resulting fires, floods, and dust storm may have killed the dinosaurs; this is called the "impact theory."

1993 "Fusion tests jolt energy record," is the worldwide headline about today's historic experiments at Princeton's Tokamak fusion reactor. Over 3 million watts of power is produced in the reactor. The new record stands less than a day; the following afternoon 5.6 million watts is produced (enough power for 1500 U.S. homes). Unfortunately, the energy-producing reactions can only be sustained for very short periods of time. If perfected, fusion holds great promise as an energy source—it is nonpolluting, inexhaustible, and produces few, low-level radioactive by-products.

DECEMBER 9

1703 English jurist-mathematician Chester Moor Hall, inventor of the first achromatic, or blur-free, lens, is born in Leigh. His studies of the human eye convinced him that such a lens was possible (despite Newton's statements to the contrary), and in 1729 he hit on a combination of two types of glass that did not blur. In 1733 he built the first refracting telescope free of "chromatic aberration."

1748 French chemist Claude-Louis Berthollet, the first to note that the completeness of chemical reactions depends on the masses of the reacting substances, is born into poverty in Talloires. His findings were the foundation of the law of mass action, a major principle in modern chemistry theory.

1868 Physical chemist Fritz Haber, Nobel laureate (in 1918) and creator of the Haber–Bosch process for producing ammonia, is born the son of a dry-salt manufacturer in Breslau, Prussia. Haber oversaw production of the first modern chemical weapons during World War I, and vainly tried to produce gold from seawater to help pay Germany's enormous reparations after the war. Ironically, he was forced to flee his beloved homeland before World War II because he was Jewish.

1884 Roller skates with ball bearings are first patented in the United States. Levant M. Richardson of Chicago receives patent No. 308,990 for his invention.

1889 The Chicago Auditorium opens. Created by Louis Sullivan, it is an architectural and technological milestone, reflecting its revolutionary acoustics and its "form follows function" design.

1907 Christmas seals (used to raise funds to fight tuberculosis) first go on sale. The main post office in Wilmington, Delaware, begins selling them in sheets of 228 seals that read either "Merry Christmas" or "Merry Christmas and a Happy New Year." The originator, designer, and producer was Emily Perkins Bissell of Wilmington. The first year's sales made $3000 for the Delaware chapter of the American National Red Cross.

1917 Physicist James Rainwater is born in Council, Idaho. He won the 1975 Nobel Prize for Physics for his "distorted nucleus" theory, in which the atomic nucleus is lumpy and asymmetric, not spherical.

1923 The night before his Nobel Prize acceptance speech, Niels Bohr is notified by telegram that the element hafnium has been found, with properties that he had predicted.

1938 Lafayette Benedict Mendel, codiscoverer of vitamin A (in 1913) and the vitamin B complex (in 1915), dies at 63 in New Haven, Connecticut. His discoveries sprang from a long-term study of rats in which he determined that pure fats, carbohydrates, and proteins are not enough to sustain life. He also determined that the nutritive value of proteins depends on their content of essential amino acids. His work helped establish the modern science of nutrition.

1962 Petrified Forest National Park is established in Arizona. It contains Native American ruins, petrified wood, and part of the Painted Desert.

1993 Astronauts Jeffrey Hoffman and Story Musgrave complete the last of five spacewalks from the shuttle *Endeavor* to repair the bus-sized $3 billion Hubble Space Telescope, originally launched into space with a host of difficulties including a misshapen mirror that blurred its view of space. The repairs restored the prestige of NASA, and enabled the Hubble to provide spectacular photos of the cosmos.

1742 Self-taught Swedish chemist-apothecary Carl Wilhelm Scheele, discoverer of oxygen, chlorine, and manganese, is born the 7th of 11 children in Stralsund. He was the first to isolate a great many substances, including some deadly poisons. He died at just 43 from overwork, chronic poor health, and habitual tasting of each new compound he produced.

1198 Muslim scholar-philosopher Averroës dies at 72 in Marrakech, Morocco. He integrated Islamic and Greek thought, while writing extensive commentaries on Aristotle, whose writings were the dominant thought in science for 20 centuries. Averroës's works were translated by European scholars into Latin, and some of these still exist in their original translations today.

1787 Thomas Hopkins Gallaudet, educator, philanthropist, and founder of the first U.S. school for the deaf, is born in Philadelphia.

1799 The metric system is made mandatory in France—the country where it was invented.

1896 Swedish inventor-philanthropist Alfred Bernhard Nobel dies of a stroke at 63 in San Remo, Italy (see October 21, 1833).

DECEMBER 10

1899 The nation's first tuberculosis hospital—providing free care to indigents on a nationwide, nonsectarian basis—opens with 58 beds in Denver, Colorado. It is the National Jewish Hospital, established by B'nai Brith.

1906 Physicist Walter Henry Zinn is born in Kitchener, Canada. It was Zinn who withdrew the control rods in Chicago in 1942 to begin history's first man-made nuclear chain reaction. Zinn specialized in the design of nuclear reactors, and in 1951 built the first experimental "breeder reactor," in Idaho.

1911 Sir Joseph Dalton Hooker, celebrated botanist and one of Darwin's earliest champions, dies at 94 in Sunningdale, England. Born the son of botanist Sir William Jackson Hooker, director of the Royal Botanic Gardens at Kew, Hooker trained as a doctor. In 1839 he combined both of these elements when he sailed as a surgeon and botanist on the Antarctic expedition of the HMS *Erebus*. It was the first of a number of foreign plant-gathering expeditions Hooker made. He was able to discover and name many new plant species. He pondered the scientific explanation of why and how such a huge variety of plants had come to exist on Earth. Darwin's theory of evolution answered those questions, and Hooker embraced it immediately. Hooker and geologist Sir Charles Lyell presided at the 1858 meeting of England's Linnaean Society at which Darwin's theory was first aired publicly.

1934 Oncologist Howard Martin Temin is born in Philadelphia. He shared the 1975 Nobel Prize for Physiology or Medicine for discovering "reverse transcriptase," the first enzyme to affect communication from RNA to DNA (which is opposite to the normal path of genetic information, DNA to RNA).

1956 Today John Bardeen becomes a corecipient of the Nobel Prize in Physics for discovering and inventing the transistor. In 1972 he won another Nobel Prize in Physics, for developing superconductivity. Bardeen is the first to win two Prizes in the same category. Ironically, the morning he won his 1972 prize he could not open his automatic garage door—although it was operated by a device that his transistor research had made possible.

1990 For the first time in history, a fisherman is imprisoned for failing to use a device to prevent endangered sea turtles from being snared in his nets. U.S. Magistrate Elizabeth A. Jenkins sentences Virgil Lynn Coleman to 30 days in jail for failing to use a TED (turtle excluder device) while trawling for shrimp off Tarpon Springs, Florida, the previous summer. The federal government estimated that 11,000 endangered sea turtles were drowned in shrimp nets annually prior to the TED law.

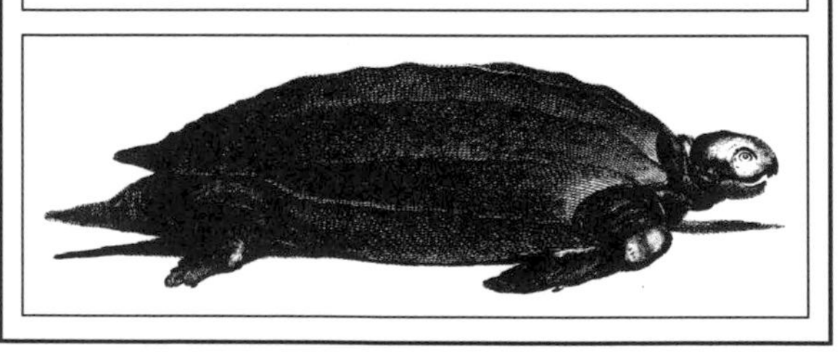

1992 One hundred thousand copies of *Penn & Teller's How to Play With Your Food* are recalled by publisher Villard Books because each copy has a fake sugar packet filled with a serious chemical irritant. The packets are nearly impossible to open, and therefore supposed to provide humor when slipped among genuine sugar packets in restaurants. Unfortunately, the fake packets contain a silica gel dyed with cobalt chloride, a potential health hazard.

DECEMBER 11

1719 The first known record of an aurora borealis (northern lights) display in America is made, in New England. An entry in the Massachusetts Historical Society Collections, Vol. II, reads, "This evening, about eight o'clock, there arose a bright and red light in the E.N.E. like the light which arises from an house when on fire (as I am told by several credible persons who saw it, when it first arose) which soon spread itself through the heavens from east to west, reaching about 43 or 44 degrees in height, and was equally broad."

1843 Robert Koch, cofounder of the science of bacteriology, is born in Clausthal, Germany, one of 13 children of a mining official. Koch's scientific contributions were enormous, foremost of which were discovering the organisms that cause cholera and tuberculosis. He won the 1905 Nobel Prize for Physiology or Medicine, but the award resulted in the greatest fiasco in Nobel history. The great scientist was awarded the prize while his antituberculosis serum was still in development; this drug started killing people six months after the award was made. The Nobel judges were denounced as idiots.

1880 Pasteur injects rabbits with the saliva of a five-year-old child who had just died of rabies. The rabbits get sick and die, and their saliva then kills other rabbits. It is Pasteur's first step in understanding the disease and how to cure it.

1882 German physicist Max Born is born the son of an anatomy professor in Breslau. In 1954 he won the Nobel Prize for his statistical/probabilistic analysis of atomic particles.

1886 The *British Medical Journal* publishes a complete physical description of "Elephant Man" Joseph Carey Merrick, famous for his horrible disfigurements and dramatic life.

1892 Psychiatrist John Augustus Larson, inventor of the polygraph, or "lie detector," is born in Shelbourne, Nova Scotia.

1919 The nation's first monument devoted to an insect is dedicated in Enterprise, Alabama. Inscribed "in profound appreciation of the Boll Weevil and what it has done as the herald of prosperity," the monument commemorates the destruction of cotton crops by the weevil, which led to crop diversification, which in turn led to an immediate tripling of profits in cotton.

1946 The Williams storage cathode-ray tube is patented. It was used in the first stored-program computer, the Manchester University Mark I (first run in 1948).

1949 The heaviest brain on record is that of a sperm whale found in the Antarctic, and weighs 20.24 pounds. The heaviest human brain on record was that of a 30-year-old Ohio man; it weighed 5 pounds 1.1 ounces.

1978 Biochemist Vincent Du Vigneaud dies in White Plains, New York. He was the sole recipient of the 1955 Nobel Prize for Chemistry for isolating and synthesizing two hormones that are critical to human existence: vasopressin—which regulates blood pressure by controlling the muscles around blood vessels—and oxytocin—which causes the uterus to contract and the breasts to secrete milk.

1990 Officials in San Mateo County, California, approve the nation's first law that requires neutering of all county cats and dogs (except those of licensed breeders).

1994 "Ecology House" is the nation's first federally subsidized residence for hyperallergic persons. Today the Associated Press reports that all eight of its tenants are sick. Although equipped with high-powered ventilation, water filtration, tile floors, metal cabinets, and no fluorescent lights, the San Rafael, California, structure causes many health problems, including headaches, pain, and sleeplessness. Unsealed concrete walls are the suspected culprit. "This was a miracle and now it's a nightmare," said one of the residents, each of whom has been diagnosed with "multiple chemical sensitivity-environmental illness."

1781 Scottish physicist Sir David Brewster, inventor of the kaleidoscope (in 1816), is born the son of a schoolmaster in Jedburgh. He also invented the stereoscope (in the 1840s) which provides three-dimensional imagery. His most important scientific work was on creating and analyzing polarized light.

1630 Swedish naturalist Olof Rudbeck, discoverer of the lymph system, is born the 10th of 11 children of a bishop in Westerås.

1731 Erasmus Darwin, one of the foremost doctors of his day, is born the son of a prominent lawyer in Elton, England. Most famous as the grandfather of Charles Darwin (by his first wife) and Francis Galton (by his second), Erasmus was a radical freethinker who wrote long, didactic poems about natural history; he published his own concepts of evolution, which were held as plausible until his grandson developed the theory of evolution.

1796 The first machine to cut and head nails is patented by George Chandler of Maryland.

1846 German chemist Eugen Baumann is born in Cannstatt. In 1896 he discovered that the thyroid gland is rich in iodine. This was the first time any animal tissue was found to contain that element, and it led others to develop treatments for thyroid disorders.

DECEMBER 12

1851 U.S. diplomat Joel R. Poinsett dies at 72 near Statesburg, South Carolina. His name was given to the poinsettia flower. An accomplished botanist, Poinsett brought the flower from Mexico to the United States. His name was also given to the Mexican concept of poinsettismo, which means intrusive, officious behavior. Poinsett was the nation's first minister to Mexico, but during his tenure (1825–1829) he became a sympathizer with revolutionaries seeking to overthrow the government; eventually he was expelled.

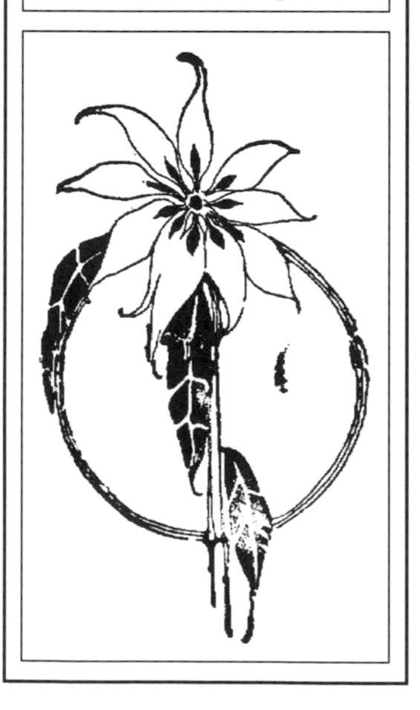

1866 Swiss chemist Alfred Werner is born the son of an ironworker in Mulhouse, France. A beloved teacher, Werner is most famous for his "coordination theory" (1891). He claims that it came to him during sleep, and he wrote it down between 2 AM and 5 AM that morning. The theory describes how atoms are organized in three-dimensional space. It established a new system for classifying inorganic compounds, and won Werner the 1913 Nobel Prize.

1899 The golf tee is patented. George F. Grant of Boston receives patent No. 638,920 for a wooden tee with a concave platform and a tapered base.

1901 A milestone in wireless communication takes place. The first radio transmission across the Atlantic is sent from Cornwall, England, to Newfoundland, Canada. The entire message is three short "dots," the Morse code signal for the letter "S." Inventor Guglielmo Marconi, who designed the system, was on the receiving end. Eight years later he received a Nobel Prize.

1912 The earliest known experiments to develop an airplane ejector seat are conducted in Issy-les-Moulineaux, France, by an Austrian inventor named Odolek. The German firm Heinkel produced the first workable seat in 1941.

1925 The first motel in the United States opens on the north side of San Luis Obispo, California. The Motel Inn was designed by architect Arthur S. Heineman, and erected on the property of Neil Cook. It could accommodate 160 guests in individual cabins with garage, bathrooms, and telephone. The sign in front of the inn flashed M and H alternatively, so it read "Motel," then "Hotel."

1953 An aircraft reaches mach 2.5 (2.5 times faster than the speed of sound) for the first time.

1955 The hovercraft is patented by Sir Christopher Sydney Cockerell.

1979 History's largest oil-tanker wreck occurs. The 354,043-ton *Energy Determination* blows up and breaks in two in the Strait of Hormuz, connecting the Persian Gulf to the Gulf of Oman.

1980 Armand Hammer pays $4.5 million at a Christie's auction for Leonardo da Vinci's 36-page Codex Leicester manuscript (compiled around 1507). At the time, it was the highest price ever for a complete manuscript. The current record is $11.4 million for the 1990 sale of the thirteenth-century *Northumberland Bestiary*, a brightly illustrated encyclopedia of real and mythical animals.

1990 In Xapuri, Brazil, the trial of rancher Darly Alves da Silva (accused of killing rain forest defender Chico Mendes) opens. Surprisingly, da Silva's son confesses to the murder. Mendes was a rubber tapper who became a leader in the battle against rain forest destruction by cattle ranchers; his murder at age 44 made him a martyr in the struggle.

DECEMBER 13

1204 Jewish philosopher-physician Maimonides dies at 69 in Cairo. He was the personal physician to Saladin, the great Muslim ruler who fought Richard the Lion-Heart during the third Crusade; Richard invited Maimonides to England to serve as his own physician, but Maimonides refused the offer. In his *Guide for the Perplexed*, Maimonides was an early opponent of astrology. He also attempted to reconcile Aristotle's works with the Bible.

1565 Natural historian-encyclopedist Conrad Gesner dies at 49 in the town of his birth, Zurich, Switzerland. He published several monumental works: a Greek–Latin dictionary (1537); a "Universal Bibliography" (1545) that listed and evaluated works by some 1800 authors; a 21-volume encyclopedia in which Gesner's aim was to survey all of the world's recorded knowledge; the elaborately illustrated "Historiae animalium" (the first of five volumes appeared in 1551) in which each book covered a portion of the animal kingdom; and a book describing some 130 languages (1551).

1577 Sir Francis Drake departs England with five ships for a three-year circumnavigation of Earth.

1642 Dutch explorer Abel Tasman becomes the first European to reach New Zealand.

1730 Scientist-diplomat Sir William Hamilton is born in Scotland. He is remembered for his studies of the volcanoes at Vesuvius and Etna, for his extensive archaeological collection (many pieces of which are now in the British Museum), and for marrying Lady Emma Hamilton, the mistress of Admiral Horatio Nelson.

1780 German chemist Johann Wolfgang Döbereiner, discoverer of catalysis (the process in which a substance speeds a reaction between other chemicals, but is unaffected itself), is born the son of a coachman in Hof. Döbereiner also discovered that several groups of elements had properties that set them apart from other elements; this was the foundation on which others created the periodic table of the elements.

1805 Astronomer Johann von Lamont is born in Braemar, Scotland. At 12 he was sent to Bavaria for an education at a Benedictine monastery, and he spent the rest of his life there. Among his scientific achievements was a catalogue of nearly 35,000 stars and discovering a 10-year cycle in the strength of Earth's magnetic field.

1878 England becomes the second country (after France) to have electric streetlighting. Twenty "Yablochkov candles" are put in operation today on London's Victoria embankment.

1920 Betelgeuse becomes the first fixed star to be measured in U.S. astronomical history. Using an inferometer designed by Albert Michelson, Dr. Francis Gladheim Pease observes the star at Mount Wilson Observatory (in Wilson, California) and estimates its diameter to be 260 million miles.

1923 Physicist Philip Warren Anderson is born in Indianapolis, Indiana. He shared the 1977 Nobel Prize for his analysis of impurities in superconductive materials.

1968 The tanker *Witwater* breaks in the Caribbean Sea, near Galeta Island, Panama, spilling oil throughout the island's mangrove swamps and coral reefs. The disaster has one positive aspect: The Smithsonian Institute maintains a research station on the island, and scientists there begin an intense and long-term study of the carnage.

1990s This day is a landmark in the history of doctor-assisted "medicide." In 1990, a judge in Clarkston, Michigan, dismissed a first-degree murder charge against Dr. Jack Kevorkian for his assistance in the death of Janet Adkins, an Alzheimer's patient. In 1993 a Circuit Court Judge in Wayne County, Michigan, declared the state's ban on assisted suicide unconstitutional. And exactly one year after that (1994), the state Supreme Court reversed this decision, making the ban legal.

1993 "The world's most expensive repair truck," the space shuttle *Endeavor* touches down at Cape Canaveral after a successful 11-day mission to repair the $3 billion Hubble Space Telescope.

1621 Furs are first exported from America. The SS *Fortune* sets sail for England from Massachusetts with a cargo, estimated at $2450, of pelts, spices, clapboards, and wainscot. The ship is later captured by the French and the cargo seized.

1546 Tycho Brahe, the last and probably the greatest naked-eye astronomer, is born in Knudstrup, Scania (now Sweden, then Denmark). He began life in turmoil: His twin brother did not survive birth, and Tycho was kidnapped by a childless uncle at age one. He entered the University of Copenhagen at age 13 to study law and politics, but in 1560 when he observed a solar eclipse his career path changed forever. He soon achieved prominence as a gifted astronomer. The fact that astronomers had predicted the 1560 eclipse made a great impression on Brahe, and he dabbled in astrology throughout life; he prepared astrological charts for Denmark's King Frederick II, who built Brahe an observatory on the island of Aven. It became a mecca for Europe's scientific elite. There Brahe compiled the most extensive, most accurate records of planets' and stars' movements that had ever been produced. It was Tycho's recalculation of the length of an Earth year that led in 1582 to the production of the now-universal Gregorian calendar. "Oh, that it may not appear I have lived in vain," he uttered on his deathbed.

1798 The nation's first screw-related patent is granted to David Wilkinson of Rhode Island for a screw-making machine.

DECEMBER 14

1873 Louis Agassiz, one of history's greatest natural scientists and educators, dies at 66 in Cambridge, Massachusetts. At 19, while still a student in Munich, he assumed the task of classifying many Amazonian fish specimens. The ensuing publication established Agassiz as a first-class scientist. He became the world's preeminent fish expert in his day, but he was also known for studies of other animal groups and for his studies of glaciers. It was Agassiz who concluded that huge ice sheets once covered many now-temperate parts of Earth. In 1847 he became a professor of zoology at Harvard where he established its famous museum of comparative zoology. Agassiz, a revered and revolutionary teacher, influenced higher education in the United States by replacing book learning with experiential learning in the study of science. Ironically, he never accepted Darwin's theory of evolution; to his death he conceived of each species as a divinely created "thought of God," rather than as the result of eons of organic interaction.

1900 Quantum physics begins. Max Planck, 42, reluctantly announces to a meeting of the German Physical Society that his experimental results can best be understood if energy exists in discrete packets called "quanta," not continuous waves.

1909 Biochemist Edward Lawrie Tatum is born the son of a pharmacology professor in Boulder, Colorado. He shared the 1958 Nobel Prize for Medicine or Physiology for formulating the "one gene, one enzyme" principle: Each biochemical reaction in a living thing is controlled by one enzyme, and production of that enzyme is controlled by one gene. This is how genes control life.

1911 Roald Amundsen and four compatriots become the first to reach the South Pole. All five held the Norwegian flag when it was planted. "Five weatherbeaten, frostbitten fists ... grasped the pole, raised the waving flag in the air, and planted it as the first at the geographical South Pole."

1940 Plutonium is discovered. The feat is accomplished at the University of California at Berkeley by Glenn Seaborg, Arthur Charles Wahl, and Joseph William Kennedy. Announcement of the discovery was postponed for six years because of security concerns.

1952 Siamese twins are successfully separated for the first time. Dr. Jac S. Geller leads the team at Mount Sinai Hospital in Cleveland, Ohio, that separates two girls joined at the breastbone.

1966 The nation's first biosatellite—a research vessel containing small life-forms to be exposed to cosmic rays, weightlessness, and the absence of a day–night rhythm—is launched from Cape Canaveral. *Biosatellite I* weighs 936.5 pounds, including a 275-pound compartment containing 13 experiments. Scheduled to return to Earth on December 17, retrofire did not occur, and *Biosatellite I* was never recovered.

1972 Gene Cernan becomes the last man to walk on the moon, so far.

1975 Zoologist Sir James Gray dies at 84 in Cambridge, England (see October 14, 1891).

1986 Dick Rutan and Jeana Yeager depart Edwards Air Force Base, California, in the experimental plane *Voyager* on history's first nonstop, nonrefueled flight around Earth.

1990 Apparently for the first time in history, a worker's compensation claim is settled for secondhand smoke causing a heart attack. Waiter Avtar Uhbi claimed his coronary resulted from the smoke he had inhaled at the bar/restaurant in Sausalito, California, where he used to work.

DECEMBER 15

1834 Astronomer Charles Augustus Young is born in Hanover, New Hampshire, where, in 1866, he succeeded his father and grandfather as a science professor at Dartmouth College. He discovered the sun's "reversing layer," or lines in the solar spectrum that are normally dark but which glow brightly during eclipses.

1852 Henri Becquerel, cowinner of the third Nobel Prize in Physics, is born in Paris (see August 25, 1908).

1854 The country's first practical street cleaning machine begins operation in Philadelphia. It is a cart containing a series of brooms attached to a cylinder that is rotated via an endless chain by movement of the cart's wheels. Ninety-seven years earlier, Philadelphia had also become the first U.S. city with a street cleaning service, when Benjamin Franklin hired "a poor industrious man" to sweep portions of the city streets twice a week.

1860 Danish physician Niels Ryberg Finsen is born in Thorshavn, the Faeroe Islands. He spent most of his life in poor health, dying at just 44, the year after winning the Nobel Prize. Finsen was especially interested in treating disease with light. He discovered that some light frequencies could kill bacteria and cure some bacterial diseases (lupus vulgaris in particular), and his research laid the groundwork for the use of X rays.

1877 Edison patents the phonograph. It was his first great success.

1896 Biochemist Carl Cori is born the son of an eminent zoologist in Prague. Cori and wife Gerty shared the 1947 Nobel Prize for their research into the pathway by which muscles receive energy.

1916 Maurice Wilkins is born the son of a physician in Pongoaroa, New Zealand. He shared the 1962 Nobel Prize for helping determine the structure of DNA.

1923 Physicist Freeman Dyson is born in Crowthorne, England. He is noted for work on quantum electrodynamics and his theories on extraterrestrial civilizations.

1965 Manned spacecraft *Gemini 6* and *Gemini 7* maneuver within ten feet of each other in orbit.

1967 President Lyndon Johnson signs the Wholesome Meat Act. The ceremony is attended by an odd cross section of campaigners for consumers' rights, including Ralph Nader, Betty Furness, and Upton Sinclair.

1970 The Soviet *Venera 7* becomes the first craft to soft-land on another planet. It reaches Venus and transmits data from the surface for 23 minutes.

1994 International wire services announce the discovery of 39 prehistoric pine trees, thought to have been extinct for 150 million years. They were found alive and healthy in an almost inaccessible section of rain forest in the Wollemi National Park outside Sydney, Australia. Said Carrick Chambers, director of the Royal Botanic Gardens, "The discovery is the equivalent of finding a small dinosaur still alive on Earth."

1994 In the interest of public health, news reports carry the story of Larry Rust of Ham Lake, Minnesota, who gives his wife Pat one of her Christmas presents early. It is a carbon monoxide detector, which goes off at 5 o'clock the next morning and saves both of their lives. "Thank goodness [the detector] wasn't wrapped up underneath the tree," said Pat.

1939 Nylon yarn is manufactured for the first time, by E.I. du Pont de Nemours & Co., Inc. in Delaware. The yarn was then sent to hosiery mills for the manufacture of women's stockings. This is the second commercial use of nylon; in February 1938, du Pont had begun manufacturing toothbrush bristles with the material.

1612 "Soon after dark" Simon Marius discovers the Great Spiral Nebula in the constellation Andromeda.

1620 The *Mayflower* anchors off Plymouth Rock. (Today is the date as recorded by the Pilgrims in the Old Style calendar; the date is December 26 in the New Style.)

1763 Davidis van Royen publishes *Incendiis Corporis Humani Spontaneis* in Amsterdam. It is the first scientific study of human spontaneous combustion.

1776 German physicist Johann Wilhelm Ritter, discoverer of ultraviolet radiation and electroplating, is born the son of a minister in Samitz, Germany (now Chojnów, Poland).

1798 Thomas Pennant, one of the renowned zoologists and great nature writers of his time, dies at 72 in the town of his birth, Downing, Wales. His 1766 *British Zoology* spurred animal research, especially in ornithology. He traveled extensively throughout Europe, mostly on horseback, and carefully recorded his observations on the flora, fauna, local peoples and antiquities.

DECEMBER 16

1826 Italian astronomer Giovanni Battista Donati, discoverer of six comets (including Donati's comet), is born in Pisa. His most important work came in 1864 when he was able to determine the spectrum of a comet before and after it was heated to glowing during a near pass of the sun; this was the first essential step in understanding the structure of comets.

1850 Hans Buchner (older brother and mentor of Nobel laureate Eduard Buchner) is born in Munich, Germany. Hans enjoyed a distinguished career, studying bacteria and the natural substances that fight them in the blood. He discovered and named alexins (1891) and pioneered research into gamma globulins, both of which are blood-borne proteins that attack invading bacteria.

1857 Astronomer Edward Emerson Barnard is born into poverty in Nashville, Tennessee. Barnard only received two months' formal education in his life. Because his father had died years before, he went to work in a photo studio to support his family at age 9. Later he obtained a number of prestigious academic positions, including the Lick Observatory and the University of Chicago, where he made important discoveries. These included Barnard's Runaway Star (the star with the fastest relative speed ever discovered) and Barnard's satellite (the fifth moon of Jupiter, which was the first Jupiter satellite discovered since Galileo, and the only satellite ever named after its discoverer). He and Maximilian Wolf were the first to realize that dark nebulae in the Milky Way are clouds of dust and gas.

1901 Anthropologist Margaret Mead is born in Philadelphia.

1899 The nation's first children's museum opens. It is the Brooklyn Children's Museum, at the Brower Park Building, Brooklyn Avenue and Park Place, in the Crown Heights section of New York City.

1920 History's deadliest landslide kills 180,000 in Kansu Province, China.

1954 A synthetic diamond is produced for the first time, by Professor H.T. Hall at the General Electric Research Laboratories.

1976 The U.S. government halts swine flu vaccinations, following reports linking the vaccine with paralysis.

1982 Environmental Protection Agency chief Anne M. Gorsuch becomes the first member of the Cabinet to be cited for contempt of Congress for refusing to submit documents requested by Congress.

1992 Astronomers at a NASA news conference in Washington report they have observed what seem to be planets forming. Through the Hubble Space Telescope, at least 156 stars in the Orion Nebula have been seen to be circled by "protoplanetary disks," bands of dust thought to form planets. "We have found a place where it is very possible that there will be planets within the next few million years," said program scientist Edward J. Weiler. "This takes us closer to the final proof that there are other planets where there could be life."

DECEMBER 17

1778 Chemist Sir Humphry Davy is born in Penzance, England, the son of a poor woodcarver who soon died and left Davy a legacy of a £1300 debt. Davy hated school, and soon he apprenticed with an apothecary. His chemistry education was largely self-taught (and risky, as he is famous for the accidental explosions in his labs and for his habit of breathing every gas he created). He discovered a number of elements, including potassium and sodium, and invented the miner's safety lamp.

1787 Czech physiologist Jan Evangelista Purkinje is born the son of an estate manager in Libochovice, Bohemia (see July 28, 1869).

1853 French bacteriologist Émile Roux is born in Confolens, France. In 1878 he was selected as an assistant to Louis Pasteur, and was instrumental in engineering anthrax vaccination strong enough to protect against the disease, but not so strong as to induce serious symptoms. Roux's greatest success was the fight against diphtheria; he proved that bacteria cause the disease, and further showed that it is specifically a toxin from the bacillus bacteria that really does the damage. Once the toxin was identified, others were able to produce a successful antitoxin.

1861 Electrical engineer Arthur Edwin Kennelly is born the son of an Irish lawyer in Bombay, India. He is known for mathematically analyzing electric currents, and for suggesting that an atmospheric layer reflected radio waves back to Earth; such a layer was eventually discovered and is now called the Kennelly–Heaviside layer.

1881 Lewis Henry Morgan, a father of scientific anthropology, dies at 63 in Rochester, New York (see November 21, 1818).

1908 Willard F. Libby, inventor of radiocarbon dating, is born in Grand Valley, Colorado. "Once you ask the question, where is the Carbon-14, and where does it go, it's like one, two, three, you have [radiocarbon] dating," Libby once said about his great innovation. The son of a farmer, Libby became a nuclear physicist, and was part of the atomic bomb project during World War II. In 1945 at the University of Chicago, Libby had his great insight about carbon dating. He was awarded the 1960 Nobel Prize for Chemistry. His technique has become standard in anthropology and fossil-dating; for example, it was used to prove that the Dead Sea Scrolls were authentic and that Piltdown man was a fake.

1938 At the Kaiser Wilhelm Institute for Chemistry in Berlin, Hahn and Strassmann discover nuclear fission (although they do not realize it at the time).

1959 *On The Beach* becomes the first motion picture presented simultaneously in major cities worldwide. It premieres in 18 cities, including at the Astor Theatre, New York City. The movie was an adaptation of the Nevil Shute novel about the aftermath of nuclear war; it was produced by Stanley Kramer and starred Fred Astaire, Ava Gardner, Gregory Peck, and Anthony Perkins.

1969 The U.S. Air Force ends Project Blue Book because it claims not to have found any valid evidence of extraterrestrial life.

1986 History's first triple transplant—heart, lung, and liver—is conducted on Mrs. Davina Thompson, 37, at Papworth Hospital in Cambridge, England. Dr. John Wallwork and Professor Sir Roy Calne lead the team of 15 that conduct the seven-hour operation.

1903 At 10:35 AM near Kill Devil Hill in Kitty Hawk, North Carolina, the Wright brothers conduct the first motor-driven, sustained flight by man. Four flights are made on this day. The first lasts 12 seconds, with a maximum altitude of 12 feet. The longest flight lasts 59 seconds and covers a distance of 852 feet.

December 18

1661 Swedish inventor Christopher Polhem is born in Visby, Gotland Island. He created a number of industrial machines, especially for mining, and in 1700 he built a water-powered tool factory. His use of division of labor was well ahead of his time, as it did not become standard procedure until the 1900s.

1796 A newspaper appears on Sunday for the first time in U.S. history. Philip Edwards publishes the Sunday *Monitor* in Baltimore. It contains four pages, 10¼ by 17 inches in size.

1856 Sir Joseph John Thomson, discoverer of the electron, is born the son of a bookseller in Cheetham Hill, near Manchester, England. Thomson's discovery (in 1897) advanced the understanding of atomic structure, and he received the 1906 Nobel Prize for Physics. Seven of his research assistants at the Cavendish Laboratory in Cambridge eventually received Nobel Prizes.

1878 English physicist Sir Joseph Swan presents the first incandescent light bulb. Unfortunately, the carbon filament had already burned out, and no light is emitted. Edison's early attempts also suffered from quick burnout; Edison patented his bulb in 1879. Swan and Edison sued each other for priority in the invention of the light bulb, and then went into business together in 1883.

1890 Edwin Howard Armstrong, inventor of FM radio, is born the son of a publisher and a schoolteacher in New York City. Among his innovations was the first amplifier for weak radio signals. His work forms the basis for modern television–radio broadcasting. Armstrong devoted much of his energy and resources to legal fights with others; toward the end of his career, before he leapt out of his apartment window in New York City at age 63, he was convinced there was a conspiracy against him.

1892 Sir Richard Owen, coiner of the word "dinosaur" (in 1842), dies at 88 in London. He obtained a medical degree at the University of Edinburgh, but soon thereafter he became an assistant in London's Museum of the Royal College of Surgeons, where he fell in love with the field of comparative anatomy. He dissected every possible animal he could. In 1852 he discovered the parathyroid gland in the throat of a rhinoceros, which was not discovered in man for many years. He undertook a massive study in the forms of animal teeth in the 1840s, which led him to the study of fossilized and extinct forms. Among his achievements were the first description of the newly extinct moa bird of New Zealand; the first description of *Archaeopteryx* (the first known fossil bird, which Owen had actually reconstructed with numerous errors); and the preparation of the first public display of reconstructed dinosaurs (done for the 1854 exhibition at the Crystal Palace in London). Owen was violently opposed to Darwin's theory of evolution; he wrote many anonymous articles attacking the theory, and secretly provided material to its opponents.

1912 Bones of "Piltdown man" are presented to the Geological Society of London by their "discoverer," Charles Dawson, and by an official of the British Museum. The bones initially cause revision of evolutionary theory, but eventually the whole affair is revealed as the greatest hoax in anthropology history.

1912 Biologist Daniel Mazia, known for exploring the chemical events that produce cell division, is born in Scranton, Pennsylvania.

1936 A giant panda first arrives in the United States. The cub Su-Lin weighs about 5 pounds as it arrives in San Francisco aboard the *President McKinley*, imported by Ruth Harkness. The baby panda was captured in China about a month before. It choked to death in a Chicago zoo on April 1, 1938, the very day Harkness departed for Asia in search of more pandas.

1957 The nation's first nuclear power plant, the Shippingport Atomic Power Station in Pennsylvania, begins operations.

1972 The term "junk food" appears in print for the first time, in *Time* magazine.

1994 The Associated Press reports that the United States is exporting massive amounts of vodka to Russia, and that many Russians prefer the Yankee drink to their own brands. This is mostly a matter of price; efficient production here often makes imported vodka the cheapest available. One of the biggest suppliers is the A. Smith Bowman Distillery in Virginia, run by direct descendants of John Adams and Robert E. Lee.

1683 John Reid, the first landscape architect in American history, arrives at Staten Island, New York. He left Aberdeen, Scotland, the previous August 28 on the *Exchange* with his wife and three daughters. In Europe he was the gardener to Sir George MacKenzie of Rosebaugh, lord advocate for King Charles II.

1809 Parasitologist Pierre-Joseph van Beneden is born in Mechlin, Belgium. He is most famous for a 15-year study of tapeworms (cestodes) that resulted in the first complete description of their complicated life cycle. His son Edouard also studied worms, but became famous for discoveries in the field of genetics. Edouard determined that each cell in an organism's body has the same number of chromosomes, that this number is a characteristic of that species, and that sperm and egg have half that number of chromosomes.

1813 Physical chemist Thomas Andrews, discoverer of critical temperature (for each gas, the temperature at which no amount of pressure can liquefy that gas) is born the son of a linen merchant in Belfast, Northern Ireland. It was a crucial discovery, because until it was made, science had been unable to liquefy some gases, called "permanent gases."

DECEMBER 19

1852 Physicist Albert Abraham Michelson is born in Strelno, Prussia. At age 4 his family emigrated to the United States. Michelson is famous for measuring the speed of light, for defining the length of a meter in terms of wavelength, and for inventing the interferometer (a device for splitting light beams). Also, his famous Michelson–Morley experiment (1887) opened the door to Einstein's relativity theory by shattering the notions of absolute space and the cosmic "ether." Michelson won the 1907 Nobel Prize for Physics, the first U.S. Nobel laureate in the sciences.

1882 The killer disease beriberi is struck a death blow. On this day the Japanese ship *Riujo* departs for a 10-month tour of the Pacific; when she returns, most of the crew has the disease. Dr. Kanekiro Takaki then sends out another ship on the exactly the same tour, except that the crew eats a diet typical of English sailors (who seldom contracted the disease). The Japanese crew hated the strange foods, but none got the disease (except for 14 who had smuggled their own rations aboard). The experiment was science's first big clue to beriberi's cause and cure, which is now known to be a lack of the vitamin thiamine.

1938 Otto Hahn writes from Berlin to his ex-research partner Lise Meitner (who had recently fled to Copenhagen to escape Nazism) about puzzling results from his latest experiment of bombarding uranium with neutrons. It is Meitner who realizes that nuclear fission has taken place and that enormous energy has been released. The experiment verified Einstein's $E = mc^2$, and led directly to the atomic bomb mechanism.

1944 Richard Leakey is born in Nairobi, Kenya, to Mary and Louis Leakey, two of history's premier anthropologists. Richard initially resisted his parental influence, but eventually became an important paleontologist and conservationist after discovering an ancient jawbone in Tanzania in 1963.

1958 The first outer-space radio broadcast is made. A tape recording of President Dwight Eisenhower's Christmas message is broadcast on 107.97 and 107.94 megacycles from a rocket revolving around Earth.

1972 *Apollo 17*, the last *Apollo* flight, splashes down in the Pacific.

1991 Science has created mice with Alzheimer's. The development is expected to be a major boon to treatment and prevention. Today's *Nature* reports work by three researchers who injected mouse embryos with human gene fragments; when mature, the mouse brain cells contained plaques, tangles, and degeneration typical of Alzheimer's.

1900 One of medicine's most repugnant, yet important experiments ends in Quemados, Cuba, when volunteers Levi E. Folk, Warren G. Jernegan, and Dr. Robert P. Cook emerge after 20 days in a tiny, hot, fetid shack that had been packed with bedding, towels, underwear, and hospital gowns that had been used by yellow fever victims. Blood, vomit, and excrement covered everything. The fact that none of the three volunteers caught the disease disproved one theory of how it spread (thereby helping Walter Reed and colleagues prove how it was spread).

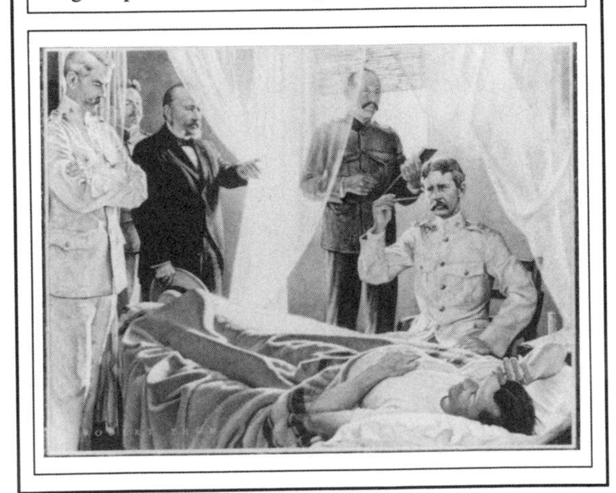

1790 The country's first mill that could spin cotton yarn begins operation in Pawtucket, Rhode Island.

1805 Thomas Graham (the father of colloid chemistry and one of the founders of physical chemistry) is born the son of a successful manufacturer in Glasgow, Scotland. Graham resisted his father's urging to go into the ministry (which caused his father to withdraw all financial support) and spent his life studying how molecules move through gases and liquids. Graham's law of diffusion describes how the size of molecules influences their speed. While examining molecule movement in solutions he discovered that particles come in two forms: crystalloid and colloid. He invented and named the process of "dialysis," and he coined the term "osmosis" to describe the movement of water and other liquids through membranes.

1820 A state bachelor tax is first levied in the United States. Effective January 1, 1821, Missouri passes a one-dollar tax "on every unmarried free white male above the age of 21 years and under 50 years."

December 20

1852 Physician-bacteriologist Shibasaburo Kitasato, the man who discovered the cause of bubonic plague, is born the son of a village mayor in Oguni, Japan. While still a student in the Berlin lab of Robert Koch, Kitasato became the first to isolate the tetanus bacillus bacteria. He then collaborated with Emil von Behring to demonstrate how dead tetanus bacteria could be made into a vaccine to provide "passive immunity" against the disease. On returning to Japan, he founded several laboratories and institutes.

1876 Astronomer Walter Sydney Adams, the first to determine the true composition of "white dwarf" stars, is born the son of U.S. missionaries in Kessab, Turkey. He specialized in spectroscopic analysis; he investigated sunspots, the rotation of the sun, the distances and movements of thousands of stars, and planetary atmospheres. He discovered that Venus's atmosphere was rich in carbon dioxide.

1879 Edison demonstrates his incandescent light bulb at Menlo Park, New Jersey.

1590 Ambroise Paré invented a new procedure for treating gunshot wounds at the battlefield of Turin, Italy, in 1536. The traditional treatment was pouring boiling oil on the wound; fortunately on this day the supply of oil expired, leading Paré to improvise with a mixture of egg yolk, rose water, and turpentine. This new compound was less traumatic, caused less infection, and produced faster healing. Paré's report of his discovery was initially ridiculed because it was published in French, not Latin. It was one of many innovations during his life. Paré was the chief surgeon to four French kings.

1890 Physical chemist Jaroslav Heyrovský is born the son of a law professor in Prague. Inspired by a question posed to him during his doctoral examination, Heyrovský constructed the first polarograph—an electrolytic device for measuring ion concentration. It won him the 1959 Nobel Prize.

1892 Alexander T. Brown and George F. Stillman of Syracuse, New York, receive patent No. 488,492; it is the first U.S. patent for a pneumatic automobile tire.

1901 Robert Jemison Van de Graaff, inventor of the Van de Graaff generator, is born in Tuscaloosa, Alabama. In the 1920s he attended lectures by Marie Curie at the Sorbonne, which turned his interests to atomic physics. He saw the need for generating enormous voltages in the study of subatomic particles, and built his first device in the United States in 1931. He worked out the principles on a model made of tin cans, a silk ribbon, and a small motor. The latest versions of Van de Graaff's machine have generated charges exceeding 30 million volts.

1938 The iconoscope, the first practical television camera is patented by U.S. physicist Vladimir Kosma Zworykin, the father of modern television.

1951 The first breeder reactor—a nuclear reactor that produces more fissionable material than it consumes—reaches an output of 100,000 watts. The machine is the EBR 1 at the National Reactor Testing Station near Idaho Falls, Idaho. It was the country's first machine that converts nuclear energy to electricity.

1620 Dr. Samuel Fuller, the first physician in New England, goes ashore from the *Mayflower* with the first exploratory party of Pilgrims to touch the New World. An original signer of the Mayflower Compact, Fuller was the only doctor in the colony for some time. He was the second doctor in colonial America, preceded by Dr. Lawrence Bohune who arrived in Virginia in 1610, but was killed two years later on board the *Margaret and John* when it was attacked by Spaniards.

1766 "May we not infer from this experiment that the attraction of electricity is subject to the same laws with that of gravitation?" This deep and farsighted insight was reached by Joseph Priestley from an experiment he begins on this date in Leeds, England. The experiment was suggested by Benjamin Franklin, and involved measuring the electricity inside an electrified metal cup.

1773 Robert Brown, botanist and discoverer of Brownian motion, is born in Montrose, Scotland, the son of an Anglican priest. Brown's keen powers of observation earned him a place in history. He recognized basic distinctions between gymnosperms (conifers) and angiosperms (flowering plants); he realized, and reported, that the nucleus is a fundamental component of plant cells (he coined the term "nucleus," which is Latin for "little nut"); and he improved plant classification by establishing new groups and using seed forms in classification. His observation skill also allowed him to detect and experiment with the irregular, minute movements of particles in solution. This action is called "Brownian motion" and is now known to be omnipresent in many molecules in the physical and living worlds.

1889 Sewell Wright, a founder of population genetics, is born in Melrose, Massachusetts. He created the concept of genetic drift, also called the Sewell Wright effect, in which the gene pool of any small, isolated population may be forever changed by random sampling processes, as well as environmental pressure.

1890 Hermann Joseph Muller, the first to induce genetic mutations with X rays, is born in New York City. The mutation experiments occurred in 1926, for which he won the Nobel Prize in 1946.

1898 Marie and Pierre Curie discover radium.

1909 The nation's first two junior high schools are authorized, in Berkeley, California. They are the McKinley and Washington "introductory high schools" for seventh, eighth, and ninth graders in buildings separate from the main high school, and established with their own curriculum and their own administration. Both schools opened on January 2, 1910. Frank Forest Bunker was the driving force behind their creation.

1956 Lewis Madison Terman, developer of the modern intelligence test, dies at 79 in Palo Alto, California. In 1916 Terman was a professor of education at Stanford University (where he spent the last 46 years of his life) when he published *The Measurement of Intelligence*, which presented the Stanford–Binet test. This was an expansion and revision of the Binet–Simon test, and it remains the premier gauge of mental capacity. At its core is "IQ" (intelligence quotient), which compares an individual's chronological and mental ages. In 1921 Terman launched a large study of "gifted" persons, which is expected to end in 2010.

1968 *Apollo 8* is launched at 7:51 AM from Cape Kennedy. Its astronauts (Lovell, Anders, and Borman) become the first humans to orbit the moon.

1987 *Soyuz TM4* is launched toward the space station *Mir*. Its astronauts (Titov and Manarov) return to Earth exactly one year later (365 days 22 hours 39 minutes 47 seconds), setting the current record for the longest spaceflight.

1993 In the interest of public safety, and $78 million in punitive damages just awarded against it, Domino's Pizza Inc. announces an end to its 30-minute-delivery-or-you-get-$3-off policy. This happens just four days after the settlement was awarded by a St. Louis jury in a 1989 case in which a Domino's driver ran a red light and struck a woman.

1666 The French Academy of Sciences holds its inaugural meeting, in the library of Louis XIV on Rue Vivienne in Paris. A commemorative coin is made, with the king on one side and Minerva—the Roman goddess of wisdom—on the other.

1823 Jean Henri Fabre, a leading entomologist in his day, and one of the first ethologists, is born in Saint-Léon, France. He wrote a number of books to popularize science, isolated the biological stain alizarin, and his practice of studying insects alive in their natural habitat (rather than mounted in display cases) was most unusual at the time.

1895 The first roetgenogram on a human is taken in Würzburg, Germany, by Wilhelm Roentgen, discoverer of X rays. He takes an image of his wife's hand; only her bones and her wedding ring are clearly seen when the image is developed. Six days later Roentgen submits his first report of X rays, and a revolution in science begins.

DECEMBER 22

1902 Psychiatrist Richard Krafft-Ebing dies at 62 near Graz, Austria. He coined the terms "paranoia," "sadism," and "masochism," and is famous as a pioneer in the scientific study of sexual aberration. His 1886 *Psychopathia Sexualis* opened the door to the theories of Freud, which came two decades later.

1903 Physiologist Haldan Keffer Hartline is born in Bloomsburg, Pennsylvania. He shared the 1967 Nobel Prize for Medicine or Physiology with George Wald and Ragnar Granit; together their work showed how light is processed by cells in the eye and turned into visual images in the brain.

1911 Grote Reber, who advanced astronomy by building the first radio telescope, is born in Wheaton, Illinois. He built the device in his backyard in 1937; it was the world's only radio telescope until after World War II. It was donated to the National Bureau of Standards in 1947.

1938 The "living fossil," the coelacanth, is first pulled up in a trawler's net from the Indian Ocean off East London, South Africa. The fish was believed to be extinct for over 60 million years. All modern amphibians, reptiles, birds, and mammals are thought to have evolved from a coelacanth relative.

1938 From Berlin, Otto Hahn and Fritz Strassmann finish and mail their paper describing man's first splitting of the atom (although they did not realize at the time what they had done).

1964 History's fastest jet (the USAF Lockheed SR-71, a reconnaissance plane) first flies in its definitive form.

1956 A gorilla is born in captivity in the United States for the first time, at a zoo in Columbus, Ohio. Colo (3¼-pounds) is the offspring of Baron (11 years old, 380 pounds) and Christiana (9 years old, 260 pounds).

1988 Ecologist Francisco "Chico" Mendes, 44, is shot and killed behind his shack in Xapuri, Brazil. For years he had been battling the destruction of rain forests by meat-and-leather ranchers.

1993 Maryland Governor William Donald Schaefer grants a full pardon to Kirk Bloodsworth, who had spent nine years in prison, and at one point was scheduled for execution for an assault that DNA evidence finally proved he did not commit. Prosecutors now admit they never would have taken Bloodsworth to court if DNA testing had been available in 1984 when the crime occurred.

1993 The Australian Senate votes to allow aborigines to pursue legal land claims to approximately one-tenth of their continent. Those who can prove unbroken connection to the land will be entitled to ownership of state and federal lands; privately owned land is exempt. Today's vote, coupled with a 1992 High Court ruling, officially ends the policy of "terra nullius," which contends that Australia was unoccupied until Europeans arrived in 1778. In fact, aborigines had been there 40,000 years before.

1722 Swedish mineralogist Axel Fredrik Cronstedt, the first to isolate the element nickel (in 1751), is born the son of an army officer in Stroepsta. Nickel was the first metal since iron—discovered 23 centuries before—known to have magnetic properties. Cronstedt was also the first to use a blowpipe to superheat minerals so as to analyze their chemical composition; this procedure became fundamental procedure for the next century.

1732 Inventor-industrialist Sir Richard Arkwright is born the youngest of 13 children in Preston, England. With little formal education, he prospered first as a traveling wig-maker, then in the 1760s he invented a thread-making machine that replaced the historic "spinning jenny" because it produced stronger thread. Arkwright's machines were first operated by animals, then waterpower, and finally steam. Arkwright has been called the "first capitalist of the newborn industrialist age."

1834 Sociologist-economist Thomas Malthus dies at 68 near Bath, England (see February 13, 1766).

1854 A tsunami—a giant tidal wave—strikes the coast of Japan; 12 hours later it hits San Francisco. Two years later A.D. Bache of the U.S. Coast and Geodetic Survey uses the information recorded about the tsunami and a newly discovered law relating depth to wave speed to provide the first scientific estimate of the depth of an ocean.

1913 The federal reserve banking system is created with the passage by Congress of the Federal Reserve Act "to provide for the establishment of Federal Reserve Banks, to furnish an elastic currency...to establish a more effective supervision of banking in the United States."

1919 The USS *Relief*, the nation's first ambulance ship designed as a hospital, is launched. She is christened by Mrs. William G. Braisted and delivered to the U.S. Navy on December 28, 1920. It is equipped with 515 beds in 14 wards.

1924 In his lab in Johannesburg, South Africa, Raymond Dart frees history's first-discovered *Australopithecus* skull from its surrounding stone. "No diamond cutter ever worked more lovingly or with such care on a priceless jewel—nor, I am sure, with such inadequate tools," he later said. The job took 73 days. *Australopithecus* (meaning "southern ape") was the name coined by Dart; the beast was one of the first "missing link" species known in man's evolution.

1947 The transistor is invented.

1954 The first successful kidney transplantation in humans is conducted by Dr. John P. Merrill at Peter Bent Brigham Hospital (now Brigham and Women's Hospital) in Boston. The patient, 23-year-old Richard Herrick, receives the kidney from his identical twin Ronald.

1986 The first nonstop round-the-world flight without refueling ends at Edwards Air Force Base in California. Dick Rutan and Jeana Yeager completed the 24,987-mile trip in 9 days 3 minutes 44 seconds, in their experimental craft *Voyager*.

1993 Chuck Fallis, spokesman for the Centers for Disease Control and Prevention, announces that the last of Earth's smallpox virus will *not* be destroyed—yet. The last of the species is held in 600 frozen vials in Moscow and Atlanta and was to be destroyed with heat on December 31. The reprieve came after dozens of researchers protested the impending destruction.

1994 Researchers report in *Science* magazine that they have discovered a cellular "fountain of youth," an enzyme called "telomerase" that enables cells to reproduce without limits. It has been called an "immortality" enzyme, and its presence might call for celebration, except that it seems to be the cause of cancer. Telomerase is found in cancer cells, not in normal cells, and the researchers report that it has been found in a variety of different cancers. Therein lies its promise: control telomerase and you control cancer.

2012 The world will end on this day according to ancient Mayan timepieces.

1834 The bellows is patented by John R. Morrison of Springfield, Ohio.

1761 French astronomer Jean Louis Pons is born into poverty in Peyres. His education was sparse, and his first job in an observatory was as a janitor. Yet he went on to discover 37 comets, including the Pons–Brookes and the Pons–Winnecke. His most notable discovery (in 1818) was the Encke comet—named for the scientist who worked out its orbit the following year—which has the shortest period of any comet known.

1800 Archaeologist Ferdinand Keller, the first to excavate Stone Age and Bronze Age lake dwellings in Europe, is born in Marthalen, Switzerland. His findings (especially after 1854 at Obermeilen on Lake Zurich) revealed much about everyday prehistoric life, and they precipitated a rush by others to find similar settlements.

1801 Passengers are carried in a motorized vehicle for the first time. It is the first passenger car that works. Inventor Richard Trevithick, 30, carries eight passengers up a hill in Camborne, Cornwall, England in his steam-powered carriage, nicknamed "Captain Dick's puffer." Bad roads and poor steam production end the demonstration.

DECEMBER 24

1818 Physicist James Joule, the first to realize that all forms of energy—mechanical, electrical, and heat—are essentially the same thing, is born the second son of a wealthy brewer. He was blessed with family wealth, but cursed with poor health; both factors led to his building a home laboratory early in life. He was a fanatic measurer. It is said that on his honeymoon he constructed a thermometer to measure heat generated by a nearby scenic waterfall. Joule showed that all forms of energy can be changed, one into another. This is the law of conservation of energy, or the first law of thermodynamics. The joule, a unit of energy or work, is named in his honor.

1851 Fire destroys 35,000 books in the Library of Congress.

1856 Geologist-theologian-science popularizer Hugh Miller dies at 54 in Edinburgh, Scotland (see October 10, 1802).

1888 Vincent van Gogh cuts off his own ear. It is history's most famous case of self-mutilation.

1889 A bicycle with a back-pedal brake is patented by Daniel C. Stover and William A. Hance of Freeport, Illinois, who are awarded patent No. 418,142.

1906 A radio program is broadcast for the first time. Physics professor Reginald Aubrey Fessenden (inventor of AM radio) sends the transmission from Brant Rock, Massachusetts. First heard on the air waves are the letters "CQ," followed by a song, the reading of verse, a violin solo, and a speech.

1927 Russian psychiatrist-neurobiologist Vladimir Bekhterev dies at 70 in Moscow (see February 1, 1857).

1936 Anthropology superstars Louis and Mary Leakey are wed in Ware, England. It causes a scandal, because Louis was just recently divorced from his first wife.

1948 The country's first house completely heated by the sun is occupied. Designed by Eleanor Raymond and built in Dover, Massachusetts, the structure traps energy in black sheet-metal collectors and stores it in "heat bins" containing a sodium compound. Fans then blow the heat when and where it is desired.

1991 Wildlife officials in the state of Assam, India, announce that rogue elephants have killed at least 31 people during the year. The pachyderms are often drunk on rice wine from the villages. Shrinking natural habitats and expanding populations (of both elephants and humans) are blamed.

1992 Called "a landmark" in the prevention of birth defects, the *New England Journal of Medicine* publishes a Hungarian study of 4753 women that proves that folic acid in the diet slashes the risk of all birth defects, especially those of the nervous system (spina bifida and anencephaly). Fruits and vegetables are the natural sources of folic acid, but it can also be taken as a vitamin supplement. Unfortunately, most women do not yet realize that they are pregnant during the formation of the embryo's nervous system, which is why the federal government urges all females of childbearing age to eat properly.

1642 Isaac Newton, "the greatest intellect who ever lived," is born prematurely to an unwed mother in Woolsthorpe, England. Coincidentally, it is the year of Galileo's death. At age 3 Newton was abandoned by his mother, and raised by grandparents. He showed no signs of exceptional ability as an adolescent. Among his ultimate achievements were the invention of calculus, discovering the nature of color and light, and explaining the movements of planets.

1758 Halley's Comet returns to Earth's vicinity, as predicted in 1705 by Edmund Halley. Never before had such a prediction come true. It was an early and startling demonstration that the heavens could be understood scientifically. Unfortunately, Halley had died 16 years before the comet's arrival.

1761 Mineralogist William Gregor, discoverer of the element titanium (in 1791), is born in Trewarthenick, England.

1780 "The electrical fluid should be considered a means to excite nervo-muscular force." Luigi Galvani records this conclusion in his lab in Bologna, Italy, after a series of experiments and his accidental discovery that muscles are operated by electrical stimulation of nerves.

1821 Legendary nurse Clarissa Harlowe Barton is born in Oxford, Massachusetts.

1871 In Newark, New Jersey, Thomas Alva Edison, 24, marries Mary Stilwell, 16, and returns to work an hour later.

1876 German chemist Adolf Windaus, the first to synthesize histamine, is born in Berlin. He also determined the structure of cholesterol, and discovered that vitamin D is a steroid molecule in which a bond is broken by sunlight. He won the 1928 Nobel Prize.

1904 Physicist Gerhard Herzberg is born in Hamburg, Germany. He won the 1971 Nobel Prize for his studies on the electronic and geometric structure of molecules, especially free radicals (atoms or molecules with unpaired electrons). He was a specialist in spectroscopic analysis of gases, including gases around stars and the outer planets.

1906 Ernst Ruska, inventor of the electron microscope, is born in Heidelberg. He won the 1986 Nobel Prize for Physics.

1914 Thyroxine—the thyroid hormone that controls metabolism in cells and tissues—is crystallized for the first time, by Edward Calvin Kendall of the Mayo Foundation in Rochester, Minnesota.

1930 The country's first bobsled run built to international specifications opens to the public. Designed by Stanislaus Zentzytzki, the run was erected in 4½ months at Mt. Van Hoevenberg at North Elba, New York, on the highway between Lake Placid and Elizabethtown, New York.

1960 The unlikely start of cryosurgery—medical operations performed with supercold instruments—occurs in Pelham, New York. Dr. Irving Cooper's wife gives him a wine-opener as a Christmas gift. The device releases small squirts of very cold gas with each operation. "I played with that gadget for hours, freezing tiny areas on the palm of my hand and watching them thaw. What impressed me most was that I could control the area of freezing without overlapping into adjacent tissue. Why couldn't the same thing be done in the brain?...Then I got the idea of developing an instrument that would be a cannula, insulated on the sides, with a cooling tip that could be put into different parts of the brain." Cooper finally developed such an instrument.

1991 Hundreds of mice, rabbits, and hamsters in Padua, Italy, receive a very special Christmas gift—their freedom. The radical group Animalistic Counter-attack breaks into the university's Institute for Experimental Surgery in the dead of night to liberate the creatures. "It's not right to make animals suffer" was written on a sign left on the gate. The same group freed 1500 mink the previous October from Italy's largest mink farm.

1993 An anonymous British woman gives birth to twins at age 59, becoming the oldest woman to bear children. Her record is in jeopardy as soon as it is set, as a 62-year-old Italian woman is already three months pregnant. Both women were implanted with eggs by the controversial Dr. Severino Antinori at his clinic in Rome. The issue of the elderly giving birth has sparked a worldwide debate. Dr. John Marks, former chair of the British Medical Association Council, called it "the Frankenstein syndrome" or science creating life in a manner unapproved by society.

1780 Physician John Fothergill dies at 68 in London. He was the first to describe coronary arteriosclerosis (hardening of the arteries that carry blood to the heart muscle itself). A very successful London physician, Fothergill first gained notoriety when he determined the difference between diphtheria and scarlet fever, in the wake of a horrific diphtheria epidemic in 1747. He also promoted the use of coffee in England and its cultivation in the West Indies.

1792 Mathematician-inventor Charles Babbage, apparently the first to conceive of an automatic digital computer, and certainly the first to begin building one, is born the son of a banker in Teignmouth, England. In addition to designing the first computer—called the "analytical engine"—Babbage helped establish England's modern postal system, cofounded the Royal Astronomical and Royal Statistical societies, invented an ophthalmoscope, compiled the first reliable actuarial tables, and invented the locomotive cowcatcher.

DECEMBER 26

1854 Wood-pulp paper is first exhibited in the United States. John Beardsley of Buffalo, New York, shows three samples of his basswood paper to the editor of the Buffalo *Democrat*. Wood pulp quickly replaced a variety of other materials (including grasses, rag, and flax), and greatly increased the speed and quantity of paper production.

1865 The coffee percolator is patented by James H. Nason of Franklin, Massachusetts.

1890 Archaeologist Heinrich Schliemann, considered the discoverer of ancient Greece (and famed for excavating Troy, Mycenae, and Tiryns), dies at 68 in Naples, Italy. Today is also the 37th birthday of Wilhelm Dörpfeld (born in Barmen, Prussia, in 1853) who worked with and then succeeded Schliemann as head of the Troy excavation in the 1890s.

1896 Emil Du Bois-Reymond, the founder of modern electrophysiology, dies at 78 in Berlin (see November 7, 1818).

1906 History's first full-length feature film opens, at the Melbourne Town Hall in Australia. The movie is *The Story of the Kelly Gang*, a biography of notorious bandit Ned Kelly (1855–1880). It was filmed in Melbourne on a budget of £450.

1194 Scholar and enlightened monarch Frederick II is born in Jesi, central Italy. A "Renaissance man" two centuries before the Renaissance, Frederick II patronized scholars, fostered all sciences and mathematics, established a zoo, and wrote an important book on falconry from his own experimental research. In 1240 at his palace in Palermo, Frederick II presented pharmacists with the first edict in Europe to separate the two disciplines and to lay out professional regulations for pharmacists.

1975 The first SST (supersonic transport) airline service begins operation. The Soviet *Tupolev-144* makes its maiden commercial flight from Moscow to Alma-Ata, Kazakhstan.

1985 Dian Fossey, 53, becomes a martyr in the battle to save the mountain gorilla. On this night she is hacked to death with a machete, probably by gorilla poachers, in her hut at her research station on Mount Visoke, Rwanda. Hired by Louis Leakey, she went to Africa in 1966 to study the primates' behavior. Over the years she became an ardent crusader for their conservation. The movie *Gorillas in the Mist* portrays her struggle.

1993 A brush fire in India's Jaldapara wildlife preserve panics a herd of 50 elephants, which flee the area and head for Calcutta. Blazing torches, thundering drums, tranquilizer guns, and a roadblock of trucks were used to quell the stampede, which lasted over a week and killed at least six people.

1993 Big Bertha, the world's oldest cow, dies at age 49 in Kenmare, Ireland. According to *The Guinness Book of World Records* Bertha also died as the world holder of the lifetime breeding record for cattle, having spawned 39 calves. She was born on St. Patrick's Day, 1944.

DECEMBER 27

1571 German astronomer Johannes Kepler is born in Weil der Stadt, the son of a professional soldier who soon deserts his family. At age 3, Kepler contracted smallpox, and was left with crippled hands and weakened eyes. It was Kepler who disproved ancient Greek astronomy by discovering that Earth and planets orbit the sun in ellipses, not circles. Kepler coined the term "satellite."

1822 Louis Pasteur is born in Dôle, France, in the foothills of the Alps. His father was a tanner who urged his son to be a good student, which young Pasteur was not. In chemistry he was graded as "mediocre." Pasteur trained to be an artist and art professor, until he heard lectures by Jean Dumas and Antoine Balard, which turned him to a life in science.

1831 Charles Darwin, 21, sets sail on his legendary voyage to the Galapagos Islands as an unpaid naturalist on the HMS *Beagle*. He is soon seasick for the first of many times during the five-year trip.

1845 Anesthesia is used for the first time in childbirth. Dr. Crawford W. Long in Jefferson, Georgia, uses ether to ease the pain of his wife's delivery of their second child, Fanny.

1892 The APA (American Psychological Society) holds its first scientific meeting, at the University of Pennsylvania in Philadelphia. Its first president was Professor Granville Stanley Hall, and Dr. Joseph Jastrow was its first secretary and treasurer. The APA was the first national psychology society in the United States, and remains its largest and most influential.

1915 Sexologist-physician William Howell Masters is born in Cleveland. *Human Sexual Response* (1966) by Masters and Virginia Johnson is considered the first comprehensive study of the anatomy and physiology of human sexual behavior under laboratory controlled conditions.

1938 Geneticist Calvin Bridges dies at just 49 in Los Angeles. His research helped prove the "chromosomal theory of heredity." Working with Thomas Hunt Morgan at Columbia University in the early 1900s, Bridges showed that inherited variations in animals were related to chromosome structure. This led to the construction of "gene maps," in which specific traits are identified with specific chromosome segments. Morgan and Bridges conducted their research on the fruit fly, making this animal one of the favorite subjects of genetics research and teaching.

1968 *Apollo 8* splashes down in the Pacific. It is the first nighttime landing of a manned satellite, and its astronauts (Borman, Lovell, and Anders) were the first Americans to orbit the moon.

1984 The first artificial comet is launched from the German satellite *IRM*, which releases barium 110,000 km above Earth. The barium sublimates, forming a fiery tail 10,000 km long. It is visible at night from the Pacific. The event is part of a research project on solar wind and solar emissions.

1994 Worldwide news services report that hundreds of residents in Bombay have protested violence and sex on TV by throwing their televisions out the window. It started one night a few months before when a devout Muslim, Safira Ali Mohammed, unplugged her set, carried it to the window of her high-rise apartment, and tossed it outside. Minutes later other TVs were thrown to the pavement, and the trend was born. Within days not one of the 1200 residents in the apartment complex in the fashionable Versova district owned a television.

1773 Sir George Cayley, founder of the science of aerodynamics, and the first to build a man-carrying glider, is born in Scarborough, England.

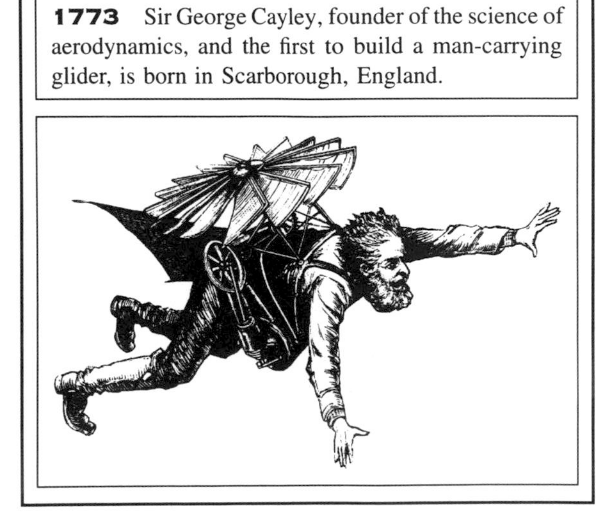

1798 Scottish astronomer Thomas Henderson, the first to measure the parallax of a star, is born the son of a tradesman in Dundee.

1814 Sir John Bennet Lawes, founder of the artificial fertilizer industry, is born in Rothamsted, England. He also founded the Rothamsted Experimental Station, the world's oldest agricultural research station.

1869 Chewing gum is patented. William Finley Semple of Mount Vernon, Ohio, receives patent No. 98,304 for a "combination of rubber with other articles, in any proportions adapted to the formation of an acceptable chewing gum."

1873 William Draper Harkins is born in Titusville, Pennsylvania, the site where oil was first obtained by drilling. He was the first to predict the existence of neutrons and heavy hydrogen, and to deduce nuclear fusion.

DECEMBER 28

1882 Astronomer Sir Arthur Stanley Eddington is born the son of a Quaker headmaster in Kendal, England. He achieved fame by deducing the internal structure of stars (published in 1926). He was also a great popularizer of science, and was the first to explain Einstein's relativity theory in lay terms for the English-speaking world. Many, including Einstein himself, have considered Eddington's presentation of the subject to be the clearest in any language. In 1919 Eddington led the famous expedition to Príncipe Island, West Africa, that observed the bending of light during a total solar eclipse; it was the first direct confirmation of Einstein's theory. Eddington, however, missed seeing the eclipse while changing photographic plates.

1895 History's first motion picture *La Sortie des ouvriers de l'usine Lumière* ("Workers Leaving the Lumière Factory") debuts publicly at the Grand Café on the Boulevard des Capucines in Paris. It was created by brothers Auguste and Louis Lumière with their camera and projector, called the Cinématographe.

1877 John Stevens of Neenah, Wisconsin, submits a patent application for the flour rolling mill, calling it a "grain crushing mill." The patent was granted on March 23, 1880. Stevens's innovation boosted flour production efficiency by 70% and produced flour of a superior quality (and higher price).

1895 Wilhelm Roentgen announces his discovery of X rays to the Würzburg Medical Society. Some historians call this event the start of the Second Scientific Revolution, the First Revolution being the result of Galileo's work. This announcement begins a flood of other discoveries that changed the face of physics forever. Within months, radioactivity was discovered; within a year, a thousand papers on X rays were published.

1929 Maarten Schmidt, the first to explain the nature of quasars ("quasi-stellar objects"), is born in Groningen, the Netherlands. In 1963, while working at the Mount Wilson and Mount Palomar Observatories (now Hale Observatories) Schmidt realized that the unusual radio emissions by quasars could be an example of an enormous redshift caused by the quasars' extreme distance and speed. This explanation is now widely accepted. Some believe that the light of quasars traveled for 15 billion years before reaching Earth, thus challenging accepted theories of the formation of the universe.

1929 Davidson Black first announces the discovery of the "missing link" Peking man, to a special session of the Geological Society of China in Peking.

1973 *The Gulag Archipelago*, an exposé of the Soviet penal system by Aleksandr Solzhenitsyn, is first published.

1981 Elizabeth Jordan Carr, the first U.S. test-tube baby, is born in Norfolk, Virginia.

1992 International news services report that a pygmy blue tongue lizard, thought to be extinct, has been discovered. Near Burra, Australia, 100 miles north of Adelaide, amateur herpetologist Graham Armstrong discovered the lizard in the stomach of a squashed snake that he found on the highway.

1994 Another mystery arises in man's struggle with AIDS. The *Journal of the American Medical Association* reports that women succumb faster than men.

December 29

1788 Archaeologist Christian Thomsen, who coined the terms "Stone Age," "Bronze Age," and "Iron Age," is born in Copenhagen. Thomsen arrived at the realization that man's prehistory could be broken up into three distinct eras during 20 years of work in the National Museum, of Denmark. Not only did he successfully organize a huge and previously chaotic collection of Scandinavian antiquities, but his tripartite system (published in 1836) organized understanding of man's development throughout the world.

1800 Charles Goodyear, inventor of vulcanized rubber, is born the son of an inventor of farm implements in New Haven, Connecticut. The father's bankruptcies and poverty helped prepare Goodyear for life as an inventor. In fact, Goodyear was in debtor's prison when he began experiments on vulcanized rubber in 1834. The final process was discovered by accident. Goodyear died in deep debt trying to defend and exploit his creation.

1813 Chemist Alexander Parkes, discoverer of the first plastic, is born in Birmingham, England. In the 1850s Parkes found that cellulose—the major component in plant cell walls—could be treated in a series of chemical steps to become a heat-malleable substance. Although he was unable to exploit the material, others after him did. Parkes had created celluloid.

1816 Carl Ludwig, the first to keep animal organs alive *in vitro* (outside the animal's body), is born in Witzenhausen, Germany. He performed this trick in 1856 by pumping a bloodlike solution through frog hearts. It was the dawn of transplant surgery. Ludwig was also the first to measure blood pressure in capillaries, discovered the mechanism in the brain that regulates blood pressure, and invented the kymograph (a device originally for measuring blood pressure, but which has now been widely adapted by many others for the measurement of many physiological processes). In 1844 Ludwig proposed the basic, now-accepted explanations of how urine and lymph are formed. He is also remembered as an exceptional teacher.

1836 German botanist-explorer Georg August Schweinfurth is born in Riga, Latvia. His search for exotic plants brought him through many unexplored parts of Africa. He was the first European to find the Uele River (which helped in the final estimate of the size of the Nile River system) and he provided the first authoritative account of Congo Pygmies.

1851 The U.S. branch of the YMCA is organized in Boston, 20 days after a branch was founded in Montreal, Canada. Both chapters were modeled on a group started in London in 1844. The first well-equipped gymnasium in a U.S. YMCA was opened in New York City in 1869.

1863 Swiss cardiologist Wilhelm His is born in Basel. Son of the renowned anatomist of the same name, His discovered (in 1893) the "bundle of His," which controls heart contractions. His was also one of the first to realize that the heartbeat originates within individual cells of the heart itself.

1923 The patent application for the iconoscope, the first practical television camera, is filed by Vladimir Kosma Zworykin.

1952 The first hearing aid to use a transistor goes on sale. Manufactured by the Sonotone Corporation of Elmsford, New York, the device weighs 3½ ounces and measures 3 by ¾ by 19⁄32 inches.

1952 The biggest gallstone in medical history is removed from an 80-year-old woman by Dr. Humphrey Arthure at Charing Cross Hospital in London. The stone weighs 13 pounds 14 ounces.

1967 The term "black hole" is coined by Professor John Archibald Wheeler at an Institute for Space Studies meeting in New York City.

1992 Former school principal Jean Harris is granted clemency by New York Governor Mario Cuomo for her 1980 shooting of Dr. Herman Tarnower, author of the best-selling *Scarsdale Diet.* "Despite her advancing age [69] and medical problems [she had a heart attack just days before], Ms. Harris consistently sought to apply her skills as a teacher and educational administrator for the benefit of other inmates and their children." Harris has consistently maintained that she shot Tarnower four times by accident.

1993 After three weeks of testing the repairs to the Hubble Space Telescope, ground scientists report, "So far, so good. We have absolutely no sign of problems." Within weeks, the enormous $3 billion Hubble starts transmitting spectacular images of some of the farthest and faintest objects in the universe.

1850 Geologist John Milne, inventor of the modern seismograph (in 1880), is born in Liverpool, England. A roving spirit carried him to Canada, Asia, and Africa before he became a professor (for 20 years) at the Imperial College of Engineering in Tokyo. Because of the high number of earthquakes in Japan, Milne had great opportunity to develop his device for measuring their strength. He established a chain of seismographs throughout Japan, marking the beginning of modern seismology.

1851 Asa Griggs Candler, founder of the Coca-Cola empire, is born on a farm near Villa Rica, Georgia. Candler was a pharmacist and owner of a successful wholesale drug business when he purchased the formula for the obscure beverage from a colleague in 1887. He improved the manufacturing and marketing procedures, and finally sold his business for $29 million in 1919. He donated much of his wealth to the improvement of Atlanta's Emory University.

1854 The first oil company in U.S. history incorporates in New York City. The Pennsylvania Rock Oil Company begins with a capital stock of $250,000 (10,000 shares at $25). George H. Bissell is the chief organizer, and he and Jonathan G. Eveleth are the principal trustees.

DECEMBER 30

1868 A letter signed by Teddy Roosevelt, J.P. Morgan, and 17 other dignitaries is sent to Andrew H. Green, the first commissioner of Central Park in New York City. The letter proposes "that a great Museum of Natural History ... be established in Central Park." The cornerstone of the American Museum of Natural History is laid six years later. It is now the world's largest nature museum.

1890 Gifford Pinchot, 25, delivers the address "Government Forestry Abroad" to a joint meeting in Washington, D.C. of the American Economic Association and the American Forestry Association. Pinchot was the first U.S. forester educated in Europe, and his address introduced the advanced techniques of continental silviculture. He was appointed the first chief of the U.S. Forest Service when it was created in 1905.

1913 Ductile tungsten is patented by Dr. William David Coolidge of the General Electric Company, Schenectady, New York. Patent No. 1,082,933 is issued for "tungsten and method of making the same, for use as filaments of incandescent electric lamps." Coolidge developed a method of superheating the metal and then drawing it out into fine threads for use in electric light bulbs.

1938 Lise Meitner and Otto Frisch deduce that their colleagues in Berlin (Otto Hahn and Fritz Strassmann) have split the atom. Hahn and Strassmann have already submitted a paper describing their experiments, but did not realize they had broken one element into two. At the time of their great insight, Meitner and Frisch are cross-country skiing at a Swedish resort, and, at the critical moment, they sit down on a fallen tree beside the trail and take out pencils and paper for calculations. Before they stand up, Meitner has determined that the release of energy is 20 million times greater than an equivalent amount of TNT would produce.

1954 Hurricane Alice strikes the Leeward Islands. Because Alice persisted until January 5 of the following year, she qualifies as both the latest and earliest Atlantic hurricane on record.

1972 History's highest measured wave is encountered in the North Atlantic by the British ship *Weather Reporter*. The wave is 86 feet high.

1872 The modern science of oceanography truly begins today. In the Atlantic Ocean off Portugal, the HMS *Challenger* dredges up a sample of the ocean floor. It is the first data collected on the ship's 3½-year voyage. No great discoveries are made on this day, but the cold mud from the ocean floor is used to chill celebratory champagne.

1994 A legally blind woman wins a Washington court battle with the investment firm of Smith Barney; from now on, she will be sent monthly statements in large print. She will also be paid $1500 in compensation. Beyond this, all present and future Smith Barney clients will be offered large-print statements. Harriet Afeld of Clearwater, Florida, had originally filed a complaint under the Americans With Disabilities Act after Smith Barney refused her request for larger print.

1994 A new strategy against cancer is reported in the journal *Cell* to shut down developing blood vessels and starve the cancer. An antibody called LM609 stops growth of new blood vessels, including those that automatically develop when a cancer takes root.

1668 Hermann Boerhaave, the "Dutch Hippocrates," is born the son of a clergyman in Voorhout, the Netherlands. Although he made few discoveries, he was the first to describe the sweat glands and to establish that smallpox is spread by contact only. He advanced the teaching of medicine by bringing his students to the bedside of patients. He has therefore been called "the founder of clinical teaching." He was the most eminent European physician since Galen, 16 centuries before, and he established the University of Leiden as an international mecca for medical students.

1679 Astronomer-physiologist Giovanni Alfonso Borelli dies at 71 in Rome (see January 28, 1608).

1808 Physicist Joseph-Louis Gay-Lussac, 30, announces Gay-Lussac's law relating temperature to gas volume.

1841 Alabama becomes the first state to pass a law regulating dental surgery. It specifies that there be "medical boards of the state to examine and to issue a license to applicants to practice dental surgery under the same rules and regulations, and subject to the same restrictions as those who apply for license to practice medicine."

1866 Gregor Mendel—then an unknown monk who had already single-handedly discovered the laws of heredity in his spare time in a small pea patch in the monastery in Brünn, Austria-Hungary—initiates an eight-year correspondence with Karl von Nägeli. Although an eminent biologist, Nägeli misunderstands the significance of Mendel's studies; his discouraging remarks and misguided advice cause Mendel to abandon his research while he is still relatively young and in good physical health.

1879 Edison conducts the first public demonstration of his incandescent light bulb, in Menlo Park, New Jersey. The Pennsylvania Railroad Company runs special trains to Edison's lab to satisfy public demand to view the device.

1955 The General Motors Corporation of Detroit becomes the nation's first company to earn more than $1 billion in one year. GM's 47th annual report, for the year ending on this date, showed a net income of $1,189,447,082.

1958 International Geophysical Year ends, 18 months after it began.

1977 A band of six poachers kill the silverback male gorilla "Digit" while he was defending his family on Mount Visoke, Rwanda. The animal was a personal favorite of primatologist Dian Fossey. Fossey is grief-stricken, then enraged. She increases her antipoaching activities, and eventually founds the Digit Fund for gorilla preservation. The death of Digit is history's most publicized ape death. Digit was one of the few remaining mountain gorillas on Earth; the poachers received about $20 for his head and hands. Exactly eight years later (1985) Fossey's funeral is held on the same mountain. She too was apparently murdered by poachers.

1993 The last known samples of the smallpox virus, once the world's deadliest disease, are scheduled to be destroyed by heat on this date; but an 11th-hour reprieve is granted, following protests by a number of scientists who want to continue research on the organism. The last of the virus is held in 600 frozen vials in Moscow and Atlanta at the Centers for Disease Control and Prevention.

1994 Arnie Wilson and Lucy Dicker spend the day skiing in Colorado. It was the 365th consecutive day the pair had been skiing. Across five continents, and not always on snow, the two skied every day of 1994, setting a record. Why did they do it? "Actually, we're both quite stubborn people," said the Briton Wilson.

1514 The great anatomist Andreas Vesalius is born in Brussels. Vesalius conducted anatomy lessons at the University of Padua, where he was a professor from 1537 to 1543. His lectures were popular among medical students, artists, government officials, and doctors. Vesalius was the first to use actual human bodies in teaching and demonstrations, and thereby revolutionized both science and philosophy. His father was court pharmacist to Emperor Charles V. Vesalius's *De Corporis Humani Fabrica* (1543) was a turning point in science, as the first accurate book on human anatomy.

NAME INDEX